IMMUNOLOGY OF RECURRENT PREGNANCY LOSS AND IMPLANTATION FAILURE

Reproductive Immunology

IMMUNOLOGY OF RECURRENT PREGNANCY LOSS AND IMPLANTATION FAILURE

VOLUME 3

Edited by

JOANNE KWAK-KIM, MD, MPH

Agnes D. Lattimer MD, Professor, Director, Reproductive Medicine and Immunology, Obstetrics and Gynecology, Clinical Sciences Department, Chicago Medical School, Rosalind Franklin University of Medicine and Science, Vernon Hills, IL, USA

Series Editor

GIL MOR

John M. Malone Jr., MD, Endowed Chair, Scientific Director, C.S. Mott Center for Human Growth and Development, Wayne State University School of Medicine, Detroit MI, USA

ELSEVIER

ACADEMIC PRESS

An imprint of Elsevier

ASRI American Society for Reproductive Immunology

Academic Press is an imprint of Elsevier
125 London Wall, London EC2Y 5AS, United Kingdom
525 B Street, Suite 1650, San Diego, CA 92101, United States
50 Hampshire Street, 5th Floor, Cambridge, MA 02139, United States
The Boulevard, Langford Lane, Kidlington, Oxford OX5 1GB, United Kingdom

Notices

Knowledge and best practice in this field are constantly changing. As new research and experience broaden our understanding, changes in research methods, professional practices, or medical treatment may become necessary.

Practitioners and researchers must always rely on their own experience and knowledge in evaluating and using any information, methods, compounds, or experiments described herein. In using such information or methods they should be mindful of their own safety and the safety of others, including parties for whom they have a professional responsibility.

To the fullest extent of the law, neither the Publisher nor the authors, contributors, or editors, assume any liability for any injury and/or damage to persons or property as a matter of products liability, negligence or otherwise, or from any use or operation of any methods, products, instructions, or ideas contained in the material herein.

ISBN: 978-0-323-90805-4

Publisher: Stacy Masucci
Acquisitions Editor: Linda Versteeg-Buschman
Editorial Project Manager: Pat Gonzalez
Production Project Manager: Sreejith Viswanathan
Cover Designer: Matthew Limbert

Typeset by MPS Limited, Chennai, India

Contents

14. Rheumatic diseases and reproductive outcomes 205

BROOKE MILLS AND BONNIE L. BERMAS

III

Repeated implantation failures

15. The role of reproductive immunology in recurrent pregnancy loss and repeated implantation failure 223

JENNY S. GEORGE, ROISIN MORTIMER AND RAYMOND M. ANCHAN

16. The embryo in reproductive failure: immunological view 241

TIA BRODEUR AND NAVID ESFANDIARI

25. Endometriosis and reproductive failures 395

GENTARO IZUMI, KAORI KOGA AND YUTAKA OSUGA

26. Laboratory approaches for reproductive failure: immunological biomarkers for reproductive failures 403

KATHARINE WOLF, KENNETH BEAMAN, SVETLANA DAMBAEVA AND ALICE GILMAN-SACHS

List of contributors

Maria Socorro L. Agcaoili-De Jesus Division of Allergy and Immunology, Department of Medicine, University of the Philippines College of Medicine, Philippine General Hospital, Manila, Philippines

Lara Theresa C. Alentajan-Aleta Division of Allergy and Immunology, Department of Medicine, University of the Philippines College of Medicine, Philippine General Hospital, Manila, Philippines

Lujain Alsubki Reproductive Medicine and Immunology, Obstetrics and Gynecology, Clinical Sciences Department, Chicago Medical School, Rosalind Franklin University of Medicine and Science, Vernon Hills, IL, United States

Raymond M. Anchan Center for Infertility and Reproductive Surgery, Brigham and Women's Hospital, Harvard Medical School, Boston, MA, United States

Petra Clara Arck Division of Experimental Feto-Maternal Medicine, Department of Obstetrics and Fetal Medicine, University Medical Center Hamburg-Eppendorf, Hamburg, Germany

Shihua Bao Department of Reproductive Immunology, Shanghai First Maternity and Infant Hospital, Tongji University School of Medicine, Shanghai, P.R. China

Ricardo Barini Department of Obstetrics and Gynecology, Campinas University (UNICAMP), Campinas, São Paulo, Brazil

Kenneth Beaman Clinical Immunology Laboratory, Faculty of Microbiology and Immunology, Center for Cancer Biology, Infection and Immunology, The Chicago Medical School, Rosalind Franklin University of Medicine and Science, Chicago, IL, United States

Bonnie L. Bermas Division of Rheumatic Diseases, University of Texas Southwestern Medical Center, Dallas, TX, United States

Elena Borodina Department of Obstetrics, Gynecology and Reproductive Sciences, Saint Petersburg State University, Saint Petersburg, Russia; Laboratory of the Mosaic of Autoimmunity, Saint Petersburg State University, Saint Petersburg, Russia

Tia Brodeur Division of Reproductive Endocrinology and Infertility, Department of Obstetrics and Gynecology and Reproductive Sciences, Larner College of Medicine at the University of Vermont, Burlington, VT, United States

Songchen Cai Shenzhen Key Laboratory for Reproductive Immunology of Peri-implantation, Shenzhen Zhongshan Institute for Reproduction and Genetics, Shenzhen Zhongshan Urology Hospital, Shenzhen, P.R. China

Marcelo Borges Cavalcante Postgraduate Program in Medical Sciences, Universidade de Fortaleza (UNIFOR), Fortaleza, Ceará, Brazil; CONCEPTUS – Reproductive Medicine, Fortaleza, Ceará, Brazil

Yuekun Chen Department of Obstetrics and Gynecology, School of Medicine, Hyogo Medical University, Nishinomiya, Hyogo, Japan

Tiffany Alexis Clinton Department of Obstetrics and Gynecology, Henry Ford Health System, Detroit, MI, United States

Svetlana Dambaeva Clinical Immunology Laboratory, Faculty of Microbiology and Immunology, Center for Cancer Biology, Infection and Immunology, The Chicago Medical School, Rosalind Franklin University of Medicine and Science, Chicago, IL, United States

Nicoletta Di Simone Department of Biomedical Sciences, Humanitas University, Pieve Emanuele, Milan, Italy; IRCCS Humanitas Research Hospital, Rozzano, Milan, Italy

Lianghui Diao Shenzhen Key Laboratory for Reproductive Immunology of Peri-implantation, Shenzhen Zhongshan Institute for Reproduction and Genetics, Shenzhen Zhongshan Urology Hospital, Shenzhen, P.R. China

Kerrilyn R. Diener Clinical and Health Sciences Unit, University of South Australia, Adelaide, SA, Australia; Robinson Research Institute and Adelaide Medical School, The University of Adelaide, Adelaide, SA, Australia

Jinli Ding Reproductive Medicine Center, Renmin Hospital of Wuhan University, Wuhan, Hubei, P.R. China

Mengyang Du Department of Reproductive Immunology, Shanghai First Maternity and Infant Hospital, Tongji University School of Medicine, Shanghai, P.R. China

Joseph Duero Department of Obstetrics, Gynecology and Reproductive Sciences, Yale School of Medicine, New Haven, CT, United States

Lyailya Kh. Dzhemlikhanova Department of Obstetrics, Gynecology and Reproductive Sciences, Saint Petersburg State University, Saint Petersburg, Russia; FSBSI "The Research Institute of Obstetrics, Gynecology and Reproductology named after D.O. Ott," Saint Petersburg, Russia

Navid Esfandiari Division of Reproductive Endocrinology and Infertility, Department of Obstetrics and Gynecology and Reproductive Sciences, Larner College of Medicine at the University of Vermont, Burlington, VT, United States

Xuhui Fang Reproductive Medicine Center, Department of Obstetrics and Gynecology, The First Affiliated Hospital of USTC, Division of Life Sciences and Medicine, University of Science and Technology of China, Hefei, Anhui, P.R. China

Atsushi Fukui Department of Obstetrics and Gynecology, School of Medicine, Hyogo Medical University, Nishinomiya, Hyogo, Japan; Fukushima Medical Center for Child and Woman, Fukushima Medical University, Fukushima, Japan

Jenny S. George Center for Infertility and Reproductive Surgery, Brigham and Women's Hospital, Harvard Medical School, Boston, MA, United States

Alice Gilman-Sachs Clinical Immunology Laboratory, Faculty of Microbiology and Immunology, Center for Cancer Biology, Infection and Immunology, The Chicago Medical School, Rosalind Franklin University of Medicine and Science, Chicago, IL, United States

Ruth Marian Guzman-Genuino Clinical and Health Sciences Unit, University of South Australia, Adelaide, SA, Australia

Alexander M. Gzgzyan Department of Obstetrics, Gynecology and Reproductive Sciences, Saint Petersburg State University, Saint Petersburg, Russia; FSBSI "The Research Institute of Obstetrics, Gynecology and Reproductology named after D.O. Ott," Saint Petersburg, Russia

Ae Ra Han Cha University, Fertility Center, Daegu, Republic of Korea

Jae Won Han Department of Obstetrics and Gynecology, Konyang University College of Medicine, Daejeon, Republic of Korea

Tetsuro Hanada Deparment of Obstetrics and Gynecology, Shiga University of Medical Science, Otsu, Shiga, Japan

Akiko Hasegawa Department of Obstetrics and Gynecology, School of Medicine, Hyogo Medical University, Nishinomiya, Hyogo, Japan

Gentaro Izumi Department of Obstetrics and Gynecology, Graduate School of Medicine, The University of Tokyo, Tokyo, Japan

Toru Kato Department of Obstetrics and Gynecology, School of Medicine, Hyogo Medical University, Nishinomiya, Hyogo, Japan

Sun Kwon Kim Department of Obstetrics and Gynecology, Division of Maternal-Fetal Medicine, Henry Ford Health System, Detroit, MI, United States

Fuminori Kimura Department of Obstetrics and Gynecology, Nara Medical University, Kashihara, Nara, Japan

Jun Kitazawa Deparment of Obstetrics and Gynecology, Shiga University of Medical Science, Otsu, Shiga, Japan

Kaori Koga Department of Obstetrics and Gynecology, Graduate School of Medicine, The University of Tokyo, Tokyo, Japan

Keiji Kuroda Division Manager of Implantation Research and Endoscopy, Center for Reproductive Medicine and Implantation Research, Sugiyama Clinic Shinjuku, Tokyo, Japan

Joanne Kwak-Kim Reproductive Medicine and Immunology, Obstetrics and Gynecology, Clinical Sciences Department, Chicago Medical School, Rosalind Franklin University of Medicine and Science, Vernon Hills, IL, United States

Maud Lansiaux Hôpital des Bluets, Centre de PMA, Paris, France

Trespidi Laura Department of Obstetrics and Gynaecology, Fondazione Ca Granda, Ospedale Maggiore Policlinico, Milan, Italy

Nathalie Lédée Hôpital des Bluets, Centre de PMA, Paris, France; MatriceLab Innove, Pépinière Paris- Santé- Cochin, Paris, France

Si won Lee Department of Obstetrics and Gynecology, Mount Sinai Medical Center, Miami Beach, FL, United States

Sung Ki Lee Department of Obstetrics and Gynecology, Konyang University College of Medicine, Daejeon, Republic of Korea

Meroni Pier Luigi Istituto Auxologico Italiano, IRCCS, Immunorheumatology Research Laboratory, Milan, Italy

Thanh Luu Reproductive Medicine and Immunology, Obstetrics and Gynecology, Clinical Sciences Department, Chicago Medical School, Rosalind Franklin University of Medicine and Science, Vernon Hills, IL, United States

Brooke Mills Division of Rheumatic Diseases, University of Texas Southwestern Medical Center, Dallas, TX, United States

Aina Morimune Deparment of Obstetrics and Gynecology, Shiga University of Medical Science, Otsu, Shiga, Japan

Roisin Mortimer Center for Infertility and Reproductive Surgery, Brigham and Women's Hospital, Harvard Medical School, Boston, MA, United States

Takashi Murakami Deparment of Obstetrics and Gynecology, Shiga University of Medical Science, Otsu, Shiga, Japan

Koji Nakagawa Director, Center for Reproductive Medicine and Implantation Research, Sugiyama Clinic Shinjuku, Tokyo, Japan

Akiko Nakamura Deparment of Obstetrics and Gynecology, Shiga University of Medical Science, Otsu, Shiga, Japan

Dariko A. Niauri Department of Obstetrics, Gynecology and Reproductive Sciences, Saint Petersburg State University, Saint Petersburg, Russia; Laboratory of the Mosaic of Autoimmunity, Saint Petersburg State University, Saint Petersburg, Russia; FSBSI "The Research Institute of Obstetrics, Gynecology and Reproductology named after D.O. Ott," Saint Petersburg, Russia

Cherie C. Ocampo-Cervantes Division of Allergy and Immunology, Department of Medicine, University of the Philippines College of Medicine, Philippine General Hospital, Manila, Philippines

Yutaka Osuga Department of Obstetrics and Gynecology, Graduate School of Medicine, The University of Tokyo, Tokyo, Japan

Jenifer R. Otadoy-Agustin Division of Allergy and Immunology, Department of Medicine, University of the Philippines College of Medicine, Philippine General Hospital, Manila, Philippines

Joon Cheol Park Department of Obstetrics and Gynecology, Keimyung University School of Medicine, Daegu, Republic of Korea

Erra Roberta Department of Obstetrics and Gynaecology, Fondazione Ca Granda, Ospedale Maggiore Policlinico, Milan, Italy

Shinichiro Saeki Department of Obstetrics and Gynecology, School of Medicine, Hyogo Medical University, Nishinomiya, Hyogo, Japan

Hiroaki Shibahara Department of Obstetrics and Gynecology, School of Medicine, Hyogo Medical University, Nishinomiya, Hyogo, Japan

Yehuda Shoenfeld Laboratory of the Mosaic of Autoimmunity, Saint Petersburg State University, Saint Petersburg, Russia; Ariel University, Ariel, Israel; Zabludowicz Center for Autoimmune Diseases; Sheba Medical Center, Tel-Aviv, Israel

Rikikazu Sugiyama CEO, Center for Reproductive Medicine and Implantation Research, Sugiyama Clinic Shinjuku, Tokyo, Japan

Ryu Takeyama Department of Obstetrics and Gynecology, School of Medicine, Hyogo Medical University, Nishinomiya, Hyogo, Japan

Reshef Tal Department of Obstetrics, Gynecology and Reproductive Sciences, Yale School of Medicine, New Haven, CT, United States

Chiara Tersigni Fondazione Policlinico Universitario A. Gemelli IRCCS, Dipartimento di Scienze della Salute della Donna e del Bambino e di Sanità Pubblica, Rome, Italy

Kristin Thiele Division of Experimental Feto-Maternal Medicine, Department of Obstetrics and Fetal Medicine, University Medical Center Hamburg-Eppendorf, Hamburg, Germany

Virginie Vaucoret Hôpital des Bluets, Centre de PMA, Paris, France

Yu Wakimoto Department of Obstetrics and Gynecology, School of Medicine, Hyogo Medical University, Nishinomiya, Hyogo, Japan

Ossola Wally Department of Obstetrics and Gynaecology, Fondazione Ca Granda, Ospedale Maggiore Policlinico, Milan, Italy

Xiao Wang Department of Reproductive Immunology, Shanghai First Maternity and Infant Hospital, Tongji University School of Medicine, Shanghai, P.R. China

Yanshi Wang Reproductive Medicine Center, Department of Obstetrics and Gynecology, The First Affiliated Hospital of USTC, Division of Life Sciences and Medicine, University of Science and Technology of China, Hefei, Anhui, P.R. China

Yiqiu Wei Reproductive Medicine Center, Renmin Hospital of Wuhan University, Wuhan, Hubei, P.R. China

Ronja Wöhrle Division of Experimental Feto-Maternal Medicine, Department of Obstetrics and Fetal Medicine, University Medical Center Hamburg-Eppendorf, Hamburg, Germany

Katharine Wolf Clinical Immunology Laboratory, Faculty of Microbiology and Immunology, Center for Cancer Biology, Infection and Immunology, The Chicago Medical School, Rosalind Franklin University of Medicine and Science, Chicago, IL, United States

Li Wu Reproductive Medicine Center, Department of Obstetrics and Gynecology, The First Affiliated Hospital of USTC, Division of Life Sciences and Medicine, University of Science and Technology of China, Hefei, Anhui, P.R. China

Ayano Yamaya Department of Obstetrics and Gynecology, School of Medicine, Hyogo Medical University, Nishinomiya, Hyogo, Japan

Tailang Yin Reproductive Medicine Center, Renmin Hospital of Wuhan University, Wuhan, Hubei, P.R. China

Yong Zeng Shenzhen Key Laboratory for Reproductive Immunology of Peri-implantation, Shenzhen Zhongshan Institute for Reproduction and Genetics, Shenzhen Zhongshan Urology Hospital, Shenzhen, P.R. China

About the editor

Dr. Joanne Kwak-Kim, MD, MS, MPH, is the Agnes D. Lattimer MD Professor, Obstetrics and Gynecology, Clinical Sciences Department, and Center for Cancer Cell Biology, Immunology and Infection, Chicago Medical School, Rosalind Franklin University of Medicine and Science, and the Director of Reproductive Medicine and Immunology, Rosalind Franklin University Health Clinic. She also served as an Assistant Chair, Department of Obstetrics and Gynecology, Chicago Medical School, and the Medical Director, Rosalind Franklin University Health System.

Dr. Kwak-Kim has spent the last three decades serving as a clinician, a researcher, and an educator. Her keen interest lies in reproductive immunology, particularly recurrent pregnancy loss and repeated implantation failures, which have been the focus of her research. Her work in natural killer cells and Th1/Th2 immune responses in women with recurrent pregnancy losses and repeated implantation failure is distinguished. The research work that she has done, which applies basic science techniques to clinical problems, is cutting edge. She has over 160 publications in peer-reviewed articles and books. She edited two special volumes, "Topics in Clinical Reproductive Immunology" and "The Clinical Reproductive Immunology," published in the *American Journal of Reproductive Immunology*. In addition, she is the editor of two books, "Endometrial gene expression; an emerging paradigm for reproductive disorder" and "Reproductive immunology: recurrent pregnancy losses and repeated implantation failures."

Dr. Kwak-Kim is the editor-in-chief of the *Journal of Reproductive Immunology*, an associate editor of the *American Journal of Reproductive Immunology*, and an editorial board member of many peer-reviewed scientific journals. She is a sought-after lecturer and has reviewed numerous national and international scientific studies. In addition, she has provided external scientific advisory services to scientific institutions worldwide.

She has mentored and continues to mentor medical students, residents, PhD students, and postdoctoral fellows. As a mentor of trainees, she is an outspoken advocate of reproductive medicine and immunology specialties. When asked, what a reproductive immunologist needs to succeed, she recommends a creative mind, a listener's ear, and a compassionate heart—attributes that have served Dr. Kwak-Kim well in her over 30 years of exemplary service.

She has been a dedicated member of the American Society for Reproductive Immunology (ASRI), the American Society for Reproductive Medicine (ASRM), the Society for Reproductive Investigation, and the International Society for Immunology of Reproduction.

Dr. Kwak-Kim has received numerous awards, including the prestigious J. Christian Herr Award for outstanding achievement in basic or applied research in reproductive immunology by the ASRI in 2002. In 2009, she won the distinguished service award from the ASRI. In 2011, she received the Lawrence R. Medoff Award from Rosalind Franklin University of Medicine and Science for outstanding contribution to Research, Teaching, and Clinical Care. Dr. Kwak-Kim's recent honor came in 2015 at the 35th Annual Meeting of ASRI when she was presented with the *American Journal of Reproductive Immunology Award* given to a senior investigator who has made outstanding contributions to the area of reproductive immunology. In 2017, she was invited as the 23rd Raymond O. Berry Memorial lecture speaker at Texas A&M University. In 2018, she received the Agnes D. Lattimer professorship from Chicago Medical School, Rosalind Franklin University of Medicine and Science. Dr. Agnes D. Lattimer overcame incredible adversity to become a respected and influential pediatrician, eventually becoming Cook County Hospital's medical director. The induction of professorship was named in honor of Dr. Lattimer's tireless service, bravery, and enduring kindness for generations to come. Dr. Joanne Kwak-Kim was recognized as an archetypal of Dr. Lattimer. In 2020, she received the "Yonsei Medical Grand Award," which is given to the Yonsei Alumni, who had a significant scholastic contribution in the field of medicine.

Dr. Kwak-Kim served as the president of the ASRI and successfully organized and chaired the 21st, the 22nd, and the 38th annual meetings of ASRI. She also served as the secretary-general and treasurer of the International Society of Immunology for Reproduction (ISIR) and currently is the president-elect, ISIR. She has been a chair, Reproductive Immunology Special Interest Group, ASRM. In 2021, she established a "Clinical Reproductive Immunology Fellowship" under the auspice of the American Society for Reproductive Immunology.

Dr. Kwak-Kim has been working as the director for Reproductive Medicine and Immunology at Rosalind Franklin University Health Clinic. The program has an emphasis on treating infertility with implantation failures, recurrent spontaneous abortion of immune etiology, and pregnancies complicated with alloimmune/autoimmune diseases. The center is well known nationally and internationally and serves patients from all over the United States and other countries. Her expertise in Reproductive Medicine and Immunology is matched only by her deep compassion for her patients and dedication to their care. She helps patients from all walks of life and around the globe, providing knowledge and solace to women who have already endured many losses by the time they reach her door. Her patients speak freely of how she turned their lives around; their years of confusion and frustration ended thanks to her consultation. A pioneer in her field, many of the treatments available in her office today were developed based on decades of her research and patient care as the reproductive immunology specialty emerged.

SECTION I

Introduction

CHAPTER

1

Clinical reproductive immunology: a window to understanding reproduction and immunology

Joanne Kwak-Kim

Reproductive Medicine and Immunology, Obstetrics and Gynecology, Clinical Sciences Department, Chicago Medical School, Rosalind Franklin University of Medicine and Science, Vernon Hills, IL, United States

Reproductive immunology has become an active field of reproductive medicine. Although many books and journals are available for scholarly works of reproductive immunology, a textbook with the systemic review of clinical reproductive immunology has not been readily available until this point. This book is part of the reproductive immunology textbook series dealing with clinical reproductive immunology, particularly focused on immunopathology, patient evaluation, diagnosis, and treatment of recurrent pregnancy losses (RPLs) and repeated implantation failures (RIFs) based on up-to-date clinical and scientific data.

Clinical translation of the reproductive immunology concept started in the 1980s [1]. The early observation of immunological significance on human reproduction was documented by William Hunter by proving complete quarantine of fetal circulation from the mother [2]. In 1871, Charles Darwin associated reduced fertility with profligacy in women, moving toward an immunological explanation for infertility [3]. In 1890, Walter Heape made an early immunological landmark by reporting a successful transfer of allogeneic blastocysts to surrogate mother rabbits [3]. The earliest documentation of reproductive immunology could be found in the literature of Clarence Cook Little in 1924, in which immune tolerance for pregnancy was stated as "in terms of the genetic uniqueness of the individual, the female mammal must be the same way in tolerance" [4]. In 1932, the intraspecies embryo transfer became successful for the first time by Raymond O. Berry, implying that the uterus is not an immune-privileged site and the placenta must be immunologically compatible with the mother [5]. In 1953, Sir Peter Medawar, a Nobel laureate, congregated reproductive

Immunology of Recurrent Pregnancy Loss and Implantation Failure
DOI: https://doi.org/10.1016/B978-0-323-90805-4.00024-9

immunology concepts by raising the following question: "How does the pregnant mother contrive to nourish within itself for many weeks or months a fetus that is an antigenically foreign body" at the Society of Experimental Biology meeting [6]. Almost at the same period, the field of immunology started to shift its focus to immunobiology, which was mainly instituted by the advancement of transplantation biology [7]. These changes influenced reproductive scientists who were interested in transplantation immunology, and in 1981, these scientists established the American Society of Reproductive Immunology [8]. A few years later, the first clinical translation study was reported by Alan E Beer, in which the first clinical study of lymphocyte immunotherapy for recurrent spontaneous abortion was reported [9]. Since then, clinical trials of various immunotherapies for women with RPL and/or RIF of immune etiology have been reported.

Reproductive immunology refers to a field of medicine that studies the interaction between the immune system and components related to the reproductive system. Its principle has been applied to understand maternal–fetal tolerance and unravel obstetrical and gynecological disorders such as Rh sensitization and neonatal alloimmune thrombocytopenia in early developmental days. In the 1980s and 1990s, reproductive immunology concepts were applied to investigate underlying immunopathologies of pregnancy losses, subfertility, and preeclampsia. Reproductive immunology has matured to develop sophisticated therapeutic approaches. However, many challenges remain in the clinical translation of research findings.

The clinical field of reproductive immunology has been recognized as an area of obstetrics and gynecology, and the clinical reproductive immunology fellowship was initiated in 2021 under the auspice of the American Society of Reproductive Immunology. Certified clinical reproductive immunologists are currently practicing reproductive immunology, although their numbers are quite small. The field of reproductive immunology requires the implementation of more clinical training programs and the development of educational materials. In this way, the time is ripe for publishing a textbook on reproductive immunology.

This book delivers the essence of gynecological and obstetrical, particularly early pregnancy-related, reproductive immune disorders. The content of this book includes autoimmune, cellular immune, and alloimmune disorders affecting human pregnancy, focusing on RPL and RIF. Autoimmune disorders, including antiphospholipid syndrome, rheumatic diseases, thyroid autoimmunity, antisperm antibodies, and cellular immune abnormalities, such as T, B, NK, and mast cell-related immunopathologies are comprehensively reviewed in relation to RPL and RIF. In addition, immune-inflammatory disorders and conditions which can affect RPL and RIF are summarized, including endometrial immune pathologies, thrombophilic conditions, ovarian autoimmunity, polycystic ovarian syndrome, endometriosis, assisted reproductive technology, metabolic disorders, and stress-induced immune deviations.

Even with the rapid advancement of immunology, clinical reproductive immunology is still in its infancy.Further studies are needed to develop new biological markers and therapeutic targets. For the clinical translation, well-designed clinical trials with systematic data collection are needed. In addition, new nomenclatures for disease entities and clinical reporting systems should be considered in the future, as well as a physician network to share up-to-date clinical knowledge. Clinical reproductive immunology is on the new horizon with the rapid technological development and the arrival of new biologics. Thus, much study is needed in the future, followed by the timely translation of reproductive immunology research findings.

I hope this book is helpful for physicians and other health care providers practicing clinical reproductive immunology. Unfortunately, due to the COVID-19 pandemic, many authors who participated in this book had difficulties preparing the manuscript. I sincerely appreciate all the authors who participated and contributed to this book, particularly Dr. Na Young Sung, who assisted in various aspects of book development. Lastly, I want to thank my husband, Dr. Joon Woo Kim, and my children, Caroline and Michael, who understood my dedication to the profession along with its busy hours and schedule.

References

[1] Beer AE, Quebbeman JF, Ayers JW, Haines RF. Major histocompatibility complex antigens, maternal and paternal immune responses, and chronic habitual abortions in humans. Am J Obstet Gynecol 1981;141 (8):987–99. Available from: https://doi.org/10.1016/s0002-9378(16)32690-4.

[2] Shippen Jr W. The hunters. N Engl J Med 1963;268(5):271–2. Available from: https://doi.org/10.1056/nejm196301312680514.

[3] Billingham RE, Beer AE. Reproductive immunology: past, present, and future. Perspect Biol Med 1984;27 (2):259–75. Available from: https://doi.org/10.1353/pbm.1984.0042.

[4] Little CC. The genetics of tissue transplantation in mammals. J Cancer Res 1924;8(1):75–95. Available from: https://doi.org/10.1158/jcr.1924.75.

[5] Kwak-Kim J, Sung N, Saab W, Fukui A. Introduction of the special issue, "clinical reproductive immunology.". Am J Reprod immunol (New York, NY: 1989) 2021;85(4):e13415. Available from: https://doi.org/10.1111/aji.13415.

[6] Medawar PB. Some immunological and endocrinological problems raised by the evolution of viviparity in vertebrates. Symp Soc Exp Biol 1953;1953(7):320–37.

[7] Kaufmann SHE. Immunology's coming of age. Review. Front Immunol 2019;10(684). Available from: https://doi.org/10.3389/fimmu.2019.00684.

[8] Officers of The American Society for the. Immunology of reproduction and the international committee for immunology of reproduction. Am J Reprod Immunol 1980;1(1):1. Available from: https://doi.org/10.1111/j.1600-0897.1980.tb00002.x.

[9] Beer AE. 10 - New Horizons in the diagnosis, evaluation and therapy of recurrent spontaneous abortion. ClObstet Gynaecol 1986;13(1):115–24. Available from: https://doi.org/10.1016/S0306-3356(21)00158-8.

SECTION II

Recurrent pregnancy losses

CHAPTER

2

Natural killer (NK) cell pathology and reproductive failure: NK cell level, NK cell cytotoxicity, and KIR/HLA-C

Svetlana Dambaeva[1], Thanh Luu[2], Lujain Alsubki[2] and Joanne Kwak-Kim[2]

[1]Clinical Immunology Laboratory, Faculty of Microbiology and Immunology, Center for Cancer Biology, Infection and Immunology, The Chicago Medical School, Rosalind Franklin University of Medicine and Science, Chicago, IL, United States [2]Reproductive Medicine and Immunology, Obstetrics and Gynecology, Clinical Sciences Department, Chicago Medical School, Rosalind Franklin University of Medicine and Science, Vernon Hills, IL, United States

1 Introduction

The presence of a large number of granulated cells in the late secretory phase endometrium and first-trimester decidua was noticed by histologists as early as the 1920s and it was suggested that these cells were originated from undifferentiated endometrial stromal cells. In the 1980s, it was discovered that they expressed lymphocyte lineage receptors and were further identified as natural killer (NK) cells [1,2]. Immunohistochemistry evaluation revealed that these endometrial granulated lymphocytes have unusual characteristics as they stained strongly with NK cell marker NKH1 (CD56), while there was no reactivity for the other NK cell marker (CD16). Since the discovery of endometrial NK (eNK) cells, the importance of eNK cells for pregnancy, their role in decidualization, embryo implantation, and placentation, and their origins have become topics of interest for reproductive biologists, obstetricians and gynecologists, and reproductive immunologists.

Immunology of Recurrent Pregnancy Loss and Implantation Failure
DOI: https://doi.org/10.1016/B978-0-323-90805-4.00012-2

2 Natural killer cells—definition

NK cells are a type of lymphoid cells that play a crucial role in innate immunity. In contrast to T and B lymphocytes, they neither undergo somatic gene rearrangement nor carry out specific immune responses defining the acquired immunity. Instead, NK cells express various receptors that allow them to identify healthy cells from virally infected cells undergoing stress or malignant transformation. Their immediate capacity to destroy cancer cells without prior sensitization was a reason why they were named NK cells. The mechanism behind NK cells' cytotoxic ability was interpreted by the missing-self concept. The basis for this concept was that cells that have lost self-HLA class I molecules would be recognized by NK cells as targets [3]. However, further research on NK cells has determined that in many cases, the missing self is not sufficient or imperative to trigger NK cell cytotoxicity, and additional signaling through activating receptors is involved in promoting the cytotoxic function [4]. Moreover, further research on NK cells has revealed that they are very potent cytokine producers in addition to their cytotoxic nature. NK cells' immunoregulatory capabilities are nonetheless important for local and systemic responses [4].

2.1 Natural killer cells receptors

NK cells express a wide range of activating and inhibitory receptors. Signaling that is transmitted upon ligation of activating receptors is strictly counterbalanced by signaling from inhibitory receptors. Therefore, to trigger cytotoxic activity, the signaling from activating receptors has to be superior to the signals received via the inhibitory receptors.

Two major families of receptors specialized in recognizing HLA class I molecules are involved in regulating NK cell cytotoxic effector function. The first family of receptors is evolutionarily conserved and includes C-type lectin-like receptors, such as NKG2A, NKG2C, and NKG2E. They form heterodimeric complexes with CD94. The CD94/NKG2A complex mediates inhibitory signaling to NK cells, while CD94/NKG2C and CD94/NKG2E mediate activating signaling. These receptors recognize the nonclassical HLA class I molecule, HLA-E. In contrast to highly polymorphic classical HLA-A, -B, -C molecules, the HLA-E has low polymorphism and a very specific spectrum of peptides that could be bound within the antigen-binding groove. Therefore, peptides only derived from other HLA class I molecules could be presented by HLA-E [5].

The second family of HLA class I specific receptors expressed by NK cells includes a large group of receptors called killer immunoglobulin-like receptors (KIRs). The KIR family is similar to CD94/NKG2 in that it contains both inhibitory and activating receptors. KIRs are encoded by highly polymorphic genes that form the KIR locus located on chromosome 19. There are nine genes for inhibitory receptors (KIR2DL1, KIR2DL2, KIR2DL3, KIR2DL4, KIR2DL5A, KIR2DL5B, KIR3DL1, KIR3DL2, and KIR3DL3) and six genes for activating receptors (KIR2DS1, KIR2DS2, KIR2DS3, KIR2DS4, KIR2DS5, and KIR3DS1) [6]. In contrast to NKG2 genes, the genes within the KIR locus are among the most recently evolved genes that rapidly expanded in primates [7]. Modern humans carry the highest number of KIR genes. Moreover, the content of allelic genes at the KIR locus varies between individuals, with some of the KIR genes being present or absent. Different combinations of these genes are known as KIR haplotypes. The haplotype "A" includes a set of six inhibitory KIRs and only one activating KIR. The content of this haplotype is fixed. All other combinations of genes within the KIR locus are termed as "B" haplotypes. An individual with two "A" haplotypes is referred to have the KIR AA genotype or "inhibitory" KIR genotype. Individuals carrying "A" and "B" or two "B" haplotypes are referred to have the KIR Bx genotype or "activating" KIR genotype. KIR Bx genotypes are characterized by a high degree of diversity in the number of KIRs due to the difference in gene content within the KIR locus. In addition, the presence of two or more activating KIRs differentiates KIR Bx genotypes from KIR AA.

KIRs recognize classical HLA class I molecules. The most recently evolved member of the HLA class I family, HLA-C, is the dominant ligand for KIRs [8]. Depending on an epitope recognized by KIRs, all HLA-C molecules as ligands could be divided into either C1 or C2. Thus, the HLA-C C1 molecules are ligands for KIR2DL2 and KIR2DL3. The HLA-C C2 molecules are ligands for KIR2DL1 and KIR2DS1. Other HLA class I molecules are also known to serve as ligands for KIRs. The HLA-G is recognized by KIR2DL4. In addition, KIR3DL1 recognizes Bw4 epitope bearing HLA-B and HLA-A. Approximately one-third of HLA-B alleles and a few HLA-A alleles (HLA-A*23, 24, 32) possess the Bw4 epitope [9]. However, no ligands are clearly identified for KIR3DL3, KIR2DL5, KIR3DS3, and only selected HLA class I alleles were shown to bind KIR3DL2 (HLA-A*03, 11), KIR2DS2 (HLA-C*01), KIR2DS5 (some HLA-C C2 alleles), and KIR3DS1 (HLA-B*51) [6].

CD16 is a non-HLA-restricted receptor used by NK cells to carry out their innate immunity task. CD16, which is a low-affinity Fc gamma receptor, recognizes cells opsonized with IgG and mediates antibody-dependent cellular cytotoxicity (ADCC) [10]. Another family of NK cell receptors regulating their effector functions are natural cytotoxicity receptors (NCRs). The NCR family is comprised of three receptors, NKp30, NKp44, and NKp46. Primarily they mediate an activating signal upon binding to their ligand. In contrast to KIRs and NKG2 receptors, the NCRs are not HLA-restricted. Their ligands are various tumor or pathogen-associated molecules, including heparan sulfate, glycosaminoglycans on malignant cells, viral hemagglutinins, protein pp65 from cytomegalovirus, and some bacterial and bacterial–fungal ligands [11].

2.2 Natural killer cell cytokines

Among key cytokines that are produced by NK cells are IFN-γ and TNF. In addition, NK cells secrete interleukins, such as IL-6 and IL-5, growth and angiogenic factors (GM-CSF, VEGF), and various chemokines (CXCL8, CCL3, CCL4, and others) [12]. Some of these cytokines directly affect the activity of other immune cells, whereas others are important for angiogenesis, vascular remodeling, and tissue homeostasis. Thus, NK cells are the predominant producer of IFN-γ, an integral element of the immune response. IFN-γ stimulates macrophages and dendritic cells, and it is critical for T cell polarization to Th1 type T cells. TNF and IL-6 are acute phase responders and cause the expression of other inflammatory mediators like prostaglandin E2. TNF activates tissue macrophages and stimulate phagocytosis and the production of oxidants. Chemokines play an active role in recruiting neutrophils and other leukocytes into tissues. GM-CSF is a growth factor that facilitates the development of myeloid cells and promotes the activation of macrophages. NK cells secrete the cytokines in response to stimulation by type-I IFNs, IL-15 or IL-18, and other cytokines. The expression of receptors to these stimulants on NK cells enables their rapid response in the production of immunoregulatory cytokines. The tissue environment also affects NK cells' reactivity, where a hypoxic condition can stimulate NK cells to secrete the angiogenic factor, VEGF-A [13].

2.3 Natural killer cell subsets

Since their discovery in the 1970s as the "null" lymphocytes that are neither T nor B cells [14], it has become clear that NK cells are an extremely diverse population in terms of their residency, phenotypes, and functional characteristics [15,16]. NK cells represent approximately 10% of peripheral blood lymphocytes. Moreover, they are also present in many tissues, including the liver, gut, tonsils, and lung. Notably, the highest number of NK cells is found in the uterus.

In peripheral blood, two subsets of NK cells are distinguished based on their phenotypes. The major population of peripheral blood NK (pNK) cells is CD16 positive and has a dim expression for CD56 (CD56dim pNK cells). About 10% of all pNK cells are negative for CD16 and have bright CD56 expression (CD56bright pNK cells). There are other phenotypic differences between these subsets. For example, CD56bright pNK cells are characterized with very low KIRs but are uniformly positive for NKp46 and NKG2A. In contrast, the CD56dim pNK cells express KIRs and show variable expression of NKp46 and NKG2A. The two subsets have distinct functional capabilities. CD56dim pNK cells display higher cytotoxic activity than CD56bright cells. In contrast, CD56bright pNK cells can rapidly produce cytokines and chemokines and are recognized as immunoregulatory cells [17]. NK cells that are present in tissues, tissue-resident NK (trNK) cells, differ from both pNK cell subsets. They are mostly CD16 negative and CD56dim. In general, trNK cells are capable of producing cytokines but are poorly cytotoxic [15].

Uterine NK (uNK) cells possess unique characteristics. Similar to CD56bright pNK cells or other trNK cells, they lack CD16 expression and cannot mediate ADCC. However, they seem to have strong cytotoxic abilities, as they are packed with granules that contain lytic proteases (granzymes) and pore-forming molecules (perforin). Gene expression studies revealed that granzyme A and B, and perforin expression are the highest in uNK cells compared to two pNK cell subsets [18]. An increased presence of inhibitory receptors might be necessary to control any potential harmful triggering of this cytotoxic potential. This finding might explain the intensive KIR expression by uNK cells. Indeed, uNK cells have the highest KIR expression over any other NK cell subset [18]. Another inhibitory receptor that recognizes HLA class I ligand, CD94/NKG2A, is also highly expressed by uNK cells [16]. Strong immunoregulatory characteristics of uNK cells play an important role in pregnancy support. Their production of IFN-γ is known to be essential for spiral arteries remodeling [19]. Moreover, uNK cells secrete chemokines, growth, and angiogenic factors that promote trophoblast migration and placental growth [20,21]. In cycling endometrium, uNK cells contribute to the clearance of senescent endometrial stromal cells. This acute senescence is induced in a fraction of endometrial stromal cells during each cycle upon increasing levels of progesterone, and it is accountable for a transient inflammatory response associated with the window of implantation [22].

2.4 Natural killer cells education and licensing

Functional competency of an individual NK cell, including cytotoxicity and cytokine production ability, is determined during its development in a process called education or licensing that is dependent on HLA class I expression in the surrounding environment. To become "licensed" for its effector potential, an NK cell has to receive an inhibitory signal triggered by ligation to self-HLA class I molecule on neighboring cells. Without this self-recognition, the NK cell cannot gain full functional competency. Thus, only NK cells that express inhibitory receptors recognizing self-HLA class I acquire full functional potential, whereas the rest of NK cells remain anergic [23]. In addition, the second type of interaction which is based on the engagement of activating receptor KIR2DS1 with its HLA class I ligand, HLA-C C2, prevents the development of harmful self-responding NK cells. This type of education is called "disarming" and explains the hyporesponsiveness of KIR2DS1-expressing NK cells from HLA-C C2 individuals [24,25]. Thus, educational signals obtained from the recognition of self-HLA class I molecules determine the functional competence of developing NK cells. Receptors involved in NK cell education include HLA-C and HLA-Bw4 binding KIRs and HLA-E binding CD94/NKG2A (Table 1). Because highly polymorphic KIRs and HLA genes are located on different chromosomes and inherited independently, there is significant variability between individuals on what types of receptor/ligand combinations they possess. Therefore, the presence/absence of certain KIRs or their ligands impacts the degree of NK cell education for each individual.

3 Natural killer cell-related pathologies

NK cell-related disorders are divided into primary deficiencies when abnormality of NK cells leads to a clinical, often life-threatening immunodeficiency or a set of conditions where NK cell functional characteristics have an impact on disease progression. Pure NK cell immunodeficiencies are rare [31]. Instead, NK cell abnormalities often accompany other primary immunodeficiencies, like severe combined immunodeficiency, when a defect in the IL2RG gene causes T cell deficiency along with the absence of NK cells. A unifying clinical presentation of these primary NK cell-related disorders includes an exceptional susceptibility to herpesviruses, which typically become evident during infancy or childhood. Beyond primary NK cell deficiencies, an increasing amount of data is accumulating on the role of NK cells, namely their educational status in infection, cancer, and autoimmune disorders.

HLA class I expression is often downregulated in virally infected cells as a mechanism to avoid T cell surveillance. Thus, cells infected with human immunodeficiency virus (HIV) have diminished HLA-B and appear as "missing-self" targets for KIR3DL1 expressing NK

TABLE 1 NK cells receptors and their ligands that are essential for natural killer cell education during their development [23,24,26].

NK cell receptor	Gene frequency[a]	Ligand	Allelic frequency[b]	Education model	NK cell functional competence
Inhibitory receptors					
KIR2DL1	95.3	HLA-C C2	0.37	Educated/Licensed	Competent
KIR2DL3	86.7	HLA-C C1	0.63	Educated/Licensed	Competent
KIR2DL2	52.9	HLA-C C1	0.63	Educated/Licensed	Competent
KIR3DL1	94.9	HLA-Bw4	see note[c]	Educated/Licensed	Competent
CD94/NKG2A[d]	100	HLA-E	see note[e]	Educated/Licensed	Competent
Activating receptors					
KIR2DS1	41.2	HLA-C C2	0.37	Disarmed	Hypofunctional

[a]*Gene frequency for KIR genes is shown for the N. American Caucasian population [27].*
[b]*Allelic frequency is shown for the N. American Caucasian population. HLA-C C1 and C2 frequencies were calculated based on declared frequencies for individual HLA-C alleles [28,29].*
[c]*HLA-Bw4 includes HLA-B and a few HLA-A alleles that are characterized by a serologically recognizable Bw4 epitope.*
[d]*Highly conserved genes.*
[e]*HLA-E expression is influenced by a sequence of a leader peptide from HLA-B [30].*

cells. A significant difference in controlling HIV was observed in individuals with a high-affinity HLA-Bw4 ligand for KIR3DL1, which ensures "education" for KIR3DL1 expressing NK cells. Individuals lacking KIR3DL1-educated NK cells were found to progress to AIDS more rapidly [25,32].

Poor functional activity of NK cells was shown to be associated with an increased risk for cancer. In a study involving 3625 individuals (40 and older) tested for NK cell cytotoxicity, a follow-up survey on cancer incidence conducted 11 years later revealed that the risk of developing cancer was lower if a participant had a high or a medium NK cell cytotoxic activity [33]. In patients treated for acute myeloid leukemia (AML), a higher cytotoxic activity of NK cells correlated with a better long-term outcome [34]. Alloreactivity of donor-derived NK cells is recognized as beneficial in hematopoietic stem cell transplantation used as a treatment for AML [6].

Either increased or impaired NK cell activity has been identified in patients with autoimmune disorders, including primary biliary cirrhosis (PBC), rheumatoid arthritis (RA), systemic lupus erythematosus (SLE), and psoriasis [35]. Thus, increased NK cell numbers and high cytotoxicity are found in patients with PBC. SLE and RA patients revealed lower than normal pNK cell count and cytotoxic activity. However, synovial fluid from RA patients contained a high number of activated NK cells [36]. Certain KIR/HLA genotype combinations that favor NK cell activation are known to be associated with an increased risk for several autoimmune disorders. They include activating receptor KIR3DS1 and HLA-B alleles with the I-80 isoform of the Bw4 epitope (HLA-Bw4-I-80). Individuals with these genes are predisposed to RA and psoriasis [37,38].

4 Natural killer cells in reproduction

4.1 Uterine natural killer cells

An abundance of NK cells in the uterine lining at the time of embryo implantation and during the first trimester of pregnancy underlines the importance of these cells in women's reproductive health. The unique characteristics of uNK cells were discussed above among other subsets of NK cells, highlighting their significance for angiogenesis and placental growth support. Immunohistochemistry analysis of endometrial samples stained with CD56 revealed an increased frequency of uNK cells in idiopathic recurrent implantation failure (RIF) cases [39]. Another study showed that endometrial samples from women with RIF substantially vary for uNK cell counts. When compared to the fertile control group, both CD56dim and bright cell counts were found in the RIF group [40]. For women with recurrent pregnancy loss (RPL), both increased [41] and decreased [42] uNK cells in comparison to control subjects were reported. Gene expression analysis demonstrated a substantially wider range of NK cell-related gene expressions in a reproductive failure of unknown etiology group compared with a control group [28]. This implies considerable variance in the size of the uNK cell population, and divergence from normality could be pathogenic or reflective of the problem that causes reproductive failure.

4.2 Peripheral blood natural killer cells

Another important aspect of evaluating NK cells is their functional characteristics. Assessment of cytotoxic abilities, degranulation, and cytokine production by NK cells requires a suspension of live cells. Indeed, initial observations on abnormality in NK cell parameters in connection with reproductive failures were made with peripheral blood testing. It was shown that pNK cells in women with RIF are often characterized with higher cytotoxicity than normal fertile controls [43,44]. Flow cytometric immunophenotyping of peripheral blood lymphocytes revealed that women with RPL [45–47] and RIF [39] have significantly higher pNK cell percentages; this parameter was shown to be the best to differentiate between patient and control groups [46]. Moreover, increased activation of pNK cells, which was assessed by CD69 expression, was demonstrated in women with RPL or infertility of unknown etiology [48,49]. Thus, evaluation of pNK cells for their frequency, the status of their activation, and functional activity reveal if alterations with NK cells could be attributed to reproductive failures and guide with therapeutic approaches.

4.3 Maternal killer immunoglobulin-like receptor/placental HLA-C combinations and pregnancy risks

Fetal-derived trophoblast cells lack HLA-A and HLA-B but highly express HLA-C molecules. In addition, they express nonclassical HLA-G and HLA-E. This pattern of HLA class I expression suits well for recognition by NK cells at the maternal–fetal interface. Indeed, uNK cells are characterized by the strong expression of KIRs (their ligands are HLA-C and HLA-G molecules) and CD94/NKG2A (HLA-E). Many KIR/ligand combinations exist due to a high gene polymorphism in both KIRs and ligands families and could be associated with pregnancy disorders. It was shown that the risk of preeclampsia is high in pregnancies where the mother has an "inhibitory" KIR AA genotype and carries a fetus with the HLA-C C2C2 genotype [50]. This combination implicates a high-affinity interaction between inhibitory KIR2DL1 on uNK cells and its ligand HLA-C2 on fetal trophoblast. Strong inhibition of uNK cells is likely responsible for the impairment in remodeling of spiral arteries, leading to the development of preeclampsia. A higher frequency of KIR AA genotype was also reported in women with RPL, where an increased frequency of HLA-C C2 alleles was observed in both women and their partners [51]. A retrospective cohort analysis of maternal KIR genes and embryonic HLA-C genes after autologous fresh IVF cycles resulting in positive beta-human chorionic gonadotropin confirmed maternal KIR AA and fetal HLA-C C2C2 combination is associated with increased risk of losing a pregnancy [52]. However, an even greater risk was revealed for the combination of maternal KIR Bx and embryonic HLA-C C1C1. This implicates that there are additional aspects of uNK cells/trophoblast recognition and interaction that could impact pregnancy. One of such aspects is a degree of functional competency of maternal NK cells, namely their "educational" status.

4.4 Natural killer cell education (maternal KIR/maternal HLA-C combinations) and pregnancy risks

Self HLA-C, HLA-Bw4, HLA-E repertoires provide education/licensing for NK cells. This process is restricted by the individual's KIR genotype and the presence/absence of KIRs that are important for education/licensing (Table 1). Thus, KIR2DL2/HLA-C C1 recognition is important for all NK cells expressing KIR2DL2. However, this KIR/ligand combination is relatively less frequent in a population, as KIR2DL2 is present in approximately 53% of individuals (data for N. American Caucasians) [27]. Assuming Hardy-Weinberg equilibrium, approximately 40% of individuals are HLA-C C1C1 (this was calculated for N. American Caucasian population) [28,29] and, correspondingly, approximately 21.2% of individuals are expected to be KIR2DL2/HLA-C C1C1. However, among KIR2DL2 positive women suffering from RPL or infertility of unknown etiology, the frequency of HLA-C- C1C1 was significantly lower (25.9%), and only 13.5% of women were KIR2DL2/HLA-C C1C1 [53]. This data points out the protective effect of KIR2DL2/HLA-C1 combination, leading to a lower risk of pregnancy failures in women with these genes. Another study revealed that the frequency of KIR2DS1/HLA-C C2C2 combination, which contributes to the development of hypofunctioning NK cells, is significantly increased among women with RPL. The frequency of KIR2DS1/HLA-C C2C2 women in the studied cohort was 11.5%, while a calculated prevalence of this combination for a corresponding population is less than 6% [28]. Thus, the presence of hypofunctioning cells among uNK cells could be related to pregnancy failures, where uNK cells activity was not sufficient to support placental growth.

Failure to educate NK cells via CD94/NKG2A receptor that recognizes HLA-E is also associated with pregnancy disorders. The expression of HLA-E is influenced by HLA-B allele-derived binding peptides that are necessary for appropriate folding and transporting of HLA-E to the cell surface. However, HLA-B alleles (identified by threonine in leader sequence, -21T variant) provide a nonfunctioning peptide. As a result, diminished HLA-E expression limits the CD94/NKG2A education of NK cells [26,30]. A genome-wide *meta*-analysis of a large cohort of mothers (>160,000 participants) revealed that a presence of the -21T variant of the HLA-B gene is associated with a 7% relative risk of preeclampsia [54]. Testing for maternal HLA-B allele that does not favor CD94/NKG2A-related NK cell education could also be relevant for other pregnancy-related disorders.

5 Evaluation

Flow cytometry analysis is a standard methodology to determine the frequency of NK cells in peripheral blood. Antibodies to detect the expression of CD56 and CD16 are used to obtain a total pNK cell percentage as well as proportions of two pNK cell subsets (Fig. 1). *Meta*-analysis of 22 studies evaluating peripheral NK cell numbers demonstrated significantly higher NK cell numbers in women with RPL compared with controls ($P < 0.00001$) but not in infertile women compared to fertile women [55]. However, in a prospective cohort study, women with RIF had significantly increased NK cell numbers and percentages, increased $CD56^{dim}$ and $CD56^{dim}CD69^{+}$ subsets [56].

To evaluate the cytotoxic ability of NK cells, a target-cell killing assay is used. Radioactive chromium (51Cr)-release assay with human erythromyelocytic leukemia cells (K562) cells as targets is the gold standard assay to measure NK cell cytotoxicity. K562 cells are labeled with 51Cr and incubated with NK cells at different effector: target ratios. Target-cell killing is determined by measuring of 51Cr released into the media from dead K562 cells. A modification of this assay is the use of a fluorescent dye to label K562 cells. A percentage of target-cell killing was measured by flow cytometry analysis using another dye that binds to dead cells (Fig. 2). Using an NK cell cytotoxicity of 17.8% as a cut-off, NK cell cytotoxicity was able to discriminate between RIF and controls with a sensitivity of 83.2% and a specificity of 84%, giving an odds ratio (OR) of 9.3 [57]. In women with RPL, using a cut-off of 17.7%, NK cell cytotoxicity was able to differentiate RPL from controls with a sensitivity of 78% and a specificity of 84%, giving an OR of 7.2 [57].

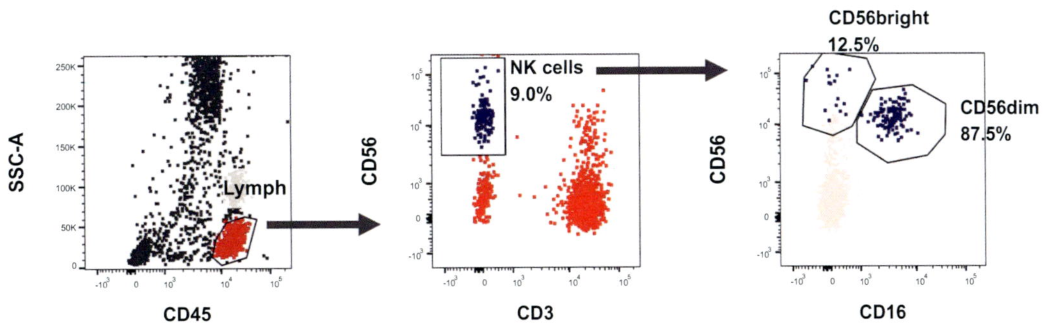

FIGURE 1 Flow cytometry analysis of peripheral blood NK cells. Lymphocytes are gated as CD45 positive/low side scatter (SSC) cells. Among lymphocytes, NK cells are gated as CD56 positive and CD3 negative cells. Finally, two subsets of NK cells can be identified on CD56 versus CD16 plot. A majority of pNK cells are CD56dim cells and CD16 positive. A minor population of pNK cells is CD56bright and CD16 negative.

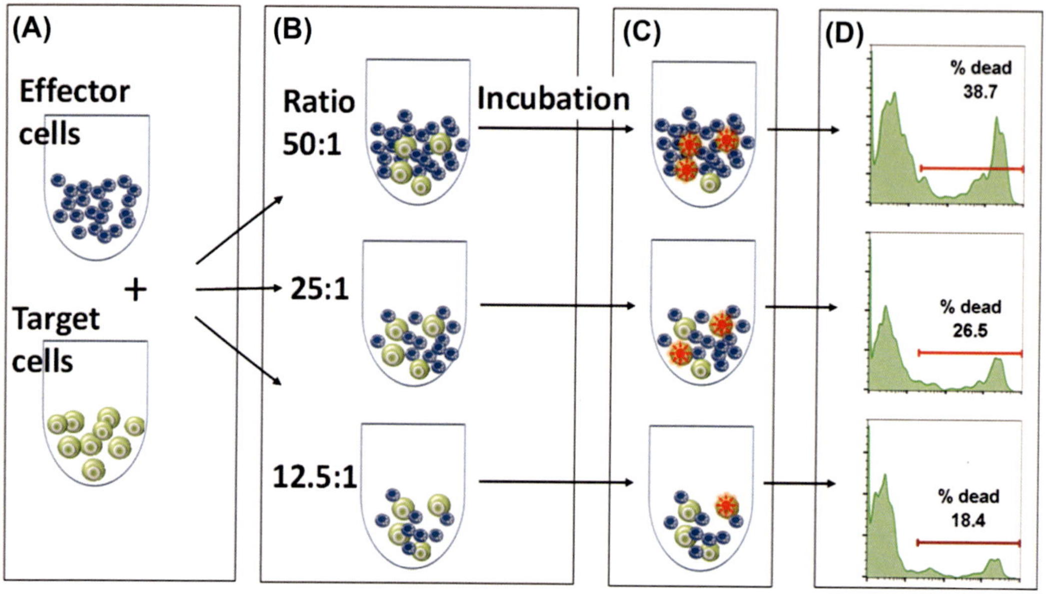

FIGURE 2 Flow cytometry-based NK cell cytotoxicity assay. (A) Effector cells are peripheral blood mononuclear cells, which are isolated from venous blood collected into sodium heparin tubes. Target cells are K562 cells labeled PKH67 green fluorescent dye. (B) Effector and target cells are combined at indicated ratios and incubated for 2 h. (C) Propidium iodide (PI) is added to the tubes to stain target cells that are killed by NK cells. (D) Samples were analyzed by flow cytometry, where K562 cells are gated by their green fluorescence and dead K562 cells further identified by PI staining. Histograms show the percentage of PI positive cells for each effector target ratio.

Other methods that are employed for the evaluation of NK cells include CD69 expression analysis and CD107a degranulation assay. These markers reflect the functional reactivity of NK cells. An incubation with K562 cells is known to cause pNK cells activation and upregulation of CD69, which can be measured by flow cytometry analysis [48,58]. Cell surface detection of CD107a is indicative of recent degranulation because, normally, CD107a is present on membranes of cytoplasmic granules. Similarly, incubation of pNK cells with K562 cells is used for the stimulation, followed by CD107a expression analysis by flow cytometry.

To assess uNK cells in endometrial biopsies, immunohistochemistry staining with antibodies for CD56 is performed [59]. Analysis of mRNA expression by molecular assays is also used for the evaluation of uNK cells. Thus, the endometrial immune profile test based on quantitative RT-PCR methodology includes CD56 mRNA evaluation [40]. Granzyme B expression as a marker for uNK cells is measured in the decidualization score assay based on targeted RNA sequencing methodology [60]. Timing of endometrial biopsy sample collection is crucial for the uNK cell assessment. The population of uNK cells undergoes dramatic changes during every cycle, with a rapid acceleration in the number of uNK cells that occurs 7–8 days after ovulation [59].

Finally, HLA and KIR genotyping provides information on KIR repertoire and HLA ligand availability. Inclusion of partner's evaluation for HLA-C groups will help predict fetal HLA-C genotypes and could guide the estimation of potential risks for following embryo transfers.

6 Reproductive outcome

Elevated NK cell proportion and cytotoxicity have been reported in women with RPL. Various medications have been reported to suppress or modulate NK cell cytotoxicities and/or their levels, including prednisone, intravenous immunoglobulin G (IVIg), intralipid infusion, etc. [61–63].

In a recent study of women with RPL and RIF of immune etiologies ($n = 197$) undergoing IVF/ET or FET cycle with personalized immunomodulatory and anticoagulation treatment, the NK cell levels, and cytotoxicities were significantly decreased at the time of embryo transfer compared to the baseline levels. The pregnancy rate, live birth rate (LBR) per embryo transfer cycle, and LBR per pregnancy (48.2%, 39.6%, and 82.1%, respectively) in women undergoing index IVF/ET or FET cycle were significantly higher than those of historical controls (33%, 1.8%, and 5.3%, respectively) ($P < 0.001$, $P < 0.0001$, and $P < 0.0001$, respectively) [64].

Regarding KIR and HLA-C genotypes, women with KIR2DL2pos and HLA-C C1C1 had significantly higher NK cytotoxicity than women with KIR2DL2pos and HLA-C C2C2, but no differences in CD56+ NK cell levels, suggesting an NK cell education effect [53]. There were no differences in the genotype frequencies of HLA-C, KIR2DL2, KIR AA, and Bx genotypes between 26 women with RPL who delivered at term and 32 women who had obstetrical complications, including 22 with first trimester losses, two with second-trimester losses, one with intrauterine growth restriction, two with preeclampsia, and five with preterm deliveries, suggesting that immunomodulation therapy improved the pregnancy outcome of KIR2DL2pos women with RPL and HLA-C mismatching [53].

7 Treatment

Multiple immunomodulatory treatments have been reported for NK cell-related reproductive failures, including prednisone, IVIg, intralipid, granulocyte-macrophage colony-stimulating factor, TNF-blockers, lymphocyte immunization therapy, and hydroxychloroquine.

7.1 Corticosteroid treatment

uNK cells express glucocorticoid receptors [65]. In women with RPL, prednisolone treatment significantly reduced uNK cell numbers ($P = 0.0004$) [66]. Additionally, prednisone treatment was reported to shift Treg/Th17 immune balance to a T reg bias and improve pregnancy outcomes in women with RIF [67]. Although well-known for its possible side effects, prednisolone treatment can be easily applied due to its convenient route of application and relatively low cost. In a double-blind, randomized placebo control study of 160 women with unexplained RPL, combined prednisolone and heparin treatment had a significantly higher success rate (70.3%) than heparin and low dose aspirin treatment only (9.2%) ($P < 0.05$) [68]. It was reported that the effect might be due to a suppressive effect of steroids on the peripheral CD16 NK cells concentration [68]. However, the study did not consider the NK cell evaluation of study participants. In the Cochran review of 14 RCTs for women undergoing IVF or ICSI, prednisone treatment did not improve the live birth rate (3 RTCs, OR 1.21) or pregnancy rate (13 RCTs, OR 1.16). However, a subgroup analysis of women undergoing IVF/ICSI (6 RCTs) revealed a significantly higher pregnancy rate using glucocorticoids (OR 1.5) [69]. Recently, another systemic review of 4 RCTs, which compared glucocorticoid supplementation during IVF stimulation versus placebo, revealed a possible increase in clinical pregnancy rate but little or no impact on the live birth rate [70]. Unfortunately, all RCTs included in these systemic reviews did not evaluate study participants for immune abnormalities, and glucocorticoid treatment was given empirically. Hence, it seems that subfertility and undergoing IVF or ICSI are not sufficient indications for empirical use of glucocorticoid treatment. Future trials should focus on specific clinical indications, such as RPL and RIF, and immune abnormalities.

7.2 Intravenous immunoglobulin G treatment

IVIg treatment has been reported to effectively suppress NK cell levels and cytotoxicity in women with RPL and RIF, favoring pregnancy outcomes [61,71,72]. Additionally, IVIg treatment increased NK cell inhibitory receptor expression and decreased the activating receptor expression [73].

In an observational study of women with RPL and NK or NKT like expansion, IVIg treatment significantly improved the live birth rate to 96.3%, compared with 30.6% in those without IVIg ($P < 0.0001$) [74]. In women with RIF, the pregnancy rate was significantly higher in the IVIg group (93.8%) than in the nontreated group (26.2%). The live birth rate was also significantly higher in the IVIg group (80.0%) than in the nontreated group (17.0%) [74]. In women with RPL, the live birth rate of women with cellular immune abnormalities (increased NK cell levels, or cytotoxicity and Th1/Th2 ratio) who were treated with IVIg was comparable to those without cellular immune abnormalities and no treatment [75], indicating clinical efficacy of IVIg in RPL with cellular immune abnormalities.

Polanski et al. conducted a systematic review in infertile women undergoing ART with elevated NK cell numbers and/or activity and reported a beneficial effect of the interventions on clinical pregnancy rates with RR of 1.63 (95% CI, 1.00–2.66) for prednisone and 3.41 (95% CI, 1.90–6.11) for IVIg, favoring intervention. However, due to significant heterogeneity, a cautious interpretation was suggested [76]. In a recent retrospective cohort study of 197 RPL and/or RIF, prednisone only or prednisone and IVIg treatment with anticoagulation treatment significantly increased the pregnancy rate (48.2% vs 33.0%, $P < 0.001$), and the live birth rate (39.6% vs 1.8%, $P < 0.001$) of IVF cycles compared to historical controls [64]. In this study, personalized immunomodulatory treatment was given to normalize NK cell levels, cytotoxicities, and Th1/Th2 cell ratios. Overall, reproductive outcomes of IVF cycles were significantly improved with immunomodulatory treatment when immune markers were better controlled.

7.3 Other treatments

Other immunotherapies have been reported to regulate NK cells in women with RPL and RIF. Lymphocyte immunotherapy was reported to effectively suppress NK cell levels and cytotoxicities in women with RPL [77]. Intralipid treatment was also reported to control peripheral blood NK cell levels and cytotoxicity and improve pregnancy outcomes [78]. A total of 50 women with abnormal NKC received intralipid treatment, either with 2 mL of intralipid 20% diluted in 250 mL saline or 4 mL of intralipid 20% diluted in 250 mL saline. 78% had suppression of NKC after one week, and 22% showed suppression but still above the normal threshold. Both animal and human studies suggested that intralipid treatment may enhance implantation and maintenance of pregnancy, and reproductive outcomes of intralipid treatment were comparable to IVIg in women displaying abnormal immune risk factors [79]. A recent systemic review of intralipid treatment in IVF/ICSI outcome was inconclusive [80]. Again, unfortunately, intralipid treatment was given empirically without NK cell evaluation.

In a randomized controlled trial of women with refractory innate immune RPL, etanercept 25 mg/week starting cycle day 1 significantly reduced TNF-a levels and NK cell activity ($P < 0.05$) than placebo, and live birth rate (89.47%) was significantly higher in the etanercept treatment group than the placebo group (72.04%) (OR 3.30, 95% CI 1.49–7.32, $P = 0.01$). Based on this study, etanercept can be an attractive therapeutic strategy for refractory innate immune RPL [81].

Various immunotherapies may enhance implantation and maintenance of pregnancy in women with abnormal NK cell levels or activities. However, the potential medications and candidates for immunomodulatory treatment should be carefully selected after a thorough evaluation and consultation by specialists in reproductive immunology.

8 Summary and recommendations

NK cells are important effector cells that contribute to host defense against infections and control over malignancies. Moreover, they play an active role at the maternal–fetal interface and are important for pregnancy support. Their activity is tightly regulated by receptor/ligand recognition, among which KIR and HLA class I interactions play an essential role in human NK cells. Various disorders of reproduction, such as RPL or RIF, have been associated with NK cell-related abnormalities that were revealed with either peripheral blood or endometrial samples testing.

- NK cells are the most abundant immune cells in the endometrium at the time of embryo implantation and during the first trimester of pregnancy.
- NK cells play an important role at the maternal–fetal interface by regulating angiogenesis and placental growth.
- Elevated NK cell proportion and cytotoxicity have been reported in women with RPL.
- Various immunotherapies, including glucocorticoid, IVIg, lymphocyte immunotherapy, intralipid, TNF-blockers, have been reported to suppress/modulate NK cell numbers and cytotoxicities.
- Well-designed clinical trials are needed in the future to investigate the efficacy of various immunotherapies.
- Immunotherapy for women with RPL and RIF should be given for women with immune abnormalities, but not empirically.

References

[1] Bulmer JN, Sunderland CA. Immunohistological characterization of lymphoid cell populations in the early human placental bed. Immunology 1984;52(2):349–57.
[2] Ritson A, Bulmer JN. Endometrial granulocytes in human decidua react with a natural-killer (NK) cell marker, NKH1. Immunology 1987;62(2):329–31.
[3] Karre K, Ljunggren HG, Piontek G, Kiessling R. Selective rejection of H-2-deficient lymphoma variants suggests alternative immune defence strategy. Nature 1986;319(6055):675–8.
[4] Long EO, Kim HS, Liu D, Peterson ME, Rajagopalan S. Controlling natural killer cell responses: integration of signals for activation and inhibition. Annu Rev Immunol 2013;31:227–58.
[5] Braud V, Jones EY, McMichael A. The human major histocompatibility complex class Ib molecule HLA-E binds signal sequence-derived peptides with primary anchor residues at positions 2 and 9. Eur J Immunol 1997;27(5):1164–9.
[6] Pende D, Falco M, Vitale M, et al. Killer Ig-like receptors (KIRs): their role in nk cell modulation and developments leading to their clinical exploitation. Front Immunol 2019;10:1179.
[7] Hao L, Nei M. Rapid expansion of killer cell immunoglobulin-like receptor genes in primates and their coevolution with MHC Class I genes. Gene 2005;347(2):149–59.
[8] Parham P, Norman PJ, Abi-Rached L, Guethlein LA. Human-specific evolution of killer cell immunoglobulin-like receptor recognition of major histocompatibility complex class I molecules. Philos Trans R Soc Lond B Biol Sci 2012;367(1590):800–11.
[9] Gumperz JE, Litwin V, Phillips JH, Lanier LL, Parham P. The Bw4 public epitope of HLA-B molecules confers reactivity with natural killer cell clones that express NKB1, a putative HLA receptor. J Exp Med 1995;181(3):1133–44.
[10] Lanier LL, Le AM, Civin CI, Loken MR, Phillips JH. The relationship of CD16 (Leu-11) and Leu-19 (NKH-1) antigen expression on human peripheral blood NK cells and cytotoxic T lymphocytes. J Immunol 1986;136 (12):4480–6.
[11] Barrow AD, Martin CJ, Colonna M. The natural cytotoxicity receptors in health and disease. Front Immunol 2019;10:909.
[12] Fauriat C, Long EO, Ljunggren HG, Bryceson YT. Regulation of human NK-cell cytokine and chemokine production by target cell recognition. Blood. 2010;115(11):2167–76.
[13] Hawke LG, Whitford MKM, Ormiston ML. The production of pro-angiogenic VEGF-A isoforms by hypoxic human NK cells is independent of their TGF-beta-mediated conversion to an ILC1-like phenotype. Front Immunol 2020;11:1903.
[14] Greenberg AH. The origins of the NK cell, or a Canadian in King Ivan's court. Clin Invest Med 1994;17 (6):626–31.
[15] Freud AG, Mundy-Bosse BL, Yu J, Caligiuri MA. The broad spectrum of human natural killer cell diversity. Immunity. 2017;47(5):820–33.
[16] Bjorkstrom NK, Ljunggren HG, Michaelsson J. Emerging insights into natural killer cells in human peripheral tissues. Nat Rev Immunol 2016;16(5):310–20.
[17] Cooper MA, Fehniger TA, Turner SC, et al. Human natural killer cells: a unique innate immunoregulatory role for the CD56(bright) subset. Blood. 2001;97(10):3146–51.
[18] Koopman LA, Kopcow HD, Rybalov B, et al. Human decidual natural killer cells are a unique NK cell subset with immunomodulatory potential. J Exp Med 2003;198(8):1201–12.
[19] Ashkar AA, Di Santo JP, Croy BA. Interferon gamma contributes to initiation of uterine vascular modification, decidual integrity, and uterine natural killer cell maturation during normal murine pregnancy. J Exp Med 2000;192(2):259–70.
[20] Lash GE, Schiessl B, Kirkley M, et al. Expression of angiogenic growth factors by uterine natural killer cells during early pregnancy. J Leukoc Biol 2006;80(3):572–80.
[21] Fu B, Zhou Y, Ni X, et al. Natural killer cells promote fetal development through the secretion of growth-promoting factors. Immunity. 2017;47(6):1100–13 e1106.
[22] Brighton PJ, Maruyama Y, Fishwick K, et al. Clearance of senescent decidual cells by uterine natural killer cells in cycling human endometrium. Elife. 2017;6.
[23] Anfossi N, Andre P, Guia S, et al. Human NK cell education by inhibitory receptors for MHC class I. Immunity. 2006;25(2):331–42.

[24] Fauriat C, Ivarsson MA, Ljunggren HG, Malmberg KJ, Michaelsson J. Education of human natural killer cells by activating killer cell immunoglobulin-like receptors. Blood. 2010;115(6):1166–74.
[25] Boudreau JE, Hsu KC. Natural killer cell education and the response to infection and cancer therapy: stay tuned. Trends Immunol 2018;39(3):222–39.
[26] Horowitz A, Djaoud Z, Nemat-Gorgani N, et al. Class I HLA haplotypes form two schools that educate NK cells in different ways. Sci Immunol 2016;1(3).
[27] Hollenbach JA, Meenagh A, Sleator C, et al. Report from the killer immunoglobulin-like receptor (KIR) anthropology component of the 15th International Histocompatibility Workshop: worldwide variation in the KIR loci and further evidence for the co-evolution of KIR and HLA. Tissue Antigens 2010;76(1):9–17.
[28] Dambaeva SV, Lee DH, Sung N, et al. Recurrent pregnancy loss in women with killer cell immunoglobulin-like receptor KIR2DS1 is associated with an increased HLA-C2 allelic frequency. Am J Reprod Immunol 2016;75(2):94–103.
[29] Wang SS, Abdou AM, Morton LM, et al. Human leukocyte antigen class I and II alleles in non-Hodgkin lymphoma etiology. Blood. 2010;115(23):4820–3.
[30] Hallner A, Bernson E, Hussein BA, et al. The HLA-B -21 dimorphism impacts on NK cell education and clinical outcome of immunotherapy in acute myeloid leukemia. Blood 2019;133(13):1479–88.
[31] Orange JS, How I. Manage natural killer cell deficiency. J Clin Immunol 2020;40(1):13–23.
[32] Boudreau JE, Mulrooney TJ, Le Luduec JB, Barker E, Hsu KC. KIR3DL1 and HLA-B density and binding calibrate nk education and response to HIV. J Immunol 2016;196(8):3398–410.
[33] Imai K, Matsuyama S, Miyake S, Suga K, Nakachi K. Natural cytotoxic activity of peripheral-blood lymphocytes and cancer incidence: an 11-year follow-up study of a general population. Lancet. 2000;356 (9244):1795–9.
[34] Carlsten M, Jaras M. Natural killer cells in myeloid malignancies: immune surveillance, NK cell dysfunction, and pharmacological opportunities to bolster the endogenous NK cells. Front Immunol 2019;10:2357.
[35] Fogel LA, Yokoyama WM, French AR. Natural killer cells in human autoimmune disorders. Arthritis Res Ther 2013;15(4):216.
[36] Dalbeth N, Callan MF. A subset of natural killer cells is greatly expanded within inflamed joints. Arthritis Rheum 2002;46(7):1763–72.
[37] Yen JH, Moore BE, Nakajima T, et al. Major histocompatibility complex class I-recognizing receptors are disease risk genes in rheumatoid arthritis. J Exp Med 2001;193(10):1159–67.
[38] Nelson GW, Martin MP, Gladman D, Wade J, Trowsdale J, Carrington M. Cutting edge: heterozygote advantage in autoimmune disease: hierarchy of protection/susceptibility conferred by HLA and killer Ig-like receptor combinations in psoriatic arthritis. J Immunol 2004;173(7):4273–6.
[39] Santillan I, Lozano I, Illan J, et al. Where and when should natural killer cells be tested in women with repeated implantation failure? J Reprod Immunol 2015;108:142–8.
[40] Ledee N, Petitbarat M, Prat-Ellenberg L, et al. Endometrial immune profiling: a method to design personalized care in assisted reproductive medicine. Front Immunol 2020;11:1032.
[41] El-Azzamy H, Dambaeva SV, Katukurundage D, et al. Dysregulated uterine natural killer cells and vascular remodeling in women with recurrent pregnancy losses. Am J Reprod Immunol 2018;80(4):e13024.
[42] Lucas ES, Vrljicak P, Muter J, et al. Recurrent pregnancy loss is associated with a pro-senescent decidual response during the peri-implantation window. Commun Biol 2020;3(1):37.
[43] Fukui A, Fujii S, Yamaguchi E, Kimura H, Sato S, Saito Y. Natural killer cell subpopulations and cytotoxicity for infertile patients undergoing in vitro fertilization. Am J Reprod Immunol 1999;41(6):413–22.
[44] Coulam CB, Roussev RG. Correlation of NK cell activation and inhibition markers with NK cytoxicity among women experiencing immunologic implantation failure after in vitro fertilization and embryo transfer. J Assist Reprod Genet 2003;20(2):58–62.
[45] Kwak JY, Beaman KD, Gilman-Sachs A, Ruiz JE, Schewitz D, Beer AE. Up-regulated expression of CD56 + , CD56 + /CD16 + , and CD19 + cells in peripheral blood lymphocytes in pregnant women with recurrent pregnancy losses. Am J Reprod Immunol 1995;34(2):93–9.
[46] King K, Smith S, Chapman M, Sacks G. Detailed analysis of peripheral blood natural killer (NK) cells in women with recurrent miscarriage. Hum Reprod 2010;25(1):52–8.
[47] Kuon RJ, Vomstein K, Weber M, et al. The "killer cell story" in recurrent miscarriage: Association between activated peripheral lymphocytes and uterine natural killer cells. J Reprod Immunol 2017;119:9–14.

[48] Ntrivalas EI, Kwak-Kim JY, Gilman-Sachs A, et al. Status of peripheral blood natural killer cells in women with recurrent spontaneous abortions and infertility of unknown aetiology. Hum Reprod 2001;16(5):855–61.
[49] Thum MY, Bhaskaran S, Abdalla HI, et al. An increase in the absolute count of CD56dimCD16 + CD69 + NK cells in the peripheral blood is associated with a poorer IVF treatment and pregnancy outcome. Hum Reprod 2004;19(10):2395–400.
[50] Hiby SE, Walker JJ, O'Shaughnessy KM, et al. Combinations of maternal KIR and fetal HLA-C genes influence the risk of preeclampsia and reproductive success. J Exp Med 2004;200(8):957–65.
[51] Hiby SE, Regan L, Lo W, Farrell L, Carrington M, Moffett A. Association of maternal killer-cell immunoglobulin-like receptors and parental HLA-C genotypes with recurrent miscarriage. Hum Reprod 2008;23(4):972–6.
[52] Morin SJ, Treff NR, Tao X, et al. Combination of uterine natural killer cell immunoglobulin receptor haplotype and trophoblastic HLA-C ligand influences the risk of pregnancy loss: a retrospective cohort analysis of direct embryo genotyping data from euploid transfers. Fertil Steril 2017;107(3):677–83 e672.
[53] Yang X, Yang E, Wang WJ, et al. Decreased HLA-C1 alleles in couples of KIR2DL2 positive women with recurrent pregnancy loss. J Reprod Immunol 2020;142:103186.
[54] Shreeve N, Depierreux D, Hawkes D, et al. The CD94/NKG2A inhibitory receptor educates uterine NK cells to optimize pregnancy outcomes in humans and mice. Immunity 2021;54(6):1231–44 e1234.
[55] Seshadri S, Sunkara SK. Natural killer cells in female infertility and recurrent miscarriage: a systematic review and meta-analysis. Hum Reprod Update 2014;20(3):429–38.
[56] Sacks G, Yang Y, Gowen E, Smith S, Fay L, Chapman M. Detailed analysis of peripheral blood natural killer cells in women with repeated IVF failure. Am J Reprod Immunol 2012;67(5):434–42.
[57] Salazar MD, Wang W, Skariah A, et al. Post-hoc evaluation of peripheral blood natural killer cell cytotoxicity in predicting the risk of recurrent pregnancy losses and repeated implantation failures. J Reprod Immunol 2022.
[58] Dons'koi BV, Chernyshov VP, Osypchuk DV. Measurement of NK activity in whole blood by the CD69 up-regulation after co-incubation with K562, comparison with NK cytotoxicity assays and CD107a degranulation assay. J Immunol Meth 2011;372(1–2):187–95.
[59] Russell P, Sacks G, Tremellen K, Gee A. The distribution of immune cells and macrophages in the endometrium of women with recurrent reproductive failure. III: further observations and reference ranges. Pathology 2013;45(4):393–401.
[60] Dambaeva S, Bilal M, Schneiderman S, et al. Decidualization score identifies an endometrial dysregulation in samples from women with recurrent pregnancy losses and unexplained infertility. F S Rep 2021;2(1):95–103.
[61] Ruiz JE, Kwak JY, Baum L, et al. Effect of intravenous immunoglobulin G on natural killer cell cytotoxicity in vitro in women with recurrent spontaneous abortion. J Reprod Immunol 1996;31(1–2):125–41.
[62] Nair MP, Schwartz SA. Immunomodulatory effects of corticosteroids on natural killer and antibody-dependent cellular cytotoxic activities of human lymphocytes. J Immunol 1984;132(6):2876–82.
[63] Austin Taylor M, Bennett M, Kumar V, Schatzle JD. Functional defects of NK cells treated with chloroquine mimic the lytic defects observed in perforin-deficient mice. J Immunol 2000;165(9):5048–53.
[64] Sung N, Khan SA, Yiu ME, et al. Reproductive outcomes of women with recurrent pregnancy losses and repeated implantation failures are significantly improved with immunomodulatory treatment. J Reprod Immunol 2021;148:103369.
[65] Henderson TA, Saunders PT, Moffett-King A, Groome NP, Critchley HO. Steroid receptor expression in uterine natural killer cells. J Clin Endocrinol Metab 2003;88(1):440–9.
[66] Quenby S, Kalumbi C, Bates M, Farquharson R, Vince G. Prednisolone reduces preconceptual endometrial natural killer cells in women with recurrent miscarriage. Fertil Steril 2005;84(4):980–4.
[67] Huang Q, Wu H, Li M, Yang Y, Fu X. Prednisone improves pregnancy outcome in repeated implantation failure by enhance regulatory T cells bias. J Reprod Immunol 2021;143:103245.
[68] Gomaa MF, Elkholy AG, El-Said MM, Abdel-Salam NE. Combined oral prednisolone and heparin vs heparin: the effect on peripheral NK cells and clinical outcome in patients with unexplained recurrent miscarriage. A double-blind placebo randomized controlled trial. Arch Gynecol Obstet 2014;290(4):757–62.
[69] Boomsma CM, Keay SD, Macklon NS. Peri-implantation glucocorticoid administration for assisted reproductive technology cycles. Cochrane Database Syst Rev 2012;6 CD005996.
[70] Kalampokas T, Pandian Z, Keay SD, Bhattacharya S. Glucocorticoid supplementation during ovarian stimulation for IVF or ICSI. Cochrane Database Syst Rev 2017;3 CD004752.

[71] Kwak JY, Kwak FM, Ainbinder SW, Ruiz AM, Beer AE. Elevated peripheral blood natural killer cells are effectively downregulated by immunoglobulin G infusion in women with recurrent spontaneous abortions. Am J Reprod Immunol 1996;35(4):363–9.
[72] Ruiz JE, Kwak JY, Baum L, et al. Intravenous immunoglobulin inhibits natural killer cell activity in vivo in women with recurrent spontaneous abortion. Am J Reprod Immunol 1996;35(4):370–5.
[73] Ahmadi M, Ghaebi M, Abdolmohammadi-Vahid S, et al. NK cell frequency and cytotoxicity in correlation to pregnancy outcome and response to IVIG therapy among women with recurrent pregnancy loss. J Cell Physiol 2019;234(6):9428–37.
[74] Ramos-Medina R, Garcia-Segovia A, Gil J, et al. Experience in IVIg therapy for selected women with recurrent reproductive failure and NK cell expansion. Am J Reprod Immunol 2014;71(5):458–66.
[75] Lee SK, Kim JY, Han AR, et al. Intravenous immunoglobulin g improves pregnancy outcome in women with recurrent pregnancy losses with cellular immune abnormalities. Am J Reprod Immunol 2016;75(1):59–68.
[76] Polanski LT, Barbosa MA, Martins WP, et al. Interventions to improve reproductive outcomes in women with elevated natural killer cells undergoing assisted reproduction techniques: a systematic review of literature. Hum Reprod 2014;29(1):65–75.
[77] Kwak JY, Gilman-Sachs A, Moretti M, Beaman KD, Beer AE. Natural killer cell cytotoxicity and paternal lymphocyte immunization in women with recurrent spontaneous abortions. Am J Reprod Immunol 1998;40(5):352–8.
[78] Roussev RG, Acacio B, Ng SC, Coulam CB. Duration of intralipid's suppressive effect on NK cell's functional activity. Am J Reprod Immunol 2008;60(3):258–63.
[79] Coulam CB, Acacio B. Does immunotherapy for treatment of reproductive failure enhance live births? Am J Reprod Immunol 2012;67(4):296–304.
[80] Zhou P, Wu H, Lin X, Wang S, Zhang S. The effect of intralipid on pregnancy outcomes in women with previous implantation failure in in vitro fertilization/intracytoplasmic sperm injection cycles: a systematic review and meta-analysis. Eur J Obstet Gynecol Reprod Biol 2020;252:187–92.
[81] Fu J, Li L, Qi L, Zhao L. A randomized controlled trial of etanercept in the treatment of refractory recurrent spontaneous abortion with innate immune disorders. Taiwan J Obstet Gynecol 2019;58(5):621–5.

CHAPTER

3

T helper cell pathology and recurrent pregnancy losses; Th1/Th2, Treg/Th17, and other T cell responses

Joon Cheol Park[1], *Jae Won Han*[2] *and Sung Ki Lee*[2]

[1]Department of Obstetrics and Gynecology, Keimyung University School of Medicine, Daegu, Republic of Korea [2]Department of Obstetrics and Gynecology, Konyang University College of Medicine, Daejeon, Republic of Korea

1 The immunologic perspective of recurrent pregnancy loss

The concept of T helper 1 (Th1) and Th2 dichotomy was proposed in 1986 by Mosmann et al. for the first time [1]. Subsequently, the Th1/Th2 paradigm was adopted to explain maternal tolerance toward fetal alloantigens during pregnancy by Wegmann et al. [2]. About a decade ago, the Th1/Th2 paradigm was further extended to Th1/Th2/Th17/regulatory T (Treg) cell paradigm [3]. Since then, other Th cell populations, including Th9, Th22, follicular Th (Tfh), and memory cells, have been introduced, and immune checkpoint molecules have been actively investigated to regulate immune effectors. Immunopathology of recurrent pregnancy loss (RPL) is one of the most extensively explored subjects in reproductive immunology, which is deeply rooted in T cell pathology. In this chapter, recent advances in Th cell immunity concerning RPL are comprehensively reviewed.

2 Definition and prevalence of recurrent pregnancy loss

Recently, RPL has been redefined as two or more pregnancy losses [4,5]. This new definition is now widely welcomed by many clinicians and researchers. A pregnancy loss or miscarriage can be defined as the spontaneous demise of a pregnancy before the fetus reaches viability, commonly 24 weeks of gestation [6]. However, there is no consensus in

DOI: https://doi.org/10.1016/B978-0-323-90805-4.00019-5

terms of the definition of pregnancy. American Society of Reproductive Medicine claimed that pregnancy indicates a clinical pregnancy documented by ultrasonography or histopathological examination [5]. While the European Society of Human Reproduction and Embryology (ESHRE) defines pregnancy as clinical and biochemical pregnancy determined by urine and serum β-human chorionic gonadotropin (hCG) test [6].

3 Prevalence

It is hard to estimate the exact prevalence of RPL. Many publications report that three or more consecutive clinical pregnancy losses before 20 weeks occur in approximately 1%–2% of couples [4,6,7]. If biochemical pregnancy is included, the prevalence increases to 2% to 3% of women [8]. Recently, an epidemiological study of all Swedish women aged 18–42 years showed that the prevalence of RPL, defined as three or more consecutive pregnancy losses before 22 weeks of gestation, increased by 74% between 2003 and 2012 [9]. Immunological, inflammatory and environmental factors were suggested as possible causes of the increase [9] However, further research is needed to explore why RPL has been increased rapidly.

Although RPL has various causes, it is estimated that cellular immune abnormality is one of the most contributing etiologies. According to our previous study of peripheral blood (PB) NK cell levels, NK cytotoxicities, and Th1/Th2 cytokine-producing Th cell ratios of nonpregnant women with RPL, the frequency of cellular immunity was 64% among all RPL women and 76% in idiopathic RPL women [10]. Moreover, abnormal Th1/Th2 cytokine-producing Th cell ratios were relatively common, accounting for 34.7% of all RPL women and 38.1% of the unexplained group.

4 Immune regulations in normal pregnancy and recurrent pregnancy loss

4.1 T cells during a menstrual cycle

Lymphocytes in PB and the uterus fluctuate during a menstrual cycle. Circulating CD3+ and CD4+ T cell proportions were significantly decreased in the luteal phase compared to the follicular phase [11]. However, CD8+ T cell proportions did not change significantly during the menstrual cycle. As a result, the Th/Tc ratio was lower in the luteal phase than in the follicular phase. In contrast to T cells, proportions of CD3-CD56+ natural killer (NK) cells increased significantly after ovulation. In CD4+ T cells, the production of tumor necrosis factor (TNF)-α, interferon (IFN)-γ, and IL-10 and the Th1/Th2 ratio did not fluctuate during the menstrual cycle [11].

In the nonpregnant endometrium, the proportion of T cells was decreased in the secretory phase compared to the proliferative phase (1.4% vs 0.8%) [12]. The percentage of CD8+ T cells among all CD3+ cells decreased significantly from 63% in the proliferative phase to 54.2% in the luteal phase. On the other hand, because CD4+ T cells slightly increased, the CD4+/CD8+ T cell ratio was increased from 1.9 in the proliferative phase to 4.2 in the luteal phase. This change is probably due to an increase in endometrial NK cells in the late secretory phase.

The proportion of PB Foxp3 + Treg cells expands during the follicular phase along with the increased serum estrogen concentration [13], and endometrial Foxp3 + cells were gradually upregulated throughout the proliferative phase [14]. Th17 cells in PB of non-pregnant women occupy a small proportion of Th cells as 1.4% [15]. However, it is unclear whether Th17 cells fluctuate during a menstrual cycle as Treg cells do [16].

4.2 T cells in human pregnancy

Endometrium prepares for the implantation of an embryo through the influx of leukocytes [17]. During a menstrual cycle, the density of NK cells and macrophages increases from the proliferative phase to the secretory phase [18]. The proportion of T cells in the decidual immune cells is as low as 10%–20% in early pregnancy but increases up to 44% as pregnancy progresses [19]. However, the numbers of CD4 + and CD8 + T cells do not change during pregnancy [19]. Among decidual CD3 + T cells, 30%–45% are CD4 + T cells, 45%–75% are CD8 + T cells, and 5% are Treg cells in early pregnancy [20]. CD8 + T cell-dominance over CD4 + T cell was confirmed in the decidua of early pregnancy by recent single-cell transcriptome analysis [21]. Comparing cytokine production of PB Th cells in three different trimesters, the absolute cell count of IL-4 + Th cells was highest in the first trimester [22]. The TNF-α/IL-10 ratio of Th cells was significantly increased toward the end of pregnancy [22]. On the other hand, it is argued that there are no differences in the percentages of PB CD4 + IFN-γ + Th1 and CD4 + IL-10 + Th2 cells during the third trimester compared to those of nonpregnant women [23]. Nevertheless, the production of Th2 cytokines was increased in the decidua, suggesting that Th2 shift is not an inherent feature of Th cells but rather regulated by the microenvironment [20].

CD4 + CD25 + Foxp3 + suppressive Treg cells in PB increase from the beginning of pregnancy, peak in the second trimester, and slowly decrease until the postpartum period [24]. The proportion of Treg cells in the term decidua parietalis was increased compared with those of nonpregnant peripheral blood mononuclear cells (PBMCs) and menstrual blood [17]. Of note, the high proportion of decidual Treg cells at term is primarily due to the increase in differentiated Treg cells rather than naïve Treg cells. In the same study, the profiles of Th1, Th2, and Th17 cells were similar between the nonpregnant endometrium and the term decidua. Likewise, circulating Th17 cells were not different between the non-pregnant period, the first, second, and third trimester of pregnancy [15,16]. On the other hand, another study demonstrated that PB Th17 cells were decreased in the third trimester of pregnancy compared to the nonpregnant period [23]. Overall, decidua has a unique immune environment to adapt to a potential pregnancy.

4.3 Three immunologic stages during a pregnancy

In the past, tissue inflammation was considered detrimental to pregnancy, and an antiinflammatory or Th2 type dominance maintains a healthy pregnancy. However, the last 20 years' studies have revealed that a normal pregnancy requires local tissue inflammation during early pregnancy, resulting in a revised concept [25]. For a successful pregnancy, the motherrsquo;s immune system must adapt to each stage of pregnancy. Gil et al. described the adaptation of maternal immunity during pregnancy as three immunological stages [25].

4.3.1 *Implantation and placentation: proinflammatory stage*

The process of implantation and early placentation is very similar to the process of tissue injury and repair. At the implantation site, many inflammatory mediators are produced by both endometrial stromal cells and local immune cells, that is, interleukin−6 (IL−6), IL−8, IL−15, granulocyte-macrophage colony-stimulating factor (GM−CSF), CXC chemokine ligand 1 (CXCL1), CC−chemokine ligand 4 (CCL4), osteopontin, and tumor necrosis factor (TNF) [25]. This proinflammatory environment in the endometrium at the implantation period increases the expression, reorganization, and affinity of adhesion molecules and removes the mucin layer. These processes are essential for endometrial receptivity and early placentation. Uterine dendritic cells (DCs) also play a role in uterine receptivity by regulating tissue remodeling and angiogenesis [26]. The uterine DCs do not migrate to the lymph nodes, which promotes the tolerance to paternal antigens [27].

4.3.2 *Fetal growth: antiinflammatory stage*

The fetal growth stage following implantation and placentation is in a Th2 type milieu. It lasts from 13 weeks to 27 weeks. The intrauterine antiinflammatory microenvironment is established at this stage by M2 macrophages, decidual NK (dNK) cells, and Treg cells [25]. M2 macrophages are essential for tissue renewal and phagocytosis of apoptotic trophoblast cells, preventing paternal antigen release [25]. In addition, dNK cells, which have low cytotoxicities, induce Treg cells through interaction with macrophages [28]. During early pregnancy, Treg cells systemically expand and accumulate in the fetomaternal interface, where they protect fetal cells from rejection and regulate Th17-mediated inflammation [25] with their antiinflammatory and antiapoptotic capacities.

4.3.3 *Labor: proinflammatory switch*

When fetal development is complete, the antiinflammatory milieu in the uterus converts into a proinflammatory environment. The proinflammatory nuclear factor-κB (NF-κB) signaling pathway plays a critical role in uterine contractions and delivery. TLR4 induces NF-κB activation by ligating fetal lung-driven-surfactant protein A and endogenous damage-associated molecular patterns (DAMPs) such as high mobility group box1 (HMGB1) [25]. During this process, immune cells invade into the uterine myometrium and resulting in uterine contraction and delivery.

In summary, Th1 dominance is needed during the implantation phase, which switches into Th2 immunity lasting most pregnancy periods after placental development. Then, when labor begins, Th2 immunity is converted back to Th1 immunity to prepare for parturition and postpartum.

4.4 Paternal and fetal antigen-specific tolerance during a pregnancy

In 1953, Sir Peter Medawar proposed three hypotheses for the "immunological paradox" of how a mother supports her semiallogenic fetus in utero for 9 months, including (1) anatomical separation of the fetus from mother by the placenta, (2) immaturity of fetal antigens, and (3) immunological inertness of the maternal immune system during pregnancy [29]. His idea has evolved into a concept of immune tolerance at the fetomaternal interface during pregnancy. However, recent studies showed that the maternal immune system exhibits a specific response to fetal antigens, thereby indicating the presence of aggression toward the fetus and immune tolerance at the same time.

Specific immunity against paternal antigens is first induced by maternal exposure to ejaculated semen. In addition, during pregnancy, continuous exposure to fetal cells and fetal cell-derived materials occurs. Even after childbirth, fetal cells continue to challenge the maternal system through microchimerism [20,30]. As a result, paternal and fetal antigen-specific memory T cells are generated. Even though the development of memory T cells can be harmful to pregnancy, delicate memory T cell balance leads to a successful pregnancy [30].

Semen-induced Treg cells prepare the endometrium for implantation and alleviate uterine inflammation, independently of placental antigen exposure [31]. Fetal antigen is indirectly presented to CD4 + T cells by DCs, an antigen-presenting cell (APC). In humans, decidual T cells recognize mismatched fetal HLA-C and induce CD4 + $CD25^{dim}$ T cell activation and functional CD4 + $CD25^{bright}$ Treg cell proliferation in the decidua [32]. However, HLA-C matching or HLA-A, B, Dr, or DQ mismatching pregnancies did not exhibit this response, suggesting that maternal T cells recognize fetal HLA-C antigen. Additionally, the presence of fetal antigen-specific effector T cells has been reported in animal models. However, effector T cells, controlled by the immune regulatory mechanism, become more robust at the end of pregnancy.

4.5 Underlying immunopathology that impairs reproduction

It is evident that an excessive proinflammatory environment, systemically and locally, is harmful to pregnancy. Th1 and Th2 immune imbalance and Th17 and Treg cell dysregulation have been robustly studied (Fig. 1).

4.5.1 The predominance of Th1 immunity over Th2 immunity in recurrent pregnancy loss

Th1 cells

Activated Th1 cells produce IL-2, TNF-α, and IFN-γ and are the main effectors of host defense [33]. The controlled Th1 predominance seems to benefit the invading extravillous trophoblasts (EVTs) during implantation [33]. TNF-α regulates excessive invasion of trophoblast [34], while promoting MMP9 expression in the preeclamptic decidua [35]. IFN-γ plays a role in early placentation, such as initiation of vascular remodeling [36], EVT apoptosis, and regulation of EVT invasion [33]. However, in human pregnancy, exaggerated Th1 immunity leads to poor reproductive outcomes. Decidual IFN-γ+ Th1 cells were increased more in miscarried women with normal fetal karyotype than in induced

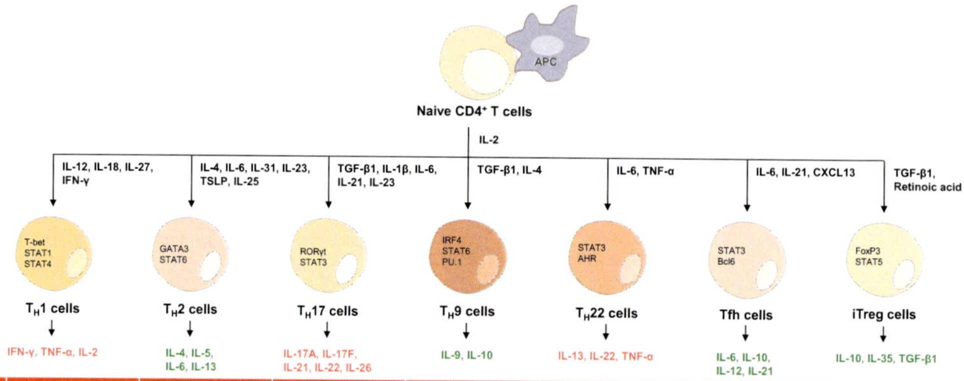

	T_H1 cells	T_H2 cells	T_H17 cells	T_H9 cells	T_H22 cells	Tfh cells	iTreg cells
Common function	• Immunity against virus and bacteria	• Immunity against helminthic infection	• Protection against extracellular bacteria and fungi. • Chronic inflammation	• Defense against helminthic infection	• Wound repair • Immunity to microbial pathogens	• B cell activation • Humoral immunity	• Regulation of all effector cells
Role in pregnancy	• Tissue remodeling • Angiogenesis • Embryo implantation	• Allograft tolerance	• Trophoblast invasion	• Unknown	• Trophoblast survival • Pregnancy maintenance	• Unknown • May be helpful for pregnancy	• Allograft tolerance

FIGURE 1 Human CD4+ T cell subsets and their roles in pregnancy. Naïve CD4 + T cells develop into activated CD4 + T cells by stimulation of antigen-presenting cells. Stimustimulated CD4 + T cells differentiate into various subsets of CD4+ T cells depending on the milieu in which they are activated. In the figure, transcriptional factors corresponding to each subset are shown inside the cells. In addition, cytokines mainly secreted by T cell subsets are presented in red and green. Red represents proinflammatory cytokines, and the green represents cytokines with regulatory functions. Finally, the general functions and roles in the pregnancy of each subset are noted.

abortion [37]. In RPL women, Th1 predominance was observed in the decidua and PB compared to women with normal pregnancies [38]. In comparing cytokine production from PBMCs, concentrations of IL-2, TNF-α, and IFN-γ in supernatants were increased in aborted RPL women compared to those of first-trimester normal pregnant women. Similarly, systemic Th1 shift was demonstrated in RPL, and repeated implantation failure (RIF) women [22,39,40]. It is well known that some autoimmune disorders, such as systemic lupus erythematosus, multiple sclerosis, scleroderma, antiphospholipid syndrome, and Hashimoto's thyroiditis, are associated with Th1 and Th17 immunity [33]. Considering that many RPL women also have these autoimmune conditions, increased Th1 immunity associated with autoimmune diseases may contribute to miscarriage

Th2 cells

Th2 cells are responsible for phagocyte-independent host defense against extracellular pathogens and produce IL-4, IL-5, IL-10, and IL-13 [20]. Local Th2 environment has been reported in a murine and human healthy pregnancy [20,41]. Th2 cells are induced by uterine DCs [42] during pregnancy and activated by paternal antigens expressed by the trophoblast [20]. In the LPS-induced abortion rat model, IL-10 or TNF-α blocker (etanercept) prevented a pregnancy loss [43]. In human pregnancy, LIF, IL-4, and IL-10 from decidual T cells were decreased in women with unexplained RPL compared to women with normal gestation [44]. In comparing IL-6 and IL-10 productions from PBMCs, both cytokines were lower in aborted RPL women than those in normal first trimester pregnant women [38].

Th1 and Th2 imbalance in recurrent pregnancy loss

The Th1 immune response plays a crucial role in rejection in organ transplants, and Th2 immunity contributes to immune tolerance. Based on this, the concept of Th1 and Th2 balance during pregnancy has been simplified as excessive Th1 immune responses to the fetus during pregnancy impairs pregnancy outcomes, and Th2 response offsets Th1 response leading to a successful pregnancy [20]. Abortion-prone mating of CBA/J female and DBA/2 male mice showed increased expression of TNF-α, IFN-γ, and IL-2 compared to control mating of CBA/J female x BALB/c male mice. The high fetal resorption rate was reversed after recombinant IL-10 injection was given to the abortion-prone mating [45].

Nonpregnant RPL women demonstrated higher ratios of TNF-α + Th/IL-10 + Th, TNF-α + Th/IL-4 + Th, and IFN-γ + Th/IL-4 + Th cells in PB than those of fertile controls [39,40]. In a study of pregnant women, Th1/Th2 cytokine ratios at various combinations were consistently increased in aborted RPL women compared to normal first-trimester pregnant women [38]. In another study that compared cytokine production from PBMCs, Th1-type/Th2-type cytokine ratios were significantly increased in pregnant RPL women who aborted in index pregnancy compared with those of normal pregnant women at the end of the first trimester [46]. When decidual T cells were cultured, production of IL-4 and IL-10 was decreased in idiopathic RPL women undergoing a miscarriage compared to women undertaking an induced abortion [44]. In this study, the production of leukemia inhibitory factor (LIF) and macrophage-colony stimulating factor (M-CSF) was also impaired in decidual Th cells from RPL women compared to normal pregnancy. In addition, decidual T cells from women with miscarriage showed higher expression of IL-2 and IFN-γ and lower expression of IL-4 and IL-10 compared with those from normal pregnant women [47]. Furthermore, the production of Th1 cytokine was positively correlated with costimulatory molecules such as CD86 and CD28 expression [47]. Decidual production of IL-2 and IFN-γ was positively correlated with the expression of CD86 and CD28 but negatively correlated with CTLA4 expression at the fetomaternal interface [47].

4.5.2 Treg and Th17 cell dysregulation in recurrent pregnancy loss

Foxp3 + Treg cells

The first study that documented the involvement of Treg cells in spontaneous abortion was reported by Sasaki et al. in 2004 [48]. They found that CD4 + $CD25^{bright}$ T cells were decreased in PB and the decidua of miscarriage group compared with those of normal pregnancy. After their report, several reports confirmed a decrease in the number and suppressive function of Foxp3 + Treg cells systemically and locally in RPL women [13,16,49]. The level of PB Treg cells peak in the late proliferative phase during a menstrual cycle and positively correlated with serum estradiol concentration [13]. Contrarily, RPL women did not show such fluctuation of PB Treg cells [13].

Th17 cells

Th17 cells play a role in protection against extracellular bacteria and fungi and are responsible for chronic inflammation, allograft rejection, and some autoimmune disorders through secretion of IL-17A, IL-17F, IL-21, IL-22, and IL-26 [20,40]. In human pregnancy, decidual stromal cells recruit Th17 cells from PB to the decidua [50]. During the first trimester of pregnancy, decidual Th17 cells promote proliferation and invasion of the trophoblast and inhibit its apoptosis via secreting IL-17 [50]. This finding suggests that Th17 cells are required in early placentation. In a study comparing Th17 cells in PB and the decidua, the percentage of decidual Th17 cells in lymphocytes was higher than that of PB Th17 cells [15]. Considering that IL-17 induces neutrophil recruitment, decidual neutrophil infiltration can be expected to increase as decidual Th17 cells increase. PB Th17 cells were increased, while PB Treg cells were decreased in nonpregnant and pregnant RPL women compared with fertile controls [40,51]. In the decidua of aborted idiopathic RPL women, an increase in Th17 cells and a decrease in Treg cells were also noted compared to normal pregnant women [51]. These findings suggest that Th17 and Treg cell dysregulation may be a critical immune mechanism involving RPL.

Furthermore, a correlation analysis between Th subsets and Treg cells in nonpregnant RPL women showed PB Th17 cell proportion was positively correlated with TNF-α + Th cells and ratios of TNF-α + CD4 + /IL-10 + CD4 + T cells and IFN-γ + CD4 + /IL-10 + CD4 + T cells [40]. In addition, Th17/Foxp3 + Treg cell ratio also positively correlated with TNF-α + Th cells, IFN-γ + Th cells, and Th1/Th2 cell ratios. These results indicate that both Th1 and Th17 immunities are involved in the immunopathology of RPL.

4.5.3 *Other Th cells in reproduction*

Th9 cells

IL-9 secreting Th cells, named Th9 cells, may play a role in allergic inflammation, autoimmune diseases, tumor immunity, and parasite infection [52,53]. In mice, Th9 cells controlled local inflammation in the uterus, while IL-6 induced to decrease decidual Th9 cells, which in turn accelerated parturition [54]. However, the role of Th9 cells in human pregnancy was not elucidated yet.

Th22 cells

IL-22 is produced by Th22, Th17, γδT, NKT, and innate lymphoid cells and has been identified in numerous tissues, including intestine, lung, liver, kidney, thymus, pancreas, skin, and placenta [55]. IL-22 plays a role in tissue regeneration and regulates host defense at barrier surfaces. In reproduction, IL-22 might be involved in the promotion of trophoblast survival and pregnancy maintenance. Women with miscarriage showed lower expression of IL-22R1 on placental villi than normal pregnant women, which indicates an association between insufficient IL-22/IL-22R signaling and miscarriage [56]. Th22 cells have marked plasticity by which they can differentiate into Th1 or Th2 cells. It is still inconclusive whether PB Th22 cells or decidual IL-22 production is increased in RPL or not. However, decidual IL-22 + IL-17 + IL-4 + Th cells were lower in RPL than in normal pregnancy. The role of Th22 cells needs to be explored further.

Follicular helper T cells

The follicular helper T (Tfh) cells play a critical role in B cell activation, antibody class switching, and germinal center formation [57]. In a mice study, CXCR5 + Tfh cells were accumulated and reached the peak in the uterus in mid-pregnancy; however, in the placenta, they are increased at late pregnancy [33]. Fetal resorption triggered by PDL1 blockade was associated with the accumulation of Tfh cells, suggesting that excessive accumulation of Tfh cells can be harmful to pregnancy. In addition, the proportion of PB Tfh cells was significantly higher in women in the third trimester than in nonpregnant women [58]. Peripheral estrogen concentration, but not progesterone, positively correlated with CXCR3 + Tfh cells [58]. Collectively, high estrogen levels during pregnancy may contribute to the increased proportion of Tfh cells, activating B cells and promoting humoral immunity. Thus, adequate Tfh cells may contribute to a successful pregnancy. Further study is required to elucidate the role of Tfh cells in RPL.

4.5.4 New types of regulatory T cells

In addition to the Foxp3 + Treg cells, new T cell subsets showing a regulatory function have been introduced. CD4 + HLA-G + $CD25^-Foxp3^-$ Treg cells can inhibit T cell proliferation by membrane-bound HLA-G1 expression and secretion of IL-10 and soluble HLA-G5 in a cell-contact dependent and independent manner [59]. CD4 + and CD8 + T cells can acquire regulatory function from APCs by cell-to-cell transfer of HLA-G1, so-called trogocytosis [60]. CD4 + HLA-G + Treg cells in PB were increased in normal pregnant women compared with those in nonpregnant women, and these cells in the decidua were more in normal pregnancy than pregnancy with preeclampsia (PE) [61].

Interestingly, Th1 cells selectively kill APCs, but not T and B cells [62]. However, CD4 + CD25-FoxP3-T regulatory cells, namely Th3 cells, produce IL-10 and IL-4 together and suppress immune cells by TGF-β and IL-10 production. Although Th3 cells were found in the decidua, their role in pregnancy remains to be explored. Contrarily, Type 1 regulatory T (Tr1) cells suppress other immune cells by producing IL-10 and TGF-β and a cell-to-cell contact manner using KIR receptors or ectoenzymes [59]. Tr1 cells have been found in PB and many tissues, including human decidua. They are induced by EVTs via HLA-G directly and have a high expression of PD-1. However, a recent mass cytometric study failed to confirm the presence of Tr1 cells in the human decidua [63]. Hence, further study is needed.

4.5.5 Memory T cells in reproduction

The immune systems of animals and humans exhibit specific immune responses to paternal and fetal antigens. Some maternal immune cells activated by the paternally driven antigens survive for a long time as memory cells. These cells show a faster response and proliferation to fetal antigens in the subsequent pregnancy and are known to play a positive role in pregnancy outcomes. To date, no reliable markers representing memory T cells have been introduced; however, $CD45RO^+$ expression or $CD45RA^-$expression are widely used as phenotypic markers [30]. The main two memory T cell subsets are the effector memory (EM) and the central memory (CM) cells (Fig. 2A). Some obstetric complications, such as PE, gestational diabetes, and preterm labor, are associated with higher proportions of total memory T cells in PB than in normal pregnancy [30].

CD4 + effector memory (T_{EM}) cells

Th subsets such as Th1, Th2, and Th17 cells can differentiate into CD4 + T_{EM} cells. The CD4 + T_{EM} cells are characterized by the lack of lymph node homing receptors CCR7 and

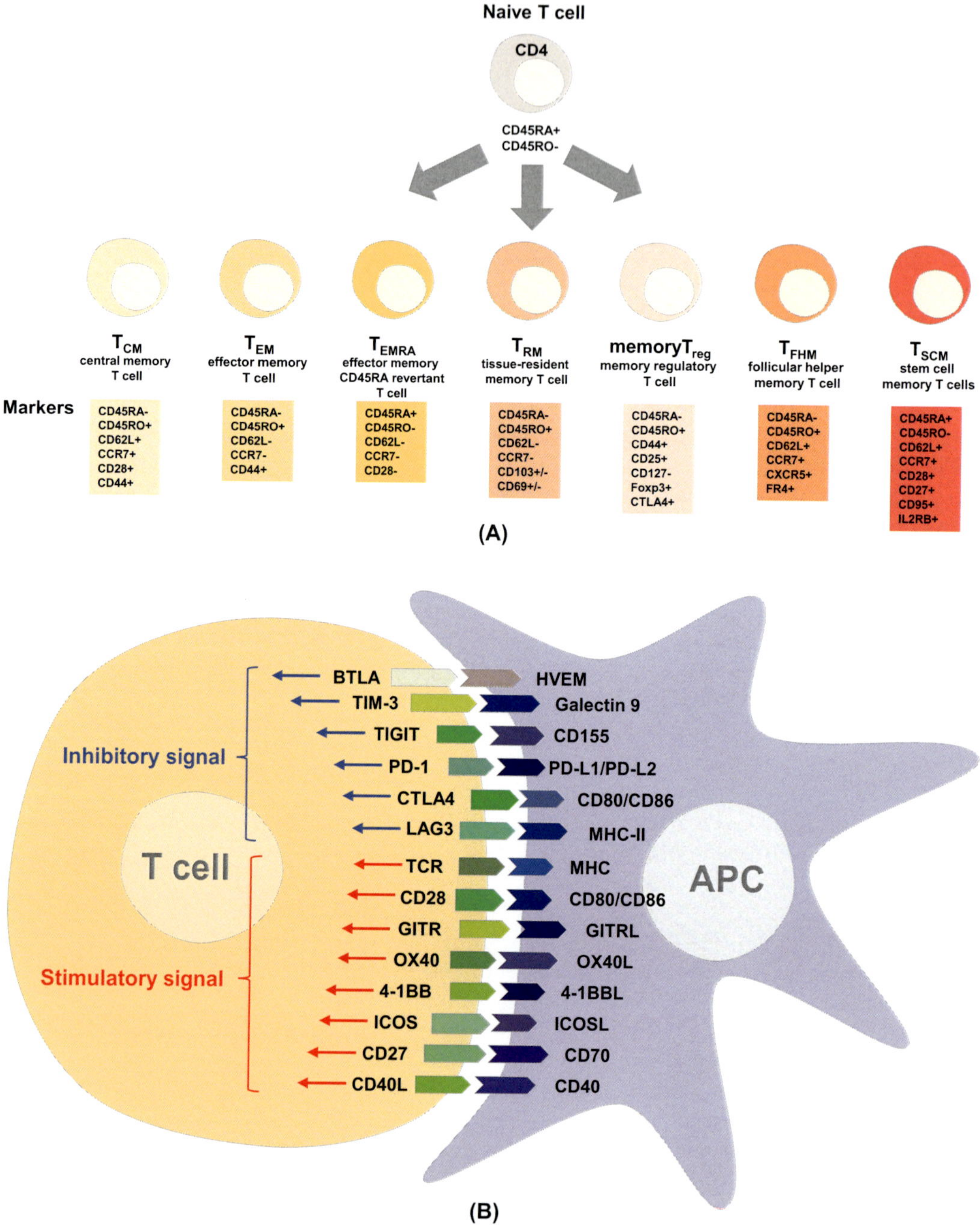

FIGURE 2 Human memory $CD4^+$ T cell subsets interact with antigen-presenting cells (APCs) through immune checkpoint molecules. (A) Human memory $CD4^+$ T cell subsets. Activated $CD4^+$ T cell subsets work as effectors, but a part of them also develop into memory cells. The phenotype of each $CD4^+$ memory cell is marked in the figure. (B) T cells receive inhibitory and stimulatory signals by APCs via immune checkpoint molecules. Blue arrows in the figure show inhibitory signaling pathways of T cells, and red arrows show the stimulatory signals.

CD62L (L-selectin) and migrate to peripheral nonlymphoid tissue [64]. Compared to the nonpregnant state, PB CD4 + T_{EM} cells increase during pregnancy and express more CD69 and PD-1. Not only during pregnancy but years after parturition, CD4 + T_{EM} cells in PB remain at increased levels as compared with those of nulligravid women [65].

CD4 + T_{EM} cells may be involved in the regulation of the fetomaternal immune environment. The expression of immune checkpoint receptors, such as PD-1, T cell immunoglobulin and mucin domain-3 (Tim-3) and CTLA-4, and lymphocyte activation gene 3 (LAG3), on decidual CD4 + T_{EM} cells was increased compared to that in the PB CD4 + T_{EM} cells [30]. In human pregnancy, most CD4 + T_{EM} cells in the decidua expressed PD-1 and Tim-3 molecules and exhibited a Th2 phenotype, producing IL-4 [66]. Furthermore, women suffering miscarriage had much lower co-expression of PD-1 and Tim-3 on decidual CD4 + T_{EM} cells as compared with normal pregnant women [66]. In mice, blockage of Tim-3 and PD-1 pathway increased fetal resorption rate. Collectively, decidual CD4 + T_{EM} cells may have a regulatory function at the fetomaternal interface and prevent miscarriage in humans [66]. However, the frequencies of CD3 + CD69 + T_{EM} cells in blood and endometrium were significantly higher in RPL women than in fertile women [67]. Further studies are required to elucidate the role of CD4 + T_{EM} cells in reproduction.

CD4 + effector memory CD45 revertant (T_{EMRA}) cells

CD4 + T_{EMRA} cells are a unique CD4 + T_{EM} subset that reexpressed CD45RA following antigen stimulation and have been implicated in protective immunity against dengue virus and cytomegalovirus infections [68]. However, the significance of this CD4 + T_{EMRA} was poorly investigated in reproduction.

CD4 + central memory (T_{CM}) cells

The CD4 + T_{EM} cells express lymph nodes homing receptors, such as CCR7 and CD62 [22]. The frequencies of CD4 + T_{CM} cells in the blood were significantly increased in PE and nonpregnant RPL compared with normal pregnancy and fertile nonpregnant controls, respectively [30].

CD4 + regulatory memory (T_{reg} memory) cells

T_{reg} memory cell subset, which survives for a long time, persists after antigen exposure [54]. No definite markers are known. T_{reg} memory cells are identified by a combination of Foxp3 +, CD25 +, CD127-, CD45RO +, and CD45RA- [30]. Interestingly, T_{reg} memory cells were increased in PE and GDM [30], but their roles were not well-evaluated in high-risk pregnancies.

CD4 + follicular helper memory (T_{FHM}) cells

T_{FHM} cells produce IL-10 and IL-21 and are characterized by CXCR5, CD62L, CCR7, and folate receptor 4. Several subsets of T_{FHM} cells can be distinguished based on the expression of PD-1, CCR7, inducible T cell costimulator (ICOS) [30]. In mice, putative T_{FHM} cells increase in the uterus and placenta toward late pregnancy, and blockage of PD-L1 (inhibition of PD-1 effect) induced fetal resorption and accumulation of putative CD4 + T_{FHM} cells, indicating that these T_{FHM} cells may involve in fetomaternal tolerance [30]. Women who experienced miscarriage showed higher PD-1 + CCR7 + and PD-

1 + ICOS + cells in the decidua but not in PB compared with normal pregnant women who had elective termination [69].

CD4 + tissue resident memory (T_{RM}) cells

CD4 + T_{RM} cells are challenging to study because they have no conclusive defining markers [30]. To date, no study was published in terms of CD4 + T_{RM} cells and reproduction.

CD4 + memory stem (T_{SCM}) cells

CD4 + T_{SCM} cells express CD45RA + , CCR7 + , CD62L + , CD28 + , CD27 + , CXCR3 + , CD95 + , and IL2RB + [30]. These cells are not studies in pregnancy yet.

4.5.6 Immune checkpoint molecules on Th cells and reproduction

wImmune checkpoint molecules are regulators of the immune system that carry costimulatory or coinhibitory signals (Fig. 2B). The most studied molecules are inhibitory immune checkpoint molecules, such as CTLA-4, PD-1, Tim-3, TIGIT, and LAG-3 [70]. CTLA-4, Tim-3, and PD-1 are key molecules to maintain immune tolerance at the fetomaternal interface.

Cytotoxic T-lymphocyte-associated protein 4 (CTLA-4)

In human pregnancy, CTLA-4 is found in decidual Treg cells. The ligands for CTLA-4, CD80, and CD86, are expressed on decidual APCs, such as DCs and macrophages [71]. Since decidual CTLA-4 expression was positively correlated with decidual Th2 immunity but negatively with Th1, CTLA-4 seems to involve intrauterine immune modulation [71]. In terms of miscarriage, the role of CTLA-4 is controversial. CTLA-4 + decidual lymphocytes were significantly decreased in women with miscarriage compared to those in normal pregnancy. However, a study presented that the ratio of CTLA-4 + Treg cells/Treg cells in the decidua was not different between normal pregnancy and miscarriage [48].

Tim-3

Tim-3 was initially identified as a Th1-specific cell surface molecule and is expressed on decidual stromal and immune cells, such as NK, macrophage, CD8 + T, Th1, Th2, Th17, Treg, γδT, and NKT cells. Tim-3 down-regulates Th1 responses by transducing apoptotic signaling through Galectin-9 engagement [72]. Tim-3 binds to two types of ligands, phosphatidylserine (PtdSer) and galectin-9. PtdSer is exposed on the surface of apoptotic cells. The ligation of Tim-3 on APCs and PtdSer induces phagocytosis of apoptotic cells, contributing to the maintenance of tolerance to self-antigens [72]. Galectin-9 is secreted by Treg, T, B, and mast cells, endometrial epithelial cells, trophoblasts, and epithelial cells of the gastrointestinal tract. Tim-3-Gal-9 pathway induces apoptosis of Th1 and Th17, expansion and resistance to apoptosis of Treg, conversion of dNK, regulatory function to macrophage, DC, and CD8 + T cells [72]. In mice, Tim-3 blockade leads to an increase in fetal resorption rate and a decrease in litter size [70]. In women with RPL, coexpression of CTLA-4 and Tim-3 on decidual CD4 + T cells was significantly lower than in normal early pregnancy [73]. In a study of infertile women who underwent embryo transfer, the proportion of Tim-3 + PD-1 + CD4 + T cells and Tim-3 + PD-1 + CD8 + T cells were reduced at the implantation period in failed pregnancy group compared with those in the success group [74].

Programmed cell death protein 1 (PD-1)

PD-1 is a transmembrane receptor and generates a potent inhibitory signal by binding to its ligands PD-L1 and PD-L2 [71]. The expression of PD-1 was found on T, B, NK, dendritic cells, and macrophages. PD-L1 is also expressed on some tissues such as the placenta, heart, and spleen [71]. In the murine abortion model, PD-1 blockade resulted in an increased abortion rate, but CTLA-4 blockage did not affect pregnancy outcome [75]. In a flow cytometry analysis of PB lymphocytes, PD-L1 expression in Th1, Th17, and Treg cells was more prevalent than PD-1 expression in RPL and control groups [76]. However, PD-1 + Th1, PD-1 + Th17, and PD-L1 + Th17 cells were significantly lower in nonpregnant RPL women than those of fertile controls [76]. In human pregnancy, PD-1 expression in decidual lymphocytes was not different between normal pregnancy and RPL, but decidual PD-L1 mRNA expression was significantly lower in RPL women than in normal pregnant women [77]. Several studies confirmed increased PD-1 + Treg cells in women with PE. A recent report demonstrated that the proportion of decidual PD-1^{high}CD8 + cells was not different between miscarriage and normal early pregnancy but decreased in PE compared with that in the third trimester of normal pregnancy [78].

5 Diagnostic approaches

5.1 Laboratory technique

5.1.1 Flow cytometry

Flow cytometric analysis is the most popular assessment method to enumerate lymphocytes and their subsets, including T, B, and NK cells in the blood of the endometrium/decidua. In clinical practice, the PB Th1/Th2 cytokine ratio has merit in several aspects: (1) venous blood is more accessible to obtain and handle than endometrial tissue and (2) PB immune cells seem to reflect endometrial immune cells since circulating immune cells are rapidly recruited into the secretory endometrium and decidua in early pregnancy. Many reports have confirmed that immunotherapy based on PB immune tests improved the clinical outcome in women with reproductive failure. Endometrial and decidual T cell immunity can also be measured by flow cytometry.

Th1/Th2 cytokine-producing cell ratios are measured as follows [39]. The PBMCs are isolated from 10 to 15 cc of PB or the endometrium and activated with phorbol myristate acetate and ionomycin. Protein transport inhibitor-containing brefeldin A is applied to inhibit cytokine secretion, and fluorochrome-conjugated monoclonal antibodies specific for cell surface antigens such as CD3, CD4, or CD8 are added. The cell membrane is permeabilized for intracellular staining, and the cells are stained with fluorochrome-conjugated monoclonal antibodies for TNF-α, IFN-γ, IL-4, or IL-10. Then the cells are analyzed by a flow cytometer [11,39].

Nevertheless, only a few laboratories worldwide have the cut-off value of Th cell subsets and apply it in clinical practice. Winger et al. reported that IVIG immunotherapy was improved the live birth rate (LBR) in women with high PB Th1/Th2 ratios or NK cell level (>30.6 for TNF-α/IL-10 ratios, >20.5 for IFN-γ/IL-10, or NK cell level >12%) compared with no immunotherapy group [79]. Lee et al. also determined the cut-off value of cytokine-producing CD4 cells: TNF-α > 70.1% of CD4+ T cells, TNF-α/IL-10 producing Th cell ratio > 36.2 [10]. For clinical practice, physicians may develop their own cut-off values of Th cell immunity or adopt the reported values.

5.1.2 Polymerase chain reaction and enzyme-linked immunosorbent assay

The gene expression of transcription factors and various cytokines can be measured by polymerase chain reaction in PB and/or endometrial/decidual tissue. Additionally, various cytokines in the PB can be measured to evaluate T cell immunity by commercial enzyme-linked immunosorbent assay kits.

5.2 Candidates for evaluation

There is no general consensus regarding the candidates for the Th cell subset study. However, women with RPL who have had two or more miscarriages, especially with one or more of the following factors, should be considered for the evaluation of T cell immunity: (1) presence of subclinical autoimmunities such as antibodies such as an antiphospholipid antibody, antinuclear antibody, anti-DNA antibody, and antithyroid antibody; (2) a history of or active autoimmune and/or rheumatic diseases; (3) a history of or active NK cell pathology; (4) maternal–fetal alloimmune disorder; (5) idiopathic RPL (Table 1).

6 Treatment and reproductive outcome

Current treatment options for T cell immunologic abnormalities in RPL women are corticosteroids, intravenous immunoglobulins (IVIGs), tacrolimus, cyclosporine, TNF-α inhibitors, and vitamin D (Table 2) [80].

6.1 Corticosteroids

Glucocorticoids can influence several key aspects of early pregnancy, including maternal immune response, embryo attachment/implantation, trophoblast outgrowth and invasion, and fetal and placental development [87]. Corticosteroids have been utilized to treat women with RPL as it is known to exert immunosuppressive effects on elevated Th1/Th2 or Treg/Th17 ratios as well as NK cell cytotoxicities [88].

In an RCT of unexplained RPL women, an addition of prednisolone (20mg/day) to heparin and low dose aspirin significantly increased the success rate (70.3%) compared to the placebo group (9.7%), and this effect might be due to a suppressive effect of corticosteroids on the peripheral CD16+ NK cells concentration [81] and Th1 cell immunity as well. Regarding the Th1/Th2 ratio imbalance, there are limited studies for the glucocorticoid effect. Corticosteroid restores the normal balance between Th1 and Th2 cytokines, such as lowering the IFN-γ and raising the IL-4. Antiinflammatory effects of corticosteroids are

TABLE 1 T helper immunity in normal pregnancy and recurrent pregnancy losses are compared in peripheral blood and endometrium/decidua.

Th subset	Menstrual cycle (secretory phase compared to follicular phase)		Normal pregnancy		Recurrent pregnancy losses (RPL)			
					Nonpregnant (RPL vs controls)		Pregnant (RPL vs controls)	
	PB	EM	PB	Decidua	PB	EM	PB	Decidua
Th1	→	?	↓ (1st trimester)	↓	→	↑	↑→	↑
Th2	→	?	↑ (1st trimester)	↑	→	?	↓	↓
Th1/Th2	→	?	↑ (3rd trimester)	?	↑	?	↑	?
Th17	?	?	→	↑	↑	?	→↑	↑
Treg	↓ (peak at late follicular phase)	↓ (higher at proliferative phase)	↑ (peak at 2nd trimester)	↑	↓	?	↓	↓
Th17/Treg	?	?	?	?	↑	?	?	?
Th9	?	?	?	?	?	?	?	?
Th22	?	?	?	?	?	?	?	↓
Tfh	?	?	?	?	?	?	?	?
CD4 + T_{EM}	?	?	?	?	?	?	?	↓
CD4 + T_{EMRA}	?	?	?	?	?	?	?	?
CD4 + T_{CM}	?	?	?	?	?	?	↑	?
Treg memory	?	?	?	?	?	?	?	?
CD4 + T_{FHM}	?	?	?	?	?	?	↑	↑
CD4 + T_{RM}	?	?	?	?	?	?	?	?
CD4 + memory stem	?	?	?	?	?	?	?	?

?, unknown; ↑, increase; ↓, decrease; →, no change.

TABLE 2 Immunomodulatory agents to control abnormal T cell immunity.

Drug	Mechanism	Dose/duration	Pregnancy outcomes	Side effects
Corticosteroids	↓Th1/Th2 ratio	Prednisolone 5–20mg daily	LBR 70.3% in Prednisolone + heparin + LDA versus 9.7% in placebo controls [81]	Preterm births, IUGR, preeclampsia, chorioamnionitis
IVIG	↓Th1/Th2 ratio ↓Th17/Treg ratio	400mg/kg in every 3–4 weeks	LBR 87.5% versus 41.6% in controls [82] LBR 47.5% versus 21.8% [83]	Headache, fever, nausea, anaphylaxis, acute renal failure
Tacrolimus	↓Th1/Th2 ratio ↓cytotoxic T cell	1–3 mg daily	LBR 60.0% versus 0.0% in controls [84]	No report associated with obstetrical and perinatal complications
Cyclosporine	↓ Th1 frequency, T-bet mRNA expression, IFN-γ, and TNFα, ↑Th2 frequency, mRNA expression of GATA binding protein, IL10	50 mg orally, three times a day, for 6 months	LBR 81.5% versus 42.1% in controls [85]	-
TNF-α blockers	↓TNFα	Adalimumab 40 mg sc every 1–2 weeks until the fetal heartbeat is confirmed by ultrasound evaluation	LBR 71.0% in humira + IVIG + heparin versus 54% IVIG + heparin [86]	Opportunistic infections, lymphomas, neurologic sequelae

Abbreviations: *Th*, T helper; *LBR*, live birth rate; *LDA*, low dose aspirin; *IUGR*, intrauterine growth restriction.

partly due to an inhibitory effect on T cell proliferation through the inhibition of IL2 production. The regulation of T cell cytokine productions by corticosteroids showed a potency-related and dose-dependent manner. The more potent corticosteroid showed more robust inhibition of IFN-γ and IL4 production from T cells. However, low concentrations, and less potent corticosteroids, such as dexamethasone, could enhance the production of IL-4 [89,90]. Sasaki et al. reported successful live births after treating women with RPL and elevated Th1/Th2 ratios with prednisolone. Th1/Th2 ratio was decreased overtime after prednisolone [91].

In women with immune disorders, the exposure of glucocorticoids in mid to late pregnancy may implicate obstetric complications, including intrauterine growth restriction, preterm labor, PE, chorioamnionitis, and effects on fetal brain development associated with alterations in prenatal and postnatal behavior [87]. Contrarily, the recent study demonstrated that treating a premature baby with steroids did not increase their risk of neurodevelopmental delay [92]. The influence of glucocorticoids in early pregnancy was less studied, and most of the available evidence comes from women undergoing assisted reproduction regarding the pregnancy rates [87]. Adverse effects of corticosteroids seem to be dose-dependent and less frequent in lower dosages of prednisolone. Exposure to the high dose of prednisone (≥10 mg/kg daily) was associated with preterm birth, while a lower dose (≤10 mg/kg daily) was not associated with preterm birth [93]. Among women with irritable bowel disease, the risk of preterm birth was not increased by oral corticosteroids exposure before day 140 but increased by exposure after day 140. However, the higher dose of oral corticosteroids in early pregnancy was associated with an increase of preterm birth as well as second trimester in women with SLE [94]. Considering these, the adverse effects of corticosteroids seem to be related to dose and exposure time during the pregnancy. In early pregnancy, low dose exposure in women with RPL might not increase the risk of preterm birth. However, RPL itself is associated with the risk of preterm birth regarding impaired trophoblast invasion and placental growth. Researches are limited due to difficulties in handling various confounding factors. Recently, the German/Austrian/Swiss Society of Obstetrics and Gynecology guideline suggests the therapeutic use of corticosteroids only for patients with preexisting autoimmune diseases. The guideline points out the possibility of significant adverse events associated with the use of corticosteroids during pregnancy, as the ESHRE guideline proposed [4,95]. However, it is noteworthy that obstetrical complications were not reported in women with RPL who were treated with low to medium dose prednisone with limited time [93]. Meticulous selection of patients for corticosteroid treatment, such as using PB Th1/Th2 cell ratios and serial follow up of T cell immunity, may increase success rate without increasing obstetrical complications in women with RPL.

6.2 Intravenous immunoglobulins

IVIG G has several immunoregulatory effects by suppressing and neutralizing autoantibodies, attenuating NK cell cytotoxicity, inhibiting complement binding, modifying cytokine production, and expanding regulatory T lymphocytes [96]. Several case-control studies and RCTs investigating IVIG efficacy in women with RPL showed conflicting results. In a Cochrane systematic review and *meta*-analysis including 8 RCTs with unexplained RPL women, IVIG did not increase the odds of live birth compared to placebo (OR 0.98, 95% CI 0.61 to 1.58) [97]. Compared to primary RPL, immunological disturbances may play a more significant role in secondary RPL. Egerup et al. showed that women with secondary RPL were more likely to obtain a potential beneficial effect from IVIG in the *meta*-analysis [98]. However, another RCT in women with secondary RPL did not demonstrate increased LBR with IVIG treatment [99]. The most recent *meta*-analysis revealed increased LBR in women with IVIG treatment compared to controls (RR 1.20, 95% CI 1.0–1.37, $p = 0.004$ by the fixed-effect method) [100]. Due to the concern of significant heterogeneity of the study population, the statistical outcome by the random-effect model (RR 1.17, 95% CI 0.95–1.44, $p = 0.14$) was primarily reported. However, it is noteworthy that a random-effects model provides poor estimates of the width of the distribution of intervention effects when the distribution pattern of the intervention effects is asymmetrical because of few studies, while the fixed effects model is less affected [101]. IVIG seems to work in specific subpopulations. Subgroup analysis revealed that preconception IVIG treatment significantly increased LBR (RR 1.69, 95% CI 1.33–2.14, $p < 0.0001$), and an IVIG treatment for secondary aborters had a borderline therapeutic effect (RR 1.24, 95% CI 0.96–1.59, $p = 0.06$) [101]. Recently, Won et al. performed a *meta*-analysis to evaluate the efficacy of immunotherapy in women specifically selected based on abnormal levels and/or activity of NK cells. *Meta*-analysis of the non-RCT IVIG studies (312 intervention, 245 controls) showed an increased live birth in IVIG treatment (RR 2.57; 95% CI 1.79–3.69; $p < 0.05$) [102].

IVIG has shown significant regulatory effects on Th1/Th2 and the Th17/Treg cell ratios in RPL. Yamada et al. reported that IVIG treatments (20 g/day for 5 days) modulated the PB cytokine levels and decreased Th1/Th2 lymphocyte ratios [103]. IVIG infusion, 400 mg/kg, IV, every 4 weeks, also significantly reduced Th1 lymphocyte frequency, transcription factor expression, and cytokine levels while significantly increased Th2 lymphocyte parameters [82]. Through a shift to Th2 responses, the LBR in IVIG-treated women was significantly higher (87.5%) than in the untreated group (41.6%) [82]. Lee et al. also reported LBR of 84.7% (94/111) with IVIG 400 mg/kg, given every 3 weeks, in women with RPL and abnormal cellular immunity (increased level, cytotoxicity of NK cells, or Th1/Th2 ratio) [104].

Kim et al. reported that IVIG treatment modulated imbalance of Th17 and Foxp3+ Treg cells in pregnant RPL women. IVIG down-regulated Th17 cells in the highest quartile, Q4 ($p = 0.001$), and upregulated CD4+ Foxp3+ T cells in the lower quartiles, Q1 and Q2 ($p = 0.025$ and 0.029, respectively) [105]. Pregnancy outcome following IVIG treatment was successful in 76% of RPL women (28 out of 37 subjects). The pregnancy success rate of RPL women with immune abnormality at the first IVIG treatment was higher (80.6%) than that of women without immune abnormality (50%). It is presumed that IVIG-mediated immune modulation of dysregulated Th17 and Treg cells is also linked with successful pregnancy outcomes in RPL women with cellular immune abnormalities during early pregnancy [105]. Ahmadi et al. confirmed that IVIG treatment showed a significant increase in Treg cell-associated parameters such as Treg cell proportion, Foxp3 expression, and cytokine mRNA levels (IL-10 and TGF-β) in women with RIF [83]. The LBR of women with RIF and IVIG treatment was 47.5% which is significantly higher than that of the control group (21.8%) [83].

Women with RPL usually tolerate IVIG well, and the adverse reactions, including headache, fever, and nausea, occur in less than 5% of patients [96]. Severe but rare complications, such as anaphylaxis, and acute renal failure, are related to IgA-deficiency and using the IVIG solution with a sugar stabilizer [106,107]. Hence, serum IgA and creatinine levels should be checked before the IVIG infusion [108]. Nowadays, the majority of IVIG preparations do not use sugar stabilizers. Prenatal exposure to drugs is postulated to cause epigenetic changes in fetal immunity, and the prevalence of autoimmune and allergic diseases was investigated in children who were in utero exposed to IVIG. However, no differences in the prevalence of any explored disease were observed. To date, there has been no report demonstrating significant side-effects in the babies whose mothers were treated with IVIG before and during pregnancy for the last 30 years in this field [108,109].

6.3 Tacrolimus

Tacrolimus is used for organ transplantation recipients and patients with autoimmune diseases such as rheumatoid arthritis and systemic lupus erythematosus, even during pregnancy or nursing. Tacrolimus inhibits lymphocyte proliferation, cytotoxic T cell generation, expression of IL-2 receptor, and the production of cytokines, such as IL-2 and IFN-γ [110]. Tacrolimus was applied for the patients with RIF and pregnancy losses showing elevated Th1/Th2 cell ratios (≥10.3). The LBR in the tacrolimus group was significantly higher than in the control group (60.0% vs 0%) [84].

Tacrolimus is classified as a class C drug by the FDA pregnancy category. Furthermore, the drug effect of tacrolimus on Th cells or other immune cells in neonates is not expected because its half-life is 7.9 ± 5.2 hours. The safety of tacrolimus was well documented for both mother and fetus/baby in several reports of female transplant recipients and a recent report from women with RIF [111,112]. From the national transplantation pregnancy registry, the incidence of congenital anomalies in the live-born appears similar to the general population, and long-term follow-up of the offspring of transplant recipients provided reassurance after 20 years of observation [113]. In a report by Nakagawa et al., tacrolimus treatment (1–4 mg/day) before and during pregnancy was not associated with obstetrical and perinatal complications [111].

6.4 Other immunosuppressants

Cyclosporine is an immunosuppressant widely used after organ transplantation and for autoimmune diseases. Some studies suggested that a low dose of cyclosporine could induce maternal-fetal immune tolerance while enhancing trophoblast invasion [114]. Cyclosporine treatment was reported to evoke a significant decrease in Th1/Th2 ratio ($p < 0.0001$) in women with RPL, by a reduction of Th1 frequency, T-bet mRNA expression, IFN-γ, and TNF-α levels, and by an increase in Th2 frequency, mRNA expression of GATA binding protein, and IL-10 level [85]. Although the study size was small ($n = 38$), women with RPL and cyclosporine treatment showed a significantly higher LBR than the controls [85,114].

Sirolimus (Rapamune), an mTOR inhibitor, can prevent allograft rejection effectively by preventing the proliferation of IL-2-stimulated T cells [115]. Additionally, sirolimus promoted the generation of CD4 + CD25 + FoxP3 + Treg cells while strongly inhibiting the differentiation of Th17 cells in peripheral tissue in humans and animal models. A significantly increased LBR (48.8%) was reported in RIF women who received sirolimus compared with a control group (21.2%) ($p < 0.0001$) [115]. The data about the impact of sirolimus on pregnancy and child-related outcomes was limited. Sirolimus is labeled Pregnancy Category C, and no apparent abnormalities were found in the babies after prenatal period exposure [116,117].

6.5 Tumor necrosis factor inhibitors

TNF-α blockers, such as Remicade (infliximab), Humira (adalimumab), and Enbrel (etanercept), are used to decrease inflammation for rheumatoid arthritis, inflammatory bowel diseases (Crohn's disease), and ankylosing spondylitis. Th1 cytokines, including IFN-γ, TNF-α, and IL-2, are known to increase fetal loss. Especially in women with elevated Th1/Th2 profile, TNF-α blockers combined with IVIG showed a significantly higher LBR. In women with RPL, the triple combination of TNF-α inhibitor, IVIG, and heparin improved the LBR compared to treatment with heparin alone or IVIG with heparin (Heparin alone 19%; Heparin + IVIG 54%; Heparin + IVIG + Humira 71%, $p = 0.0026$) [86].

TNF-α blockers are a relatively safe drug. However, these drugs may cause several serious adverse events, such as *Myobacterium tuberculosis* reactivation, opportunistic infections, lymphoma, neurologic sequelae like demyelinating diseases [118]. Additionally, TNF-α blockers can be used during the implantation period and pregnancy. Breastfeeding is also permitted. However, physicians should be aware of the potential risk of, although it is low, embryo-fetal toxicity of these drugs during pregnancy [119].

6.6 Vitamin D

In women with RPL and vitamin D deficiency, the risk of auto- and cellular immune abnormalities is increased. After supplementing vitamin D to T cell cultures, TNF-α/IL-10 and INF-γ/IL-10 expressing CD3 + /CD4 + cell ratios were significantly reduced. Besides, vitamin D supplement to NK cell culture suppressed secretion of type I cytokines such as IFN-γ and TNF-α, and increased type II cytokine such as IL-10 [120]. Vitamin D promotes a more favorable environment for pregnancy through the Th2 and NK2 shift. Moreover, vitamin D3 reduced the frequency of PB Th17 cells and the Th17/Treg ratio in women with RPL [121].

7 Summary and recommendations

- The mother's immune system differentially adapts to each stage of pregnancy for a successful pregnancy, including implantation, fetal development, labor, and postpartum.
- An excessive proinflammatory environment caused by Th cell dysregulation in circulating blood and fetomaternal interface is harmful to pregnancy.
- A marked predominance of Th1 over Th2 immunity during pregnancy provides a disadvantageous environment for implantation, leading to miscarriage and RIFs.
- Various subsets of Th cells and their roles in reproductive failures have been studied. Th17 cell/Treg cell imbalance also contributes to the development of RPL.
- The Th cell subset study can be considered for RPL women who had two or more miscarriages and additional risk factors, such as subclinical autoimmune abnormality, active autoimmune disorders, NK cell pathology, and idiopathic RPL.
- Immunomodulatory agents, such as corticosteroids, IVIGs, tacrolimus, and TNF inhibitors, may regulate an immune imbalance and improve pregnancy outcomes in women with RPL and T cell immune abnormalities.
- Immunomodulatory treatment is more effective if given preconceptionally to women with RPL and immune abnormalities, such as increased NK cell levels and cytotoxicities, increased Th1/Th2 cell ratios, autoimmune abnormalities, and idiopathic RPL.

References

[1] Mosmann TR, Cherwinski H, Bond MW, Giedlin MA, Coffman RL. Two types of murine helper T cell clone. I. Definition according to profiles of lymphokine activities and secreted proteins. Comparative study. J Immunol 1986;136(7):2348–57 Apr 1.

[2] Wegmann TG, Lin H, Guilbert L, Mosmann TR. Bidirectional cytokine interactions in the maternal-fetal relationship: is successful pregnancy a TH2 phenomenon? ImmunolToday 1993;14(7):353–6. Available from: https://doi.org/10.1016/0167-5699(93)90235-D Jul.

[3] Saito S, Nakashima A, Shima T, Ito M. Th1/Th2/Th17 and regulatory T-cell paradigm in pregnancy. Am J Reprod Immunol 2010;63(6):601–10. Available from: https://doi.org/10.1111/j.1600-0897.2010.00852.x Jun.

[4] EGGo RPL, Bender Atik R, Christiansen OB, et al. ESHRE guideline: recurrent pregnancy loss. Hum Reprod Open 2018;2018(2):hoy004. Available from: https://doi.org/10.1093/hropen/hoy004.

[5] Practice Committee of American Society for Reproductive M. Definitions of infertility and recurrent pregnancy loss: a committee opinion. Fertil Steril 2013;99(1):63. Available from: https://doi.org/10.1016/j.fertnstert.2012.09.023 Jan.

[6] Group EEPGD. Recurrent pregnancy loss. https://www.eshre.eu/Guidelines-and-Legal/Guidelines/Recurrent-pregnancy-loss/.

[7] van Dijk MM, Kolte AM, Limpens J, et al. Recurrent pregnancy loss: diagnostic workup after two or three pregnancy losses? A systematic review of the literature and meta-analysis. Hum Reprod Update 15 2020;26(3):356–67. Available from: https://doi.org/10.1093/humupd/dmz048 Apr.

[8] Larsen EC, Christiansen OB, Kolte AM, Macklon N. New insights into mechanisms behind miscarriage. BMC Med 2013;11:154. Available from: https://doi.org/10.1186/1741-7015-11-154 Jun 26.

[9] Rasmark Roepke E, Matthiesen L, Rylance R, Christiansen OB. Is the incidence of recurrent pregnancy loss increasing? A retrospective register-based study in Sweden. Acta Obstet Gynecol Scand 2017;96(11):1365–72. Available from: https://doi.org/10.1111/aogs.13210 Nov.

[10] Lee SK, Na BJ, Kim JY, et al. Determination of clinical cellular immune markers in women with recurrent pregnancy loss. Am J Reprod Immunol 2013;70(5):398–411. Available from: https://doi.org/10.1111/aji.12137 Nov.

[11] Lee S, Kim J, Jang B, et al. Fluctuation of peripheral blood T, B, and NK cells during a menstrual cycle of normal healthy women. J Immunol 2010;185(1):756–62. Available from: https://doi.org/10.4049/jimmunol.0904192 Jul 1.

[12] Flynn L, Byrne B, Carton J, Kelehan P, Orsquo;Herlihy C, Orsquo;Farrelly C. Menstrual cycle dependent fluctuations in NK and T-lymphocyte subsets from nonpregnant human endometrium. Am J Reprod Immunol 2000;43(4):209–17. Available from: https://doi.org/10.1111/j.8755-8920.2000.430405.x Apr.

[13] Arruvito L, Sanz M, Banham AH, Fainboim L. Expansion of CD4 + CD25 + and FOXP3 + regulatory T cells during the follicular phase of the menstrual cycle: implications for human reproduction. J Immunol 2007;178(4):2572–8 Feb 15.

[14] Berbic M, Hey-Cunningham AJ, Ng C, et al. The role of Foxp3 + regulatory T-cells in endometriosis: a potential controlling mechanism for a complex, chronic immunological condition. Hum Reprod 2010;25(4):900–7. Available from: https://doi.org/10.1093/humrep/deq020 doi:deq020 [pii].

[15] Nakashima A, Ito M, Yoneda S, Shiozaki A, Hidaka T, Saito S. Circulating and decidual Th17 cell levels in healthy pregnancy. Am J Reprod Immunol 2010;63(2):104–9. Available from: https://doi.org/10.1111/j.1600-0897.2009.00771.x doi:AJI771 [pii].

[16] Lee SK, Kim JY, Lee M, Gilman-Sachs A, Kwak-Kim J. Th17 and regulatory T cells in women with recurrent pregnancy loss. Am J Reprod Immunol 2012;67(4):311–18. Available from: https://doi.org/10.1111/j.1600-0897.2012.01116.x Apr.

[17] Feyaerts D, Benner M, van Cranenbroek B, van der Heijden OWH, Joosten I, van der Molen RG. Human uterine lymphocytes acquire a more experienced and tolerogenic phenotype during pregnancy. Sci Rep 2017;7(1):2884. Available from: https://doi.org/10.1038/s41598-017-03191-0 Jun 6.

[18] Lee SK, Kim CJ, Kim DJ, Kang JH. Immune cells in the female reproductive tract. Immune Netw 2015;15(1):16–26. Available from: https://doi.org/10.4110/in.2015.15.1.16 Feb.

[19] Williams PJ, Searle RF, Robson SC, Innes BA, Bulmer JN. Decidual leucocyte populations in early to late gestation normal human pregnancy. J Reprod Immunol 2009;82(1):24–31. Available from: https://doi.org/10.1016/j.jri.2009.08.001 Oct.

[20] Piccinni MP, Lombardelli L, Logiodice F, Kullolli O, Romagnani S, Le Bouteiller P. T helper cell mediated-tolerance towards fetal allograft in successful pregnancy. Clin Mol Allergy 2015;13(1):9. Available from: https://doi.org/10.1186/s12948-015-0015-y.
[21] Vento-Tormo R, Efremova M, Botting RA, et al. Single-cell reconstruction of the early maternal-fetal interface in humans. Nature 2018;563(7731):347–53. Available from: https://doi.org/10.1038/s41586-018-0698-6 Nov.
[22] Ng SC, Gilman-Sachs A, Thaker P, Beaman KD, Beer AE, Kwak-Kim J. Expression of intracellular Th1 and Th2 cytokines in women with recurrent spontaneous abortion, implantation failures after IVF/ET or normal pregnancy. Am J Reprod Immunol 2002;48(2):77–86. Available from: https://doi.org/10.1034/j.1600-0897.2002.01105.x Aug.
[23] Santner-Nanan B, Peek MJ, Khanam R, et al. Systemic increase in the ratio between Foxp3+ and IL-17-producing CD4+ T cells in healthy pregnancy but not in preeclampsia. J Immunol 2009;183(11):7023–30. Available from: https://doi.org/10.4049/jimmunol.0901154 doi:jimmunol.0901154 [pii].
[24] Somerset DA, Zheng Y, Kilby MD, Sansom DM, Drayson MT. Normal human pregnancy is associated with an elevation in the immune suppressive CD25+ CD4+ regulatory T-cell subset. Research support, Non-U.S. Govrsquo;t. Immunology 2004;112(1):38–43. Available from: https://doi.org/10.1111/j.1365-2567.2004.01869.x May.
[25] Mor G, Aldo P, Alvero AB. The unique immunological and microbial aspects of pregnancy. Nat Rev Immunol 2017;17(8):469–82. Available from: https://doi.org/10.1038/nri.2017.64 Aug.
[26] Plaks V, Birnberg T, Berkutzki T, et al. Uterine DCs are crucial for decidua formation during embryo implantation in mice. J Clin Invest 2008;118(12):3954–65. Available from: https://doi.org/10.1172/jci36682 Dec.
[27] Morelli AE, Thomson AW. Tolerogenic dendritic cells and the quest for transplant tolerance. Nat Rev Immunol 2007;7(8):610–21. Available from: https://doi.org/10.1038/nri2132 Aug.
[28] Vacca P, Cantoni C, Vitale M, et al. Crosstalk between decidual NK and CD14+ myelomonocytic cells results in induction of Tregs and immunosuppression. Proc Natl Acad Sci U S A 2010;107(26):11918–23. Available from: https://doi.org/10.1073/pnas.1001749107 Jun 29.
[29] Billington WD. The immunological problem of pregnancy: 50 years with the hope of progress. A tribute to Peter Medawar. J Reprod Immunol 2003;60(1):1–11. Available from: https://doi.org/10.1016/s0165-0378(03)00083-4 Oct.
[30] Kieffer TEC, Laskewitz A, Scherjon SA, Faas MM, Prins JR. Memory T cells in pregnancy. Front Immunol 2019;10:625. Available from: https://doi.org/10.3389/fimmu.2019.00625.
[31] Robertson SA, Prins JR, Sharkey DJ, Moldenhauer LM. Seminal fluid and the generation of regulatory T cells for embryo implantation. Am J Reprod Immunol 2013;69(4):315–30. Available from: https://doi.org/10.1111/aji.12107 Apr.
[32] Tilburgs T, Scherjon SA, van der Mast BJ, et al. Fetal-maternal HLA-C mismatch is associated with decidual T cell activation and induction of functional T regulatory cells. J Reprod Immunol 2009;82(2):148–57. Available from: https://doi.org/10.1016/j.jri.2009.05.003 Nov.
[33] Wang W, Sung N, Gilman-Sachs A, Kwak-Kim JT. Helper (Th) cell profiles in pregnancy and recurrent pregnancy losses: Th1/Th2/Th9/Th17/Th22/Tfh cells. Front Immunol 2020;11:2025. Available from: https://doi.org/10.3389/fimmu.2020.02025.
[34] Todt JC, Yang Y, Lei J, et al. Effects of tumor necrosis factor-alpha on human trophoblast cell adhesion and motility. Am J Reprod Immunol 1996;36(2):65–71. Available from: https://doi.org/10.1111/j.1600-0897.1996.tb00141.x Aug.
[35] Lockwood CJ, Oner C, Uz YH, et al. Matrix metalloproteinase 9 (MMP9) expression in preeclamptic decidua and MMP9 induction by tumor necrosis factor alpha and interleukin 1 beta in human first trimester decidual cells. Biol Reprod 2008;78(6):1064–72. Available from: https://doi.org/10.1095/biolreprod.107.063743 Jun.
[36] Ashkar AA, Di Santo JP, Croy BA. Interferon gamma contributes to initiation of uterine vascular modification, decidual integrity, and uterine natural killer cell maturation during normal murine pregnancy. J Exp Med 2000;192(2):259–70. Available from: https://doi.org/10.1084/jem.192.2.259 Jul 17.
[37] Ebina Y, Shimada S, Deguchi M, Maesawa Y, Iijima N, Yamada H. Divergence of helper, cytotoxic, and regulatory T cells in the decidua from miscarriage. Am J Reprod Immunol 2016;76(3):199–204. Available from: https://doi.org/10.1111/aji.12546 Sep.

[38] Makhseed M, Raghupathy R, Azizieh F, Omu A, Al-Shamali E, Ashkanani L. Th1 and Th2 cytokine profiles in recurrent aborters with successful pregnancy and with subsequent abortions. Hum Reprod 2001;16 (10):2219–26 Oct.

[39] Kwak-Kim JY, Chung-Bang HS, Ng SC, et al. Increased T helper 1 cytokine responses by circulating T cells are present in women with recurrent pregnancy losses and in infertile women with multiple implantation failures after IVF. Hum Reprod 2003;18(4):767–73 Apr.

[40] Lee SK, Kim JY, Hur SE, et al. An imbalance in interleukin-17-producing T and $Foxp3^+$ regulatory T cells in women with idiopathic recurrent pregnancy loss. Hum Reprod 2011;26(11):2964–71. Available from: https://doi.org/10.1093/humrep/der301 Nov.

[41] Lin H, Mosmann TR, Guilbert L, Tuntipopipat S, Wegmann TG. Synthesis of T helper 2-type cytokines at the maternal-fetal interface. J Immunol 1993;151(9):4562–73 Nov 1.

[42] Fallon PG, Jolin HE, Smith P, et al. IL-4 induces characteristic Th2 responses even in the combined absence of IL-5, IL-9, and IL-13. Immunity 2002;17(1):7–17. Available from: https://doi.org/10.1016/s1074-7613(02)00332-1 Jul.

[43] Renaud SJ, Cotechini T, Quirt JS, Macdonald-Goodfellow SK, Othman M, Graham CH. Spontaneous pregnancy loss mediated by abnormal maternal inflammation in rats is linked to deficient uteroplacental perfusion. J Immunol 2011;186(3):1799–808. Available from: https://doi.org/10.4049/jimmunol.1002679 Feb 1.

[44] Piccinni MP, Beloni L, Livi C, Maggi E, Scarselli G, Romagnani S. Defective production of both leukemia inhibitory factor and type 2 T-helper cytokines by decidual T cells in unexplained recurrent abortions. Nat Med 1998;4(9):1020–4. Available from: https://doi.org/10.1038/2006 Sep.

[45] Chaouat G, Assal Meliani A, Martal J, et al. IL-10 prevents naturally occurring fetal loss in the CBA x DBA/2 mating combination, and local defect in IL-10 production in this abortion-prone combination is corrected by in vivo injection of IFN-tau. J Immunol 1995;154(9):4261–8 May 1.

[46] Raghupathy R, Makhseed M, Azizieh F, Omu A, Gupta M, Farhat R. Cytokine production by maternal lymphocytes during normal human pregnancy and in unexplained recurrent spontaneous abortion. Hum Reprod 2000;15(3):713–18 Mar.

[47] Jin LP, Fan DX, Zhang T, Guo PF, Li DJ. The co-stimulatory signal upregulation is associated with Th1 bias at the maternal-fetal interface in human miscarriage. Am J Reprod Immunol 2011;66(4):270–8. Available from: https://doi.org/10.1111/j.1600-0897.2011.00997.x Oct.

[48] Sasaki Y, Sakai M, Miyazaki S, Higuma S, Shiozaki A, Saito S. Decidual and peripheral blood CD4 + CD25 + regulatory T cells in early pregnancy subjects and spontaneous abortion cases. Mol Hum Reprod 2004;10 (5):347–53. Available from: https://doi.org/10.1093/molehr/gah044 May.

[49] Mei S, Tan J, Chen H, Chen Y, Zhang J. Changes of CD4 + CD25high regulatory T cells and FOXP3 expression in unexplained recurrent spontaneous abortion patients. Fertil Steril 2010;94(6):2244–7. Available from: https://doi.org/10.1016/j.fertnstert.2009.11.020 doi:S0015-0282(09)04049-7 [pii].

[50] Wu HX, Jin LP, Xu B, Liang SS, Li DJ. Decidual stromal cells recruit Th17 cells into decidua to promote proliferation and invasion of human trophoblast cells by secreting IL-17. Cell Mol Immunol 2014;11(3):253–62. Available from: https://doi.org/10.1038/cmi.2013.67 May.

[51] Wang WJ, Hao CF, Yi L, et al. Increased prevalence of T helper 17 (Th17) cells in peripheral blood and decidua in unexplained recurrent spontaneous abortion patients. J Reprod Immunol 2010;84(2):164–70. Available from: https://doi.org/10.1016/j.jri.2009.12.003 doi:S0165-0378(10)00009-4 [pii].

[52] Kaplan MH. Th9 cells: differentiation and disease. Immunol Rev 2013;252(1):104–15. Available from: https://doi.org/10.1111/imr.12028 Mar.

[53] Licona-Limon P, Henao-Mejia J, Temann AU, et al. Th9 cells drive host immunity against gastrointestinal worm infection. Immunity 2013;39(4):744–57. Available from: https://doi.org/10.1016/j.immuni.2013.07.020 Oct 17.

[54] Gomez-Lopez N, Olson DM, Robertson SA. Interleukin-6 controls uterine Th9 cells and CD8(+) T regulatory cells to accelerate parturition in mice. Immunol Cell Biol 2016;94(1):79–89. Available from: https://doi.org/10.1038/icb.2015.63 Jan.

[55] Dudakov JA, Hanash AM, van den Brink MR. Interleukin-22: immunobiology and pathology. Annu Rev Immunol 2015;33:747–85. Available from: https://doi.org/10.1146/annurev-immunol-032414-112123.

[56] Wang Y, Xu B, Li MQ, Li DJ, Jin LP. IL-22 secreted by decidual stromal cells and NK cells promotes the survival of human trophoblasts. Int J Clin Exp Pathol 2013;6(9):1781–90.

[57] Jogdand GM, Mohanty S, Devadas S. Regulators of Tfh cell differentiation. Front Immunol 2016;7:520. Available from: https://doi.org/10.3389/fimmu.2016.00520.

[58] Monteiro C, Kasahara TM, Castro JR, et al. Pregnancy favors the expansion of circulating functional follicular helper T cells. J Reprod Immunol 2017;121:1–10. Available from: https://doi.org/10.1016/j.jri.2017.04.007 Jun.
[59] Krop J, Heidt S, Claas FHJ, Eikmans M. Regulatory T cells in pregnancy: it is not all about FoxP3. Front Immunol 2020;11:1182. Available from: https://doi.org/10.3389/fimmu.2020.01182.
[60] LeMaoult J, Caumartin J, Daouya M, et al. Immune regulation by pretenders: cell-to-cell transfers of HLA-G make effector T cells act as regulatory cells. Blood 2007;109(5):2040–8. Available from: https://doi.org/10.1182/blood-2006-05-024547 Mar 1.
[61] Hsu P, Santner-Nanan B, Joung S, Peek MJ, Nanan R. Expansion of CD4(+) HLA-G(+) T cell in human pregnancy is impaired in preeclampsia. Am J Reprod Immunol 2014;71(3):217–28. Available from: https://doi.org/10.1111/aji.12195 Mar.
[62] Magnani CF, Alberigo G, Bacchetta R, et al. Killing of myeloid APCs via HLA class I, CD2 and CD226 defines a novel mechanism of suppression by human Tr1 cells. Eur J Immunol 2011;41(6):1652–62. Available from: https://doi.org/10.1002/eji.201041120 Jun.
[63] van der Zwan A, van Unen V, Beyrend G, et al. Visualizing dynamic changes at the maternal-fetal interface throughout human pregnancy by mass cytometry. Front Immunol 2020;11:571300. Available from: https://doi.org/10.3389/fimmu.2020.571300.
[64] Reinhardt RL, Khoruts A, Merica R, Zell T, Jenkins MK. Visualizing the generation of memory CD4 T cells in the whole body. Nature 2001;410(6824):101–5. Available from: https://doi.org/10.1038/35065111 Mar 1.
[65] Kieffer TE, Faas MM, Scherjon SA, Prins JR. Pregnancy persistently affects memory T cell populations. J Reprod Immunol 2017;119:1–8. Available from: https://doi.org/10.1016/j.jri.2016.11.004 Feb.
[66] Wang S, Zhu X, Xu Y, et al. Programmed cell death-1 (PD-1) and T-cell immunoglobulin mucin-3 (Tim-3) regulate CD4+ T cells to induce Type 2 helper T cell (Th2) bias at the maternal-fetal interface. Hum Reprod 2016;31(4):700–11. Available from: https://doi.org/10.1093/humrep/dew019 Apr.
[67] Ramhorst R, Garcia V, Agriello E, et al. Intracellular expression of CD69 in endometrial and peripheral T cells represents a useful marker in women with recurrent miscarriage: modulation after allogeneic leukocyte immunotherapy. Am J Reprod Immunol 2003;49(3):149–58. Available from: https://doi.org/10.1034/j.1600-0897.2003.00021.x Mar.
[68] Tian Y, Babor M, Lane J, et al. Unique phenotypes and clonal expansions of human CD4 effector memory T cells re-expressing CD45RA. Nat Commun 2017;8(1):1473. Available from: https://doi.org/10.1038/s41467-017-01728-5 Nov 13.
[69] Luan X, Kang X, Li W, Dong Q. An investigation of the relationship between recurrent spontaneous abortion and memory T follicular helper cells. Am J Reprod Immunol 2017. Available from: https://doi.org/10.1111/aji.12714 Jun 21.
[70] Zhang YH, Sun HX. Immune checkpoint molecules in pregnancy: focus on regulatory T cells. Eur J Immunol 2020;50(2):160–9. Available from: https://doi.org/10.1002/eji.201948382 Feb.
[71] Miko E, Meggyes M, Doba K, Barakonyi A, Szereday L. Immune checkpoint molecules in reproductive immunology. Front Immunol 2019;10:846. Available from: https://doi.org/10.3389/fimmu.2019.00846.
[72] Hu XH, Tang MX, Mor G, Liao AH. Tim-3: expression on immune cells and roles at the maternal-fetal interface. J Reprod Immunol 2016;118:92–9. Available from: https://doi.org/10.1016/j.jri.2016.10.113 Nov.
[73] Wang S, Chen C, Li M, et al. Blockade of CTLA-4 and Tim-3 pathways induces fetal loss with altered cytokine profiles by decidual CD4(+)T cells. Cell Death Dis 2019;10(1):15. Available from: https://doi.org/10.1038/s41419-018-1251-0 Jan 8.
[74] Zhang T, Zhu W, Zhao Y, et al. Early transient suppression of immune checkpoint proteins T-cell immunoglobulin mucin-3 and programmed cell death-1 in peripheral blood lymphocytes after blastocyst transfer is associated with successful implantation. Fertil Steril 2020;114(2):426–35. Available from: https://doi.org/10.1016/j.fertnstert.2019.12.022 Aug.
[75] Wafula PO, Teles A, Schumacher A, et al. PD-1 but not CTLA-4 blockage abrogates the protective effect of regulatory T cells in a pregnancy murine model. Research support, non-U.S. Govrsquo;t. Am J Reprod Immunol 2009;62(5):283–92. Available from: https://doi.org/10.1111/j.1600-0897.2009.00737.x Nov.

[76] Wang WJ, Salazar Garcia MD, Deutsch G, et al. PD-1 and PD-L1 expression on T-cell subsets in women with unexplained recurrent pregnancy losses. Am J Reprod Immunol 2020;83(5):e13230. Available from: https://doi.org/10.1111/aji.13230 May.

[77] Li G, Lu C, Gao J, et al. Association between PD-1/PD-L1 and T regulate cells in early recurrent miscarriage. Int J Clin Exp Pathol 2015;8(6):6512–18.

[78] Morita K, Tsuda S, Kobayashi E, et al. Analysis of TCR repertoire and PD-1 expression in decidual and peripheral CD8(+) T cells reveals distinct immune mechanisms in miscarriage and preeclampsia. Front Immunol 2020;11:1082. Available from: https://doi.org/10.3389/fimmu.2020.01082.

[79] Winger EE, Reed JL, Ashoush S, El-Toukhy T, Ahuja S, Taranissi M. Elevated preconception CD56 + 16 + and/or Th1:Th2 levels predict benefit from IVIG therapy in subfertile women undergoing IVF. Am J Reprod Immunol 2011;66(5):394–403. Available from: https://doi.org/10.1111/j.1600-0897.2011.01018.x Nov.

[80] Vomstein K, Feil K, Strobel L, et al. Immunological risk factors in recurrent pregnancy loss: guidelines vs current state of the art. J Clin Med 2021;10(4). Available from: https://doi.org/10.3390/jcm10040869 Feb 20.

[81] Gomaa MF, Elkholy AG, El-Said MM, Abdel-Salam NE. Combined oral prednisolone and heparin vs heparin: the effect on peripheral NK cells and clinical outcome in patients with unexplained recurrent miscarriage. A double-blind placebo randomized controlled trial. Arch Gynecol Obstet 2014;290(4):757–62. Available from: https://doi.org/10.1007/s00404-014-3262-0 Oct.

[82] Ahmadi M, Abdolmohammadi-Vahid S, Ghaebi M, et al. Effect of Intravenous immunoglobulin on Th1 and Th2 lymphocytes and improvement of pregnancy outcome in recurrent pregnancy loss (RPL). Biomed Pharmacother 2017;92:1095–102. Available from: https://doi.org/10.1016/j.biopha.2017.06.001 Aug.

[83] Ahmadi M, Abdolmohammadi-Vahid S, Ghaebi M, et al. Regulatory T cells improve pregnancy rate in RIF patients after additional IVIG treatment. Syst Biol Reprod Med 2017;63(6):350–9. Available from: https://doi.org/10.1080/19396368.2017.1390007 Dec.

[84] Nakagawa K, Kwak-Kim J, Ota K, et al. Immunosuppression with tacrolimus improved reproductive outcome of women with repeated implantation failure and elevated peripheral blood TH1/TH2 cell ratios. Am J Reprod Immunol 2015;73(4):353–61. Available from: https://doi.org/10.1111/aji.12338 Apr.

[85] Azizi R, Ahmadi M, Danaii S, et al. Cyclosporine A improves pregnancy outcomes in women with recurrent pregnancy loss and elevated Th1/Th2 ratio. J Cell Physiol 2019;234(10):19039–47. Available from: https://doi.org/10.1002/jcp.28543 Aug.

[86] Winger EE, Reed JL. Treatment with tumor necrosis factor inhibitors and intravenous immunoglobulin improves live birth rates in women with recurrent spontaneous abortion. Am J Reprod Immunol 2008;60 (1):8–16 Jul.

[87] Michael AE, Papageorghiou AT. Potential significance of physiological and pharmacological glucocorticoids in early pregnancy. Hum Reprod Update 2008;14(5):497–517. Available from: https://doi.org/10.1093/humupd/dmn021 Sep-Oct.

[88] Kemp MW, Newnham JP, Challis JG, Jobe AH, Stock SJ. The clinical use of corticosteroids in pregnancy. Hum Reprod Update 2016;22(2):240–59. Available from: https://doi.org/10.1093/humupd/dmv047 Mar-Apr.

[89] Snijdewint FG, Kapsenberg ML, Wauben-Penris PJ, Bos JD. Corticosteroids class-dependently inhibit in vitro Th1- and Th2-type cytokine production. Immunopharmacology 1995;29(2):93–101. Available from: https://doi.org/10.1016/0162-3109(94)00048-k Mar.

[90] Milburn HJ, Poulter LW, Dilmec A, Cochrane GM, Kemeny DM. Corticosteroids restore the balance between locally produced Th1 and Th2 cytokines and immunoglobulin isotypes to normal in sarcoid lung. Clin Exp Immunol 1997;108(1):105–13. Available from: https://doi.org/10.1046/j.1365-2249.1997.d01-979.x Apr.

[91] Sasaki A, Kuroda K, Hisano M, Ozawa N, Sago H, Yamaguchi K. Successful treatment for a recurrent pregnancy loss woman with high Th1/Th2 ratio using medium-dose corticosteroids. J Obstet Gynaecol 2017;37 (5):685–7. Available from: https://doi.org/10.1080/01443615.2017.1285873 Jul.

[92] Harmon HM, Jensen EA, Tan S, et al. Timing of postnatal steroids for bronchopulmonary dysplasia: association with pulmonary and neurodevelopmental outcomes. J Perinatol 2020;40(4):616–27. Available from: https://doi.org/10.1038/s41372-020-0594-4 Apr.

[93] Palmsten K, Bandoli G, Vazquez-Benitez G, et al. Oral corticosteroid use during pregnancy and risk of preterm birth. Rheumatol (Oxf) 2020;59(6):1262–71. Available from: https://doi.org/10.1093/rheumatology/kez405 Jun 1.

[94] Palmsten K, Bandoli G, Watkins J, Vazquez-Benitez G, Gilmer TP, Chambers CD. Oral corticosteroids and risk of preterm Birth in the California medicaid program. J Allergy Clin Immunol Pract 2021;9(1):375–84. Available from: https://doi.org/10.1016/j.jaip.2020.07.047 e5.

[95] Toth B, Wurfel W, Bohlmann M, et al. Recurrent miscarriage: diagnostic and therapeutic procedures. guideline of the DGGG, OEGGG and SGGG (S2k-Level, AWMF registry number 015/050). Geburtshilfe Frauenheilkd 2018;78(4):364–81. Available from: https://doi.org/10.1055/a-0586-4568 Apr.

[96] Schwab I, Nimmerjahn F. Intravenous immunoglobulin therapy: how does IgG modulate the immune system? Nat Rev Immunol 2013;13(3):176–89. Available from: https://doi.org/10.1038/nri3401 Mar.

[97] Wong LF, Porter TF, Scott JR. Immunotherapy for recurrent miscarriage. Cochrane Database Syst Rev 2014;10:CD000112. Available from: https://doi.org/10.1002/14651858.CD000112.pub3 Oct 21.

[98] Egerup P, Lindschou J, Gluud C, Christiansen OB. ImmuRe MIPDSG. the effects of intravenous immunoglobulins in women with recurrent miscarriages: a systematic review of randomised trials with meta-analyses and trial sequential analyses including individual patient data. PLoS One 2015;10(10):e0141588. Available from: https://doi.org/10.1371/journal.pone.0141588.

[99] Christiansen OB, Larsen EC, Egerup P, Lunoee L, Egestad L, Nielsen HS. Intravenous immunoglobulin treatment for secondary recurrent miscarriage: a randomised, double-blind, placebo-controlled trial. BJOG. 2015;122(4):500–8. Available from: https://doi.org/10.1111/1471-0528.13192 Mar.

[100] Christiansen OB, Kolte AM, Krog MC, Nielsen HS, Egerup P. Treatment with intravenous immunoglobulin in patients with recurrent pregnancy loss: An update. J Reprod Immunol 2019;133:37–42. Available from: https://doi.org/10.1016/j.jri.2019.06.001 Jun.

[101] Saab W, Seshadri S, Huang C, Alsubki L, Sung N, Kwak-Kim J. A systemic review of intravenous immunoglobulin G treatment in women with recurrent implantation failures and recurrent pregnancy losses. Am J Reprod Immunol 2021;85(4):e13395. Available from: https://doi.org/10.1111/aji.13395 Apr.

[102] Woon EV, Day A, Bracewell-Milnes T, Male V, Johnson M. Immunotherapy to improve pregnancy outcome in women with abnormal natural killer cell levels/activity and recurrent miscarriage or implantation failure: a systematic review and meta-analysis. J Reprod Immunol 2020;142:103189. Available from: https://doi.org/10.1016/j.jri.2020.103189 Nov.

[103] Yamada H, Morikawa M, Furuta I, et al. Intravenous immunoglobulin treatment in women with recurrent abortions: increased cytokine levels and reduced Th1/Th2 lymphocyte ratio in peripheral blood. Am J Reprod Immunol 2003;49(2):84–9 Feb.

[104] Lee SK, Kim JY, Han AR, et al. Intravenous immunoglobulin G improves pregnancy outcome in women with recurrent pregnancy losses with cellular immune abnormalities. Am J Reprod Immunol 2016;75(1):59–68. Available from: https://doi.org/10.1111/aji.12442 Jan.

[105] Kim DJ, Lee SK, Kim JY, et al. Intravenous immunoglobulin G modulates peripheral blood Th17 and Foxp3 (+) regulatory T cells in pregnant women with recurrent pregnancy loss. Am J Reprod Immunol 2014;71(5):441–50. Available from: https://doi.org/10.1111/aji.12208 May.

[106] Rachid R, Bonilla FA. The role of anti-IgA antibodies in causing adverse reactions to gamma globulin infusion in immunodeficient patients: a comprehensive review of the literature. J Allergy Clin Immunol 2012;129(3):628–34. Available from: https://doi.org/10.1016/j.jaci.2011.06.047 Mar.

[107] Zhang R, Szerlip HM. Reemergence of sucrose nephropathy: acute renal failure caused by high-dose intravenous immune globulin therapy. South Med J 2000;93(9):901–4 Sep.

[108] Han AR, Lee SK. Immune modulation of i.v. immunoglobulin in women with reproductive failure. Reprod Med Biol 2018;17(2):115–24. Available from: https://doi.org/10.1002/rmb2.12078 Apr.

[109] Triggianese P, Lattavo G, Chimenti MS, et al. Reproductive outcomes 20 years after the intravenous immunoglobulin treatment in women with recurrent pregnancy losses. Am J Reprod Immunol 2020;83(4):e13224. Available from: https://doi.org/10.1111/aji.13224 Apr.

[110] Kino T, Hatanaka H, Miyata S, et al. FK-506, a novel immunosuppressant isolated from a Streptomyces. II. Immunosuppressive effect of FK-506 in vitro. J Antibiot (Tokyo) 1987;40(9):1256–65. Available from: https://doi.org/10.7164/antibiotics.40.1256 Sep.

[111] Nakagawa K, Kwak-Kim J, Hisano M, et al. Obstetric and perinatal outcome of the women with repeated implantation failures or recurrent pregnancy losses who received pre- and post-conception tacrolimus treatment. Am J Reprod Immunol 2019;82(2):e13142. Available from: https://doi.org/10.1111/aji.13142 Aug.

[112] Nakagawa K, Kwak-Kim J, Kuroda K, Sugiyama R, Yamaguchi K. Immunosuppressive treatment using tacrolimus promotes pregnancy outcome in infertile women with repeated implantation failures. Am J Reprod Immunol 2017;78(3). Available from: https://doi.org/10.1111/aji.12682 Sep.
[113] Coscia LA, Constantinescu S, Moritz MJ, et al. Report from the national transplantation pregnancy registry (NTPR): outcomes of pregnancy after transplantation. Clin Transpl 2010;65–85.
[114] Ling Y, Huang Y, Chen C, Mao J, Zhang H. Low dose Cyclosporin A treatment increases live birth rate of unexplained recurrent abortion—initial cohort study. Clin Exp Obstet Gynecol 2017;44(2):230–5.
[115] Ahmadi M, Abdolmohamadi-Vahid S, Ghaebi M, et al. Sirolimus as a new drug to treat RIF patients with elevated Th17/Treg ratio: a double-blind, phase II randomized clinical trial. Int Immunopharmacol 2019;74:105730. Available from: https://doi.org/10.1016/j.intimp.2019.105730 Sep.
[116] Perez-Garcia LF, Dolhain R, Vorstenbosch S, et al. The effect of paternal exposure to immunosuppressive drugs on sexual function, reproductive hormones, fertility, pregnancy and offspring outcomes: a systematic review. Hum Reprod Update 2020;26(6):961–1001. Available from: https://doi.org/10.1093/humupd/dmaa022 Nov 1.
[117] Shen L, Xu W, Gao J, et al. Pregnancy after the diagnosis of lymphangioleiomyomatosis (LAM). Orphanet J Rare Dis 2021;16(1):133. Available from: https://doi.org/10.1186/s13023-021-01776-7 Mar 17.
[118] Steeland S, Libert C, Vandenbroucke RE. A new venue of TNF targeting. Int J Mol Sci 2018;19(5). Available from: https://doi.org/10.3390/ijms19051442 May 11.
[119] Alijotas-Reig J, Esteve-Valverde E, Ferrer-Oliveras R, Llurba E, Gris JM. Tumor necrosis factor-alpha and pregnancy: focus on biologics. an updated and comprehensive review. Clin Rev Allergy Immunol 2017;53(1):40–53. Available from: https://doi.org/10.1007/s12016-016-8596-x Aug.
[120] Ota K, Dambaeva S, Han AR, Beaman K, Gilman-Sachs A, Kwak-Kim J. Vitamin D deficiency may be a risk factor for recurrent pregnancy losses by increasing cellular immunity and autoimmunity. Hum Reprod 2014;29(2):208–19. Available from: https://doi.org/10.1093/humrep/det424 Feb.
[121] Rafiee M, Gharagozloo M, Ghahiri A, et al. Altered Th17/Treg ratio in recurrent miscarriage after treatment with paternal lymphocytes and vitamin D3: a double-blind placebo-controlled study. Iran J Immunol. 2015;12(4):252-262. doi:IJIv12i4A3.

CHAPTER

4

B cell pathology and recurrent pregnancy loss

Ruth Marian Guzman-Genuino[1] *and Kerrilyn R. Diener*[1,2]

[1]Clinical and Health Sciences Unit, University of South Australia, Adelaide, SA, Australia

[2]Robinson Research Institute and Adelaide Medical School, The University of Adelaide, Adelaide, SA, Australia

1 Introduction

A successful pregnancy requires the maternal immune system to perform contrasting functions: it must be immunologically tolerant of the semiallogeneic conceptus while maintaining effective immunity against foreign pathogens. Immune dysfunction can contribute to pregnancy disorders such as recurrent pregnancy loss (RPL), preeclampsia, and immune-based infertility. Therefore, defining and understanding the functional role of the various immune, endometrial and fetal cells and their complex interactions are critical for developing successful treatments for pregnancy pathologies.

B cells play a central role in classical humoral immunity because of their ability to produce pathogen-specific antibodies. Antibody production is important during pregnancy, as the production of protective asymmetric antibodies against paternal antigens allows the maternal immune system to remain tolerant of the growing fetus. Research has confirmed that B cells perform crucial regulatory roles during pregnancy and undergo a wide range of modifications in proliferation, subtype switching, antibody production, cytokine production, and immune cell regulation [1,2].

Abnormal B cell subsets are increasingly being reported as factors in infertility and poor obstetric outcomes. In mice, the lack of mature B cells was associated with reduced litter size, decreased embryo size, and higher susceptibility to prenatal infections [3]. Immune profiling of women with RPL and preeclampsia has revealed altered induction and function of memory B cells [4,5]. Therefore, the developmental and functional changes in B cell subsets are considered major contributors to the immune alteration that occurs during pregnancy [2].

Imbalance within the immune system is often attributed as the cause of adverse reproductive outcomes, especially in idiopathic cases. Dysregulated immune indicators include

Immunology of Recurrent Pregnancy Loss and Implantation Failure
DOI: https://doi.org/10.1016/B978-0-323-90805-4.00016-X

abnormalities in lymphocyte quantity, phenotype, and functionality, such as dysregulated cytokines; the presence of antiphospholipid (aPL) antibodies or other autoantibodies; and imbalanced T helper (Th)1 and Th2 cell ratio [6]. Inadequate immune adaptation results in implantation failure, while suboptimal immune adaptation increases the risk of pregnancy pathologies, ranging from early RPL to intrauterine growth restriction (IUGR), preeclampsia, and preterm birth.

The current treatment options for pregnancy loss resulting from inadequate or suboptimal immune adaptation are general immunotherapies that modulate women's immune systems, including intravenous immunoglobulin (IVIg) therapy and lymphocyte immunization therapy (LIT). However, recent reviews and *meta*-analyses suggest that immunotherapy does not improve pregnancy outcomes in all cases of idiopathic RPL; this suggests that immunological treatments can only be effective if the association between the abnormality and the treatment is known [6,7].

B cell pathology related to RPL is a relatively underexplored field of reproductive immunology. The dearth of information means that few specific therapies that target B cells as an etiological factor for RPL have been identified. This chapter presents information about the role of B cells in RPL and other immune-associated pregnancy complications and discusses potential therapeutic approaches that target B cells.

2 B cell function during pregnancy

2.1 Humoral immunity

B cell function is primarily associated with humoral immunity, and B cells are thought to act via antibody production in pregnancy. The earliest investigations showed that the immunoglobulin (Ig) G fraction of maternal serum prevented the cytotoxic effects of maternal effector T cells against allogeneic trophoblast tissue in an in vitro model of human trophoblast antigenicity [8]. Increased levels of both IgG- and IgM-secreting cells in the uterine-draining lymph nodes in pregnant mice further supported the notion that antibodies played a suppressive role in supporting and maintaining the fetal allograft [9]. Recent murine and clinical studies have reinforced this concept. The levels of IgM and IgA in serum samples from pregnant mice were significantly higher than in nonpregnant controls, and there was increased production of natural IgM, IgA, IgG3, and IgG4 in serum from healthy pregnant women compared to age-matched nonpregnant women [10–12]. In a study of the human memory B cell subset during pregnancy, the level of expression of the memory marker CD27 directly affected the profile of Ig isotypes being produced: up regulated CD27 expression was associated with increased IgG and IgA production [13]. These results suggest that B cell activation and antibody production are necessary steps toward achieving a healthy pregnancy.

2.2 Asymmetric antibodies and autoantibodies

Asymmetric antibodies are a subpopulation of IgG antibodies that bear a carbohydrate residue in only one Fab region of the molecule, rendering them functionally univalent and incapable of triggering immune effector functions, such as complement fixation, phagocytosis, and cytotoxicity [14]. Studies have shown higher levels of asymmetric antibodies (most with paternal antigen specificity) in serum from healthy pregnant women than nonpregnant women, while women experiencing recurrent spontaneous abortions had significantly lower proportions of serum asymmetric antibodies [15–17]. Therefore, asymmetric antibodies were considered protective during pregnancy and referred to as "protective maternal antibodies," despite the mechanisms underpinning this response remaining unclear.

Conversely, autoantibodies have been associated with adverse pregnancy outcomes. Several aPLs—lupus anticoagulant, anticardiolipin, and anti-β-2 glycoprotein I—have been associated with RPL and infertility, while antibodies to thyroid antigens (e.g., thyroglobulin and thyroid peroxidase), nuclear antigens, antilaminin, antiprothrombin, and anti-*Saccharomyces cerevisiae* have also been implicated in pregnancy complications [18]. Recent work has focused on the effect of aPL, and agonistic autoantibodies on the angiotensin II type I receptor (AT1-AA) [19–21]. High levels of serum aPL can occur in the autoimmune condition antiphospholipid syndrome (APS). In pregnancy, APS leads to compromised reproductive outcomes, such as fetal loss and severe preeclampsia [19,22,23]. AT1-AA autoantibodies also play a significant role in the pathology and possibly the initiation of preeclampsia. These agonistic autoantibodies mimic the natural ligand of the AT1-AA and induce downstream production of the antiangiogenic factors, such as soluble fms-like tyrosine kinase 1 (s-Flt1) and endoglin; this leads to the onset of preeclamptic symptoms and results in placental abnormalities and embryonic defects [24,25].

2.3 B cell lymphocytopenia

The number, distribution, and function of B cell subsets can fluctuate during pregnancy. The earliest studies described a general reduction in B cell numbers during pregnancy and conditional responsiveness of B cells toward endocrine factors, such as estrogen and human chorionic somatomammotropin [26,27]. Mouse models of allogeneic pregnancy showed a significant reduction in the number of pre-B cells, pro-B cells, immature B cells in bone marrow, and splenic follicular B cells during the latter half of pregnancy [10,11,28]. A general reduction in B cells was also demonstrated in clinical studies, and peripheral $CD19^+$ B cells were significantly lower during the second and third trimesters, and on the day of delivery, in low-risk pregnant women than in nonpregnant women [29–32]. Additional investigations identified the specific B cell subsets undergoing lymphopenia, which included IgD^+CD38^{hi} transitional B cells, $CD24^{hi}CD38^{hi}$ transitional B cells, IgD^+CD38^- unswitched memory B cells, IgD^-CD38^- resting memory B cells, IgD^-CD38^{hi} plasmablasts, and $CD5^+$ B cells [12,31,33,34]. In addition to lower levels, there is also evidence of suppressed B cell function during human pregnancy [35]. In contrast, several murine studies have shown that some B cell subsets expand during pregnancy; these include splenic marginal zone B cells, peritoneal B1 and B2 cells, B cells within uterine-draining lymph nodes, and the uterus, and B cells in the decidua [11,28,36,37]. Further, the percentage of circulating IgD^+CD38^+ naive B cells increased in the late stages of pregnancy [31].

The decrease in circulating B cells may be due to decreased B cell lymphopoiesis or redistribution of B cells in the growing placenta [10,27,36]. Histological examination of the immune infiltrates from first trimester placentas showed that B cells were extant in the decidua of normal pregnancies and were significantly increased in decidual tissue exhibiting hydatidiform moles [38]. Placental intervillous blood was found to be rich in the chemokine ligand CCL20, which attracts mature naive B cells that express the nonpromiscuous receptor CCR6 [39]. It is speculated that B cells within the placental tissue may safeguard term pregnancies via the production of progesterone-induced blocking factor (PIBF) 1 [29,39,40].

Changes in the B cell profile have been associated with RPL and infertility. Women with recurrent miscarriages had a significantly higher number of endometrial $CD20^+$ B cells compared to controls [41,42]. Similarly, a patient with a history of RPL and obstetric complications had elevated levels of $CD19^+$ B cells in peripheral blood and low levels of $CD27^+IgD^-$ and $CD27^+IgD^+$ memory B cells [43]. However, another study found a reduction in peripheral $CD19^+IL\text{-}10^+$ B cells and mRNA levels of interleukin (IL)-10 and programmed death-ligand 1 (PD-L1) genes in women with RPL [44,45]. The relationships between B cells and pregnancy pathologies require further investigation.

2.4 Regulatory B cells

Any subset of B cell that exhibits an immune-suppressive function is termed a regulatory B cell (Breg) [1]. In murine pregnancy, Bregs include $IL\text{-}10^+$ B cells, $CD5^+CD1d^+$ B cells, $CD80^+CD86^+$ B cells, $CD80^+CD86^+CD27^+IL\text{-}10^+$ B cells, $IL\text{-}35^+$ B cells and progesterone immunomodulatory binding factor 1^+ ($PIBF1^+$) choriodecidual B cells. In human pregnancy, identified Bregs are $IL\text{-}10^+$ B cells, $CD24^{hi}CD27^+$ B cells, $CD24^{hi}CD38^{hi}$ transitional B cells, $CD27^+IgM^+$ memory B cells, $CD38^{hi}CD27^{hi}$ plasmablasts, marginal zone B cells, and $IL\text{-}35^+$ B cells. The ability of B cells to produce IL-10 and, to a lesser extent, transforming growth factor beta (TGFβ) remains a unifying hallmark of Bregs across species. Additional markers, such as IL-35 and PIBF1, have also been shown to promote a regulatory capacity in B cells during pregnancy [40,46–49].

Pregnancy induces the expansion of marginal zone B cell populations in the spleen of normal mice, whereas this expansion does not occur in mice with pregnancy complications. In a successful pregnancy, marginal zone B cells produce higher levels of IgM and IgA, which is suspected to aid, or be indicative of, alterations in the cellular immune response from a Th1-like profile to a Th2-like profile. During the murine peri-implantation period, IL-10-producing B cells in the uterine tissue exhibit regulatory properties and display significantly upregulated levels of the costimulatory molecules CD80 and CD86 [28]. Similarly, splenic B cells also upregulate costimulatory expression during this period [50–52], which may allow both local and systemic B cells to modulate regulatory T cell populations and cytokine production in preparation for successful embryo implantation.

Bregs found in the uterine-draining lymph nodes, and spleen of pregnant mice and peripheral blood of pregnant women have demonstrated the ability to produce and respond to IL-35 and contribute to establishing maternal immune tolerance and maintenance of a healthy pregnancy [47,49]. Choriodecidual B cells have also been classified as Bregs in murine pregnancy, as IL-33-mediated B cell production of PIBF1 in late gestation protected against preterm labor; these effects were enhanced by therapeutic administration of IL-33 [40]. Moreover, choriodecidual B cells expressed elevated levels of the activated memory B cell and plasma cell-associated molecules, such as CD11c, CD27, CD38, CD70, CD80, CD86, CD95, CD138, and B cell maturation antigen, indicating higher activation status, class switching, memory and plasmacytoid differentiation [40].

Bregs have also been shown to participate in the peri-implantation period in mice. Toll-like receptor 9 and CD86 (Breg-associated markers) were upregulated 3.5 days postcoitus in splenic B cells, while uterine B cells expanded and the B cell phenotype changed between 2.5 and 8.5 days postcoitus [28,50]. Uterine B cells effectively suppressed the proliferation and activation of syngeneic $CD4^+$ effector T cells [28], and trophoblast education in the earliest phase of pregnancy may have contributed to changes in the B cell phenotype [46]. Lower levels of IL-10 mRNA in peripheral B cells and significantly lower levels of IL-10 secretion on ex vivo restimulation have been found in women with repeated implantation failure compared to healthy controls [45], which suggests that impaired B cell production of IL-10 may be a key mechanism in repeated implantation failure. These studies suggest that Bregs support appropriate embryo implantation in early pregnancy.

3 Immunopathologies related to reproduction

3.1 B cell immune dysregulation in pregnancy

The coordinated interaction of correctly functioning maternal immune cells in early pregnancy is crucial to the physiological changes required for a successful, healthy, full-term pregnancy. Immune dysregulation, which includes B cell pathologies, has been proposed as a potential etiology in unexplained RPL. Maternal immune tolerance is characterized by an induction of regulatory T cells and an antiinflammatory Th2 profile. Without these conditions, an aberrant maternal alloimmune response may result in pregnancy loss.

Immunological alterations in B cell subsets may be involved in the etiopathogenesis of unexplained RPL. For instance, the percentage of B-1 $CD19^+CD5^+$ lymphocytes was significantly higher in women with RPL compared to healthy women [53]. Moreover, CD83 expression on B cells is locally regulated and differentially expressed in disturbed murine pregnancy [54]. Therefore, dysregulation in B cell number and function may contribute to the pathogenesis of APS, suboptimal implantation, IUGR, preeclampsia, and preterm birth (see Fig. 1).

3.2 Antiphospholipid syndrome

APS is a humoral autoimmune disorder characterized by the presence of aPLs that are directed against membrane phospholipids. Clinical features of APS relate to a hypercoagulable state and may lead to obstetric complications via thrombosis and impaired endothelial function in pregnant women [21,55].

As APS is an autoantibody-mediated disease, B cells are thought to play a crucial role in its pathogenesis; however, few studies have addressed the dysregulation of B cells and the characteristics of aPL-producing clones. Disturbance of B cell homeostasis during B cell differentiation and B cell tolerance checkpoints may lead to alterations in the B cell

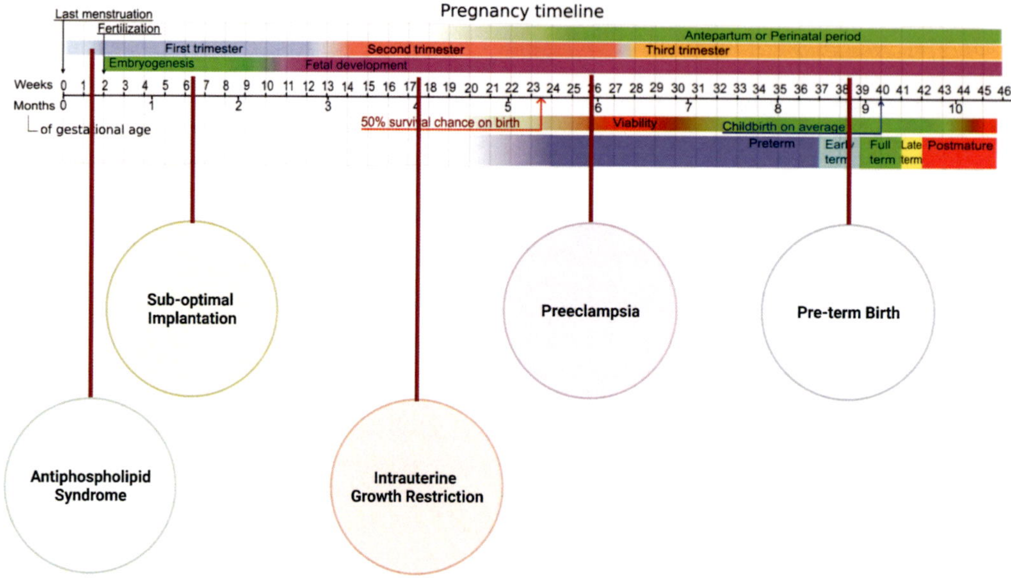

FIGURE 1 The potential contribution of B cell dysregulation to conditions during pregnancy. Source: *Created with BioRender.com adapted from Medical gallery of Mikael Häggström 2014.*

repertoire and the generation of autoantibody-producing cells. In obstetric APS patients, perturbations to B cell activation and differentiation have been linked to a bias toward plasma cell differentiation that resulted in fewer memory B cells [4,56]. The reduction in memory B cells and the consequent increase in circulating naive cells were associated with lower concentrations of serum C4 and concomitant increase in IgG anticardiolipin and anti-β-2 glycoprotein I antibodies in patients with thrombotic complications [4]. A study that profiled the lymphocyte subsets of 11 patients with thrombotic primary APS supported this finding; increased proportions of B1 cells, transitional B cells ($CD24^{+}CD38^{+}$), and naive B cells ($CD27^{-}IgD^{+}$), and fewer memory B cells ($CD27^{+}IgD^{+}$ and $CD27^{+}IgD^{-}$) [57]. However, contradictory findings were reported in a recent paper, where lower percentages of naive B cells were observed in thrombotic and obstetric APS patients compared to healthy controls [56].

To date, the debate continues on the role of naive B cells in APS pathogenesis. Recent findings reveal that, in general, APS patients have further deviations in B cell homeostasis, demonstrated by fewer total memory B cells, preswitched memory B cells, postswitched memory B cells, and IL-10-producing Bregs [58]. Furthermore, $CD20^{-}$ B cells are a major source of aPL, while $CD20^{+}$ B cells partially contribute to anticardiolipin antibody production [58]. Another underresearched soluble mediator that could dysregulate B cell survival and differentiation is B cell-activating factor belonging to the tumor necrosis factor (TNF) family (BAFF). In APS, soluble BAFF level is increased in obstetric patients with either primary or secondary APS and correlated with a higher risk of thrombotic events [59]. BAFF may be developed as a biomarker or therapeutic target for obstetric APS patients in the future.

3.3 Suboptimal implantation

Failure in implantation leads to pregnancy loss. The underlying reasons may be related to problems with gametes or the embryo or suboptimal endometrial conditions. In the latter, assuming that the blastocyst is intact and fully competent to undergo implantation, impairment in the recruitment, induction, or activation of immune components at the fetal–maternal interface may lead to erroneous implantation [60]. The immune network at implantation is complex and poorly understood; however, B cells may play a role during early pregnancy. In murine models, the proportion of uterine B cells increased during the peri-implantation period and B cells exhibited increased IL-10 production and a regulatory capacity over T cells in vitro [28]. Likewise, the proportion of peritoneal IL-10-producing Bregs gradually increased during early pregnancy and decreased during the second trimester [61].

Although few studies have examined the functionality of B cells during the early stages of pregnancy, it appears that B cells contribute to successful implantation. Recent reports show an increase in decidual B cells during the first and second trimesters of pregnancy, specifically in $CD24^{hi}CD27^{+}$ B cells and $CD24^{hi}CD38^{hi}$ B cells, and decidual B cells have a higher production capacity of IL-10 compared to peripheral blood B cells [36]. Moreover, decidual B cells were observed to colocalize with T cells, possibly exerting regulatory control that supports regulatory T cell induction and maintenance while inhibiting the development of naive T cells into Th1 or Th17 cells [36]. Finally, an in vitro study has shown that B cells develop a regulatory phenotype in response to trophoblast presence, providing immunologic protection against detrimental T cell-mediated inflammation [46]. These findings support the emerging view that B cells are intimately involved in promoting the optimal conditions for implantation, and alterations or disturbances in B cell homeostasis may contribute to spontaneous abortion. Indeed, a clinical study of Breg function during pregnancy demonstrated that inadequate levels of $CD24^{hi}CD27^{+}$ B cells at pregnancy onset are correlated with spontaneous abortion [48], while adoptive transfer of IL-10-producing Bregs can restore pregnancy tolerance in abortion-prone mice [62].

3.4 Intrauterine growth restriction

IUGR is defined as the failure of a fetus to reach its genetic growth potential and is associated with increased perinatal mortality and morbidity. IUGR is the common result of maternal, placental, fetal, or genetic factors or a combination of these factors. The leading cause of IUGR is placental insufficiency, which affects 5%–10% of all pregnancies [63]. As placental insufficiency occurs when the decidual cellular immunity limits trophoblastic invasion, thereby curtailing microcirculation and metabolism in the placenta, partial rejection of the fetal allograft may be an immunological etiology for IUGR. A recent study found that severe abnormalities in the immune system led to adaptation disorders in pregnant women, including placental insufficiency [63]. However, the association and mechanisms between IUGR and disturbances in the immune network at the fetal–maternal interface remain unclear.

Several studies that evaluated lymphocyte subsets in patients with IUGR reported that dysregulation in B cell frequency was significantly negatively correlated with birth weight

[64,65]. Women with IUGR were shown to have a significantly higher number of circulating lymphoid cells positive for cytoplasmic IgM, IgG, and IgA, as well as a higher absolute number and percentage of B cells. Functionally, the expanded B cell subset may be correlated to the increased production of antibodies and chronic activation of the lymphoid system observed in patients with IUGR [64,66]. A cohort study of patients with IUGR found a positive correlation between the percentage and number of B lymphocytes in umbilical cord blood and infant weight, length, and head circumference at birth [65]. Comparison of cytokine profiles of peripheral blood showed an increase in the levels of pro-inflammatory cytokines IL-8, IL-6, IL-12, and TNF in women with IUGR compared with healthy pregnant women. At the same time, IL-10 was lower in women with IUGR and placental insufficiency compared to women with IUGR alone [67–69]. As these cytokines are produced by a variety of immune cells, including B cells, it is unclear how B cells affect these results. However, these findings collectively indicate that B cells are involved in the pathophysiology of IUGR.

3.5 Preeclampsia

Preeclampsia is characterized by elevated blood pressure, a decline in renal function, IUGR, chronic immune activation, and multiorgan dysfunction that begins in the latter half of pregnancy. Preeclampsia places a woman and her fetus at considerable risk of stroke, preterm birth, and death. Like IUGR, the disease etiology is attributed to inadequate placental vascular remodeling and poor placental circulation, resulting in placental hypoxia and ischemia [70]. Evidence for the role of B cells in hypertension is limited; however, B cells are heavily implicated in preeclampsia via their memory immune response and production of autoantibodies. Dysregulation in B cell frequency and function was observed in one study that found higher percentages of circulating memory B cells ($CD27^{+}CD38^{-}$ and $CD27^{+}CD38^{+}$) and increased B cell proliferation in preeclamptic women compared to normal controls [71]. Further, preeclamptic placentas stained positive for markers of B1 B lymphocytes ($CD5^{+}$), and this B cell subset demonstrated a higher capacity to produce IgG autoantibodies specific to AT1-AA [72]. The role of B cells in the pathophysiology of hypertension arising from placental ischemia has been further supported by observations in rat models of reduced uterine perfusion pressure. Adoptive transfer of patient-derived AT1-AA was shown to induce preeclampsia in pregnant rats, while depletion of circulating B cells using the atnihuman CD20 antibody rituximab resulted in the attenuation of preeclampsia [25,73].

3.6 Preterm birth

Preterm birth is defined as a birth that occurs prior to 37 weeks of gestation. Preterm birth is a complex phenomenon that is the leading cause of infant death and predisposes surviving infants to long-term illnesses [74]. Intrauterine and systemic infection and the ensuing inflammation are the most common pathophysiological mechanism in preterm labor, which commonly results in preterm birth.

B cells have been found in both the decidua basalis and decidua parietalis prior to parturition and at the onset of labor at term, but B cells were expanded only in the decidua basalis [75]. Phenotypical and functional characterization of B cells found in the amniotic fluid of murine pregnancies reveals an abundance of B1 B cells that spontaneously produce IgM [76]. Increased proportions of total B cells in the decidua parietalis were apparent in women who underwent spontaneous preterm labor, while no such changes were observed in women who delivered at term [40,77]. Preterm samples of placentas and attached chorioamniotic membranes also exhibited higher frequency and enhanced activation of B1 cells that secrete autoantibodies and polyreactive antibodies [40].

Recent findings highlight the regulatory role of B cells during labor. In a mouse model of lipopolysaccharide (LPS)-induced preterm labor, a lack of B cells promoted uterine inflammation and a higher incidence of preterm labor and neonatal mortality [3,40]. In both spontaneous term and preterm labor, B cells in the human choriodecidual stroma were shown to have phenotypic and functional profiles that indicate higher activation, class switching, memory, and plasmacytoid differentiation; however, in preterm labor, the choriodecidua contained B cells that had altered functions (i.e., aberrant expansion, higher activation, and enhanced antibody production) [40]. Mouse models of LPS-induced preterm birth showed that decidual B cell production of PIBF1 mediated by IL-33 safeguards term pregnancy, and this mechanism seems to be defective in preterm labor [40,78].

The mouse model of LPS-induced preterm birth has demonstrated the essential role of IL-10 in term pregnancy [79]. IL-10-producing Bregs have emerged as a key player in maintaining pregnancy well-being as a recent study showed that adoptive transfer of IL-10-producing Bregs reduced fetal loss and restored immune balance in vivo [3].

4 Recurrent pregnancy loss

RPL occurs in approximately 1%–3% of women [6]. Primary RPL is defined as having suffered two or more consecutive failed clinical pregnancies without previous live birth with the same partner, while secondary RPL is defined as consecutive recurrent miscarriages, usually after one or more live births [80,81]. Pregnancy loss can be attributed to fetal chromosomal abnormalities, parental chromosomal defects, infections, endocrine dysregulation, or uterine abnormalities [6,80,82]. More than half of all cases of RPL are idiopathic, although the likely cause is a lack of, or inappropriate initiation of, maternal tolerance due to immune dysregulation [6,7,80]. The prevalence of RPL from B cell pathologies is difficult to estimate because of limited research on immunological causes of RPL and the lack of standardized diagnostic procedures.

5 Diagnosis

The intricate coordination of different immune cell subsets necessary for a successful pregnancy creates a wide range of potential etiologies for RPL. Currently, there is no consensus on the tests that indicate abnormal alloimmune response to pregnancy or that are necessary to evaluate idiopathic RPL. Therefore, the diagnosis of RPL is most often determined by exclusion.

Minimum diagnostic workup for RPL includes complete medical, surgical, genetic, and family history and physical examination. After excluding genetic etiologies, immune indicators are assessed via hysterosalpingography, sonohysterography, or hysteroscopy, and measurement of IgG and IgM anticardiolipin antibodies, lupus anticoagulant, and autoantibodies. Further immunological tests use biomarkers and clinical assays, such as immune cell profiling, human leukocyte antigen typing, mixed lymphocytotoxic antibody tests, and mixed lymphocyte culture reactions. These diagnostic tests are the foundation for determining an immune diagnosis of RPL [7].

Indicators of B cell pathology that may be predictive of pregnancy loss include low levels of IL-10^+ B cells and high numbers of circulating $CD5^+$ B cells [48,72]. However, these indicators have only been validated as predictors of pregnancy outcome for research purposes and have not yet been confirmed as pre-pregnancy clinical tests to predict the risk of pregnancy loss.

6 Treatments

Several immunological treatments have been advocated by some investigators for idiopathic RPL, based on evidence that pre-transplant blood transfusions decreased rejection of organ allografts. The most popular attempts at maternal immunomodulation are paternal leukocyte transfusion prior to conception and passive immunization with IVIg during pregnancy [82]. Other interventions include third-party lymphocyte immunization and paternal antigen desensitization, although these methods have not been proven efficacious [6,82] (see Fig. 2).

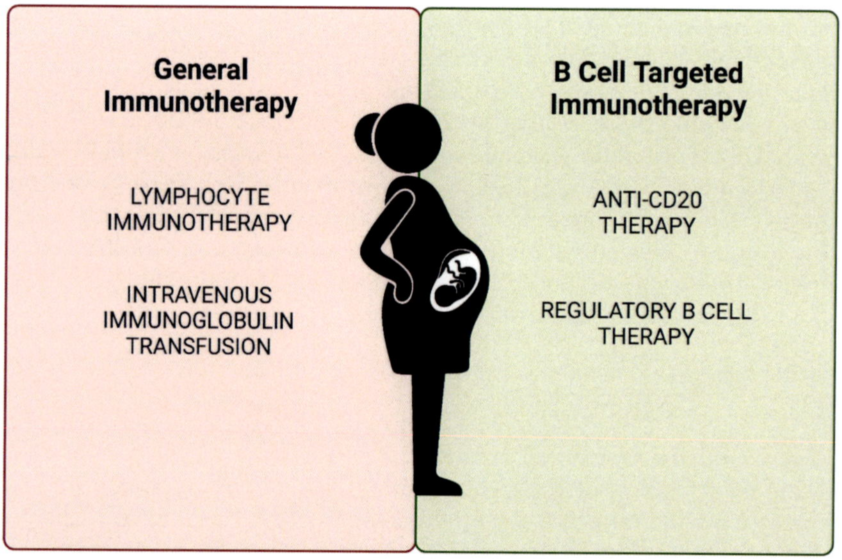

FIGURE 2 Immunotherapy options for immune-mediated etiopathology of recurrent pregnancy loss. Source: *Created with BioRender.com.*

6.1 Lymphocyte immunotherapy

Immunization with lymphocytes is considered for cases of idiopathic RPL. LIT was developed based on the theory of acclimatizing the prospective mother's body to foreign cells to improve the immunologic tolerance of the paternal cells. For RPL, LIT treatment involves extracting lymphocytes from the donor and injecting them into the prospective mother. LIT induces changes in the maternal immune system to preserve pregnancy via several potential mechanisms: expression of antipaternal cytotoxic antibodies, PIBF, antiidiotypic antibodies, and mixed lymphocyte reaction blocking antibodies, reducing the Th1/Th2 ratio, and modifying the cytokine production pattern [83].

There is no consensus regarding the efficacy and safety of LIT. Some studies report efficacy rates ranging from 38% to 70% in clinical studies in patients with RPL; however, a 2006 Cochrane systematic review found that immunotherapy does not have any beneficial effect for unexplained RPL [6,82–84]. Reactions at the paternal lymphocyte application site have been reported that may persist for up to 15 days, but LIT did not increase the risk of autoimmune complications [84]. More studies are needed to establish a consensus on the clinical use of LIT.

6.2 Intravenous immunoglobulin therapy

After immunological abnormalities were discovered in women with RPL in clinical practice, several immunomodulatory therapies were introduced to maintain the immune balance at the fetal–maternal interface. Immunosuppressive therapy with IVIg administration was proposed for the various immunomodulating effects of the principal component, the IgG antibodies. Immunomodulatory effects include antibody-mediated immune damage blockade, T cell–receptor blockade, inhibition of natural killer cell activity, inhibition of Th1 cytokine secretion, Fc receptor blockade, complement inactivation, downregulation of B cell responsiveness, and enhanced suppressor T cell function [85,86].

Reported side effects of IVIg treatment include headache, skin rash, and predisposition to thrombosis in APS without significant risk of congenital malformation. A 20-year follow-up of IVIg used during spontaneous pregnancy reported no changes in thrombotic state or development of new autoimmune disease in the mother and lower rates of preeclampsia and preterm deliveries, which may be attributed to improved placentation [87].

Several studies on the therapeutic effect of IVIg in RPL patients have produced contradictory results. While some studies have shown that IVIg can significantly improve the pregnancy outcome in women with RPL, other studies suggested that IVIg has no obvious therapeutic effect [6,82,85,86]. For this reason, IVIg is not included in clinical treatment guidelines for RPL. This may change in the future as the mechanisms of IVIg in immunologic RPL become clearer.

6.3 B cell-targeting therapies

Immunotherapies that target B cells remain experimental; however, anti-CD20 antibody therapy (rituximab) has been tested in women with autoantibody-mediated pathologies, based on its use in cancer patients. Women with APS have benefited from rituximab treatment—some patients exhibited normalization or reduction of aPL titers and significantly reduced clinical indicators, such as thrombocytopenia [88]. How this may translate to successful pregnancy remains unknown.

Treatments that harness the regulatory power of B cells have been proposed for pregnancy pathologies, particularly those of known immune origin, to potentially develop and protect the immune alterations required to support implantation. Hypothetically, B cells obtained from women could be expanded into Breg populations and reinfused to stimulate the appropriate immune alterations required to maintain the pregnancy.

The fact that these treatments remain in the experimental phase highlights the need for a greater understanding of the mechanisms that underlie B cell-mediated pathophysiology. A thorough understanding of the intricate cellular interplay with networks that dictate the health and well-being of pregnancy is essential to develop and test the next generation of therapies.

7 Summary and recommendations

Targeted approaches for disrupting B cell distribution and function to improve pregnancy outcomes are in the initial stages of the investigation. B cell depletion via monoclonal anti-CD20 antibodies and the use of inflammatory mediators specific to B cell cytokines may, in principle, inhibit deleterious inflammation; however, these therapies need to be refined to avoid restricting the functionality of certain B cell subsets deemed helpful to pregnancy health.

Therefore, to progress any type of immune therapy for the treatment of pregnancy pathologies, it is recommended to

- extend research aimed at deciphering the mechanisms of maternal tolerance induction,
- concurrently assess clinical samples to determine the validity of such proposed mechanisms,
- identify further immune parameters to examine in cases of unexplained fertility and RPL,
- widen assessment of immune cell subsets and function in other pregnancy pathologies, and
- Continue to explore new treatment options for use in early pregnancy to set or correct aberrant immune adaptations.

References

[1] Guzman-Genuino RM, Diener KR. Regulatory B cells in pregnancy: lessons from autoimmunity, graft tolerance, and cancer. Front Immunol 2017;8:172.

[2] Guzman-Genuino RM, Hayball JD, Diener KR. Regulatory B cells: dark horse in pregnancy immunotherapy? J Mol Biol 2021;433(1):166596.

[3] Busse M, Campe KJ, Nowak D, et al. IL-10 producing B cells rescue mouse fetuses from inflammation-driven fetal death and are able to modulate T cell immune responses. Sci Rep 2019;9(1):9335.

[4] Carbone J, Gallego A, Lanio N, et al. Quantitative abnormalities of peripheral blood distinct T, B, and natural killer cell subsets and clinical findings in obstetric antiphospholipid syndrome. J Rheumatol 2009;36(6):1217–25.

[5] Zeng B, Kwak-Kim J, Liu Y, Liao AH. Treg cells are negatively correlated with increased memory B cells in preeclampsia while maintaining suppressive function on autologous B-cell proliferation. Am J Reprod Immunol 2013;70(6):454–63.
[6] Achilli C, Duran-Retamal M, Saab W, Serhal P, Seshadri S. The role of immunotherapy in in vitro fertilization and recurrent pregnancy loss: a systematic review and meta-analysis. Fertil Steril 2018;110(6):1089–100.
[7] Beaman KD, Ntrivalas E, Mallers TM, Jaiswal MK, Kwak-Kim J, Gilman-Sachs A. Immune etiology of recurrent pregnancy loss and its diagnosis. Am J Reprod Immunol 2012;67(4):319–25.
[8] Taylor PV, Hancock KW. Antigenicity of trophoblast and possible antigen-masking effects during pregnancy. Immunology. 1975;28(5):973–82.
[9] Carter J, Dresser DW. Pregnancy induces an increase in the number of immunoglobulin-secreting cells. Immunology. 1983;49(3):481–90.
[10] Muzzio DO, Soldati R, Ehrhardt J, et al. B cell development undergoes profound modifications and adaptations during pregnancy in mice. Biol Reprod 2014;91(5):115.
[11] Muzzio DO, Ziegler KB, Ehrhardt J, Zygmunt M, Jensen F. Marginal zone B cells emerge as a critical component of pregnancy wellbeing. Reproduction. 2016;151(1):29–37.
[12] Ziegler KB, Muzzio DO, Matzner F, et al. Human pregnancy is accompanied by modifications in B cell development and immunoglobulin profile. J Reprod Immunol 2018;129:40–7.
[13] Grimsholm O, Piano Mortari E, Davydov AN, et al. The interplay between CD27(dull) and CD27(bright) B cells ensures the flexibility, stability, and resilience of human B cell memory. Cell Rep 2020;30(9):2963–77 e2966.
[14] Margni RA, Paz CB, Cordal ME. Immunochemical behavior of sheep non-precipitating antibodies isolated by immunoadsorption. Immunochemistry. 1976;13(3):209–14.
[15] Barrientos G, Fuchs D, Schrocksnadel K, et al. Low levels of serum asymmetric antibodies as a marker of threatened pregnancy. J Reprod Immunol 2009;79(2):201–10.
[16] Malan Borel I, Gentile T, Angelucci J, et al. IgG asymmetric molecules with antipaternal activity isolated from sera and placenta of pregnant human. J Reprod Immunol 1991;20(2):129–40.
[17] Zenclussen AC, Gentile T, Kortebani G, Mazzolli A, Margni R. Asymmetric antibodies and pregnancy. Am J Reprod Immunol 2001;45(5):289–94.
[18] Carp HJ, Meroni PL, Shoenfeld Y. Autoantibodies as predictors of pregnancy complications. Rheumatol (Oxf) 2008;47(Suppl 3):iii6–8.
[19] Di Prima FA, Valenti O, Hyseni E, et al. Antiphospholipid syndrome during pregnancy: the state of the art. J Prenat Med 2011;5(2):41–53.
[20] Herse F, LaMarca B. Angiotensin II type 1 receptor autoantibody (AT1-AA)-mediated pregnancy hypertension. Am J Reprod Immunol 2013;69(4):413–18.
[21] Salmon JE, Girardi G. Antiphospholipid antibodies and pregnancy loss: a disorder of inflammation. J Reprod Immunol 2008;77(1):51–6.
[22] Heilmann L, Schorsch M, Hahn T, Fareed J. Antiphospholipid syndrome and preeclampsia. Semin Thromb Hemost 2011;37(2):141–5.
[23] Vinatier D, Dufour P, Cosson M, Houpeau JL. Antiphospholipid syndrome and recurrent miscarriages. Eur J Obstet Gynecol Reprod Biol 2001;96(1):37–50.
[24] Robinson CJ, Johnson DD, Chang EY, Armstrong DM, Wang W. Evaluation of placenta growth factor and soluble Fms-like tyrosine kinase 1 receptor levels in mild and severe preeclampsia. Am J Obstet Gynecol 2006;195(1):255–9.
[25] Zhou CC, Zhang Y, Irani RA, et al. Angiotensin receptor agonistic autoantibodies induce preeclampsia in pregnant mice. Nat Med 2008;14(8):855–62.
[26] Christiansen JS, Andersen AR, Osther K, Peitersen B, Bach-Mortensen N, Lebech PE. The relationship between pregnancy, HCS and B lymphocytes. Acta Pathol Microbiol Scand C 1976;84(4):313–18.
[27] Medina KL, Smithson G, Kincade PW. Suppression of B lymphopoiesis during normal pregnancy. J Exp Med 1993;178(5):1507–15.
[28] Guzman-Genuino RM, Eldi P, Garcia-Valtanen P, Hayball JD, Diener KR. Uterine B cells exhibit regulatory properties during the peri-implantation stage of murine pregnancy. Front Immunol 2019;10:2899.
[29] de Carvalho Schettini JA, Gomes TV, Junior C, Heraclio SA, Coelho I, Torres LC. Evaluation of immunological parameters in pregnant women: low levels of B and NK cells. Rev Bras Ginecol Obstet 2019;41(4):213–19.
[30] Kraus TA, Engel SM, Sperling RS, et al. Characterizing the pregnancy immune phenotype: results of the viral immunity and pregnancy (VIP) study. J Clin Immunol 2012;32(2):300–11.

[31] Lima J, Martins C, Leandro MJ, et al. Characterization of B cells in healthy pregnant women from late pregnancy to post-partum: a prospective observational study. BMC Pregnancy Childbirth 2016;16(1):139.
[32] Mahmoud F, Abul H, Omu A, Al-Rayes S, Haines D, Whaley K. Pregnancy-associated changes in peripheral blood lymphocyte subpopulations in normal Kuwaiti women. Gynecol Obstet Invest 2001;52(4):232–6.
[33] Bhat NM, Mithal A, Bieber MM, Herzenberg LA, Teng NN. Human CD5 + B lymphocytes (B-1 cells) decrease in peripheral blood during pregnancy. J Reprod Immunol 1995;28(1):53–60.
[34] Watanabe M, Iwatani Y, Kaneda T, et al. Changes in T, B, and NK lymphocyte subsets during and after normal pregnancy. Am J Reprod Immunol 1997;37(5):368–77.
[35] Birkeland SA, Kristoffersen K. Lymphocyte transformation with mitogens and antigens during normal human pregnancy: a longitudinal study. Scand J Immunol 1980;11(3):321–5.
[36] Benner M, Feyaerts D, Garcia CC, et al. Clusters of tolerogenic B cells feature in the dynamic immunological landscape of the pregnant uterus. Cell Rep 2020;32(13):108204.
[37] Carter J, Newport A, Keeler KD, Dresser DW. FACS analysis of changes in T and B lymphocyte populations in the blood, spleen and lymph nodes of pregnant mice. Immunology. 1983;48(4):791–7.
[38] Hussein MR, Abd-Elwahed AR, Abodeif ES, Abdulwahed SR. Decidual immune cell infiltrate in hydatidiform mole. Cancer Invest 2009;27(1):60–6.
[39] Solders M, Lundell AC, Gorchs L, Gidlof S, Tiblad E, Kaipe H. Mature naive B cells are retained in the placental intervillous blood and positively associate with specific chemokines in full-term healthy pregnancy. Am J Reprod Immunol 2019;82(3):e13154.
[40] Huang B, Faucette AN, Pawlitz MD, et al. Interleukin-33-induced expression of PIBF1 by decidual B cells protects against preterm labor. Nat Med 2017;23(1):128–35.
[41] Harrity C, Bereir MM, Walsh DJ, Marron KD. Moving from peripheral blood to local uterine immunophenotype analysis in patients with poor reproductive history: pilot study of a novel technique. Ir J Med Sci 2019;188(3):893–901.
[42] Marron K, Walsh D, Harrity C. Detailed endometrial immune assessment of both normal and adverse reproductive outcome populations. J Assist Reprod Genet 2019;36(2):199–210.
[43] Sung N, Byeon HJ, Garcia MDS, et al. Deficiency in memory B cell compartment in a patient with infertility and recurrent pregnancy losses. J Reprod Immunol 2016;118:70–5.
[44] Danaii S, Ghorbani F, Ahmadi M, et al. IL-10-producing B cells play important role in the pathogenesis of recurrent pregnancy loss. Int Immunopharmacol 2020;87:106806.
[45] Koushaeian L, Ghorbani F, Ahmadi M, et al. The role of IL-10-producing B cells in repeated implantation failure patients with cellular immune abnormalities. Immunol Lett 2019;214:16–22.
[46] Guzman-Genuino RM, Dimova T, You Y, et al. Trophoblasts promote induction of a regulatory phenotype in B cells that can protect against detrimental T cell-mediated inflammation. Am J Reprod Immunol 2019;82(6):e13187.
[47] Liu J, Chen X, Hao S, et al. Human chorionic gonadotropin and IL-35 contribute to the maintenance of peripheral immune tolerance during pregnancy through mediating the generation of IL-10(+) or IL-35(+) Breg cells. Exp Cell Res 2019;383(2):111513.
[48] Rolle L, Memarzadeh Tehran M, Morell-Garcia A, et al. Cutting edge: IL-10-producing regulatory B cells in early human pregnancy. Am J Reprod Immunol 2013;70(6):448–53.
[49] Slawek A, Lorek D, Kedzierska AE, Chelmonska-Soyta A. Regulatory B cells with IL-35 and IL-10 expression in a normal and abortion-prone murine pregnancy model. Am J Reprod Immunol 2020;83(3):e13217.
[50] Lorek D, Kedzierska AE, Slawek A, Chelmonska-Soyta A. Expression of Toll-like receptors and co-stimulatory molecules in splenic B cells in a normal and abortion-prone murine pregnancy model. Am J Reprod Immunol 2019;82(2):e13148.
[51] Maj T, Slawek A, Chelmonska-Soyta A. CD80 and CD86 co-stimulatory molecules differentially regulate OT-II CD4(+) T lymphocyte proliferation and cytokine response in cocultures with antigen-presenting cells derived from pregnant and pseudopregnant mice. Mediators Inflamm 2014;2014:769239.
[52] Slawek A, Maj T, Chelmonska-Soyta A. CD40, CD80, and CD86 co-stimulatory molecules are differentially expressed on murine splenic antigen-presenting cells during the pre-implantation period of pregnancy, and they modulate regulatory T-cell abundance, peripheral cytokine response, and pregnancy outcome. Am J Reprod Immunol 2013;70(2):116–26.
[53] Darmochwal-Kolarz D, Leszczynska-Gorzelak B, Rolinski J, Oleszczuk J. The immunophenotype of patients with recurrent pregnancy loss. Eur J Obstet Gynecol Reprod Biol 2002;103(1):53–7.

[54] Einenkel R, Packhauser KRH, Ehrhardt J, Tungler A, Zygmunt M, Muzzio DO. CD83 is locally regulated and differentially expressed in disturbed murine pregnancy. Reproduction. 2019;158(4):323–33.
[55] Satta R, Biondi G. Antiphospholipid syndrome and pregnancy. G Ital Dermatol Venereol 2019;154(3):277–85.
[56] Alvarez-Rodriguez L, Riancho-Zarrabeitia L, Calvo-Alen J, Lopez-Hoyos M, Martinez-Taboada V. Peripheral B-cell subset distribution in primary antiphospholipid syndrome. Int J Mol Sci 2018;19:2.
[57] Simonin L, Pasquier E, Leroyer C, et al. Lymphocyte disturbances in primary antiphospholipid syndrome and application to venous thromboembolism follow-up. Clin Rev Allergy Immunol 2017;53(1):14–27.
[58] Hisada R, Kato M, Sugawara E, et al. Circulating plasmablasts contribute to antiphospholipid antibody production, associated with type I interferon upregulation. J Thromb Haemost 2019;17(7):1134–43.
[59] Li XY, Duan HJ, Liu XY, Deng XL. Change of serum B-cell activating factor level in patients with positive antiphospholipid antibodies and previous adverse pregnancy outcomes and its significance. Chin Med J (Engl) 2020;133(19):2287–94.
[60] Mor G, Cardenas I, Abrahams V, Guller S. Inflammation and pregnancy: the role of the immune system at the implantation site. Ann N Y Acad Sci 2011;1221:80–7.
[61] Schumacher A, Ehrentraut S, Scharm M, et al. Plasma cell alloantigen 1 and IL-10 secretion define two distinct peritoneal B1a B cell subsets with opposite functions, PC1(high) cells being protective and PC1(low) cells harmful for the growing fetus. Front Immunol 2018;9:1045.
[62] Jensen F, Muzzio D, Soldati R, Fest S, Zenclussen AC. Regulatory B10 cells restore pregnancy tolerance in a mouse model. Biol Reprod 2013;89(4):90.
[63] Bekmukhambetov Y, Mamyrbayev A, Dzharkenov T, et al. Metabolic and immunologic aspects of fetoplacental insufficiency. Am J Reprod Immunol 2016;76(4):299–306.
[64] Bartha JL, Comino-Delgado R. Lymphocyte subpopulations in intrauterine growth retardation in women with or without previous pregnancies. Eur J Obstet Gynecol Reprod Biol 1999;82(1):23–7.
[65] Xiong F, Tong Y, You Y, et al. Prospective cohort study about the lymphocyte subpopulations' change and impact on the pregnancy outcome in fetal growth restriction. J Matern Fetal Neonatal Med 2012;25(12):2773–7.
[66] Selvaggi L, Lucivero G, Iannone A, et al. Analysis of mononuclear cell subsets in pregnancies with intrauterine growth retardation. Evidence of chronic B-lymphocyte activation. J Perinat Med 1983;11(4):213–17.
[67] Al-Azemi M, Raghupathy R, Azizieh F. Pro-inflammatory and anti-inflammatory cytokine profiles in fetal growth restriction. Clin Exp Obstet Gynecol 2017;44(1):98–103.
[68] Bartha JL, Romero-Carmona R, Comino-Delgado R. Inflammatory cytokines in intrauterine growth retardation. Acta Obstet Gynecol Scand 2003;82(12):1099–102.
[69] Raghupathy R, Al-Azemi M, Azizieh F. Intrauterine growth restriction: cytokine profiles of trophoblast antigen-stimulated maternal lymphocytes. Clin Dev Immunol 2012;2012:734865.
[70] LaMarca B, Cornelius D, Wallace K. Elucidating immune mechanisms causing hypertension during pregnancy. Physiol (Bethesda) 2013;28(4):225–33.
[71] Liao AH, Liu LP, Ding WP, Zhang L. Functional changes of human peripheral B-lymphocytes in preeclampsia. Am J Reprod Immunol 2009;61(5):313–21.
[72] Jensen F, Wallukat G, Herse F, et al. CD19 + CD5 + cells as indicators of preeclampsia. Hypertension. 2012;59(4):861–8.
[73] Laule CF, Odean EJ, Wing CR, et al. Role of B1 and B2 lymphocytes in placental ischemia-induced hypertension. Am J Physiol Heart Circ Physiol 2019;317(4):H732–42.
[74] Quinn JA, Munoz FM, Gonik B, et al. Preterm birth: case definition & guidelines for data collection, analysis, and presentation of immunisation safety data. Vaccine. 2016;34(49):6047–56.
[75] Rinaldi SF, Makieva S, Saunders PT, Rossi AG, Norman JE. Immune cell and transcriptomic analysis of the human decidua in term and preterm parturition. Mol Hum Reprod 2017;23(10):708–24.
[76] Bommer I, Juriol L, Muzzio D, et al. Characterization of murine amniotic fluid B cells in normal pregnancy and in preterm birth. Reproduction. 2019;158(4):369–76.
[77] Leng Y, Romero R, Xu Y, et al. Are B cells altered in the decidua of women with preterm or term labor? Am J Reprod Immunol 2019;81(5):e13102.
[78] Valeff N, Juriol L, Quadrana F, et al. Expression of IL-33 receptor is significantly up-regulated in B cells during pregnancy and in the acute phase of preterm birth in mice. Front Immunol 2020;11:446.
[79] Robertson SA, Skinner RJ, Care AS. Essential role for IL-10 in resistance to lipopolysaccharide-induced preterm labor in mice. J Immunol 2006;177(7):4888–96.

[80] Zarnani AH. Recurrent pregnancy loss through the lens of immunology. J Reprod Infertil 2015;16(2):59–60.
[81] Pillarisetty LS, Gupta N. Recurrent Pregnancy Loss. Treasure Island (FL): StatPearls; 2021.
[82] Wong LF, Porter TF, Scott JR. Immunotherapy for recurrent miscarriage. Cochrane Database Syst Rev 2014;10:CD000112.
[83] Hajipour H, Nejabati HR, Latifi Z, et al. Lymphocytes immunotherapy for preserving pregnancy: mechanisms and challenges. Am J Reprod Immunol 2018;80(3):e12853.
[84] Cavalcante MB, Sarno M. Araujo Junior E, Da Silva Costa F, Barini R. Lymphocyte immunotherapy in the treatment of recurrent miscarriage: systematic review and meta-analysis. Arch Gynecol Obstet 2017;295 (2):511–18.
[85] Palmeira P, Quinello C, Silveira-Lessa AL, Zago CA, Carneiro-Sampaio M. IgG placental transfer in healthy and pathological pregnancies. Clin Dev Immunol 2012;2012:985646.
[86] Sung N, Han AR, Park CW, et al. Intravenous immunoglobulin G in women with reproductive failure: The Korean Society for reproductive immunology practice guidelines. Clin Exp Reprod Med 2017;44(1):1–7.
[87] Woon EV, Day A, Bracewell-Milnes T, Male V, Johnson M. Immunotherapy to improve pregnancy outcome in women with abnormal natural killer cell levels/activity and recurrent miscarriage or implantation failure: a systematic review and meta-analysis. J Reprod Immunol 2020;142:103189.
[88] Dieudonne Y, Guffroy A, Poindron V, et al. B cells in primary antiphospholipid syndrome: review and remaining challenges. Autoimmun Rev 2021;20(5):102798.

CHAPTER

5

Mast cell pathology and reproductive failures

Maria Socorro L. Agcaoili-De Jesus[1], *Lara Theresa C. Alentajan-Aleta*[1], *Cherie C. Ocampo-Cervantes*[1], *Jenifer R. Otadoy-Agustin*[1] *and Joanne Kwak-Kim*[2]

[1]Division of Allergy and Immunology, Department of Medicine, University of the Philippines College of Medicine, Philippine General Hospital, Manila, Philippines [2]Reproductive Medicine and Immunology, Obstetrics and Gynecology, Clinical Sciences Department, Chicago Medical School, Rosalind Franklin University of Medicine and Science, Vernon Hills, IL, United States

1 Introduction

Mast cells (MCs) have long been recognized for their participation in allergic diseases. However, more than being a key cell mediator in allergic diseases, MCs can modulate physiological and pathological processes in the body, including reproduction. Their vast distribution in the male and female reproductive tracts has emphasized their active roles in immune responses with both positive and negative implications on fertility and the establishment and maintenance of pregnancy. Activation of MCs leads to degranulation, releasing various preformed and/or newly synthesized inflammatory mediators. In this chapter, the function of MCs and MC granules, their distribution in the reproductive tract, and their involvement in numerous reproductive processes are discussed in depth. Moreover, reproductive and obstetrical disorders and reproductive outcomes arising from abnormal MC density and pathology are explored.

2 Mast cells biology

MCs are multifunctional tissue-resident cells, particularly residing in the mucosa and epithelial tissues, making them part of the innate immune system and the body's first line

Immunology of Recurrent Pregnancy Loss and Implantation Failure
DOI: https://doi.org/10.1016/B978-0-323-90805-4.00015-8

of defense. MCs were first discovered by Paul Ehrlich in 1878 and were initially named *"mastzellen"* due to the characteristic metachromatic staining of their secretory granules [1]. The German word "mast" denotes a "fattening" or "suckling" function, which was based on his belief that they had a nutritional function because of their granules. Around 60 years later, the connection between MCs and histamine was established. Since then, MCs have been strongly associated with allergic diseases and parasitic infections [2].

However, MCs are present in all classes of vertebrates, and histamine storage of MCs in vertebrates and its use as an inflammatory mediator was estimated to be established approximately 276 million years ago in primitive reptiles. It suggests that MC mediates essential functions for life rather than just representing a vestigial remnant of the immune system. Furthermore, no MC-deficient human has been identified so far, indicating that MCs have an important biological role [3].

2.1 Origin of mast cells

MCs are derived from pluripotent hematopoietic bone marrow stem cells called the common myeloid progenitor (CMP) cell that are released into the systemic circulation as undifferentiated CD34 receptor, tyrosine kinase Kit-positive (CD117+) mononuclear cells [4]. They are the only immune cells that leave the bone marrow in an immature state, which travel to connective tissues and mucosal linings before they mature. There are inter-species variations in terms of MC numbers and distribution. MCs are numerous in man, monkey, mouse, rat, and fish but few in the rabbit. MCs are largely present in the dermal and subcutaneous tissue of the skin. In the digestive system, the omentum, intestines, and tongue have large numbers of MCs, but few MCs are found in the liver, spleen, and adrenal glands because of the paucity of connective tissue in these organs. Recently, it was reported that MCs occur in large numbers in the myometrium but are only scanty in the endometrium, where they tend to be confined to the stratum basalis [5]. There seem to be no apparent differences in the number of MCs between the normal proliferative and secretory phase endometrium; however, endometrial MCs are considerably reduced during the immediate premenstrual phase of the menstrual cycle [5]. It is presumed that this, almost exclusive, presence of MCs in the basal layer of the endometrial matrix, combined with the discharge of their cytoplasmic granules towards the end of the cycle, may be related to the mechanism of menstruation by producing ischemic spasm of the spiral arteries, preceding menstruation. On the other hand, the relative paucity of MCs in the functional layer may contribute to the immune tolerance of the gestational endometrium to the implantation of the blastocyst.

2.2 Differentiation and development of mast cells

MCs develop from pluripotent hematopoietic bone marrow stem cells that are released into the peripheral circulation. These cells subsequently migrate into a tissue, potentially proliferate, and then mature under the influence of the local cytokine milieu, tissue matrix, and resident cells such as fibroblasts that profoundly influence their phenotype. The interaction between the stem cell factor (SCF) and c-kit, and the subsequent signaling that follows, is crucial for the growth and development of MCs.

Cytokines that promote MC development were studied. Adding SCF, IL-3, IL-4, IL-5, or IL-6 to the cultures in both serum-containing and serum-free medium prolonged MC survival in a dose-dependent manner [6]. SCF is an obligatory growth factor together with IL-6, and their removal from culture media was shown to cause MC apoptosis. Several cofactors enhance or inhibit the effects of SCF depending on the MC origin and its level of maturity. Nerve growth factor, IL-3, IL-6, IL-9, and IL-10 potentially enhance SCF-dependent MC growth and/or promote survival [7,8]. On the other hand, granulocyte macrophage-colony stimulating factor (GM-CSF), retinoids, and transforming growth factor β (TGF-β) inhibit MC growth and differentiation [8].

2.3 Mast cell activation

The best-studied mechanism of MC activation (as it is the most relevant to allergic disease) is the one mediated through the high-affinity IgE receptor FcεRI by harmless allergens thru cross-linking IgEs on the surface of the MCs. However, in a physiological setting, other receptors, such as KIT or CD 117, Toll-like receptors, and receptors for TNF-α might also markedly influence the release of mediators by MCs [6]. Other pathways of MC activation include proteases such as tryptase, neuropeptides (particularly skin MC), pollutants, occupational agents, cell-cell contact, and hyperosmolarity.

The activation of innate immune responses can be due to tissue injury or in response to pathogen recognition. Due to MCs distribution and their strategic location at portals of foreign substance entry into the body, they are the body's first line of defense. Its activation has been reported to be initiated by various etiologies, such as parasitic, bacterial, and viral pathogens [6].

2.4 Types of mast cells

Variation in MCs' morphologic, biochemical, and functional characteristics has been reported, and this phenomenon is referred to as MC heterogeneity. The MCs' ultrastructure, receptor expression, mediator content, immunologic and nonimmunologic activation, and pharmacologic responsiveness make them heterogeneous. MCs differ depending on species, between different organs, and even within the same organ. The factors leading to this heterogeneity are multifactorial, including interactions with the tissue matrix and resident cells such as fibroblasts. Although it is also possible that progenitors are committed to a particular phenotype early on in their development, marked plasticity in protease expression is also evident. There are two groups of MCs in humans, based on their protease content; containing tryptase only (MC_T) and containing tryptase, chymase, carboxypeptidase A, and cathepsin G (MC_{TC}). The MC_T phenotype is typically found at mucosal surfaces such as the nasal and lower airway epithelium. The MC_{TC} phenotype, on the other hand, is found in connective tissues such as healthy skin. The heterogeneity of MCs also extends to their cytokine content. IL-4 and IL-13 are predominantly found in MC_{TC}, whereas IL-5 and IL-6 are almost exclusively present in MC_T, suggesting unique functions for these phenotypes [9].

2.5 Mast cell granules

As primary effector cells of the immediate hypersensitivity reactions in humans, MCs have been shown to respond to stimuli within minutes of exposure. Upon activation via the FcεRI, preformed and newly formed mediators are secreted. This process is also known as degranulation. The major preformed mediators are histamine, tryptase [10], chymase, and heparin [9]. In addition to the stored granule-derived mediators, newly formed metabolites of arachidonic acid, leukotriene C4 and D4, prostaglandin D2, and platelet-activating factor are also released from MCs after IgE-dependent activation. These granules can be released in a non-IgE mediated way as part of the innate immune response to viruses, parasites, and bacteria. Recently, it was noted that MCs' granules are rich in zinc, a potent caspase inhibitor, and NF-kB translocation modulator and have been associated with the MCs' role in homeostasis [11].

Once MCs are activated, degranulation begins within seconds of antigen-binding [12]. Granule contents include short-lived histamine, serine esterases, and proteases such as chymase and tryptase [13]. Histamine and heparin are toxic mediators that help to clear parasitic infestations, increase vascular permeability, cause smooth muscle contraction to expel the parasite, and promote anticoagulation. Tryptase, chymase, cathepsin G, and carboxypeptidase are essential mediators in fibrotic processes and help tissue remodeling postinflammation [13]. Tryptase has a proliferative effect on fibroblasts and acts as a chemoattractant for those cells. It induces collagen synthesis in human fibroblasts. The cytokines, chemokines, and lipid mediators that MCs synthesize after activation contribute to the sustained proinflammatory state, which can be beneficial or detrimental, depending on the site of inflammation.

3 Role of mast cells in reproductive system

3.1 Male reproductive tract

MCs play an important role in male reproductive organs [14]. Both MC_{TC} containing tryptase and chymase and MC_T containing tryptase only are found in the human testis and epididymis, but MC_{TC} is the predominant population in the human testis. MC number varies in the different developmental stages of life, with a slight but notable increase in the testis and epididymis during infancy followed by decreasing their number during childhood and increasing again during adolescence [15]. During adulthood, these cells appear to reduce progressively in the testicular connective tissue. The average MC in the testis in healthy men is approximately $4.8 \pm 3.1 \times 10^6$ μm [16].

MCs are present in the lamina propria of seminiferous tubules in the testis, epididymis, and excurrent ducts. The majority of them are located in the interstitial compartment, and a small portion of MCs are in proximity to the seminiferous tubule or in the lamina propria itself [16]. MCs are also described in the epididymis and prostate as well. In the epididymis, they are spread evenly throughout the head, body, and tail. Their number also changes during different stages of life. MCs were also found in the semen of fertile men [17,18].

3.2 Female reproductive system

MCs have been described to be involved in numerous processes in the female reproductive tract. For example, they have significant roles in ovulation, implantation, tissue remodeling during early pregnancy, decidualization, placentation, fetal development, and delivery.

3.2.1 Ovulation

Tryptase-positive MCs are greater in number than chymase-positive MCs in the interstitial cortex and medulla of cyclic and polycystic ovaries than in postmenopausal women. In cyclic ovaries, areas with a moderate density of nerve fibers contained many MCs. However, as the nerve fiber density increases in polycystic ovaries, the number of MCs decreases. MCs were not seen in the postmenopausal ovaries, suggesting that decreased MC number and increased nerve fiber density indicate a change in neuroimmune communication [19].

3.2.2 Implantation and remodeling during early pregnancy

MCs express receptors for estrogen and progesterone, and these hormones can attract MCs to the uterus [20]. Uterine MCs in humans vary in number and structure in response to the menstrual cycle, indicating that the changes are hormone-dependent. The number of MCs remains constant during the proliferative and secretory endometrium but decreases during the premenstrual phase of the menstrual cycle [5]. In the nonpregnant uterus, tryptase-chymase-positive-MCs were dominant, while in the pregnant myometrium, the tryptase-positive MCs were dominant [21]. MCs containing chymase are usually found surrounding the blood vessels in the myometrium, placenta, and umbilical cord of pregnant women [22].

MCs produce histamine, which may be involved in blastocyst implantation. However, in murine studies, blastocyst implantation occurs even in the setting of MC deficiency [23], suggesting that other cells such as endothelial cells or decidual cells may be the source of histamine [24]. Activated MCs release both preformed and newly formed mediators such as tryptase, chymase, histamine, and TGF-β, which may be involved in blastocyst implantation. During implantation, endometrial tissues secrete inducible nitric oxide synthase, which produces nitric oxide. Nitric Oxide plays a vital role in endometrial functions such as endometrial receptivity, implantation, and menstruation. MCs secrete IL-8, which triggers neutrophil migration into the stroma, and elastase, which breaks down the Type III collagen, generating spaces in the stroma. Interaction with hyaluronic acid promotes water retention in the stromal cells, leading to edema observed during implantation [25].

3.2.3 *Decidualization and placentation*

The endometrial epithelium and the decidual cells secrete chemokines during blastocyst attachment to the endometrium and trophoblast invasion into the decidua. Chemokines such as CX3CL1, CCL14, and CCL4 are abundantly expressed by the endometrial cells and stromal fibroblasts. Their corresponding receptors, CX3R1, CCR1, CCR2, and CCR5, on the other hand, are expressed in the trophoblast cells [26]. Secretion of chemokines promotes trophoblast migration, underlying the importance of chemokines in maternal–fetal communication [27].

In the endometrium, MCs surround the endometrial glands and spiral arteries and express tryptase, chymase, and carboxypeptidase A3. In addition, endometrial MCs secrete CCL8, which promotes the migration of endometrial epithelial and stromal cells. The overexpression of CCL8 and its receptor CCR1 was reported in the ovarian endometrioma and ectopic endometrium, contributing to endometriosis [28]. Biopsies from patients with deeply infiltrating endometriosis showed abundant activated and degranulated MCs surrounding the nerve fibers. The strong association of MCs with neuronal cells suggests the critical role of MCs in the development of pain and hyperalgesia in endometriosis [29].

3.2.4 *Modulation of the fetal and maternal immune response*

Decidual MCs secrete vascular endothelial growth factor-A (VEGF-A), and arginase-1, thereby promoting angiogenesis [30]. Murine studies show that MCs localized in the decidua secrete IL-9, which contributes to the Treg-dependent allograft tolerance at the fetal–maternal interface [31]. Recent studies suggest that MCs in the decidua express killer cell Ig-like receptor (KIR) 2DL4, which serves as a receptor for HLA-G expressed by the trophoblasts. Interaction of HLA-G with the KIR2DL4 on NK cells inhibits the cytotoxic activity of NK cells, further supporting maternal tolerance [32].

MCs play a role in placentation through their influence on spiral artery remodeling. Examination of placentation sites in MC deficient mice showed a decrease in the size of the placenta and impaired remodeling of the spiral arteries [33]. Placental MC-derived histamine is also involved in the different steps of placental development. Histamine enhances human cytotrophoblast invasion through the histamine type 1 receptor [34]. Histamine increases $\alpha v\beta 3$ integrin expression by the trophoblast [35] and influences cytotrophoblast apoptotic activity via interaction with H1 receptors [36].

3.2.5 *Fetal development and delivery*

The generation of strong myometrial contractions in nonpregnant and pregnant uterine tissues when exposed to an allergen and histamine and pregnant uterus with the abundance of MCs suggest that MCs are involved in uterine contractility during pregnancy [21]. Stretching the myometrial tissues to accommodate the growing fetus releases mediators such as SCF and other chemokines that attract MCs to the myometrium. Chemokines such as CXCL-1, -2, -3, -5, and -8 CXC are dominant during term labor in the myometrium, cervix, and fetal membranes. The receptors to these chemokines, CXCR-1 and -2, are present in both neutrophils and MCs, leading to their recruitment into the uterus [37].

4 Mast cell-related immunopathology in reproduction

4.1 Male infertility

4.1.1 Mast cell-related male infertility

The increased density of MCs has been reported in the ejaculates of infertile males with idiopathic oligozoospermia, asthenozoospermia, or azoospermia. Quantitative analysis of MCs in testicular fine needle aspiration biopsy smears showed a striking increase of MCs. Only Sertoli cells and increased numbers of MCs (a mean value of 19.5 MCs/20 field) were present in idiopathic male infertility cases. MCs were round to oval with heavily loaded with purple-colored granules by H and E staining. This finding was highly significant compared to cases with normal spermatogenesis (0.66 MCs/20 field; $p < 0.001$). Similarly, the number of MCs in late spermatogenic maturation arrest was also significantly higher (8 MCs/20 field; $p < 0.01$) than normal spermatogenesis cases. In early maturation arrest cases, the mean number of MCs/20 field was 3.5; however, no statistical analysis was made in this group due to a small number of cases [38].

4.1.2 Prevalence

MCs were present in 64.5% of semen samples from males with idiopathic asthenozoospermia and 68.8% of semen samples from males with asthenozoospermia and scrotal varicocele, compared with only 34.8% semen samples from normal fertile controls [18].

4.1.3 Immunopathology

MCs have been shown to affect male fertility. Studies with immunohistological techniques have shown that the majority of fertility disorders in men are associated with a high number of testicular MCs. MCs are a crucial player in inflammatory processes as well as fibrosis which can lead to organ dysfunction. Pathophysiological mechanisms leading to infertility are the following [39,40]: (1) MC-derived tryptase and histamine induce fibrosis, which can affect normal spermatogenesis; (2) tryptase and chymase induce a mitogenic effect on fibroblast, leading to testicular fibrosis and collagen deposition; (3) tubular wall thickening disrupts the exchange of fluids between the tubular and interstitial regions; (4) MC-derived matrix metalloproteinases (especially MMP-9) induce fibrosis; (5) decorin interferes with the signaling of growth factor, including epidermal growth factor (EGF). MC-derived tryptase may induce the production of decorin by testicular peritubular cells. Decorin regulates collagen fibrillogenesis and can bind to TGF-β, and platelet-derived growth factor (PDGF) [41], and (6) MC-derived chymase generates angiotensin-II acting on peritubular cells producing proinflammatory factors.

A significantly increased seminal MCs were seen among idiopathic asthenozoospermic and varicocele-associated asthenospermic infertile patients compared to fertile controls [18,42]. Moreover, obstructive azoospermia patients have increased MC count and MC_{TC} to total MC ratios. In patients with varicocele, the average number of MC_{TC} per seminiferous tubules was significantly increased compared to controls [43]. Elevated seminal MCs have a negative correlation with sperm concentration, progressive sperm motility, and total sperm count, suggesting underlying pathophysiology of varicocele-induced infertility [17]. MC-derived mediators such as tryptase in the seminal fluid reduce sperm motility via activation of mitogen-activated protein kinase pathway [14,44]. Seminal MCs are significantly increased in patients with high varicocele grade and laterality [45] and seminal oxidative stress has been documented in varicocele-associated infertility [46]. Varicocele grade determines the extent of alteration to semen quality, and as it increases, the level of seminal reactive oxygen species (ROS) also increases, while the sperm concentration decreases. ROS produced in the seminal fluid can stimulate MC degranulation, followed by the release of MC mediators such as tryptase and histamine. Indeed, ROS have been associated with abnormal forms of sperm, affected sperm motility, DNA integrity, and accelerated apoptosis.

4.1.4 Evaluation and candidates for evaluation

The presence of MCs in the ejaculates of infertile men can be evaluated by semen slides fixed through air drying and stained with 1% toluidine blue-pyronine. The stained slides are examined under an immersion objective [18]. Flow cytometry using c-kit (CD117) expression can be utilized as well [47].

Infertile men with any following conditions may need the MC evaluation: Varicocele-associated, and idiopathic asthenozoospermia [18], nonobstructive azoospermia [48], smokers [18], and age above 40 years old [18].

4.1.5 Therapeutic options

Pretreatment with either antiprotease-activated receptor 2 (anti-PAR-2) antiserum and anti-tryptase antibody or washing with human tubular fluid medium (HTFM) could reverse negative effects of tryptase on sperm mobility [40,49].

Given the role of MCs in male infertility as described above, the use of MC stabilizers and antihistamines has been investigated as a potential therapeutic option. The use of ketotifen 1 mg orally every 12 hours has significantly improved pregnancy rates, sperm counts, and motility among infertile male patients [50,51] and without varicocele [52]. Tranilast 300 mg/day likewise has improved pregnancy rates, sperm count, and sperm motility [53,54]. Among the antihistamines, Ebastine 10 mg/day was beneficial in improving pregnancy rates and sperm quality [55], while fexofenadine did not show benefit [56]. A common side-effect of the medications observed for these three drugs was mild drowsiness. Table 1 summarizes the various studies using ketotifen, tranilast, ebastine, and fexofenadine among patients with oligozoospermia and asthenozoospermia.

On the contrary, the potential adverse effects of antihistamines have been reported in male reproductive function since these drugs may affect testicular homeostasis, reduce sperm viability, and decrease human sexual behavior and penile erection [57].

4.2 Endometriosis

4.2.1 Mast cell and endometriosis

Endometriosis is a common estrogen-dependent inflammatory disorder among women of reproductive age. It is characterized by endometrial tissue implants outside the uterus, which can cause pelvic scarring, pain, and infertility [58]. MCs have recently been considered to have a role in endometriosis and infertility through its mediators by directly

TABLE 1 Summary of studies on mast cell stabilizers and antihistamines in male infertility.

Author, Year	Population	Study Type	Intervention	Outcome
Schill [52]	17 men with idiopathic oligozoospermia $<20 \times 10^6$ sperm/mL 22 men with idiopathic asthenozoospermia	Prospective cohort	Ketotifen 1 mg/tab × 2 tabs/day × 3 months	Oligozoospermia: improvement of sperm count & total sperm output; 9% pregnancy rate after 1 year of treatment Asthenozoospermia: increased progressive motility, total motile, progressive motile sperm count; 20% pregnancy rate after 1 year of treatment
Azadi [50]	103 infertile men who underwent varicocelectomy Ketotifen group: $n = 51$ Control group: $n = 52$	RCT	Ketotifen 1 mg every 12 hours after varicocelectomy	Pregnancy rate 9 months postsurgery: Ketotifen group: 41.17% Control group: 21.15% $p < 0.05$ Significant improvement in sperm count, protamine content & DNA integrity after 3 months
Zaazaa [51]	120 infertile men with varicocele	Prospective cohort	Group 1: varicocelectomy Group 2: Ketotifen 1 mg BID × 3 months Group 3: varicocelectomy + Ketotifen 1 mg BID × 3 months	Varicocelectomy followed by ketotifen significantly improved sperm DNA fragmentation index versus varicocelectomy or ketotifen alone
Yamamoto [54]	50 men with severe oligozoospermia $<5 \times 10^6$ sperm/mL	Placebo-controlled single-blind	Tranilast 300 mg/day × 3 months	Pregnancy rate Tranilast group: 28.6% Placebo group: 0% $p < 0.05$ The Tranilast group had significantly higher levels of sperm density, sperm motility, and total motile sperm count
Hibi [53]	17 men with severe oligozoospermia $<10 \times 10^6$ sperm/mL	Prospective cohort	Tranilast 300 mg/day × ≥ 12 weeks	Increase in sperm count in 41.1% of patients 18% pregnancy rate
Matsuki [55]	15 men with idiopathic oligozoospermia	Prospective cohort	Ebastine 10 mg/day × 3 months	66.7% with a definite improvement of sperm quality 20% pregnancy rate within 6 months
Cayan [56]	16 men with idiopathic azoospermia or oligozoospermia	Prospective cohort	Fexofenadine 180 mg/day × 4–9 months	No benefit

suppressing sperm motility. MC-related endometriosis can be defined as the presence of increased numbers of MCs in the endocervical brushings and endocervical smears among patients with endometriosis.

TABLE 2 Comparison of mast cell population morphology and numbers. [59].

	With endometriosis (mean ± 1 S.D.)	Without endometriosis (mean ± 1 S.D.)
Ectopic endometrial lesion	Granulated: 0.69 ± 0.72 Degranulated: 10.97 ± 3.71 Total: 11.66 ± 4.07 MCs exhibited degranulation. Scattered granules were observed.	
Eutopic endometrium	Granulated: 0.82 ± 0.97 Degranulated: 0.18 ± 0.40 Total: 1.00 ± 1.19	Granulated: 0.95 ± 0.96 Degranulated: 0.47 ± 0.73 Total: 1.42 ± 1.37
Normal serosa samples collected by the laparoscopic surgery	Granulated: 0.54 ± 0.63 Degranulated: 0.16 ± 0.37 Total: 0.70 ± 0.86	Granulated: 0.52 ± 0.65 Degranulated: 0.16 ± 0.42 Total: 0.68 ± 0.84

TABLE 3 Mast cell density in peritoneal endometriosis [60].

	Black lesions	Red lesions	p-value
Mast cell density	15.6% ± 4.9%	3.1% ± 1.5%	0.0002
Number of MCs per 0.13 mm^2 of stroma	0–12	0–26	

Sugamata et al. compared the morphology and numbers of degranulated MCs between endometriotic lesions, eutopic endometrium, and normal uterine serosa of patients with or without endometriosis (Table 2). In areas with endometriotic lesions, MCs exhibiting degranulation and scattered granules were noticed. In eutopic endometrium and normal serosa samples from patients with and without endometriosis, both degranulated and granulated MCs were rarely detected [59]. Matsuzaki et al. investigated MC density in peritoneal endometriosis [60]. MCs were detected in the peritoneal endometriotic lesions, and MC density in the stroma of black lesions was significantly higher than that in red lesions (Table 3). Additionally, there was a significant difference in MC density between red peritoneal lesions and eutopic endometrium of patients with endometriosis. In peritoneal endometriosis, there was a marked heterogeneous distribution of MCs.

4.2.2 *Prevalence*

As of this writing, there is no data on the prevalence of increased numbers of MCs among patients with endometriosis. However, in the study of Tsuji et al., all patients with dysmenorrhea had increased numbers of MCs in their endocervical brushings and endocervical smears compared to patients without dysmenorrhea [61].

4.2.3 *Immunopathology*

In the menstrual process of the normal endometrium, MCs facilitate tissue desquamation and attract other immune effectors into the tissue to participate in this process.

However, in the eutopic endometrium of women with endometriosis, MCs fluctuate across the cycle, suggesting an aberrant MC regulation or function [58], although conflicting studies have been reported [58–60]. Increased number of MCs in endometriosis lesions exacerbate inflammation by the release of proinflammatory factors, which increase infiltration of other immune effectors to the lesion [58]. The presence of MCs, and their primary mediator, tryptase, has been proven among patients with endometriosis. MCs-sperm interaction with the presence of peritoneal fluid from endometriosis patients induced the degranulation of the MC, which could directly modulate sperm function [62]. Contrarily, MCs may have a role in developing endometriosis. Crosstalk between endometrial cells and MCs via CCL8/CCR1 has been demonstrated in the development of endometriosis [28].

4.2.4 *Evaluation*

Patients presenting with symptoms of endometriosis, such as dysmenorrhea and dyspareunia, may be candidates for evaluation.

Endocervical brushing and endocervical smears of patients with dysmenorrhea and presumed endometriosis may be examined for MCs using methylene blue staining. A significant difference in the number of MCs between the dysmenorrhea group and the control group (median 35, interquartile range 17–58; and median 2, interquartile range 0–6, respectively) has been noted [61]. Peritoneal fluid may also be examined for the presence of MCs [62] and IL-25 levels [63].

4.2.5 *Therapeutic options*

Animal studies on inhibitors targeting MCs, such as zafirlukast, cromolyn sodium, the JAK2 inhibitor tyrphostin, and ultra micronized palmitoylethanolamide in endometriosis models have shown promising results. Ketotifen significantly suppresses hyperalgesia by modulating MC activity and, therefore, reducing peripheral sensitization due to noxious signals from endometriotic lesions in rats [64]. Small clinical trials (consisting of 4–30 participants) on palmitoylethanolamide and polydatin and the levonorgestrel intrauterine system have shown promising results in decreasing the numbers of MCs and the symptoms of dysmenorrhea, dyspareunia, and pelvic pain [65].

4.3 Recurrent pregnancy loss

4.3.1 *Mast cells and recurrent pregnancy loss*

Accumulation of MCs throughout layers of the endometrium has been reported in women with a history of pregnancy losses [40]. Additionally, increased endometrial MCs in the mid-luteal phase have been reported in women with RPL. Per Derbala et al., MC-related RPL was described as women with two or more consecutive pregnancy losses with endometrial tissue SCF >0.0418, tryptase >0.0035, matrix metalloproteinases (MMP2) >0.1534, and heparan sulfate >0.0121 in the endometrial gene expression assay [66]. However, the study has not been confirmed by others yet.

4.3.2 *Prevalence*

There are no data regarding the prevalence of MC pathology in women with RPL to date.

4.3.3 Immunopathology

While MCs play a positive role in tissue remodeling during implantation by releasing histamine, it is speculated that these cells have a negative role during the later stages of pregnancy since an increased MC activation is associated with preterm labor and delivery [33]. MC deficiency in mice has been associated with severely impaired implantation, which could be blocked by systemic or local transfer of MCs from the bone marrow [67]. It also has been associated with decreased blood flow due to impaired spiral artery remodeling, leading to intrauterine growth restriction [68]. Additionally, abortion-prone mice have insufficient counts of uterine MCs, and the adoptive transfer of Tregs prevented pregnancy losses by enhancing the remodeling of spiral arteries and the development of the placenta [69].

MCs are present throughout the different layers of the endometrium. In the endometrium of women with RPL, MC counts and the mRNA expression of various mediators, such as SCF, tryptase, heparan sulfate, and matrix metalloproteinases (MMP-2) are significantly increased as compared with those of normal fertile controls. The increase in MC counts and activity was noted during the mid-luteal phase, the implantation window [66]. In addition, women with first-trimester miscarriage have been observed to have a deranged organization of the collagen framework, reactive fibroblasts, and an increased number of activated MCs in the cervix. These suggest that an inflammatory-like response takes place in the cervix in women with miscarriage [70].

Stress can trigger the secretion of the corticotropin-releasing hormone (CRH), which activates the hypothalamic–pituitary–adrenal axis, and can cause spontaneous abortion. Aside from the hypothalamus, CRH and its analog urocortin (Ucn) are also secreted locally outside the brain, where they can activate MCs, resulting in inflammation. Ucn, tryptase, which is a mediator secreted by MCs, and IL-8 have been noted to be significantly increased in the products of a conception from women with at least two spontaneous abortions compared with those with only one spontaneous abortion and those with elective abortion. Thus, high levels of CRH and Ucn due to high-stress levels in women with RPL may activate uterine MCs, secreting the abortogenic tryptase and IL-8 [71].

A positive correlation has been observed between perceived stress scores assessed by the questionnaire and the number of decidua basalis tryptase$^+$ MCs (MC_T) and TNF-α producing CD8$^+$ T cells [72]. Aside from the secretion of inflammatory cytokines, the release of tryptase from MCs may pose an additional danger to the pregnancy. Tryptase cleaves proteinase-activated receptor 2 (PAR-2) and induces widespread inflammation. PAR-2 expressing primary spinal afferent neurons contain pro-inflammatory neuropeptide, substance P. Tryptase prompts neurons to stimulate the release of such neuropeptides, which mediate inflammatory edema. These mechanisms contribute to the stress-triggered pro-inflammatory effects of MCs in the uterine decidua of women with miscarriage [73].

4.3.4 Evaluation

Women with two or more failed clinical pregnancies, confirmed by ultrasound or biochemical testing, with no causes identified on initial screening, such as chromosomal abnormalities, anatomic defects, endocrine disorders, autoimmune disease, and thrombophilia (inherited or acquired) in the presence of personal or family history of venous thromboembolism, may be considered for the evaluation of MC-related RPL (based on the study done by Derbala et al. [66]). However, no particular assay has been reported for clinical use.

Based on the study of Derbala et al., endometrial MC can be evaluated by endometrial biopsy, which is performed 5–7 days postovulation or the mid-luteal phase. Serial sections are stained to detect surface (c-kit) and activation (tryptase) markers for MCs using c-kit monoclonal antibody and MC tryptase antibody, respectively. Immunostaining intensity grading and the number of positive cells per mm^2 are noted. This is based on a comparative study of Derbala among patients with RPL against fertile controls [66].

4.3.5 *Therapeutic options*

There is no clinical trial reported so far, and further study is warranted.

4.4 Mast cells and in vitro fertilization failures

Histamine, a mediator released by MCs, may have a role in implantation failures after in-vitro fertilization and embryo transfer (ET) and frozen ET cycles. An excessive level of histamine may enhance the Th1 response and suppresses the Th2 response before implantation, leading to immune rejection and failure of implantation [40]. However, studies on the association of MCs and implantation failures after assisted reproductive technology cycles are limited.

4.5 Mast cell activation disease

4.5.1 *Definition*

Mast cell activation disease (MCAD) refers to a collection of disorders characterized by accumulation of pathological MCs in potentially any or all organs and tissues and/or aberrant release of variable MC mediators. MCAD includes mastocytosis, mast cell activation syndrome (MCAS), and mast cell leukemia (MCL).

Mastocytosis is a spectrum of a rare medical condition characterized by a clonal MC proliferation and accumulation in organs. In cutaneous mastocytosis (CM), there is an increased dermal accumulation of MCs, and involvement is limited to the skin. On the other hand, extracutaneous tissue involvement results in systemic mastocytosis (SM) resulting from the release of MC mediators provoked by several stimuli such as stress, infections, medications, and others. MCAS is a rare condition defined by acute and/or chronic symptoms of MC activation with an elevation of MC mediators at baseline or during an acute episode without MC hyperplasia [74]. Patients suffering not only anaphylaxis but also other inflammatory and/or allergic comorbidities, and with laboratory evidence of MC activation but no other evident unifying diagnosis such as mastocytosis, are diagnosed with MCAS. MCL is an aggressive MC neoplasm that is defined by increased numbers of MCs in bone marrow smears (>20%) and the presence of circulating MCs. Mastocytosis involving the reproductive tract, particularly the uterus and cervix, has been reported [75,76].

4.5.2 *Prevalence*

Few data are available concerning the prevalence of MCAD. Both SM and MCL are rare disorders. The exact incidence and prevalence of mastocytosis are not known due to its rarity. In the western European population, the incidence is ~5–10/1,000,000, and prevalence is ~5–13/100,000 [77,78]. CM outnumbers SM by ~10:1. In all of its forms, the estimated prevalence is 1 per 10,000 persons affecting both sexes but with a slight predominance in females in adulthood. In contrast, MCAS seems to be a more common disorder. Only preliminary data have been reported for MCAS. Evidence has been presented that MCAS may be an underlying cause of various clinical presentations. Hence, the prevalence of MCAS is likely to lie within the single-digit percentage range. A 3:1 female preponderance is observed in MCAS.

The European Competence Network on Mastocytosis followed 23 patients with mastocytosis and recorded 5 spontaneous abortions and 4 preterm deliveries, including one due to preeclampsia at 26 weeks, resulting in neonatal death [79].

4.5.3 *Immunopathology*

Patients with MC disorders such as MCAS can manifest reproductive failures with endometriosis, infertility, and miscarriages. These are most likely due to inflammatory and/or fibrogenic mediators released by MCs [66,80].

Around 20% of pregnant women with mastocytosis reported worsening disease-related symptoms, 30% experienced a clinical improvement in the first trimester, and the other 50% had unchanged MC-mediator-related symptoms. Worsening of symptoms was also observed during the first or third trimester when T helper cell 1 (TH1) -mediated pro-inflammatory conditions dominate [81]. Many physiological changes during pregnancy and labor also trigger multiple stressors that can worsen mastocytosis symptoms. In turn, elevated levels of histamine may result in uterine contractions, leading to an increased risk of preterm labor and other obstetrical complications in mastocytosis patients. Spontaneous abortions, preterm deliveries, neonatal deaths, and rare cases of fetal demise due to anaphylactic reaction have been reported in mastocytosis patients [79,82,83]. One case of neonatal CM was also reported from the mother with mastocytosis [79].

4.5.4 *Evaluation*

Patients with a history of miscarriage, infertility, and endometriosis along with signs and symptoms of MC activation, such as recurrent unexplained anaphylaxis, flushing, osteoporosis, ulcerative gastrointestinal disease, chronic abdominal cramping, idiopathic dyspareunia, dysfunctional uterine bleeding, genital tract inflammation, or multisystem inflammatory/allergic/growth issues, need to be screened for MCAD.

Due to the propensity of MCAD to affect the immune system, including induction of autoimmunity, idiopathic miscarriages in MCAD patients should prompt testings for anti-phospholipid antibody syndrome (MCAS-rooted antiphospholipid antibody syndrome) [84]. Inspite of MC heparin content, MCAD-induced coagulopathy may spur placental microthrombi, prevent placental nidation, disrupt embryonic blood supply and threaten miscarriages [78]. The diagnosis and classification of mastocytosis are based on criteria set by the World Health Organization as outlined in Tables 4 and 5.

TABLE 4 The World Health Organization classification of mastocytosis [85,86].

1. Cutaneous mastocytosis (CM):
a. Urticaria pigmentosa (UP)/ Maculopapular cutaneous mastocytosis (MPCM)
b. Diffuse cutaneous mastocytosis
c. Solitary mastocytoma of skin
2. Indolent systemic mastocytosis (ISM)
– Meets criteria for systemic mastocytosis (SM).[a] No "C" findings.[a] No evidence of associated hematological neoplasm.
– Isolated bone marrow mastocytosis[b] As above (ISM), but with bone marrow involvement and no skin involvement, generally low-burden of MC.
3. Smoldering systemic mastocytosis (SSM)
– As above (ISM), but with 2 or more "B" findings, and no "C" findings [1], generally high-burden of MC.
4. SM with an associated hematological neoplasm (SM-AHN)
– Meets criteria for SM and criteria for AHN as a distinct entity per the WHO classification
5. Aggressive systemic mastocytosis (ASM)
– Meets criteria for SM. One or more "C" findings.[a] No evidence of MCL.
6. Mast cell leukemia (MCL)
– Meets criteria for SM. Bone marrow biopsy shows diffuse infiltration, usually dense, by atypical, immature MCs. BM aspirate smears show ≥20% MCs. In classic cases, MCs account for ≥10% of peripheral blood white cells. Aleukemic MCL variant (<10% circulating MCs).
7. Mast cell sarcoma (MCS)
– No evidence of SM. Generally localized destructive growth pattern. High-grade cytology.

[a] *See Table 5 for diagnostic criteria for SM and definition of "B" and "C" findings.*
[b] *Provisional categories.*

4.5.5 Therapeutic options

Treatment of MCAD involves trigger identification and avoidance and control of MC mediator production and activity. A blockade of elevated mediators can help. Acute and chronic immunosuppressants can be considered. Treatment with H1 antihistamines, cromolyn sodium, omalizumab, steroids, leukotriene receptor antagonists, and cytoreductive therapies such as interferon-alpha 2b can reduce such complications [82,87].

Comprehensive reports describe histamine H1 and H2 receptor blockers as safe in pregnancy [84], but there have been no drug safety studies specifically in the gravid MCAD population. Glucocorticoids can help control flares, but chronic toxicities limit their use. The anti-IgE monoclonal antibody Omalizumab has been described in multiple case reports as safe and effective in MCAD [88]. Cytoreductive drugs are prescribed in mastocytosis with an aggressive clinical course. These include interferon-α (IFN-α) and cladribine, commonly with glucocorticoids [89–91]. MC-targeted treatments were successfully reported in 5 MCAS patients presenting with chronic idiopathic dyspareunia, dysfunctional uterine bleeding, and genital infections [92].

TABLE 5 World Health Organization diagnostic criteria for systemic mastocytosis [85,86].

The diagnosis of SM can be made when the major criterion and one minor criterion or at least three minor criteria are present

Major criterion

Multifocal, dense infiltrates of MCs (≥15 MCs in aggregates) were detected in sections of bone marrow and/or other extracutaneous organs

Minor criteria

a. In biopsy sections of bone marrow or other extracutaneous organs, >25% of the MCs in the infiltrate are spindle-shaped or have atypical morphology or, of all MCs in bone marrow aspirate smears, >25% are immature or atypical.
b. Detection of an activating point mutation at codon 816 of KIT in bone marrow, blood, or other extracutaneous organs.
c. MCs in bone marrow, blood, or other extracutaneous organs express CD25 with/without CD2 in addition to normal MC markers.[a]
d. Serum total tryptase persistently exceeds 20 ng/mL (unless there is an associated myeloid neoplasm, in which case this parameter is not valid).

"B" findings

1. High MC burden is shown on BM biopsy: >30% infiltration of cellularity by MCs (focal, dense aggregates) and serum total tryptase level >200 ng/mL
2. Signs of dysplasia or myeloproliferation in the non-MC lineage(s), but insufficient criteria for definitive diagnosis of an associated hematological neoplasm (AHN), with normal or only slightly abnormal blood counts.
3. Hepatomegaly without impairment of liver function, palpable splenomegaly without hypersplenism, and/or lymphadenopathy on palpation or imaging.

"C" findings

1. Bone marrow dysfunction caused by neoplastic MC infiltration, manifested by ≥1 cytopenia(s) (ANC $<1.0 \times 10^9$/L, Hgb <10 g/dL, and/or platelet count $<100 \times 10^9$/L).
2. Palpable hepatomegaly with impairment of liver function, ascites and/or portal hypertension.
3. Skeletal involvement with large osteolytic lesions with/without pathological fractures (pathological fractures caused by osteoporosis do not qualify as a "C" finding).
4. Palpable splenomegaly with hypersplenism.
5. Malabsorption with weight loss due to gastrointestinal MC infiltrates.

[a]*MC CD25 is the more sensitive marker by both flow cytometry and immunohistochemistry.*
ANC, absolute neutrophil count; Hgb, hemoglobin; ng, nanograms; mL, milliliter; L, liter; dL, deciliter.

Pregnant mastocytosis patients are often required to continue the intake of medications, including antihistamines. However, due to fetal safety concerns, medication doses are often reduced. This reduction in medication, sometimes coupled with an irregular intake of medications, can contribute to the worsening of mastocytosis symptoms. Severe pregnancy complications, including fetal demise, can result from undiagnosed and inappropriately treated mastocytosis [93].

Parturient mastocytosis patients who undergo a cesarian delivery is a particular challenge. Life-threatening complications may occur during the labor brought about by the risk of anaphylactoid reactions triggered by anesthesia. Therefore, emergency medications such as epinephrine, antihistamines, and glucocorticoids should always be available during the labor and postpartum period.

TABLE 6 Treatment algorithm for systemic mastocytosis.

Indolent/smoldering systemic mastocytosis	Aggressive systemic mastocytosis	Systemic mastocytosis with an associated hematological neoplasm
Avoid triggers of MC degranulation (*e.g., aspirin, narcotics, alcohol, contrast dye, anesthetics*)	Clinical trial (*potent, selective mutant KIT inhibitor—e.g., avapritinib*)	Integrate clinical, histologic, and molecular data to assess which disease component (i.e., SM or AHN) warrants immediate treatment
Symptoms of MC degranulation (*symptom burden assessment, treatment options include epinephrine, corticosteroids, histamine H1/H2-blockers, sodium cromolyn, leukotriene inhibitors, topical agents, aspirin, ketotifen, omalizumab, MC cytoreductive therapy considered in severe/refractory cases*)	Midostaurin or Cladribine (*Cladribine preferred if rapid MC debulking is indicated, midostaurin role as maintenance post-transplant?*)	For aggressive AHN (with low-burden or incidentally discovered SM) (*e.g., AML or poor-risk CMML—treat the AHN as per standard of care, e.g., allogeneic stem cell transplant for AML, with symptom management of SM as indicated*)
Osteoporosis/osteopenia (*Bone mineral density assessment, calcium & vitamin D supplementation, bisphosphonates, denosumab, interferon-α, vertebroplasty/kyphoplasty*)	Imatinib (*eosinophilia with FIP1L1-PDGFRA or KITD816V -negative*)	For SM causing organopathy (with indolent AHN) (*treat as aggressive SM, indolent AHN such as PV or ET - observation or treatment as per standard of care*)
Perioperative management (*refer to specialized texts, consult with anesthesia and surgical teams, review prior anesthetic records, use "safer" agents*)	Interferon-α (*pegylated forms likely better tolerated, with/without prednisone*)	Disease progression (*re-stage to assess dominant component of progression—SM vs. AHN, appropriate salvage therapy including allogeneic stem cell transplant as indicated, a molecular assessment may guide targeted therapy*)
	Allogeneic stem cell transplant (*refractory/relapsed disease*)	

Aside from being highly individualized, treatment of SM in adults is also symptom-directed. Table 6 lists the various therapeutic options [86].

5 Summary and Recommendation

- MCs, aside from their role in allergic diseases, have a significant role in reproduction.
- MC-released preformed and newly generated mediators are involved in numerous processes, such as steroidogenesis and spermatogenesis in the male reproductive tract and ovulation, implantation, decidualization, placentation, fetal development, and delivery in the female reproductive tract.
- An increased number of MCs, higher levels of MC-released tryptase in testis, and seminal plasma of males are associated with male infertility and sperm motility disorders.
- The use of MC stabilizers and antihistamines has been studied as a potential therapeutic option for male fertility disorders.
- MCs have a positive role during implantation, and MC deficiency has been associated with severely impaired implantation and impaired spiral artery remodeling, leading to intrauterine growth restriction.

- Increased MC activation is associated with preterm delivery.
- Elevated MC count and increased mRNA expression of various MC inflammatory mediators were observed in women with RPLs.
- Excessive histamine levels may have a role in in vitro fertilization (IVF) failures.
- Patients with MC disorders (e.g., mastocytosis, MCAS) may have clinical manifestations of reproductive failures, such as endometriosis, infertility, preterm delivery, and spontaneous abortions.
- Treatment with H1 antihistamine, cromolyn sodium, omalizumab, steroids, leukotriene receptor antagonists, and cytoreductive therapies can reduce pregnancy complications in patients with MC disorders.

References

[1] Beaven MA. Our perception of the mast cell from Paul Ehrlich to now. Eur J Immunol 2009;39(1):11–25. Available from: https://ncbi.nlm.nih.gov/pubmed/19130582 Accessed April 21, 2021.

[2] Yong L. The mast cell: origin, morphology, distribution, and function. Exp Toxicol Pathol 1997;49(6):409–24. Available from: https://doi.org/10.1016/s0940-2993(97)80129-7.

[3] Abraham St S, John A. Mast cell-orchestrated immunity to pathogens. Nat Rev Immunol 2010;10(6):440–52. Available from: https://doi.org/10.1038/nri2782.

[4] Dahlin JS, Hallgren J. Mast cell progenitors: origin, development and migration to tissues. Molecular Immunology. Mol Immunol 2015;63(1):9–17. Available from: https://sciencedirect.com/science/article/pii/s0161589014000340 Accessed April 24, 2021.

[5] Sivridis E, Giatromanolaki A, Agnantis E, Anastasiadis P. Mast cell distribution and density in the normaluterus Ð metachromatic staining using lectins. Eur J Obstet Gynecol Reprod Biol 2001;98:109–13. Available from: https://doi.org/10.1016/s0301-2115(00)00564-9.

[6] Gilfillan A.M., Tkaczyk C. Integrated signalling pathways for mast-cell activation. Nat Rev Immunol. 6(3):218–230. https://doi.org/10.1038/nri1782.

[7] Razin E, Ihle J, Seldin D, Menchia-Huerta J. Interleukin 3: a differentiation and growth factor for the mouse mast cell that contains chondroitin sulfate E proteoglycan. J Immunol 1984;132(3):1479–86. Available from: http://www.jimmunol.org/cgi/pmidlookup?view = long&pmid = 6198393 Accessed April 24, 2021.

[8] Abbas AK, Lichtman AH, Pillai S. Cellular and Molecular Immunology. Elsevier; 2011.

[9] Bradding PA. Biology of mast cells and their mediators. Middleton's Allergy Principles and Practice. 9th (ed.) Edinburgh London New York Oxford Philadelphia St Louis Sydney: Elsevier; 2020. p. 215–35.

[10] Hogan AD, Schwartz L.B. Markers of mast cell degranulation. Methods. 13(1):45–52. https://doi.org/10.1006/meth.1997.0494.

[11] Ho LH, Ruffin RE, Murgia C, Li L, Krilis SA, Zalewski PD. Labile zinc and zinc transporter ZnT4 in mast cell granules: role in regulation of caspase activation and NF-kappaB translocation. J Immunol 2004;172(12):7750–68. Available from: https://doi.org/10.4049/jimmunol.172.12.7750.

[12] Murphy K.M. In: Janeway's Immunobiology. https://books.google.com/?id = WDMmAgAAQBAJ&dq = janeway + immunobiology.

[13] Duan L., Mukherjee E. Janeway's immunobiology. 9th (ed.). Yale Journal of Biology and Medicine; 2016. http://europepmc.org/articles/pmc5045153 [accessed 25.04.21].

[14] Haidl G, Duan Y, Chen S, Kohn F, Schuppe H, Allam JP. The role of mast cells in male infertility. Exp Rev Clin Immunol 2011;7(5):627–34. Available from: https://doi.org/10.1586/eci.11.57.

[15] Nistal M, Santamaria L, Paniagua R. Mast cells in the human testis and epididymis from birth to adulthood. Acta Anat 1984;119:155–60.

[16] Jezek D, Banek L, Hittmair A, Pezerovic-Panijan R, Goluza T, Schulze W. Mast cells in testicular biopsies of infertile men with 'mixed atrophy' of seminiferous tubules. Andrologia. 1999;31:2013–210.

[17] Cincik M, Sezen SC. The mast cells in semen: their effects on sperm motility. Arch Androl 2003;49(4):307–11.

[18] El-Karasky A, Mostafa T, Shaeer O, Bahgat D, Samir N. Seminal mast cells in infertile asthenozoospermic males. Andrologia. 2007;39(6):244–7.

[19] Heider U, Pedal I, Spanel-Borowski K. Increase in nerve fibers and loss of mast cells in polycystic and post-menopausal ovaries. Fertil Steril 2001;75(6):1141–7. Available from: https://doi.org/10.1016/S0015-0282(01)01805-2.

[20] Jensen F, Woudwyk M, Teles A, et al. Estradiol and progesterone regulate the migration of mast cells from the periphery to the uterus and induce their maturation and degranulation. PLoS ONE 2010;5(12). Available from: https://doi.org/10.1371/journal.pone.0014409.

[21] Garfield RE, Irani AM, Schwartz LB, Bytautiene E, Romero R. Structural and functional comparison of mast cells in the pregnant vs nonpregnant human uterus. Am J Obstet Gynecol 2006;194(1):261–7. Available from: https://doi.org/10.1016/j.ajog.2005.05.011.

[22] Mitani R, Maeda K, Fukui R, et al. Production of human mast cell chymase in human myometrium and placenta in cases of normal pregnancy and preeclampsia. Eur J Obstet Gynecol Reprod Biol 2002;101(12):155–60. Available from: https://doi.org/10.1016/S0301-2115(01)00546-2.

[23] Robinson-White A, Beaven MA. Presence of histamine and histamine-metabolizing enzyme in rat and guinea-pig microvascular endothelial cells. J Pharmacol Exp Ther 1982;223(2):440–5.

[24] Schrey MP, Hare AL, Ilson SL, Walters MP. Decidual histamine release and amplification of prostaglandin F2α production by histamine in interleukin-1 β-primed decidual cells: potential interactive role for inflammatory mediators in uterine function at term. J Clin Endocrinol Metab 1995;80(2):648–53. Available from: https://doi.org/10.1210/jcem.80.2.7531717.

[25] Okada Y., Asahina T., Kobayashi T., Goto J., Terao T. Studies on the mechanism of edematous changes at the endometrial stroma for implantation. Semin Thromb Hemost. 27(2):67–68. https://doi.org/10.1055/s-2001-14063.

[26] Hannan NJ, Jones RL, White CA, Salamonsen LA. The chemokines, CX3CL1, CCL14, and CCL4, promote human trophoblast migration at the feto-maternal interface. Biol Reprod 2006;74(5):896–904. Available from: https://doi.org/10.1095/biolreprod.105.045518.

[27] Red-Horse K, Drake PM, Fisher SJ. Human pregnancy: the role of chemokine networks at the fetal-maternal interface. Expert Rev Mol Med 2004;6(11):1–14. Available from: https://doi.org/10.1017/S1462399404007720.

[28] Li T, Wang J, Guo X, et al. Possible involvement of crosstalk between endometrial cells and mast cells in the development of endometriosis via CCL8/CCR1. Biomed Pharmacother 2020;129(110476). Available from: https://doi.org/10.1016/j.biopha.2020.110476.

[29] Anaf V, Chapron C, El Nakadi I, De Moor V, Simonart T, Noël J-C. Pain, mast cells, and nerves in peritoneal, ovarian, and deep infiltrating endometriosis. Fertil Steril 2006;86(5):1336–43. Available from: https://doi.org/10.1016/j.fertnstert.2006.03.057.

[30] Amsalem H, Kwan M, Hazan A, et al. Identification of a novel neutrophil population: proangiogenic granulocytes in second-trimester human decidua. J Immunol 2014;193(6):3070–9. Available from: https://doi.org/10.4049/jimmunol.1303117.

[31] Popovic M, Teles A, Stancic O, Volk HD, Zenclussen AC. Mast-cell-associated genes Thp1, Mcpt1 and Mcpt5 are up-regulated after Treg-induced tolerance at the fetal–maternal interface: new role for mast cells in tolerance? J Reprod Immunol 2007;75(1):A16. Available from: https://doi.org/10.1016/j.jri.2007.06.043.

[32] Ueshima C, Kataoka TR, Hirata M, et al. Possible involvement of human mast cells in the establishment of pregnancy via killer cell Ig-like receptor 2DL4. Am J Pathol 2018;188(6):1497–508. Available from: https://doi.org/10.1016/j.ajpath.2018.02.012.

[33] Woidacki K, Jensen F, Zenclussen AC. Mast cells as novel mediators of reproductive processes. Front Immunol 2013;4(FRB):29. Available from: https://doi.org/10.3389/fimmu.2013.00029.

[34] Liu Z, Kilburn B, Leach R, Romero R, Paria B, Armant DR. Histamine enhances cytotrophoblast invasion by inducing intracellular calcium transients through the histamine type-1 receptor. Mol Reprod Dev 2004;68:345–53. Available from: https://doi.org/10.1002/mrd.20082.

[35] Szewczyk G, Pyzlak M, Smierta W, Klimkiewicz J, Szukiewicz D. Histamine stimulates av-b3 integrin expression of the human trophoblast through the H1 receptor. Inflamm Res 2006;55:79–80.

[36] Pyzlak M, Szewczyk G, Szukiewicz D. Histamine influence on apoptosis in trophoblast cell cultures. Inflamm Res 2010;59:213–15. Available from: https://doi.org/10.1007/s00011-009-0133-4.

[37] Bollopragada S, Youssef R, Jordan F, Greer I, Norman J, Nelson S. Term labor is associated with a core inflammatory response in human fetal membranes, myometrium, and cervix. Am J Obstet Gynecol 2009;200(1):104.e1–104.e11. Available from: https://doi.org/10.1016/j.ajog.2008.08.032.

[38] Dey P, Ray R. Quantitative analysis of mast cells in testicular aspiration cytology smears in azoospermic males. Diagn Cytopathol 1993;9(6):685–6. Available from: https://doi.org/10.1002/dc.2840090617.
[39] Albrecht M, Ramsch R, Kohn F, Schwarzer J, Mayerhofer A. Isolation and cultivation of human testicular peritubular cells: a new model for the investigation of fibrotic processes in the human testis and male infertility. J Clin Endocrinol Metab 2006;91(5):1956–69. Available from: https://doi.org/10.1210/jc.2005-2169.
[40] Elieh Ali Komi D, Shafaghat F, Haidl G. Significance of mast cells in spermatogenesis, implantation, pregnancy, and abortion: cross talk and molecular mechanisms. Am J Reprod Immunol N Y N 1989 2020;83(5): e13228. Available from: https://doi.org/10.1111/aji.13228.
[41] Adam M, Schwarzer J, Kohn F, Strauss L, Poutanen M, Mayerhofer A. Mast cell tryptase stimulates production of decorin by human testicular peritubular cells: possible role of decorin in male infertility by interfering with growth factor signaling. Hum Reprod Oxf Engl 2011;26(10):2613–25. Available from: https://doi.org/10.1093/humrep/der245.
[42] Agrawal S, Alvin J. Mast cells and idiopathic male infertility. Int J Fertil Int J Fertil 1987;32:283–6.
[43] Yamanaka K, Fujisawa M, Tanaka H, Okada H, Arakawa S, Kamidono S. Significance of human testicular mast cells and their subtypes in male infertility. Hum Reprod 2000;15:1543–7. Available from: https://doi.org/10.1093/humrep/15.7.1543.
[44] Weidinger S, Mayerhofer A, Kunz L. Tryptase inhibits motility of human spermatozoa mainly by activation of the mitogen-activated protein kinase pathway. Hum Reprod 2005;20:456–61. Available from: https://doi.org/10.1093/humrep/deh618.
[45] Mostafa R, Abol-Magd R, Younis S, Dessouki O, Azab M, Mostafa T. Assessment of seminal mast cells in infertile men with varicocele after surgical repair. Andrologia. 2017;49(3):e12625. Available from: https://doi.org/10.1111/and.13756.
[46] Mostafa T, Rashed L, Nabil N, Osman I, Mostafa R, Farag M. Seminal miRNA relationship with apoptotic markers and oxidative stress in infertile men with varicocele. Biomed Res Int 2016;2016. Available from: https://doi.org/10.1155/2016/4302754.
[47] Allam J-P, Langer M, Fathy A, et al. Mast cells in the seminal plasma of infertile men as detected by flow cytometry. Andrologia. 2009;41(1):1–6. Available from: https://doi.org/10.1111/j.1439-0272.2008.00879.x.
[48] Abdel-Hamid AAM, Atef H, Zalata KR, Abdel-Latif A. Correlation between testicular mast cell count and spermatogenic epithelium in non-obstructive azoospermia. Int J Exp Pathol 2018;99(1):22–8. Available from: https://doi.org/10.1111/iep.12261.
[49] Weidinger S, Mayerhofer A, Frungieri MB, Meineke V, Ring J, Kohn FM. Mast cell-sperm interaction: evidence for tryptase and proteinase-activated receptors in the regulation of sperm motility. Hum Reprod Oxf Engl 2003;18(12):2519–24. Available from: https://doi.org/10.1093/humrep/deg476.
[50] Azadi L, Abbasi H, Deemeh MR, et al. Zaditen (Ketotifen), as mast cell blocker, improves sperm quality, chromatin integrity and pregnancy rate after varicocelectomy. Int J Androl 2011;34(5 Pt 1):446–52. Available from: https://doi.org/10.1111/j.1365-2605.2010.01112.x.
[51] Zaazaa A, Adel A, Fahmy I, Elkhiat Y, Awaad AA, Mostafa T. Effect of varicocelectomy and/or mast cells stabilizer on sperm DNA fragmentation in infertile patients with varicocele. Andrology 2018;6(1):146–50. Available from: https://doi.org/10.1111/andr.12445.
[52] Schill WB, Schneider J, Ring J. The use of ketotifen, a mast cell blocker, for treatment of oligo- and asthenozoospermia. Andrologia. 1986;18(6):570–3. Available from: https://doi.org/10.1111/j.1439-0272.1986.tb01831.x.
[53] Hibi H, Kato K, Mitsui K, et al. The treatment with tranilast, a mast cell blocker, for idiopathic oligozoospermia. Arch Androl 2001;47(2):107–11. Available from: https://doi.org/10.1080/014850101316901307.
[54] Yamamoto M, Hibi H, Miyake K. New treatment of idiopathic severe oligozoospermia with mast cell blocker: results of a single-blind study. Fertil Steril 1995;64(6):1221–3. Available from: https://doi.org/10.1016/s0015-0282(16)57992-8.
[55] Matsuki S, Sasagawa I, Suzuki Y. The use of ebastine, a mast cell blocker, for treatment of oligozoospermia. Arch Androl 2000;44(2):129–32. Available from: https://doi.org/10.1080/014850100262290.
[56] Cayan S, Apa DD, Akbay E. Effect of fexofenadine, a mast cell blocker, in infertile men with significantly increased testicular mast cells. Asian J Androl 2002;4(4):291–4.
[57] Mondillo C, Varela ML, Abiuso AMB, Vázquez R. Potential negative effects of anti-histamines on male reproductive function. Reprod Camb Engl 2018;155(5):R221–7. Available from: https://doi.org/10.1530/REP-17-0685.

[58] Vallvé-Juanico J, Houshdaran S, Giudice LC. The endometrial immune environment of women with endometriosis. Hum Reprod Update 2019;25(5):564–91. Available from: https://doi.org/10.1093/humupd/dmz018.
[59] Sugamata M, Ihara T, Uchiide I. Increase of activated mast cells in human endometriosis. Am J Reprod Immunol 2005;53(3):120–5. Available from: https://doi.org/10.1111/j.1600-0897.2005.00254.x.
[60] Matsuzaki S, Canis M, Darcha C, Fukaya T, Yajima A, Bruhat MA. Increased mast cell density in peritoneal endometriosis compared with eutopic endometrium with endometriosis. Am J Reprod Immunol 1998;40(4):291–4. Available from: https://doi.org/10.1111/j.1600-0897.1998.tb00420.x.
[61] Tsuji S, Tsuji K, Otsuka H, Murakami T. Increased mast cells in endocervical smears of women with dysmenorrhea. CytoJournal. 2018;15:27. Available from: https://doi.org/10.4103/cytojournal.cytojournal_54_17.
[62] Borelli V, Martinelli M, Luppi S, et al. Mast cells in peritoneal fluid from women with endometriosis and their possible role in modulating sperm function. Front Physiol 2019;10:1543. Available from: https://doi.org/10.3389/fphys.2019.01543.
[63] Bungum HF, Nygaard U, Vestergaard C, Martensen PM, Knudsen UB. Increased IL-25 levels in the peritoneal fluid of patients with endometriosis. J Reprod Immunol 2016;114:6–9. Available from: https://doi.org/10.1016/j.jri.2016.01.003.
[64] Zhu T-H, Zou G, Ding S-J, et al. Mast cell stabilizer ketotifen reduces hyperalgesia in a rodent model of surgically induced endometriosis. J Pain Res 2019;12:1359–69. Available from: https://doi.org/10.2147/JPR.S195909.
[65] Binda MM, Donnez J, Dolmans M-M. Targeting mast cells: a new way to treat endometriosis. Expert Opin Ther Targets 2017;21(1):67–75. Available from: https://doi.org/10.1080/14728222.2017.1260548.
[66] Derbala Y, Elazzamy H, Bilal M, et al. Mast cell-induced immunopathology in recurrent pregnancy losses. Am J Reprod Immunol N Y N 1989 2019;82(1):e13128. Available from: https://doi.org/10.1111/aji.13128.
[67] Woidacki K, Popovic M, Metz M, et al. Mast cells rescue implantation defects caused by c-kit deficiency. Cell Death Dis 2013;4:e462. Available from: https://doi.org/10.1038/cddis.2012.214.
[68] Meyer N, Schüler T, Zenclussen AC. Simultaneous ablation of uterine natural killer cells and uterine mast cells in mice leads to poor vascularization and abnormal doppler measurements that compromise fetal well-being. Front Immunol 2017;8:1913. Available from: https://doi.org/10.3389/fimmu.2017.01913.
[69] Woidacki K, Meyer N, Schumacher A, Goldschmidt A, Maurer M, Zenclussen AC. Transfer of regulatory T cells into abortion-prone mice promotes the expansion of uterine mast cells and normalizes early pregnancy angiogenesis. Sci Rep 2015;5:13938. Available from: https://doi.org/10.1038/srep13938.
[70] Radulovic NV, Ekerhovd E, Abrahamsson G, Norström A. Cervical tissue changes in women with miscarriage: a morphological and biochemical investigation. Acta Obstet Gynecol Scand 2010;89(1):54–64. Available from: https://doi.org/10.3109/00016340903390737.
[71] Madhappan B, Kempuraj D, Christodoulou S, et al. High levels of intrauterine corticotropin-releasing hormone, urocortin, tryptase, and interleukin-8 in spontaneous abortions. Endocrinology. 2003;144(6):2285–90. Available from: https://doi.org/10.1210/en.2003-0063.
[72] Arck PC, Rose M, Hertwig K, Hagen E, Hildebrandt M, Klapp BF. Stress and immune mediators in miscarriage. Hum Reprod Oxf Engl 2001;16(7):1505–11. Available from: https://doi.org/10.1093/humrep/16.7.1505.
[73] Arck PC. Stress and pregnancy loss: role of immune mediators, hormones and neurotransmitters. Am J Reprod Immunol 2001;46(2):117–23. Available from: https://doi.org/10.1111/j.8755-8920.2001.460201.x.
[74] Castells M, Butterfield J. Mast cell activation syndrome and mastocytosis: initial treatment options and long-term management. J Allergy Clin Immunol Pract 2019;7(4):1097–106. Available from: https://doi.org/10.1016/j.jaip.2019.02.002.
[75] Jia Q, Fang K, Liu Y. Rare lesion in the uterine cervix with irregular vaginal bleeding. Am J Obstet Gynecol 2019;220(6):598–9. Available from: https://doi.org/10.1016/j.ajog.2018.11.1104.
[76] Zhao D, Zhang L, Li W, Duan M, Zhuang J, Zhou D. Well-differentiated systemic mastocytosis with KIT K5091 mutation and uterus infiltration in an Asian woman with good response to imatinib. Chin Med JEngl 2019;132(16):2002–3. Available from: https://doi.org/10.1097/CM9.0000000000000355.
[77] Brockow K. Epidemiology, prognosis, and risk factors in mastocytosis. Immunol Allergy Clin North Am 2014;34(2):283–95. Available from: https://doi.org/10.1016/j.iac.2014.01.003.
[78] Cohen S, Skovbo S, Vestergaard H, Kristensen T, Moller M, Bindslev-Jensen C. Epidemiology of systemic mastocytosis in denmark. Br J Haematol 2014;166:521–8. Available from: https://doi.org/10.1111/bjh.12916.

[79] Ciach K, Niedoszytko M, Abacjew-Chmylko A, et al. Pregnancy and delivery in patients with mastocytosis treated at the polish center of the european competence network on mastocytosis (ECNM). PLoS ONE 2016;11(1):e0146924. Available from: https://doi.org/10.1371/journal.pone.0146924.

[80] Murray DB, editor. Mast Cells: Phenotypic Features, Biological Functions and Role in Immunity. New York: Nova Biomedical; 2013.

[81] Woidacki K, Zenclussen AC, Siebenhaar F. Mast cell-mediated and associated disorders in pregnancy: a risky game with an uncertain outcome? Fronti Immunol 2014;5(231). Available from: https://doi.org/10.3389/fimmu.2014.00231.

[82] Watson KD, Arendt KW, Watson WJ, Volcheck GW. Systemic mastocytosis complicating pregnancy. Obstet Gynecol 2012;119:486–9. Available from: https://doi.org/10.1097/AOG.0b013e318242d3c5.

[83] Matito A, Álvarez-Twose I, Morgado JM, Sánchez-Muñoz L, Orfao A, Escribano L. Clinical impact of pregnancy in mastocytosis: a study of the Spanish network on mastocytosis (REMA) in 45 cases. Int Arch Allergy Immunol 2011;156(1):104–11. Available from: https://doi.org/10.1159/000321954.

[84] Afrin L, Butterfield J, Raithel M, Molderings G. Often seen, rarely recognized: mast cell activation disease-a guide to diagnosis and therapeutic options. Ann Med 2016;48(3):190–201. Available from: https://doi.org/10.3109/07853890.2016.1161231.

[85] Horny HP, Metcalfe DD, Akin C. WHO Classification of Tumors of Hematopoietic and Lymphoid Tissues. Lyon, France: International Agency for Research and Cancer (IARC); 2017.

[86] Pardanani A. Systemic mastocytosis in adults: 2019 update on diagnosis, risk stratification and management. Am J Hematol 2019;94(3):363–77. Available from: https://doi.org/10.1002/ajh.25371.

[87] Lei D, Akin C, Kovalszki A. Management of mastocytosis in pregnancy: a review. J Allergy Clin Immunol Pract 2017;5(5):1217–23. Available from: https://doi.org/10.1016/j.jaip.2017.05.021.

[88] Molderings G, Raithel M, Kratz F, Azemar M, Haenisch B, Harzer S. Omalizumab treatment of systemic mast cell activation disease: experiences from four cases. Intern Med 2011;50:1–5. Available from: https://doi.org/10.2169/internalmedicine.50.4640.

[89] Lim K, Tefferi A, Lasho T, Finke C, Patnaik M, Butterfield J. Systemic mastocytosis in 342 consecutive adults: survical studies and prognostic factors. Blood. 2009;113:5727–36. Available from: https://doi.org/10.1182/blood-2009-02-205237.

[90] Valent P, Sperr W, Akin C. How I treat patients with advance systemic mastocytosis. Blood. 2010;116:5812–17. Available from: https://doi.org/10.1182/blood-2010-08-292144.

[91] Butterfield J. Interferon treatment for hypereosinophilic syndromes and systemic mastocytosis. Acta Haematol 2005;114:26–40. Available from: https://doi.org/10.1159/000085560.

[92] Afrin L., Dempsey T., Rosenthal L., Dorff S. Successful mast-cell-targeted treatment of chronic dyspareunia, vaginitis, and dysfunctional uterine bleeding. J Obstet Gynaecol. https://www.tandfonline.com/action/showCitFormats?doi = 10.1080/01443615.2018.1550475

[93] Watson K., Arendt K., Watson W., Volcheck G. Systemic mastocytosis complicating pregnancy. Obstetr Gynecol. 119 (2Pt2):486-910. Available from: https://doi.org/10.1097/AOG.0b013e318242d3c5.

CHAPTER

6

Role of human leukocyte antigen in the pathogenesis of recurrent pregnancy loss

Chiara Tersigni[1] *and Nicoletta Di Simone*[2,3]

[1]Fondazione Policlinico Universitario A. Gemelli IRCCS, Dipartimento di Scienze della Salute della Donna e del Bambino e di Sanità Pubblica, Rome, Italy [2]Department of Biomedical Sciences, Humanitas University, Pieve Emanuele, Milan, Italy [3]IRCCS Humanitas Research Hospital, Rozzano, Milan, Italy

1 Introduction

Human pregnancy can be defined as nature's perfect allograft. This phenomenon is allotted by the placenta and not the fetus. This extraordinary organ adopts a complex strategy of immune escape from the maternal immune system that we can summarize in three main mechanisms: (1) anatomical separation between maternal and fetal tissues, (2) modified expression of human leukocyte antigen (HLA) molecules, and (3) suppression of maternal cell-mediated immunity. Suppression of HLA transcription in placental tissues prevents classical T-cell responses to fetal antigens. Intriguingly, fetal somatic cells that fully express HLA molecules are anatomically separated from maternal tissues by the placental trophoblast barrier, preventing direct contact of fetal antigens with maternal T cells. As an additional protective mechanism to the fetus, maternal cellular immunity undergoes a hyporeactive shift. In this chapter, we focus on the physiologic modification of HLA expression in the human placenta occurring in normal pregnancy and the potential role of the abnormal placental expression of HLA molecules in the occurrence of recurrent pregnancy loss (RPL).

2 Recurrent pregnancy loss and human leukocyte antigen studies

Pregnancy loss is defined as the spontaneous end of a pregnancy before the fetus has reached viability and encompasses all losses from conception to 20–23 weeks gestation [1]. It occurs in about 15%–20% of all clinically recognized pregnancies. Guidelines from

Immunology of Recurrent Pregnancy Loss and Implantation Failure
DOI: https://doi.org/10.1016/B978-0-323-90805-4.00025-0

TABLE 1 Main causes of recurrent pregnancy loss and related prevalence (%).

Unexplained (including non-APS thrombophilias)	**40–50**
Autoimmune	20
Endocrine factors	17 –20
Anatomic factors	10–15
Genetic factors	2–5

the European Society of Human Reproduction and Embryology and the American Society for Reproductive Medicine (ASRM) define RPL as the failure of two or more clinically recognized pregnancies. [1,2] RPL is an obstetrical syndrome with heterogeneous etiologies with a prevalence of 3%–5% in reproductive-age women [1,2]. The main causes of RPL are summarized in Table 1. Unfortunately, about 40% of cases of RPL are diagnosed as unexplained, and no etiologies can be found. Among those unexplained cases, dysregulated expression and/or function of HLA molecules at the feto-maternal interface have been reported to play a role [3–12].

3 Human leukocyte antigen class I and II molecules

The HLA system is a human gene complex encoding the major histocompatibility complex (MHC) proteins, or HLA molecules. These are cell-surface proteins mediating interactions between human cells and the immune system [13]. HLA class I (A, B, and C) molecules (Ia) are composed of a glycosylated heavy chain and a light chain β2-microglobulin. The heavy chain has three extracellular domains (α1, α2, and α3), a transmembrane region, and an intracytoplasmic domain. The amino acid sequences of α1 and α2 domains determine the antigenic specificities of the HLA class I molecules (Fig. 1A). Class I HLA molecules are expressed on the cell surface of all nucleated eukaryotic cells. They are also present on platelets but not on red blood cells. Their role is to protect healthy "self" cells from immune cell recognition. According to the "missing self" hypothesis, the absence or altered expression of HLA class I molecules render target cells susceptible to natural killer (NK) cell attack [13]. Furthermore, another function of HLA class Ia molecules is to display fragments of proteins from damaged or infected cells to cytotoxic T cells that trigger an immediate response from the immune system against not-self antigens [14].

HLA class II molecules (DP, DM, DO, DQ, and DR) are heterodimers of two noncovalently associated glycosylated polypeptide chains: α and β (Fig. 1B). The α and β chains are transmembrane polypeptides that have the same overall structure. An extracellular portion composed of two domains (α1 and α2, or β1 and β2) is anchored on the membrane. Polymorphisms of class II molecules occur in the first amino-terminal β1 domain. HLA class II molecules are expressed constitutively on professional antigen-presenting cells (APCs), such as dendritic cells, mononuclear phagocytes, and B cells. The antigens presented by class II peptides are of extracellular origin and loaded onto HLA class II

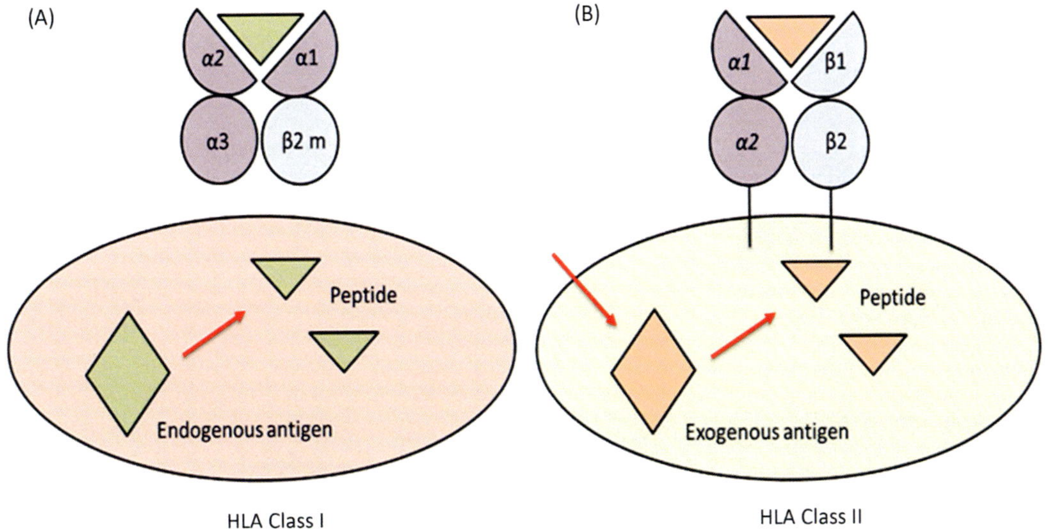

FIGURE 1 Molecular structure of human leukocyte antigen class I (a) and II (b) molecules in human cells. β2 m: β2 microglobulin.

molecules by phagocytosis. These antigens stimulate the proliferation of T-helper cells, which in turn elicit B-cells to secrete antigen-specific antibodies [13]. In particular, activation of T cells requires two signals. The first signal, which is antigen-specific, is provided through the T cell receptor (TCR), which interacts with peptide-HLA class II molecules on the membrane of APC. The second signal, the costimulatory signal, is antigen nonspecific and is provided by the interaction between costimulatory molecules expressed on the membrane of APC (i.e., CD80 and CD86) and the T cell (i.e., CD28).

3.1 Expression of human leukocyte antigen class I and II molecules in the human placenta

In human pregnancy, tight control of HLA class I and II expression in chorionic and extra-villous trophoblast (EVT) is essential for a successful outcome [15]. In humans, the villous syncytiotrophoblast does not express any HLA class I or II molecules [16]. Hence, T cells are theoretically unable to recognize and bind to the main placental barrier. EVT expresses nonclassical HLA class I molecules, such as HLA-G and -E, and classical HLA class I molecule HLA-C, but not HLA class II molecules [17,18]. The lack of classical class I HLA-A and B and the general lack of HLA class II expression on EVT prevent maternal T cell alloimmune responses against paternally-derived HLA antigens. Furthermore, the sole expression of the classical HLA-C and the atypical class I antigens (Ib), such as HLA-G and HLA-E molecules, is thought to protect EVT from immune attack by decidual natural killer (dNK) and T cells [19].

During implantation and early placental development, EVT invades maternal decidua, in which maternal leukocytes are composed of dNK cells (80%), macrophages (10%), and

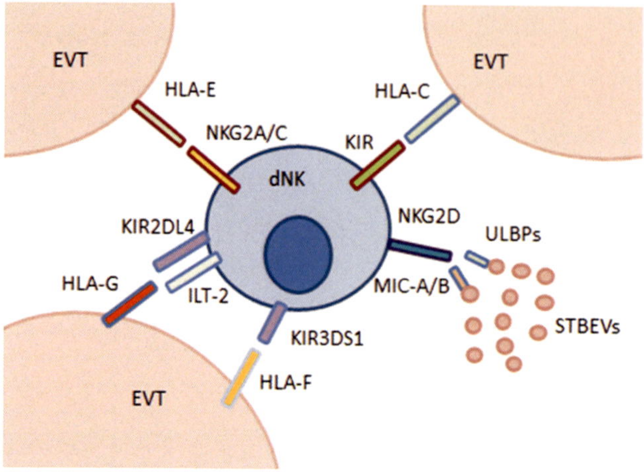

FIGURE 2 Main receptor-mediated interactions between dNK, nonclassical human leukocyte antigen class I molecules expressed by extra-villous trophoblast, and immunomodulatory soluble receptor carried by syncytiotrophoblast-derived extracellular vesicles. *NKG2*: natural killer group 2 receptor; *KIR*: killer cell immunoglobulin-like receptors; *ILT-2*: immunoglobulin-like transcript 2 receptor; *MIC*: MHC class I chain-related proteins; *ULBPs*: UL-16 binding proteins.

T cells (10%) [20]. Thus, NKs are relatively abundant in maternal decidua during the first half of pregnancy compared to other maternal immune cells. A central mechanism of maternal immune adaptation to pregnancy is that dNKs show a tolerogenic phenotype by losing cytotoxic activity and acquiring immunomodulatory properties. Low cytotoxicity of dNK cells is achieved by a particular receptor expression profile. They express the killer cell immunoglobulin-like receptors (KIR), which identify polymorphic HLA-C among other ligands, the natural killer group 2 receptor (NKG2), which recognizes HLA-E, and the immunoglobulin-like transcript 2 receptor (ILT-2), which recognizes HLA-G (Fig. 2) [21].

HLA-C expression on EVT cell surface plays a central role in modulating maternal NK and T cells and influences pregnancy outcomes. In general, it regulates cellular immune responses to viral-, allo- and self-antigens [20]. NK cells specifically recognize two groups of HLA-C allotypes, HLA-C1 and C2. They differ in the amino acid at position 80 of the HLA-C heavy chain. HLA-C1 has an asparagine, while HLA-C2 has a lysine [20]. The expression profile of KIRs gives dNK cells the ability to recognize the HLA allotypes on EVT specifically. The HLA-C1 ligates to the inhibitory KIR2DL2/3 receptors and HLA-C2 to the inhibitory KIR2DL1 or the activating KIR2DS1 receptors [17,20,22]. Thus, the interaction of dNK cells with invading EVT differs in each pregnancy and is determined by the combination of maternal KIR haplotypes, maternal and fetal HLA-C allotypes, and NK cell education.

Finally, in addition to HLA-C–KIR interaction at the maternal–trophoblast interface, the human trophoblast has been shown to adopt another mechanism of inhibiting dNK cytotoxicity. Particularly, it is known that the cytotoxicity of NK cells is commonly induced by the interaction of the NK cell receptor NKG2D with its ligands, namely the MHC class I chain-related (MIC) proteins A and B and the UL-16 binding proteins (ULBPs). This receptor–ligand interaction activates a perforin-mediated cytotoxic pathway, eliminating damaged and/or infected cells. Interestingly, MIC and ULBPs are constitutively expressed in the human placenta, and these proteins are released as soluble forms by syncytiotrophoblast-derived extracellular vesicles during pregnancy [22]. Placenta-derived soluble MIC and ULBPs have been shown to downregulate the NKG2D receptors and indirectly downregulate the NKG2D-mediated cytotoxicity (Fig. 2) on peripheral blood mononuclear cells, purified NK, CD8 + , and γδT cells [23,24].

Additionally, maternal decidual CD8 + T cells directly recognize allogenic HLA-C molecules of paternal origin. This immunological cross-talk is a key process in the development of the immune tolerance of the fetus during pregnancy. Interestingly, decidual CD8 + T cells show high expression of coinhibitory molecules, such as cytotoxic T-lymphocyte associated protein-4 (CTLA4), programmed cell death 1 (PD1), and lymphocyte activation gene-3 (LAG3), and low expression of cytolytic molecules. This uterine microenvironment functions to inhibit CD8 + effector T cell activity and maintain placental tolerance [25]. However, it has been shown that decidual CD8 + T cells retain the ability to respond to pro-inflammatory events, as with stimulation in vitro, they produce interferon-γ, tumor necrosis factor-α, perforin and granzyme B [25]. Additionally, maternal regulatory T cells (Treg) play a pivotal role in establishing immune tolerance to invading EVT and preventing inflammatory responses to placental antigens [19,26]. Indeed, Tregs are highly represented in the human decidua [16]. It is extraordinary how in vitro coculture of naïve CD4 + T cells with EVT directly increases the proportion of T cells spontaneously by differentiating into CD4 + FOXP3 + Tregs [27,28]. Intriguingly, some authors have proposed that Treg differentiation would occur after specific recognition of fetal HLA-C by maternal T cells,[20] but the precise mechanisms are still a matter of investigation. In conclusion, we can summarize that the HLA-C molecule has a central role in modulating the activity of the leading players of the cell-mediated maternal tolerance to the fetus, namely NK, CD8 + , and Treg cells.

In contrast to a classical HLA-C, HLA class Ia and Ib molecules have different polymorphisms and express different numbers of alleles. Class Ia molecule is highly polymorphic, with a high number of alleles encoding various functional proteins, whereas HLA class Ib molecules display low polymorphism, with a limited number of alleles encoding a small number of proteins. Similar to HLA Ia molecules, HLA Ib molecules can also bind peptides generated through a proteasome-driven degradation of cytosolic proteins and present them to specific subpopulations of CD8 + T cells [19]. The main function of these molecules is to modulate the immune responses in both physiological and pathological conditions [29]. In particular, HLA-G has limited polymorphic characters and forms a homodimer that can bind with high avidity to LILRB1, an inhibitory receptor expressed by all decidual APC, KIR2DL4 expressed by NK cells, [30] and CD160 expressed by T lymphocytes, NK cells, and endothelial cells [31]. These interactions deviate immune responses toward a tolerogenic rather than an immunogenic response. In particular, upon interaction with these receptors, HLA-G (i) inhibits T cells proliferation and cytotoxicity [32] and elicit the expansion of regulatory T cells [33]; (ii) inhibits differentiation, proliferation, and cytokine production in B cells [34]; (iii) inhibits proliferation and cytotoxicity of peripheral blood and uterine NK cells [35]; (iv) inhibits chemotaxis of different T, B, and NK cell populations by downregulating chemokine receptor expression on their surface [36]; (v) inhibits phagocytosis and production of reactive oxygen species in neutrophils [37]. This extraordinary and selective immunological interaction between mother and fetus has the added benefit of only occurring when there is direct physical contact between HLA-G + EVT and uterine APC, allowing maternal APC elsewhere in her body to function normally [20]. Finally, NK–EVT interactions involve HLA-E and NKG2A/C as well as HLA-F and KIR3DS1 interactions that may lead to degranulation and release of perforin (PRF) and granzymes (GZMs) (Fig. 2), [20] eliciting immunomodulatory effects that are still a matter of investigation. Interestingly, mesenchymal stromal cells derived from gestational tissue (in particular derived from the cord blood) are poorly immunogenic, which is related to the coexpression of HLA-G and HLA-E on their cell surface [38]. Similarly, induced pluripotent stem cells (iPSCs) express low levels of classical HLA-class I molecules but express high levels of HLA-G and HLA-E and are able to avoid the recognition of HLA-restricted cytotoxic T-cells, which become anergic when cocultured with iPSCs [39].

3.2 Abnormal expression of human leukocyte antigen molecules in RPL

Abnormal expression and/or function of HLA molecules at the maternal–fetal interface have been proposed as mechanisms potentially involved in the pathogenesis of unexplained RPL. HLA-G is the most investigated HLA molecule in the context of RPL. Miscarriage tissues collected from women with a history of RPL show reduced expression of HLA-G [3,4]. Consistently, HLA-G expression was found to be decreased in decidua basalis from women with RPL [5]. On the other hand, HLA-G was found to be increased in term placentas of women with a previous history of RPL compared to women without previous adverse pregnancy outcomes [6]. In vitro studies on JEG-3 cells have suggested that overexpression of miR-133a might be one of the mechanisms responsible for decreased HLA-G expression by reducing the translation of HLA-G, which could be involved in the pathogenesis of unexplained RPL [3]. Interestingly, it has been shown that HLA-G expressed in cultured trophoblast cells of RPL women can be upregulated upon exposure with both dexamethasone and hydrocortisone [7]. Hence, glucocorticoids might potentially modulate HLA-G expression in vitro and should be further examined for their therapeutic applicability in RPL.

The importance of the proper expression of the fetal HLA-C molecule for a successful pregnancy outcome has been indirectly suggested by the evidence that women with RPL show higher serum levels of anti-HLA-C antibodies in the first trimester of pregnancy than women with an uneventful pregnancy. This suggests a hypothesis that anti-HLA-C antibodies could be involved in the pathogenesis of unexplained RPL [8].

Several studies have also investigated that HLA genetic profiles and polymorphism are associated with the higher risk of RPL. In particular, 15 polymorphisms at the protein level and two null alleles have been identified to be associated with RPL. Most studies have focused on the null allele G*01:05 N [9–11]. The discovery of this allele led to speculations on whether fetal homozygosity of G*01:05 N could be an etiological factor in couples experiencing RPL, although some studies failed to confirm this theory [40–44]. Thus, further research is needed to define the precise role of HLA-G polymorphisms in the pathogenesis of RPL.

Although there are no available studies investigating endometrial/trophoblast expression of HLA class II molecules in women experiencing RPL, several cohort studies profiling HLA haplotypes have been conducted. The higher frequencies of DRB1*0101, DQA1*0101, DQB1*0501; DRB1*0102, DQA1*0101, DQB1*0501; DRB1*0103, DQA1*0101, DQB1*0501; or DRB1*0301, DQA1*0501, DQB1*0201 haplotypes were found in women with four or more fetal losses compared with controls [45]. In an Indian cohort study, the association of

DQB1*03:03:02 allele with RPL cases has been suggested [46]. Consistently, an association of haplotype DPB1*04:01:01-DQB1*02:01:01-DRB1*07:01:01 with RPL has been proposed in a more recent Lebanese case-control study [47]. Although very heterogeneous in the results, these studies suggest a pivotal role of HLA in determining pregnancy outcomes and highlight the immunological nature of RPL.

Aberrant expression of HLA-Dr molecules has been shown in syncytiotrophoblast of women developing preeclampsia [12]. It confirms that reduced immunological tolerance of placental paternal-derived antigens from the maternal immune system might be a mechanism leading to reduced trophoblast invasion and placenta development in the pathogenesis of preeclampsia However, to the authors' knowledge, there are no studies investigating the aberrant expression of HLA class II molecules in the trophoblast tissues of women with spontaneous miscarriage.

4 Evaluation of human leukocyte antigen antigens in assisted reproduction

Evaluation of soluble HLA-G in follicular fluids has been proposed as a possible tool to select oocytes for in vitro fertilization [48]. Several studies have pointed to the expression of HLA-G antigen in early embryos as a possible marker of elevated implantation rate [49–51]. Indeed, HLA-G positive embryo-derived vesicles are highly contained in embryo culture secretome and are uptaken by endometrial epithelial and stromal cells, suggesting EV-exchange as a way of communication at the maternal–fetal interface. Further studies are needed to validate the predictive value of expression levels of HLA-G in follicular fluids and/or embryos in terms of a successful pregnancy.

5 Treatment

There are no available medical treatments for aberrant expression of HLA antigen in RPL or other pregnancy complications, that is, preeclampsia or preterm delivery. In vitro studies suggest that glucocorticoid could upregulate HLA-G expression in trophoblast cells of RPL women [7]. Further studies are needed to examine the clinical effect of glucocorticoids administration in women with RPL and abnormal expression of HLA antigens.

6 Conclusions

Proper EVT expression and function of HLA class Ib molecules and complete suppression of HLA class II molecules in trophoblast tissues are crucial immunological mechanisms for successful embryo implantation and placentation. Several studies have suggested an association between the genetic polymorphisms of HLA class I and II molecules and RPL. Reduced expression of HLA-G has been shown in trophoblast tissues and decidua of women with RPL. However, studies investigating the abnormal expression of HLA class Ib other than HLA G or HLA class II molecules at the feto-maternal interface in RPL are limited. Therefore, this is a substantially unexplored field that must be further investigated.

7 Summary and recommendations

- HLA class Ib (HLA-C, -E, -F, -G) molecules are expressed by EVT.
- STB does not express HLA class Ib molecules.
- In normal pregnancy, HLA class II molecules are not expressed by trophoblast cells.
- Reduced expression of HLA-G has been shown in women with RPL.
- HLA class I and II genetic polymorphisms have been associated with RPL.

References

[1] Bender Atik, R. et al. ESHRE guideline: recurrent pregnancy loss. An in-depth evaluation of the evidence base that underpins the use of diagnostic tests and treatments in the contemporary management of recurrent pregnancy loss. Hum. Reprod. Open 2018, hoy004 (2018).
[2] Practice Committee of the American Society for Reproductive Medicine. Evaluation and treatment of recurrent pregnancy loss: a committee opinion. Fertil Steril 2012;98:1103–11.
[3] Wang X, Li B, Wang J, Lei J, Liu C, Ma Y, et al. Evidence That miR-133a Causes Recurrent Spontaneous Abortion by Reducing HLA-G Expression. Reprod Biomed Online 2012;25:415–24.
[4] Mosaferi E, Alizadeh Gharamaleki N, Farzadi L, Majidi J, Babaloo Z, Kazemi T, et al. The study of HLA-G gene and protein expression in patients with recurrent miscarriage. Adv Pharm Bull 2019;9(1):70–5.
[5] Papamitsou T, Toskas A, Papadopoulou K, Sioga A, Lakis S, Chatzistamatiou M, et al. Immunohistochemical study of immunological markers: HLAG, CD16, CD25, CD56 and CD68 in placenta tissues in recurrent pregnancy loss. Histol Histopathol 2014;29(8):1047–55.
[6] Craenmehr MHC, Nederlof I, Cao M, Drabbels JJM, Spruyt-Gerritse MJ, Anholts JDH, et al. Increased HLA-G expression in term placenta of women with a history of recurrent miscarriage despite their genetic predisposition to decreased HLA-G levels. Int J Mol Sci 2019;20:625.
[7] Akhter A, Faridi RM, Das V, Pandey A, Naik S, Agrawal S. In vitro up-regulation of HLA-G using dexamethasone and hydrocortisone in first-trimester trophoblast cells of women experiencing recurrent miscarriage. Tissue Antigens 2012;80(2):126–35.
[8] Meuleman T, van Beelen E, Kaaja RJ, van Lith JM, Claas FH, Bloemenkamp KW. HLA-C antibodies in women with recurrent miscarriage suggests that antibody mediated rejection is one of the mechanisms leading to recurrent miscarriage. J Reprod Immunol 2016;116:28–34.
[9] Pfeiffer KA, Fimmers R, Engels G, van der Ven H, van der Ven K. Implantation and pregnancy. The HLA-G genotype is potentially associated with idiopathic recurrent spontaneous abortion. Mol Hum Reprod 2001;7:373–8.
[10] Aldrich CL, Stephenson MD, Karrison T, Odem RR, Branch DW, Scott JR, et al. HLA-G genotypes and pregnancy outcome in couples with unexplained recurrent miscarriage. Mol Hum Reprod 2001;7:1167–72.
[11] Hviid TV, Hylenius S, Hoegh AM, Kruse C, Christiansen OB. HLA-G polymorphisms in couples with recurrent spontaneous abortions. Tissue Antigens 2002;60:122–32.
[12] Tersigni C, Redman CW, Dragovic R, Tannetta D, Scambia G, Di Simone N, et al. HLA-DR is aberrantly expressed at feto-maternal interface in pre-eclampsia. J Reprod Immunol 2018;129:48–52.
[13] Rock KL;, Reits E;, Neefjes J. Present yourself! By MHC class I and MHC class II molecules. Trends Immunol 2016;37:724–37.
[14] Blees A, Januliene D, Hofmann T, Koller N, Schmidt C, Trowitzsch S, et al. Structure of the human MHC-I peptide-loading complex. Nature 2017;551:525–8.
[15] Moffett A, Chazara O, Colucci F. Maternal allo-recognition of the fetus. Fertil Steril 2017;107:1269–72.
[16] Apps R, Murphy SP, Fernando R, Gardner L, Ahad T, Moffett A. Human leucocyte antigen (HLA) expression of primary trophoblast cells and placental cell lines, determined using single antigen beads to characterize allotype specificities of anti-HLA antibodies. Immunology 2009;127:26–39.
[17] Parham P, Norman PJ, Abi-Rached L, Hilton HG, Guethlein LA. Review: immunogenetics of human placentation. Placenta 2012;33:71–80.
[18] Burton G, Jauniaux E. What is the placenta? Am J Obstet Gynecol 2015;213:6–8.

[19] Datema G, Van Meir CA, Kanhai HH, Van Den Elsen PJ. Pre-term birth and severe pre-eclampsia are not associated with altered expression of HLA on human trophoblasts. Am J Reprod Immunol 2003;49:193–201.

[20] Papúchová H, Meissner TB, Li Q, Strominger JL, Tilburgs T. The dual role of HLA-C in tolerance and immunity at the maternal-fetal interface. Front Immunol 2019;10:2730.

[21] Olmos-Ortiz A, Flores-Espinosa P, Mancilla-Herrera I, Vega-Sánchez R, Díaz L, Zaga-Clavellina V. Innate immune cells and toll-like receptor-dependent responses at the maternal-fetal interface. Int J Mol Sci 2019;20:3654.

[22] Sharkey AM, Gardner L, Hiby S, Farrell L, Apps R, Masters L, et al. Killer Ig-like receptor expression in uterine NK cells is biased toward recognition of HLA-C and alters with gestational age. J Immunol 2008;181:39–46.

[23] Mincheva-Nilsson L, Nagaeva O, Chen T, Stendahl U, Antsiferova J, Mogren I, et al. Placenta-derived soluble MHC class I chain-related molecules down-regulate NKG2D receptor on peripheral blood mononuclear cells during human pregnancy: a possible novel immune escape mechanism for fetal survival. J Immunol 2006;176:3585–92.

[24] Hedlund M, Stenqvist AC, Nagaeva O, Kjellberg L, Wulff M, Baranov V, et al. Human placenta expresses and secretes NKG2D ligands via exosomes that down-modulate the cognate receptor expression: evidence for immunosuppressive function. J Immunol 2009;183:340–51.

[25] Van der Zwan A, Bi K, Norwitz ER, Crespo AC, Claas FHJ, Strominger JL, et al. Mixed signature of activation and dysfunction allows human decidual CD8(+) T cells to provide both tolerance and immunity. Proc Natl Acad Sci USA 2018;115:385–90.

[26] Tilburgs T, Roelen DL, van der Mast BJ, van Schip JJ, Kleijburg C, de Groot-Swings GM, et al. Differential distribution of CD4 + CD25bright and CD8 + CD28 – T-cells in decidua and maternal blood during human pregnancy. Placenta 2006;27:47–53.

[27] Tilburgs T, Crespo ÂC, van der Zwan A, Rybalov B, Raj T, Stranger B, et al. Human HLAG + extravillous trophoblasts: immune-activating cells that interact with decidual leukocytes. Proc Natl Acad Sci USA 2015;112:7219–24.

[28] Svensson-Arvelund J, Mehta RB, Lindau R, Mirrasekhian E, Rodriguez-Martinez H, Berg G, et al. The human fetal placenta promotes tolerance against the semi-allogeneic fetus by inducing regulatory T cells and homeostatic M2 macrophages. J Immunol 2015;194:1534–44.

[29] Reeves E, James E. Tumour and placenta establishment: the importance of antigen processing and presentation. Placenta 2017;56:34–9.

[30] Persson G, Jørgensen N, Nilsson LL, Andersen LHJ, Hviid TVF. A role for both HLA-F and HLA-G in reproduction and during pregnancy? Hum Immunol 2020;81:127–33.

[31] Le Page ME, Goodridge JP, John E, Christiansen FT, Witt CS. Killer Ig-like receptor 2DL4 does not mediate NK cell IFN-γ responses to soluble HLA-G preparations. J Immunol 2014;192:732–40.

[32] Pistoia V, Morandi F, Wang X, Ferrone S. Soluble HLAG: are they clinically relevant? Seminars in Cancer Biology, Volume 17. Cambridge, MA, USA: Academic Press; 2007. p. 469–79.

[33] Fournel S, Aguerre-Girr M, Huc X, Lenfant F, Alam A, Toubert A, Le Bouteiller P. Cutting edge: soluble HLA-G1 triggers CD95/CD95 ligand-mediated apoptosis in activated CD8 + cells by interacting with CD8. J Immunol 2000;164:6100–4.

[34] Naji A, Menier C, Morandi F, Agaugué S, Maki G, Ferretti E, et al. Binding of HLA-G to ITIM-bearing Ig-like transcript 2 receptor suppresses B cell Responses. J Immunol 2014;192:1536–46.

[35] Morandi F, Rizzo R, Fainardi E, Rouas-Freiss N, Pistoia V. Recent advances in our understanding of HLA-G biology: lessons from a wide spectrum of human diseases. J Immunol Res 2016;4326495.

[36] Morandi F, Rouas-Freiss N, Pistoia V. The emerging role of soluble HLA-G in the control of chemotaxis. Cytokine Growth Factor Rev 2014;25:327–35.

[37] Baudhuin J, Migraine J, Faivre V, Loumagne L, Lukaszewicz AC, Payen D, et al. Exocytosis acts as a modulator of the ILT4-mediated inhibition of neutrophil functions. Proc Natl Acad Sci USA 2013;110:17957–62.

[38] Stubbendorff M, Deuse T, Hua X, Phan TT, Bieback K, Atkinson K, et al. Immunological properties of extraembryonic human mesenchymal stromal cells derived from gestational tissue. Stem Cell Dev 2013;22:2619–29.

[39] Kim EM, Manzar G, Zavazava N. Human iPS cell-derived hematopoietic prog-enitor cells induce T-cell anergy in in vitro-generated alloreactive CD8(+) T cells. Blood 2013;121:5167–75.

[40] Penzes M, Rajczy K, Gyodi E, Reti M, Feher E, Petranyi G. HLA-G gene polymorphism in the normal population and in recurrent spontaneous abortion in Hungary. Transpl Proc 1999;31:1832–3.

[41] Yamashita T, Fujii T, Tokunaga K, Tadokoro K, Hamai Y, Miki A, et al. Analysis of human leukocyte antigen-G polymorphism including intron 4 in Japanese couples with habitual abortion. Am J Reprod Immunol 1999;41:159–63.
[42] Yan WH, Fan LA, Yang JQ, Xu LD, Ge Y, Yao FJ. HLA-G polymorphism in a Chinese Han population with recurrent spontaneous abortion. Int J Immunogenet 2006;33:55–8.
[43] Yan WH, Lin A, Chen XJ, Dai MZ, Gan LH, Zhou MY, et al. Association of the maternal 14-bp insertion polymorphism in the HLA-G gene in women with recurrent spontaneous abortions. Tissue Antigens 2006;68:521–3.
[44] Abbas A, Tripathi P, Naik S, Agrawal S. Analysis of human leukocyte antigen (HLA)-G polymorphism in normal women and in women with recurrent spontaneous abortions. Eur J Immunogenet 2004;31:275–8.
[45] Christiansen OB, Rasmussen KL, Jersild C, Grunnet N. HLA class II alleles confer susceptibility to recurrent fetal losses in Danish women. Tissue Antigens 1994;44(4):225–33.
[46] Aruna M, Nagaraja T, Andal Bhaskar S, Tarakeswari S, Reddy AG, Thangaraj K, et al. Novel alleles of HLA-DQ and -DR loci show association with recurrent miscarriages among South Indian women. Hum Reprod 2011;26(4):765–74.
[47] Aimagambetova G, Hajjej A, Malalla ZH, Finan RR, Sarray S, Almawi WY. Maternal HLA-DR, HLA-DQ, and HLA-DP loci are linked with altered risk of recurrent pregnancy loss in Lebanese women: a case-control study. Am J Reprod Immunol 2019;82(4):e13173.
[48] Rizzo R, Fuzzi B, Stignani M, Criscuoli L, Melchiorri L, Dabizzi S, et al. Soluble HLA-G molecules in follicular fluid: a tool for oocyte selection in IVF? J Reprod Immunol 2007;74(1–2):133–42.
[49] Desai N, Filipovits J, Goldfarb J. Secretion of soluble HLA-G by day 3 human embryos associated with higher pregnancy and implantation rates: assay of culture media using a new ELISA kit. Reprod Biomed Online 2006;13(2):272–7.
[50] Sher G, Keskintepe L, Fisch JD, Acacio BA, Ahlering P, Batzofin J, et al. Soluble human leukocyte antigen G expression in phase I culture media at 46 hours after fertilization predicts pregnancy and implantation from day 3 embryo transfer. Fertil Steril 2005;83(5):1410–13.
[51] Díaz RR, Zamora RB, Sánchez RV, Pérez JG, Bethencourt JCA. Embryo sHLA-G secretion is related to pregnancy rate. Zygote 2019;27(2):78–81.

CHAPTER

7

Stress-induced immune deviations and reproductive failure

Ronja Wöhrle, Petra Clara Arck and Kristin Thiele

Division of Experimental Feto-Maternal Medicine, Department of Obstetrics and Fetal Medicine, University Medical Center Hamburg-Eppendorf, Hamburg, Germany

1 Immune-endocrine adaptations during normally progressing pregnancies

More than six decades ago, Sir Peter Medawar discovered that organ rejection from unrelated individuals is based on immunological principles. This quickly led to the recognition that pregnancy is an immunological enigma since the fetus exhibits paternal antigens foreign to the mother, but rejection does not occur. Medawar proposed three mechanisms for aiming to explain this enigma: (1) the anatomical separation by the placenta, (2) the immaturity of fetal antigens preventing a maternal immune response, and (3) immunological inertness of the maternal immune system [1]. These attempted explanations have sparked intense research endeavors and firmly established the field of reproductive immunology. Nowadays, it is generally accepted that the maternal immune adaptation to pregnancy is a very complex and fine-tuned interplay of hormones and various immune cells of the innate and adaptive immune system to enable and sustain fetal growth and development [2,3].

Maternal immune adaptation to pregnancy has been recognized to occur during the spontaneous onset and normal progression of pregnancy until term, including the cyclical renewal of the endometrium, that is, the recurrent formation of an immunological capacity that allows receptivity for implantation [4]. After fertilization, the zygote continuously divides and develops into a blastocyst. Its outer layer, called the trophoblast, can be divided into the villous and the extravillous trophoblasts (EVT). While the former express neither major histocompatibility complex (MHC) class I nor class II molecules herewith preventing a maternal immune response, the latter, in contrast, expresses polymorphic human leukocyte antigen (HLA)-C molecules and nonpolymorphic HLA-E, HLA-G, and HLA-F [5]. Via HLA-C, EVT cells can address uterine natural killer (uNK) cells, $CD8^+$ T effector cells, and $CD4^+$ T cells [6,7]. Subsequently, immune cells accumulate around the

Immunology of Recurrent Pregnancy Loss and Implantation Failure
DOI: https://doi.org/10.1016/B978-0-323-90805-4.00013-4

differentiating trophoblast, characterized by a proinflammatory cytokine and chemokine response, mirrored by the presence of interleukin (IL)-6, IL-15, tumor necrosis factor (TNF), CXCL10, CXCL8, CCL2 [8]. Inflammation is proposed to be necessary for the differentiation and invasion of fetal trophoblast cells into the decidua [9].

Shortly after this surge of inflammation, an urgent need for a tolerogenic milieu is developing to ensure fetal tolerance. Unlike previous assumptions of a constant suppression of the immune system, this immune tolerance toward the fetus is gained through active recruitment of immune cells to the feto-maternal interface [9]. These immune cells take on distinct phenotypes and are restricted in their effector functions. Shortly summarized, NKs, macrophages, and dendritic cells (DCs) exhibit a tolerogenic phenotype differing from their peripheral counterparts. NK cells are less cytotoxic, and macrophages stay immature. Both act antiinflammatory and are of importance for decidual and vascular remodeling [10,11]. Also, DCs accumulate in higher numbers than in the periphery and exhibit immature, tolerogenic features [12]. Further, these innate immune cells contribute to changes in $CD4^+$ T cell subsets. T helper (Th)1/Th2- and the Th17/regulatory T (Treg)-balance are both shifted toward the antiinflammatory T cell types. Hence, rather Th2 and Treg cells accumulate in the decidua than Th1 and Th17 phenotypes [13]. After this long period of tolerance, parturition is initiated, representing another surge of gestational inflammation. All phenotypical and functional changes underlying the maternal immune adaptation to pregnancy are mediated by hormones, for example, human chorionic gonadotropin (hCG), progesterone, glucocorticoids, and estrogens [14]. Their effects on distinct immune cell populations are summarized in Table 1.

TABLE 1 Summary of pregnancy hormones and their effect on the immune system.

Hormone	Effect on the innate immune system	Effect on the adaptive immune system
hCG	– Induction of uterine NK cells [15] – Monocytes produce chemo-attractive IL-8 [16] – Promotes shift of macrophages toward M2 phenotype [17] – Induction of tolerogenic DCs [18]	– Inhibition of Th1 cells [19] – Increase of Treg [17] and Bregs [20] – Recruitment of Th17 to decidua [17]
Progesterone	– Induction of tolerogenic DCs [21] – Promotes shift of macrophages toward M2 phenotype [22] – NK cells: apoptosis and reduced cytotoxicity [23]	– Accumulation of Th2 [24] – Increase of Tregs [25] – Depression of Th1 [24] and Th17 cells [26] – Less cytotoxic $CD8^+$ T cells [27]
Estrogen	– Promotes M2 phenotype [28] – Maturation of DCs [21]	– High concentrations promote Th2 [29] and Treg cells [30] – Low concentrations promote Th1 cells [29]
Glucocorticoids	– Immature DCs [31] – Antiinflammatory Macrophages [32] – Decrease of uNK [33]	– Th2-shift instead of Th1 [34,35] – Induction of Tregs [36]

This complex and fine-tuned maternal immune adaptation to pregnancy is highly vulnerable to disruption, induced by, that is, medication, infection, nutrition, and high-stress perception. The latter represents the focus of the present chapter.

1.1 Distinct immunological features of maternal immune adaptation to pregnancy

1.1.1 Innate immunity

Natural killer (NK) cells make up the vast majority of immune cells during the early stage of pregnancy. In contrast to their peripheral counterparts, uterine NKs are characterized by $CD56^{bright}CD16^{dim}$ expression, a higher production of cytokines, and less cytotoxicity. They contribute to vascular and decidual stromal remodeling via the secretion of cytokines, chemokines, and angiogenic growth factors. Additionally, the uterine NKs promote the induction of $CD4^{+}$ Treg cells while dampening Th17 cells via IFN-γ secretion and secrete Galectin-1, facilitating a Th2-shift and the apoptosis of active cytotoxic T cells [10,37–40].

Monocytes, especially intermediate monocytes ($CD14^{high}CD16^{intermediate}$), increase during pregnancy leading to an elevated TNF-α production, while phagocytosis is dampened. They preferentially develop into an immunosuppressive M2-macrophage phenotype supporting the trophoblastic invasion, spiral artery formation, and remodeling [11,41].

Dendritic cells (DCs) are significantly higher in the decidua than in peripheral blood and characterized by low production of inflammatory cytokines like IL-12, suppressed antigen presentation, and reduced costimulation signals. Instead, these immature, tolerogenic DCs express galectin-1 and produce IL-10 promoting $CD4^{+}$ Treg expansion and a shift to Th2 cells [12,39,42].

1.1.2 Adaptive immunity

T helper (Th) 1 cells and their associated cytokines IFN-γ, IL-2, IL-12, IL-18 are generally responsible for an appropriate immune response toward pathogens. During pregnancy, they are downregulated to the benefit of *Th2 cells*. These cells mediate a humoral immune reaction via their typical cytokines IL-4, IL-5, IL-10, and IL-13 which support a receptive, antiinflammatory response [43].

Regulatory T (Treg) cells, a major player for the establishment and maintenance of fetal tolerance, express CD4, CD25, and Foxp3 and support a tolerogenic microenvironment by their production of IL-10, prostaglandin E_2 (PGE_2), and TGF-β. Their counterpart *Th17 cells* are proinflammatory and characterized by their IL-17 secretion, which promotes trophoblastic invasion. However, their general occurrence in the decidua is reduced via NK cell regulation [13,37,43–45].

$\gamma\delta$ T cells are a minor population in blood and decidua, although an expansion is seen during pregnancy. They exhibit low cytotoxicity, and their IL-10 and TGF-β secretion contribute to a tolerogenic milieu [46].

Regulatory B (Breg) cells increase during pregnancy and produce IL-10 and hereby dampen TNF-α production by T cells. Further, it has been suggested that Bregs support the arrest of DCs in an immature state, thereby promoting $CD4^{+}$ Treg expansion [20,47].

2 Maternal stress perception

Stress is usually understood to be a psychosocial strain. It describes situations that are able to cause disturbances, irritations, and anxiety and contributes to the impairment of the psychological and physical well-being of a person. Further, it is multifaceted, and consequently, a universal definition and its measurement are difficult due to individual perception, different cultural, social, and environmental settings, as well as the diversity of stressors [48,49]. However, different ways of describing stress have emerged:

2.1 Stressors

Stressors are highly heterogenous and understood as events or situations, people, or environmental stimuli that generate stress according to a person's own assessment. Generally, it can be distinguished between internal and external stressors, and both can be perceived as either positive (Eustress) or negative (Distress). Internal stressors are a result of a person's character and the way of raised during childhood and include high standards and self-expectations, perfectionism, unfulfilled wishes, and low load capacity. Depending on the efficiency of coping, these conditions can be incentive or overextending. External stressors are summarized as all kinds of environmental factors such as noise, great cold or heat, medical treatment or pain, financial problems, or the loss of a family member or a good friend leading to a high strain with various symptoms. During pregnancy, the expecting mother is facing a completely new situation that is often associated with uncertainty and anxiety. The existence of this "pregnancy-specific stress" is scientifically accepted and encompasses specifically health concerns for mother and child, the fear of labor, and impending parenthood [49,50]. In addition, personal life-changing events such as relocations, divorce, loss of a job, or dire environmental catastrophes like earthquakes or the collapse of the World Trade Center in 9/11 can be stressful events negatively impacting the health of pregnant women and subsequently the development of the baby (Fig. 1) [51–53].

2.2 The strength of stress

The strength of the impact on the psychological and physiological well-being of the mother is strongly depending on the deployed coping strategies and existing social support, for example, partner or other family members (Fig. 1) [54]. Further, the perception of a stressful event is subject to interindividual differences [48]. Pregnancy homeostasis can be disturbed by major and minor stressors, and the severity of prenatal stress ranges from severe traumas to moderate life-event changes to mild daily hassles [55]. The perception of stress is further dependent on the time point during pregnancy. It has been suggested that the same stressor is perceived as more severe if experienced early in pregnancy, suggesting a shrinking impact with advancing pregnancy [52,56].

FIGURE 1 **Stress perception during pregnancy:** Stressors can be highly heterogeneous, including various environmental factors, psychosocial strains, and pregnancy-specific concerns. Health effects due to prenatal stress are dependent on the severity of the stressor, the duration, and existing coping strategies and social support.

2.3 The severity of a stressor

The severity of a stressor and subsequently its impact on maternal and fetal well-being is strongly dependent on the span of exposure [57]. Hence, it can be distinguished between acute and chronic stress. While acute stress results from a specific situation normally part of our everyday life, chronic stress is defined as repeated exposure to the same or different stressors for an extended time frame. Both have the release of stress hormones such as cortisol in common. However, in acute stress situations, this release may be beneficial to increase one's awareness of a specific situation or solve a problem. In contrast, the constant release of cortisol due to chronic stress may negatively affect various body functions, for example, the cardiovascular system. Especially problematic may be the combination of both. It has been proposed that in women with spontaneous abortion, not solely the acute stress itself, but a history of chronic stress might be the crucial factor that makes acute stressors in pregnancy matter [58].

3 The prevalence of stress

It is rather difficult to determine the prevalence of stress, as there is neither a clear definition nor a diagnostical marker. Further, stress and the way it is perceived and coped with are rather subjective. It has been reported that 84 of 100 pregnant women are exposed to any sort of prenatal stress, mostly of a mild to moderate nature [59]. In addition, almost 11 of 100 pregnancies

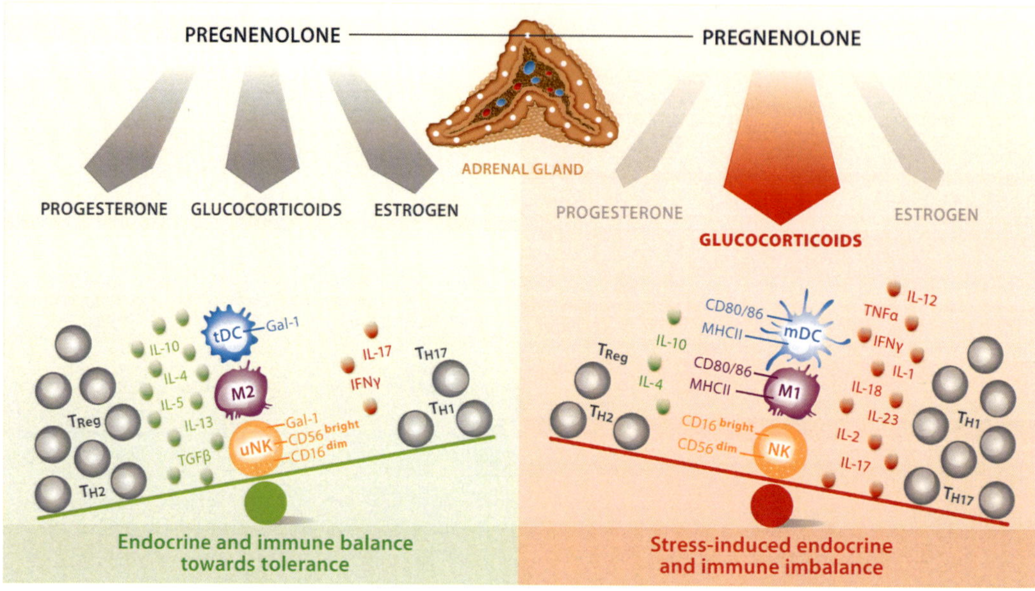

FIGURE 2 **Endocrine and immune balance and stress-induced imbalance during pregnancy:** During normal progressing pregnancy, a hormonal equilibrium favors innate immune cells with a tolerogenic phenotype, including tolerogenic dendritic cells (tDCs) expressing Galectin-1 (Gal-1), type 2 macrophages (M2), and uterine natural killer (uNK) cells characterized by $CD56^{bright}$, $CD16^{dim}$ and Gal-1 expression. That supports a shift toward antiinflammatory T cell types, T helper (Th) 2 and T regulatory (Treg) cells, and their associated cytokines interleukin (IL)-4, IL-5, IL-10, IL-13 and transforming growth factor β (TGFβ). In contrast, a stress-induced increased need for cortisol leads to a competition about pregnenolone, produced in the adrenal gland, victimizing the production of progesterone and estrogen. As a consequence, costimulatory molecules MHCII and CD80/86 are upregulated, inducing the maturation of dendritic cells (mDC) and macrophages (M1), which supports the formation of a proinflammatory milieu with Th1 and Th17 cells, including the secretion of proinflammatory cytokines IL-1, IL-2, IL-12, IL-18, IL-23, tumor necrosis factor α (TNF-α), and interferon gamma (IFNγ).

may even be affected by chronic stress in terms of depression and increased anxiety [60]. The risk of suffering from general prenatally perceived stress is reported, especially in women with one or more children [61]. Further, observations suggest that stress decreases over pregnancy, especially in primipara women, as reported by the National Institute of Child Health and Human Development fetal growth study, as anxiety is predominantly seen in the first trimester [61]. Furthermore, pregnancy-specific stress was increasingly observed in, for example, younger women, primiparous women, and women with unplanned pregnancies or lower socioeconomic statuses [50]. Nevertheless, this pregnancy-specific stress increases again in the third trimester when the time of birth is approaching.

4 Neuro-endocrine-immune imbalances and pregnancy

Early studies of stress-related pathologies during pregnancy already started last century, for example, birth weight reduction after stress exposure as reviewed in [48]. With this, impaired pregnancy progression was correlated with recent adverse life events [62]. Nowadays, the variety of known impacts of prenatal stress on pregnancy outcomes reaches from early implantation failure, recurrent abortion, or miscarriage to preterm birth and defects in fetal immune and cognitive development [49,62].

The physiologic stress response is very complexly controlled by the autonomic nervous system and the hypothalamic–pituitary–adrenal (HPA) axis. For an acute adaptation to stress, the sympathoadrenal medullary axis activates a "fight-or-flight" response by dedicating energy to the required body systems by inducing the release of adrenaline from the adrenal gland into the bloodstream. Afterward, the body's homeostasis is restored by the parasympathetic nervous system. The HPA axis, activated by chronic stress exposure, regulates the release of cortisol from the adrenal gland. Essential precondition is the release of corticotropin-releasing hormone (CRH) from the hypothalamus, which triggers the release of adrenocorticotropic hormone from the pituitary gland [55]. Typical symptoms of the stress response are increased heart rate, blood pressure, and breathing rate, and the release of glucose and fats from temporary storage sites to facilitate energy availability to the brain, muscle, and heart. Hence, especially chronic stress has effects on cardiovascular, immune, and metabolic systems [55,63].

The tolerance of the fetal antigen is strongly dependent on fine-tuned endocrine and immune adaptational processes, which are disrupted by prenatal stress perception leading to various pregnancy complications, for example, spontaneous abortion, miscarriage, or preterm birth. This includes a decrease in serum progesterone levels. Subsequently, a progesterone-induced blocking factor also decreases, leading the maternal immune system toward an inflammatory Th1-state [64]. In addition, hCG secretion is also inhibited by stress which further aggravated progesterone synthesis. Even more important is the competition about pregnenolone as a common progenitor for progesterone, estrogen, and the glucocorticoid cortisol. Since the activation of the HPA axis requires an increase in cortisol production, the production of progesterone and estrogen is simultaneously victimized leading to a shift of this hormonal equilibrium (Fig. 2). Hence, excess glucocorticoids manipulate the production of important pregnancy-protective hormones and disrupt their suppressive capacity [65]. Interestingly, this overflow of glucocorticoids is also present in the fetus despite the fact that 80%–90% of the glucocorticoids are inactivated by the placenta enzyme 11β-hydroxysteroid dehydrogenase type 2 [66]. In addition to glucocorticoids, CRH increases over gestation and is produced not only in the hypothalamus but also by the placenta. It acts as an inducer of labor and, if overexpressed, leads to premature birth, as reviewed in Ref. [67]. Further, higher levels of CRH are associated with miscarriage [68].

This stress-induced imbalance of essential pregnancy hormones disrupts the immune cell and cytokine milieu [69]. Hence, although glucocorticoids are immune-suppressive and antiinflammatory, their excess presence due to stress perception negatively affects levels of progesterone, estrogen, and hCG. In general, leukocytes decrease at the fetal-maternal interface in response to stress affecting the innate and adaptive arm of immunity [70]. Only NK cells, especially $CD56^{dim}CD16^{+}CD69^{+}$ NK cells, increase in the endometrium and in the blood resulting in a lower $CD56^{bright}/CD56^{dim}$ ratio which has been associated with recurrent pregnancy loss and infertility due to increased cytotoxicity [71–73].

However, other studies did not find differences in NK cell numbers in the decidua, only suggesting CD56^{bright} dNK cell being downregulated [74].

Prenatal stress further induced DC maturation mirrored by upregulated expression of CD11c, MHC class II and ICAM-1 in uterine draining lymph nodes and skin [75,76]. This increased maturation state of DCs is associated with the etiology of spontaneous abortion.

The CD4$^+$ T cell compartment undergoes proportional and phenotypic changes in stressed women. Hereby, Th1 immunity is promoted at the feto-maternal interface at the expense of Th2 cells [69], including the secretion of proinflammatory cytokines, such as TNF-α, IFN-γ, without being opposed by Th2 cytokines like IL-4, IL-10, or TGFβ. Proinflammatory cytokines are reported to affect cervical ripening and, therefore, preterm delivery [77]. Further, TNF-α is said to be abortogenic due to depressing pregnancy-protective transforming growth factor p2, which exhibits immunosuppressive activity [78]. Additionally, proportions of CD4$^+$ Tregs in decidua were significantly reduced in spontaneous abortion compared to induced abortion but were also associated with miscarriage and other complications [79,80]. In contrast, in chronic stress, an increase of Treg numbers was observed, which might be mediated by trophoblast cells in response to an inflammatory process [81].

Further, the CD8$^+$ T cell population was demonstrated to be present in higher frequencies in response to prenatal stress [74] Hereby, an increase of cytotoxic CD8$^+$ T cells was reported along with a reduction of tolerogenic CD8$^+$CD122$^+$ Treg cells, which was causally linked to reduced progesterone levels and placental heme oxygenase 1 expression [27] These stress-induced changes led to intrauterine growth restriction, as demonstrated in mice [27].

As the role of B cells for fetal tolerance is still widely open, their modification due to prenatal stress is even less well investigated. However, B cells were lower in stressed females [82], and decreased Breg cells along with impaired IL-10 production was observed in spontaneous abortion [47].

Also, leukocyte homing is important for inducing tolerance during pregnancy. Physiologically, antiinflammatory cell subsets are recruited and able to permeate the vascular endothelium [83]. This active process of cell transfer from the vasculature through the endothelium into the tissue is mediated by adhesion molecules, for example, intercellular adhesion molecule-1 (ICAM-1) and its ligand lymphocyte function-associated antigen-1 (LFA-1). It has been shown in a murine model that ICAM-1 at the epithelium and LFA-1 expression on antigen-presenting cells like macrophages and DCs are augmented in stressed mothers experiencing pregnancy loss [84]. Hence, leukocytes are attracted to migrate into uterine tissue and induce inflammation. Proinflammatory cytokines such as IL-1, TNF-α, and IFN-γ, already showing a higher density in stress, increase even further, leading to maturation of antigen presenting cells, for example, DCs leave their immature state and promote polarization toward Th1-type. Further, ICAM-1/LFA-1 signal pathway is shown to induce apoptosis of trophoblast, leading to abortion which could be prevented by blocking ICAM-1 adhesion [84].

Another adhesion molecule is the vascular cell adhesion protein (VCAM) 1. Equally to ICAM-1, it is expressed in higher amounts in inflammatory settings, again attracting leukocytes. VCAM-positive vessels were significantly higher in high-stress murine models [85]. Interaction with its counterpart $\alpha_4\beta_1$-integrin enables lymphocytes, monocytes, eosinophils, and NK cells to permeate into the tissue, favoring the adhesion of inflammatory cells and consequently their recruitment to the feto-maternal interphase [83].

The adhesion molecule P-selectin is involved in the formation of thrombi and has also been measured in increased numbers in the context of stress-induced abortions [84,86]. Herewith, decidual expression of P-selectin was upregulated via thrombin, histamine, and platelet-activating factor due to a rise in Th1 cells [86,87].

5 How to assess stress perception?

The evaluation of stress constitutes some difficulties due to various ways of definition and different investigation tools creating a great diversity and quantity [88]. However, to investigate the impact and pathology of stress, scientists started to find stressed cohorts after major events, for example, catastrophes, and examined people with a diagnosis like anxiety or other mental health problems. Later, daily hardships and personal life changes were made measurable by questionnaires. Hereby, the Perceived Stress Scale (PSS) and, especially for pregnancy, the Prenatal Life Events Scale are of significance [48,89].

The PSS was developed by Cohen et al. [90] and represented a self-report measure based on the psychological conceptualization of stress focusing on the daily hassles. It measures the extent to which situations were appraised as stressful in the past month, including unpredictability and overload. It is currently one of the best tools available to routinely and standardized assess daily hassles, and this questionnaire has also been validated for pregnancy [61,90].

Another commonly used questionnaire is the Prenatal Distress Questionnaire (PDQ). It has been designed in 1999 by Yali and Lobel and was edited by Lobel and Canella in 2008 (NUPDQ) [50]. This questionnaire does not only include typical stressors but also ones that are specific for pregnancy [55]. The Prenatal Life Events Scale inquires 22 items about personal life-changing events, such as relocations, divorce, loss of job, and violent crimes that have either happened to the woman herself or a close family member [89].

As already described above, the perceived severity of stress is strongly associated with social support, including social stability and participation [91]. For this purpose, there is also a questionnaire addressing 22 items about several dimensions of perceived social support, including emotional support, instrumental support, and social integration. Additionally, it assesses the woman's satisfaction with her social support network and considers several trusted individuals in a person's life [91].

6 How to treat high-stress perception?

First of all, it is particularly important to raise the awareness of gynecologists and obstetricians when supervising pregnant women throughout their pregnancy. Next to the imminent medical concerns regarding the unborn child, they need to expand their view to the women as a whole and closely observe their patients regarding signs of stress beyond the normal.

The best option to prevent the negative consequences of prenatal stress would be the avoidance of that in the first place. Stressors should be reduced, and resilience can be strengthened through stress training and social support, an effective modulator of maternal welfare [91]. Stress interventions and coping strategies like relaxation training, meditation, and yoga have been tested and claimed for options of birth outcome improvement [61,92,93].

Additionally, environmental enrichment (EE) is known as a noninvasive stimulation of sensory pathways by social and physical surroundings. Hence, it has been shown that EE modulates metabolism and acts antiinflammatory, thereby preventing preterm birth in mice [94]. Also, in human studies, it was proven that pregnancy-related impairments were reduced by sufficient leisure time, physical activity, and antistress treatments [95]. In contrast, evasive strategies only seem to enhance stress [61].

Psychological stress like anxiety might also be treatable by psychotherapy, already showing positive results in women who previously experienced recurrent spontaneous abortion [96].

Certainly, there is also the option to discuss medical interventions aiming to antagonize stress-induced obstetric complications. In women experiencing recurrent pregnancy loss, supplementation of vitamin D has shown positive results on the immunological status, making it a potential candidate for treatment [97]. Also, the exogenous administration of progesterone was shown to have an antiabortogenic effect in stressed mice [98]. Reduced abortion rates are also reported following hCG treatment of abortion-prone female mice showing increased $CD4^+$ Treg numbers [99].

7 Consequences of maternal stress on reproductive outcome

The altered hormonal situation in the mother due to stress leads to changes in the immune cell composition, as explained above. Dependent on the specific time point of stress exposure, the consequences for pregnancy outcome as well as the offspring's long-term health can be severe and of huge variety.

High cortisol levels have been associated with irregularities of the menstrual cycle influencing the ability to conceive [100] Stress during early pregnancy is strongly associated with the etiology of miscarriage and abortion [101,102]. Later in pregnancy, an active HPA axis and a resulting Th1 overflow in the mother hamper the secretion of growth hormones and elevate the blood pressure causing placental perfusion problems. This was shown to lead to low birth weight, false gut microbiome, and immunity problems of the fetus [103]. In general, an imbalance of glucocorticoids and their receptors was reported to contribute to intrauterine growth retardation [104]. While the fetus is normally protected from excessive glucocorticoids, the capacity to prevent an unwanted inundation of glucocorticoids exceeds concentrations present by stress. It was shown that glucocorticoid excess in the fetus exerts a neurotoxic effect. Brain development, especially from the hippocampal system, is impaired, leading to increased frequency of anxiety behavior and decelerated cognitive development [105,106]. In addition, a shift in the fetal HPA axis disrupts neuro-immune communication, gut maturation, and microbiota colonization with consequences for behavioral, immunological, and brain development [106–108].

An influence of maternal stress on immune development in the offspring can also be caused by direct effects on the immune cell ontogenesis. Th1 cytokine levels were reported to be lower, for example, TNF-α production. Hence, children of stressed mothers exhibit a delayed shift from Th2 to Th1 immunity that physiologically occurs within the first 2 years of life [108,109]. As a consequence, neonatal immunity is lowered with increased susceptibility to infections. Later in life, these children have an increased risk of developing allergies and asthma [110].

Especially late gestational, chronic stress might cause problems regarding passive immunity transferred from mother to fetus via IgG-antibodies [111]. Furthermore, other disadvantages like conotruncal heart defects, neural tube defects and isolated cleft lip, and mental health disorders have been summarized [112]. However, not just the stress itself but also the lack of social support is a risk factor for the fetus leading to birth weight reduction by an average of 200 g [91].

8 Summary and recommendations

- A delicate endocrine and immune balance is essential to support a tolerogenic microenvironment and ensure fetal growth and development.
- This complex and fine-tuned maternal immune adaptation to pregnancy is highly vulnerable to disruption, that is, stress perception.
- Prenatal stress affects the hormonal equilibrium, including an increase of glucocorticoids and a simultaneous decrease of serum progesterone.
- This hormonal shift disrupts the immune cell composition and cytokine milieu toward inflammation.
- The perception of prenatal stress is dependent on the severity of the stressor, the duration, and existing coping strategies.
- For the evaluation of stress, various questionnaires are available, including the PSS, the Prenatal Life Events Scale, and the PDQ.
- The consequences of prenatal stress for pregnancy outcomes can be severe and of huge variety, including an increased risk for obstetric complications such as abortion, intrauterine growth retardation, and preterm birth.
- Prenatal stress also affects offspring's long-term health due to developmental impairment leading to decelerated cognitive development and attenuated immunity.
- First-line defense to prevent negative consequences of prenatal stress is the reduction of potential stressors and a reliable social support.
- For medical options to antagonize stress-induced obstetric complications, the supplementation of vitamin D, progesterone, and hCG are discussed.

References

[1] Rendell V, Bath NM, Brennan TV. Medawar's paradox and immune mechanisms of fetomaternal tolerance. OBM Transpl 2020;4(1). Available from: https://doi.org/10.21926/obm.transplant.2001104.

[2] Arck PC, Hecher K. Fetomaternal immune cross-talk and its consequences for maternal and offspring's health. Nat Med 2013;19(5):548–56. Available from: https://doi.org/10.1038/nm.3160.

[3] Solano ME. Decidual immune cells: guardians of human pregnancies. Best Pract Res Clin Obstet Gynaecol 2019;60:3–16. Available from: https://doi.org/10.1016/j.bpobgyn.2019.05.009.
[4] Kitazawa J, Kimura F, Nakamura A, et al. Endometrial immunity for embryo implantation and pregnancy establishment. Tohoku J Exp Med 2020;250(1):49–60. Available from: https://doi.org/10.1620/tjem.250.49.
[5] Saito S, Nakabayashi Y, Nakashima A, Shima T, Yoshino O. A new era in reproductive medicine: consequences of third-party oocyte donation for maternal and fetal health. Semin Immunopathol 2016;38 (6):687–97. Available from: https://doi.org/10.1007/s00281-016-0577-x.
[6] Papuchova H, Meissner TB, Li Q, Strominger JL, Tilburgs T. The dual role of HLA-C in tolerance and immunity at the maternal-fetal interface. Front Immunol 2019;10:2730. Available from: https://doi.org/10.3389/fimmu.2019.02730.
[7] Moffett A, Loke C. Immunology of placentation in eutherian mammals. Nat Rev Immunol 2006;6(8):584–94. Available from: https://doi.org/10.1038/nri1897.
[8] PrabhuDas M, Bonney E, Caron K, et al. Immune mechanisms at the maternal-fetal interface: perspectives and challenges. Nat Immunol 2015;16(4):328–34. Available from: https://doi.org/10.1038/ni.3131.
[9] Mor G, Cardenas I, Abrahams V, Guller S. Inflammation and pregnancy: the role of the immune system at the implantation site. Ann N Y Acad Sci 2011;1221:80–7. Available from: https://doi.org/10.1111/j.1749-6632.2010.05938.x.
[10] Hanna J, Goldman-Wohl D, Hamani Y, et al. Decidual NK cells regulate key developmental processes at the human fetal-maternal interface. Nat Med 2006;12(9):1065–74. Available from: https://doi.org/10.1038/nm1452.
[11] Heikkinen J, Mottonen M, Komi J, Alanen A, Lassila O. Phenotypic characterization of human decidual macrophages. Clin Exp Immunol 2003;131(3):498–505. Available from: https://doi.org/10.1046/j.1365-2249.2003.02092.x.
[12] Bachy V, Williams DJ, Ibrahim MA. Altered dendritic cell function in normal pregnancy. J Reprod Immunol 2008;78(1):11–21. Available from: https://doi.org/10.1016/j.jri.2007.09.004.
[13] Mjosberg J, Berg G, Jenmalm MC, Ernerudh J. FOXP3 + regulatory T cells and T helper 1, T helper 2, and T helper 17 cells in human early pregnancy decidua. Biol Reprod 2010;82(4):698–705. Available from: https://doi.org/10.1095/biolreprod.109.081208.
[14] Polese B, Gridelet V, Araklioti E, Martens H, Perrier d'Hauterive S, Geenen V. The endocrine milieu and CD4 T-lymphocyte polarization during pregnancy. Front Endocrinol (Lausanne) 2014;5:106. Available from: https://doi.org/10.3389/fendo.2014.00106.
[15] Kane N, Kelly R, Saunders PT, Critchley HO. Proliferation of uterine natural killer cells is induced by human chorionic gonadotropin and mediated via the mannose receptor. Endocrinology 2009;150(6):2882–8. Available from: https://doi.org/10.1210/en.2008-1309.
[16] Kosaka K, Fujiwara H, Tatsumi K, et al. Human chorionic gonadotropin (HCG) activates monocytes to produce interleukin-8 via a different pathway from luteinizing hormone/HCG receptor system. J Clin Endocrinol Metab 2002;87(11):5199–208. Available from: https://doi.org/10.1210/jc.2002-020341.
[17] Furcron AE, Romero R, Mial TN, et al. Human chorionic gonadotropin has anti-inflammatory effects at the maternal-fetal interface and prevents endotoxin-induced preterm birth, but causes dystocia and fetal compromise in mice. Biol Reprod 2016;94(6):136. Available from: https://doi.org/10.1095/biolreprod.116.139345.
[18] Wan H, Versnel MA, Leijten LM, et al. Chorionic gonadotropin induces dendritic cells to express a tolerogenic phenotype. J Leukoc Biol 2008;83(4):894–901. Available from: https://doi.org/10.1189/jlb.0407258.
[19] Khil LY, Jun HS, Kwon H, et al. Human chorionic gonadotropin is an immune modulator and can prevent autoimmune diabetes in NOD mice. Diabetologia 2007;50(10):2147–55. Available from: https://doi.org/10.1007/s00125-007-0769-y.
[20] Rolle L, Memarzadeh Tehran M, Morell-Garcia A, et al. Cutting edge: IL-10-producing regulatory B cells in early human pregnancy. Am J Reprod Immunol 2013;70(6):448–53. Available from: https://doi.org/10.1111/aji.12157.
[21] Xiu F, Anipindi VC, Nguyen PV, et al. High physiological concentrations of progesterone reverse estradiol-mediated changes in differentiation and functions of bone marrow derived dendritic cells. PLoS One 2016;11 (4):e0153304. Available from: https://doi.org/10.1371/journal.pone.0153304.
[22] Tsai YC, Tseng JT, Wang CY, Su MT, Huang JY, Kuo PL. Medroxyprogesterone acetate drives M2 macrophage differentiation toward a phenotype of decidual macrophage. Mol Cell Endocrinol 2017;452:74–83. Available from: https://doi.org/10.1016/j.mce.2017.05.015.
[23] Arruvito L, Giulianelli S, Flores AC, et al. NK cells expressing a progesterone receptor are susceptible to progesterone-induced apoptosis. J Immunol 2008;180(8):5746–53. Available from: https://doi.org/10.4049/jimmunol.180.8.5746.

[24] Piccinni MP, Giudizi MG, Biagiotti R, et al. Progesterone favors the development of human T helper cells producing Th2-type cytokines and promotes both IL-4 production and membrane CD30 expression in established Th1 cell clones. J Immunol 1995;155(1):128–33.
[25] Mao G, Wang J, Kang Y, et al. Progesterone increases systemic and local uterine proportions of CD4 + CD25 + Treg cells during midterm pregnancy in mice. Endocrinology 2010;151(11):5477–88. Available from: https://doi.org/10.1210/en.2010-0426.
[26] Xu L, Dong B, Wang H, et al. Progesterone suppresses Th17 cell responses, and enhances the development of regulatory T cells, through thymic stromal lymphopoietin-dependent mechanisms in experimental gonococcal genital tract infection. Microbes Infect 2013;15(12):796–805. Available from: https://doi.org/10.1016/j.micinf.2013.06.012.
[27] Solano ME, Kowal MK, O'Rourke GE, et al. Progesterone and HMOX-1 promote fetal growth by CD8 + T cell modulation. J Clin Invest 2015;125(4):1726–38. Available from: https://doi.org/10.1172/JCI68140.
[28] Villa A, Rizzi N, Vegeto E, Ciana P, Maggi A. Estrogen accelerates the resolution of inflammation in macrophagic cells. Sci Rep 2015;5:15224. Available from: https://doi.org/10.1038/srep15224.
[29] Priyanka HP, Krishnan HC, Singh RV, Hima L, Thyagarajan S. Estrogen modulates in vitro T cell responses in a concentration- and receptor-dependent manner: effects on intracellular molecular targets and antioxidant enzymes. Mol Immunol 2013;56(4):328–39. Available from: https://doi.org/10.1016/j.molimm.2013.05.226.
[30] Prieto GA, Rosenstein Y. Oestradiol potentiates the suppressive function of human CD4 CD25 regulatory T cells by promoting their proliferation. Immunology 2006;118(1):58–65. Available from: https://doi.org/10.1111/j.1365-2567.2006.02339.x.
[31] Piemonti L, Monti P, Allavena P, et al. Glucocorticoids affect human dendritic cell differentiation and maturation. J Immunol 1999;162(11):6473–81.
[32] Russo-Marie F. Macrophages and the glucocorticoids. J Neuroimmunol 1992;40(2-3):281–6. Available from: https://doi.org/10.1016/0165-5728(92)90144-a.
[33] Cooper S, Laird SM, Mariee N, Li TC, Metwally M. The effect of prednisolone on endometrial uterine NK cell concentrations and pregnancy outcome in women with reproductive failure. A retrospective cohort study. J Reprod Immunol 2019;131:1–6. Available from: https://doi.org/10.1016/j.jri.2018.10.001.
[34] Xu B, Makris A, Thornton C, Hennessy A. Glucocorticoids inhibit placental cytokines from cultured normal and preeclamptic placental explants. Placenta 2005;26(8-9):654–60. Available from: https://doi.org/10.1016/j.placenta.2004.09.011.
[35] Elenkov IJ. Glucocorticoids and the Th1/Th2 balance. Ann N Y Acad Sci 2004;1024:138–46. Available from: https://doi.org/10.1196/annals.1321.010.
[36] Engler JB, Kursawe N, Solano ME, et al. Glucocorticoid receptor in T cells mediates protection from autoimmunity in pregnancy. Proc Natl Acad Sci U S A 2017;114(2):E181–90. Available from: https://doi.org/10.1073/pnas.1617115114.
[37] Fu B, Li X, Sun R, et al. Natural killer cells promote immune tolerance by regulating inflammatory TH17 cells at the human maternal-fetal interface. Proc Natl Acad Sci U S A 2013;110(3):E231–40. Available from: https://doi.org/10.1073/pnas.1206322110.
[38] Vacca P, Mingari MC, Moretta L. Natural killer cells in human pregnancy. J Reprod Immunol 2013;97(1):14–19. Available from: https://doi.org/10.1016/j.jri.2012.10.008.
[39] Blois SM, Ilarregui JM, Tometten M, et al. A pivotal role for galectin-1 in fetomaternal tolerance. Nat Med 2007;13(12):1450–7. Available from: https://doi.org/10.1038/nm1680.
[40] Zhang J, Dunk C, Croy AB, Lye SJ. To serve and to protect: the role of decidual innate immune cells on human pregnancy. Cell Tissue Res 2016;363(1):249–65. Available from: https://doi.org/10.1007/s00441-015-2315-4.
[41] Tang MX, Hu XH, Liu ZZ, Kwak-Kim J, Liao AH. What are the roles of macrophages and monocytes in human pregnancy? J Reprod Immunol 2015;112:73–80. Available from: https://doi.org/10.1016/j.jri.2015.08.001.
[42] Miyazaki S, Tsuda H, Sakai M, et al. Predominance of Th2-promoting dendritic cells in early human pregnancy decidua. J Leukoc Biol 2003;74(4):514–22. Available from: https://doi.org/10.1189/jlb.1102566.
[43] Saito S, Nakashima A, Shima T, Ito M. Th1/Th2/Th17 and regulatory T-cell paradigm in pregnancy. Am J Reprod Immunol 2010;63(6):601–10. Available from: https://doi.org/10.1111/j.1600-0897.2010.00852.x.
[44] Wu HX, Jin LP, Xu B, Liang SS, Li DJ. Decidual stromal cells recruit Th17 cells into decidua to promote proliferation and invasion of human trophoblast cells by secreting IL-17. Cell Mol Immunol 2014;11(3):253–62. Available from: https://doi.org/10.1038/cmi.2013.67.

[45] Nakashima A, Ito M, Yoneda S, Shiozaki A, Hidaka T, Saito S. Circulating and decidual Th17 cell levels in healthy pregnancy. Am J Reprod Immunol 2010;63(2):104–9. Available from: https://doi.org/10.1111/j.1600-0897.2009.00771.x.

[46] Nagaeva O, Jonsson L, Mincheva-Nilsson L. Dominant IL-10 and TGF-beta mRNA expression in gammadeltaT cells of human early pregnancy decidua suggests immunoregulatory potential. Am J Reprod Immunol 2002;48(1):9–17. Available from: https://doi.org/10.1034/j.1600-0897.2002.01131.x.

[47] Jensen F, Muzzio D, Soldati R, Fest S, Zenclussen AC. Regulatory B10 cells restore pregnancy tolerance in a mouse model. Biol Reprod 2013;89(4):90. Available from: https://doi.org/10.1095/biolreprod.113.110791.

[48] Lobel M. Conceptualizations, measurement, and effects of prenatal maternal stress on birth outcomes. J Behav Med 1994;17(3):225–72. Available from: https://doi.org/10.1007/BF01857952.

[49] Coussons-Read ME. Effects of prenatal stress on pregnancy and human development: mechanisms and pathways. Obstet Med 2013;6(2):52–7. Available from: https://doi.org/10.1177/1753495X12473751.

[50] Ibrahim SM, Lobel M. Conceptualization, measurement, and effects of pregnancy-specific stress: review of research using the original and revised prenatal distress questionnaire. J Behav Med 2020;43(1):16–33. Available from: https://doi.org/10.1007/s10865-019-00068-7.

[51] Orr ST, James SA, Casper R. Psychosocial stressors and low birth weight: development of a questionnaire. J Dev Behav Pediatr 1992;13(5):343–7.

[52] Glynn LM, Wadhwa PD, Dunkel-Schetter C, Chicz-Demet A, Sandman CA. When stress happens matters: effects of earthquake timing on stress responsivity in pregnancy. Am J Obstet Gynecol 2001;184(4):637–42. Available from: https://doi.org/10.1067/mob.2001.111066.

[53] Engel SM, Berkowitz GS, Wolff MS, Yehuda R. Psychological trauma associated with the World Trade Center attacks and its effect on pregnancy outcome. Paediatr Perinat Epidemiol 2005;19(5):334–41. Available from: https://doi.org/10.1111/j.1365-3016.2005.00676.x.

[54] Chen T, Laplante DP, Elgbeili G, et al. Coping during pregnancy following exposure to a natural disaster: the QF2011 Queensland flood study. J Affect Disord 2020;273:341–9. Available from: https://doi.org/10.1016/j.jad.2020.03.165.

[55] Coussons-Read M. Stress and immunity in pregnancy. The Oxford Handbook of Psychoneuroimmunology. Publisher: Oxford Universitry Press; 2012. p. 3–17. Available from: https://doi.org/10.1093/oxfordhb/9780195394399.013.0001.

[56] Glynn LM, Schetter CD, Wadhwa PD, Sandman CA. Pregnancy affects appraisal of negative life events. J Psychosom Res 2004;56(1):47–52. Available from: https://doi.org/10.1016/S0022-3999(03)00133-8.

[57] Bowers SL, Bilbo SD, Dhabhar FS, Nelson RJ. Stressor-specific alterations in corticosterone and immune responses in mice. Brain Behav Immun 2008;22(1):105–13. Available from: https://doi.org/10.1016/j.bbi.2007.07.012.

[58] Frazier T, Hogue CJR, Bonney EA, Yount KM, Pearce BD. Weathering the storm: a review of pre-pregnancy stress and risk of spontaneous abortion. Psychoneuroendocrinology. 2018;92:142–54. Available from: https://doi.org/10.1016/j.psyneuen.2018.03.001.

[59] Woods SM, Melville JL, Guo Y, Fan MY, Gavin A. Psychosocial stress during pregnancy. Am J Obstet Gynecol 2010;202(1):61. Available from: https://doi.org/10.1016/j.ajog.2009.07.041 e1-7.

[60] Bennett HA, Einarson A, Taddio A, Koren G, Einarson TR. Prevalence of depression during pregnancy: systematic review. Obstet Gynecol 2004;103(4):698–709. Available from: https://doi.org/10.1097/01.AOG.0000116689.75396.5f.

[61] Goletzke J, Kocalevent RD, Hansen G, et al. Prenatal stress perception and coping strategies: insights from a longitudinal prospective pregnancy cohort. J Psychosom Res 2017;102:8–14. Available from: https://doi.org/10.1016/j.jpsychores.2017.09.002.

[62] Neugebauer R, Kline J, Stein Z, Shrout P, Warburton D, Susser M. Association of stressful life events with chromosomally normal spontaneous abortion. Am J Epidemiol 1996;143(6):588–96. Available from: https://doi.org/10.1093/oxfordjournals.aje.a008789.

[63] Joseph DN, Whirledge S. Stress and the HPA axis: balancing homeostasis and fertility. Int J Mol Sci 2017;18(10). Available from: https://doi.org/10.3390/ijms18102224.

[64] Joachim R, Zenclussen AC, Polgar B, et al. The progesterone derivative dydrogesterone abrogates murine stress-triggered abortion by inducing a Th2 biased local immune response. Steroids 2003;68(10-13):931–40. Available from: https://doi.org/10.1016/j.steroids.2003.08.010.

[65] Solano ME, Arck PC. Steroids, pregnancy and fetal development. Front Immunol 2019;10:3017. Available from: https://doi.org/10.3389/fimmu.2019.03017.

[66] Gitau R, Cameron A, Fisk NM, Glover V. Fetal exposure to maternal cortisol. Lancet 1998;352(9129):707–8. Available from: https://doi.org/10.1016/S0140-6736(05)60824-0.

[67] Kalantaridou SN, Makrigiannakis A, Zoumakis E, Chrousos GP. Reproductive functions of corticotropin-releasing hormone. Research and potential clinical utility of antalarmins (CRH receptor type 1 antagonists). Am J Reprod Immunol 2004;51(4):269–74. Available from: https://doi.org/10.1111/j.1600-0897.2004.00155.x.

[68] Arck PC, Rucke M, Rose M, et al. Early risk factors for miscarriage: a prospective cohort study in pregnant women. Reprod Biomed Online 2008;17(1):101–13. Available from: https://doi.org/10.1016/s1472-6483(10)60300-8.

[69] Blois SM, Joachim R, Kandil J, et al. Depletion of CD8 + cells abolishes the pregnancy protective effect of progesterone substitution with dydrogesterone in mice by altering the Th1/Th2 cytokine profile. J Immunol 2004;172(10):5893–9. Available from: https://doi.org/10.4049/jimmunol.172.10.5893.

[70] Antonson AM, Evans MV, Galley JD, et al. Unique maternal immune and functional microbial profiles during prenatal stress. Sci Rep 2020;10(1):20288. Available from: https://doi.org/10.1038/s41598-020-77265-x.

[71] Schedlowski M, Jacobs R, Stratmann G, et al. Changes of natural killer cells during acute psychological stress. J Clin Immunol 1993;13(2):119–26. Available from: https://doi.org/10.1007/BF00919268.

[72] Murphy SP, Fast LD, Hanna NN, Sharma S. Uterine NK cells mediate inflammation-induced fetal demise in IL-10-null mice. J Immunol 2005;175(6):4084–90. Available from: https://doi.org/10.4049/jimmunol.175.6.4084.

[73] King K, Smith S, Chapman M, Sacks G. Detailed analysis of peripheral blood natural killer (NK) cells in women with recurrent miscarriage. Hum Reprod 2010;25(1):52–8. Available from: https://doi.org/10.1093/humrep/dep349.

[74] Arck PC, Rose M, Hertwig K, Hagen E, Hildebrandt M, Klapp BF. Stress and immune mediators in miscarriage. Hum Reprod 2001;16(7):1505–11. Available from: https://doi.org/10.1093/humrep/16.7.1505.

[75] Blois SM, Alba Soto CD, Tometten M, Klapp BF, Margni RA, Arck PC. Lineage, maturity, and phenotype of uterine murine dendritic cells throughout gestation indicate a protective role in maintaining pregnancy. Biol Reprod 2004;70(4):1018–23. Available from: https://doi.org/10.1095/biolreprod.103.022640.

[76] Joachim RA, Handjiski B, Blois SM, Hagen E, Paus R, Arck PC. Stress-induced neurogenic inflammation in murine skin skews dendritic cells towards maturation and migration: key role of intercellular adhesion molecule-1/leukocyte function-associated antigen interactions. Am J Pathol 2008;173(5):1379–88. Available from: https://doi.org/10.2353/ajpath.2008.080105.

[77] Zhang W, Wang L, Zhao Y, Kang J. Changes in cytokine (IL-8, IL-6 and TNF-alpha) levels in the amniotic fluid and maternal serum in patients with premature rupture of the membranes. Zhonghua Yi Xue Za Zhi (Taipei) 2000;63(4):311–15.

[78] Arck PC, Merali FS, Manuel J, Chaouat G, Clark DA. Stress-triggered abortion: inhibition of protective suppression and promotion of tumor necrosis factor-alpha (TNF-alpha) release as a mechanism triggering resorptions in mice. Am J Reprod Immunol 1995;33(1):74–80. Available from: https://doi.org/10.1111/j.1600-0897.1995.tb01141.x.

[79] Sasaki Y, Sakai M, Miyazaki S, Higuma S, Shiozaki A, Saito S. Decidual and peripheral blood CD4 + CD25 + regulatory T cells in early pregnancy subjects and spontaneous abortion cases. Mol Hum Reprod 2004;10(5):347–53. Available from: https://doi.org/10.1093/molehr/gah044.

[80] Winger EE, Reed JL. Low circulating CD4(+) CD25(+) Foxp3(+) T regulatory cell levels predict miscarriage risk in newly pregnant women with a history of failure. Am J Reprod Immunol 2011;66(4):320–8. Available from: https://doi.org/10.1111/j.1600-0897.2011.00992.x.

[81] Ronaldson A, Gazali AM, Zalli A, et al. Increased percentages of regulatory T cells are associated with inflammatory and neuroendocrine responses to acute psychological stress and poorer health status in older men and women. Psychopharmacol (Berl) 2016;233(9):1661–8. Available from: https://doi.org/10.1007/s00213-015-3876-3.

[82] Stefanski V, Raabe C, Schulte M. Pregnancy and social stress in female rats: influences on blood leukocytes and corticosterone concentrations. J Neuroimmunol 2005;162(1-2):81–8. Available from: https://doi.org/10.1016/j.jneuroim.2005.01.011.

[83] Kruse A, Martens N, Fernekorn U, Hallmann R, Butcher EC. Alterations in the expression of homing-associated molecules at the maternal/fetal interface during the course of pregnancy. Biol Reprod 2002;66(2):333–45. Available from: https://doi.org/10.1095/biolreprod66.2.333.

[84] Blois S, Tometten M, Kandil J, et al. Intercellular adhesion molecule-1/LFA-1 cross talk is a proximate mediator capable of disrupting immune integration and tolerance mechanism at the feto-maternal interface in murine pregnancies. J Immunol 2005;174(4):1820–9. Available from: https://doi.org/10.4049/jimmunol.174.4.1820.

[85] Prados MB, Solano ME, Friebe A, Blois S, Arck P, Miranda S. Stress increases VCAM-1 expression at the fetomaternal interface in an abortion-prone mouse model. J Reprod Immunol 2011;89(2):207–11. Available from: https://doi.org/10.1016/j.jri.2011.01.021.
[86] Zenclussen AC, Fest S, Sehmsdorf US, Hagen E, Klapp BF, Arck PC. Upregulation of decidual P-selectin expression is associated with an increased number of Th1 cell populations in patients suffering from spontaneous abortions. Cell Immunol 2001;213(2):94–103. Available from: https://doi.org/10.1006/cimm.2001.1877.
[87] Auvinen K, Jalkanen S, Salmi M. Expression and function of endothelial selectins during human development. Immunology 2014;143(3):406–15. Available from: https://doi.org/10.1111/imm.12318.
[88] Nast I, Bolten M, Meinlschmidt G, Hellhammer DH. How to measure prenatal stress? A systematic review of psychometric instruments to assess psychosocial stress during pregnancy. Paediatr Perinat Epidemiol 2013;27(4):313–22. Available from: https://doi.org/10.1111/ppe.12051.
[89] Lobel M, Dunkel-Schetter C, Scrimshaw SC. Prenatal maternal stress and prematurity: a prospective study of socioeconomically disadvantaged women. Health Psychol 1992;11(1):32–40. Available from: https://doi.org/10.1037//0278-6133.11.1.32.
[90] Cohen S, Kamarck T, Mermelstein R. A global measure of perceived stress. J Health Soc Behav 1983;24(4):385–96.
[91] Elsenbruch S, Benson S, Rucke M, et al. Social support during pregnancy: effects on maternal depressive symptoms, smoking and pregnancy outcome. Hum Reprod 2007;22(3):869–77. Available from: https://doi.org/10.1093/humrep/del432.
[92] Narendran S, Nagarathna R, Narendran V, Gunasheela S, Nagendra HR. Efficacy of yoga on pregnancy outcome. J Altern Complement Med 2005;11(2):237–44. Available from: https://doi.org/10.1089/acm.2005.11.237.
[93] Khianman B, Pattanittum P, Thinkhamrop J, Lumbiganon P. Relaxation therapy for preventing and treating preterm labour. Cochrane Database Syst Rev 2012;8:CD007426. Available from: https://doi.org/10.1002/14651858.CD007426.pub2.
[94] Schander JA, Aisemberg J, Correa F, et al. The enrichment of maternal environment prevents pre-term birth in a mice model. Reproduction 2020;159(4):479–92. Available from: https://doi.org/10.1530/REP-19-0572.
[95] Hegaard HK, Hedegaard M, Damm P, Ottesen B, Petersson K, Henriksen TB. Leisure time physical activity is associated with a reduced risk of preterm delivery. Am J Obstet Gynecol 2008;198(2):180. Available from: https://doi.org/10.1016/j.ajog.2007.08.038 e1-5.
[96] Stray-Pedersen B, Stray-Pedersen S. Etiologic factors and subsequent reproductive performance in 195 couples with a prior history of habitual abortion. Am J Obstet Gynecol 1984;148(2):140–6. Available from: https://doi.org/10.1016/s0002-9378(84)80164-7.
[97] Goncalves DR, Braga A, Braga J, Marinho A. Recurrent pregnancy loss and vitamin D: a review of the literature. Am J Reprod Immunol 2018;80(5):e13022. Available from: https://doi.org/10.1111/aji.13022.
[98] Haas DM, Hathaway TJ, Ramsey PS. Progestogen for preventing miscarriage in women with recurrent miscarriage of unclear etiology. Cochrane Database Syst Rev 2019;2019(11). Available from: https://doi.org/10.1002/14651858.CD003511.pub5.
[99] Schumacher A, Heinze K, Witte J, et al. Human chorionic gonadotropin as a central regulator of pregnancy immune tolerance. J Immunol 2013;190(6):2650–8. Available from: https://doi.org/10.4049/jimmunol.1202698.
[100] Nepomnaschy PA, Welch K, McConnell D, Strassmann BI, England BG. Stress and female reproductive function: a study of daily variations in cortisol, gonadotrophins, and gonadal steroids in a rural Mayan population. Am J Hum Biol 2004;16(5):523–32. Available from: https://doi.org/10.1002/ajhb.20057.
[101] Qu F, Wu Y, Zhu YH, et al. The association between psychological stress and miscarriage: a systematic review and meta-analysis. Sci Rep 2017;7(1):1731. Available from: https://doi.org/10.1038/s41598-017-01792-3.
[102] O'Hare T, Creed F. Life events and miscarriage. Br J Psychiatry 1995;167(6):799–805. Available from: https://doi.org/10.1192/bjp.167.6.799.
[103] Knackstedt MK, Hamelmann E, Arck PC. Mothers in stress: consequences for the offspring. Am J Reprod Immunol 2005;54(2):63–9. Available from: https://doi.org/10.1111/j.1600-0897.2005.00288.x.
[104] Hutter S, Hepp P, Hofmann S, et al. Glucocorticoid receptors alpha and beta are modulated sex specifically in human placentas of intrauterine growth restriction (IUGR). Arch Gynecol Obstet 2019;300(2):323–35. Available from: https://doi.org/10.1007/s00404-019-05189-7.
[105] Jia N, Yang K, Sun Q, et al. Prenatal stress causes dendritic atrophy of pyramidal neurons in hippocampal CA3 region by glutamate in offspring rats. Dev Neurobiol 2010;70(2):114–25. Available from: https://doi.org/10.1002/dneu.20766.

[106] Chagas LA, Batista TH, Ribeiro A, et al. Anxiety-like behavior and neuroendocrine changes in offspring resulting from gestational post-traumatic stress disorder. Behav Brain Res 2021;399:113026. Available from: https://doi.org/10.1016/j.bbr.2020.113026.

[107] Bailey MT, Lubach GR, Coe CL. Prenatal stress alters bacterial colonization of the gut in infant monkeys. J Pediatr Gastroenterol Nutr 2004;38(4):414–21. Available from: https://doi.org/10.1097/00005176-200404000-00009.

[108] Coe CL, Kramer M, Kirschbaum C, Netter P, Fuchs E. Prenatal stress diminishes the cytokine response of leukocytes to endotoxin stimulation in juvenile rhesus monkeys. J Clin Endocrinol Metab 2002;87(2):675–81. Available from: https://doi.org/10.1210/jcem.87.2.8233.

[109] Yabuhara A, Macaubas C, Prescott SL, et al. TH2-polarized immunological memory to inhalant allergens in atopics is established during infancy and early childhood. Clin Exp Allergy 1997;27(11):1261–9.

[110] von Hertzen LC. Maternal stress and T-cell differentiation of the developing immune system: possible implications for the development of asthma and atopy. J Allergy Clin Immunol 2002;109(6):923–8. Available from: https://doi.org/10.1067/mai.2002.124776.

[111] Coe CL, Crispen HR. Social stress in pregnant squirrel monkeys (Saimiri boliviensis peruviensis) differentially affects placental transfer of maternal antibody to male and female infants. Health Psychol 2000;19(6):554–9.

[112] Carmichael SL, Shaw GM. Maternal life event stress and congenital anomalies. Epidemiology 2000;11(1):30–5. Available from: https://doi.org/10.1097/00001648-200001000-00008.

CHAPTER

8

Antiphospholipid syndrome and recurrent pregnancy losses

Erra Roberta[1], *Trespidi Laura*[1], *Ossola Wally*[1] *and Meroni Pier Luigi*[2]

[1]Department of Obstetrics and Gynaecology, Fondazione Ca Granda, Ospedale Maggiore Policlinico, Milan, Italy [2]Istituto Auxologico Italiano, IRCCS, Immunorheumatology Research Laboratory, Milan, Italy

1 Introduction

1.1 Definition

Antiphospholipid syndrome (APS) is a systemic autoimmune disease characterized either by recurrent pregnancy morbidity or vascular thrombosis associated with the persistent presence of antibodies reacting with negatively charged phospholipid-binding proteins (aPL). The classification of APS can be formally made when the patient fulfills at least one clinical and one laboratory criterion according to the Sydney-Sapporo criteria (Table 1) [1]. The process of revision for the classification criteria is still a matter of discussion and regular updates [2].

The clinical classification criteria include the two major manifestations: venous and/or arterial thrombosis and pregnancy morbidity. In particular, three patterns of pregnancy morbidity are identified:

1. Greater than or equal to three unexplained consecutive preembryonic or embryonic miscarriages before 10 weeks of gestation, with maternal and paternal factors (such as anatomical, hormonal or chromosomal abnormalities) excluded;
2. Greater than or equal to one unexplained death of a morphologically normal fetus beyond 10 weeks of gestation;
3. Greater than or equal to one premature delivery of a morphologically normal fetus <34 weeks of gestation because of severe preeclampsia, eclampsia, or recognized features of placental insufficiency [1,3,4].

Immunology of Recurrent Pregnancy Loss and Implantation Failure
DOI: https://doi.org/10.1016/B978-0-323-90805-4.00008-0

TABLE 1 Revised Sydney Sapporo criteria for the antiphospholipid syndrome [1].

Clinical criteria	
Vascular thrombosis	One or more episodes of arterial, venous, or small vessel thrombosis in any tissue or organ.
Pregnancy morbidity	≥3 unexplained consecutive miscarriages before 10 weeks of gestation, with maternal and paternal factors (such as anatomical, hormonal or chromosomal abnormalities) excluded
	≥1 unexplained death of a morphologically normal fetus beyond 10 weeks of gestation.
	≥1 premature delivery of a morphologically normal fetus <34 weeks of gestation because of severe preeclampsia (PE), eclampsia, or recognized features of placental insufficiency.
Laboratory criteria[a]	
Lupus anticoagulant present in plasma	
Anticardiolipin antibodies (IgG or IgM)	
Anti-B2-glycoprotein I antibodies (IgG or IgM)	

[a] *Autoantibodies are relevant only when at medium-high titers and have to be confirmed on two or more occasions at least 12 weeks apart. Classification of APS should be avoided if the results of positive aPL assays are obtained less than 12 weeks or more than 5 years from the clinical manifestations.*

The aPL can be detected by three formal laboratory classification criteria: anticardiolipin (aCL), anti-β2-glycoprotein I (anti-β2GPI), and the lupus anticoagulant (LA). Their testing must be performed twice at least 12 weeks after the first detection in order to rule out transient positivities (see also paragraph 2.1).

The antiphospholipid syndrome can manifest as isolated (primary APS) or in association with another autoimmune disease, most frequently full-blown systemic lupus erythematosus (SLE) or SLE-like disorders. Frequently APS displays both vascular and obstetric events, but pure vascular or obstetric variants have been described [5].

1.2 Prevalence

There are a few studies on the prevalence of APS in the general population, and the inclusion criteria in the reports are quite heterogeneous, making the final conclusions on the real prevalence of the syndrome questionable. A large population-based study recently reported that APS occurred in approximately 2 persons per 100,000 population per year with an estimated prevalence of 50 per 100,000 population [6]. The same study did not specify the prevalence of isolated obstetric variants. In particular, APS affects women of childbearing age, and aPL are the most common cause of adverse pregnancy outcomes (APOs) [3]. The prevalence of APO was 27% and 19% in the Euro Phospholipid Project (188 pregnancies) and in the PROMISSE study (144 pregnancies), respectively [7,8]. However, The EURAPS registry reported 65.1% of APO in 5229 aPL positive pregnancies by using less stringent inclusion criteria [9]. In fact, the EURAPS registry included both early and late miscarriages, despite the debate whether or not early miscarriages should be excluded because of their less strong association with APS [4]. Another review of the literature reported 9% of pregnancy losses, 8% of severe preeclampsia, 4% of preeclampsia, 16.5% of eclampsia, and 3% of hemolysis, elevated liver enzymes, and low platelet count (HELLP) syndrome occur in women with positive aPL [10]. A recent report by the 16th Antiphospholipid Meeting Task Force on obstetric APS stated that fewer than 5% of women with recurrent miscarriages and no other causes display aPL results fulfilling the international consensus criteria [4]. On the other hand, unexplained fetal death and early delivery for severe preeclampsia and/or placental insufficiency seem to be more specific features of obstetric APS. The presence of positive IgG/IgM aPL may be associated with up to fivefold odds of stillbirth [4].

1.3 Underlying immunopathology

While thrombotic events are closely associated with vascular APS manifestations, the role of coagulation in the pathogenesis of obstetric APS is apparently much less important. Thrombotic lesions can be detected in placental tissues from APS patients but are not characteristic since they can be found in conditions unrelated to aPL. On the other hand, vascular remodeling of uterine vessels and/or decidual hypercoagulability in the absence of thrombotic manifestation have been reported in some APS [11]. There is growing evidence that a variety of pathogenic mechanisms not involving coagulation are crucial for the defective placentation responsible for APO in APS [5,11]. β2-glycoprotein-I (β2GPI) is physiologically present on the decidual endothelium and villous and extravillous trophoblasts and can be available for maternal aPL. β2GPI-dependent aPL reacts with β2GPI on trophoblasts affecting their proliferation and differentiation and inducing apoptosis. The same antibodies interfere with angiogenesis and spiral artery development and induce local inflammatory responses when bound to decidual cells and extravillous trophoblasts (Fig. 1) [11,12].

Antiphospholipid antibodies can activate complement, which in turn may interact with the coagulation system, favoring fibrin deposition and placenta vessel thrombosis. The deposition of complement components has been reported in the human placenta from APS women. Additionally, complement deficient mice (or mice treated with complement blocking molecules) are protected from fetal loss induced by passively infused human IgG aPL [13,14]. On the other hand, complement activation contributes to trigger a placental inflammatory response, as shown in pregnant naïve mice passively infused with large amounts of human IgG aPL [15]. However, an inflammatory signature in the human placental tissue was not reported in all studies [11,14]. In addition, the passive transfer of small amounts of IgG aPL still induced abortion in another model of aPL-mediated fetal loss without manifesting inflammation [16]. In conclusion, while the role of complement activation in obstetric APS is widely accepted, its exact role in fueling placental inflammation or in favoring coagulation is still a matter of research. Finally, it is unclear whether these pathogenic pathways occur throughout the pregnancy or are involved at different times. For example, an inflammatory response that occurred at the beginning of pregnancy and then resolved could explain its importance in placenta damage and the lack of histological inflammatory signs in term placentas at the same time. The effect of aPL on trophoblast proliferation and differentiation may explain early miscarriages. On the other

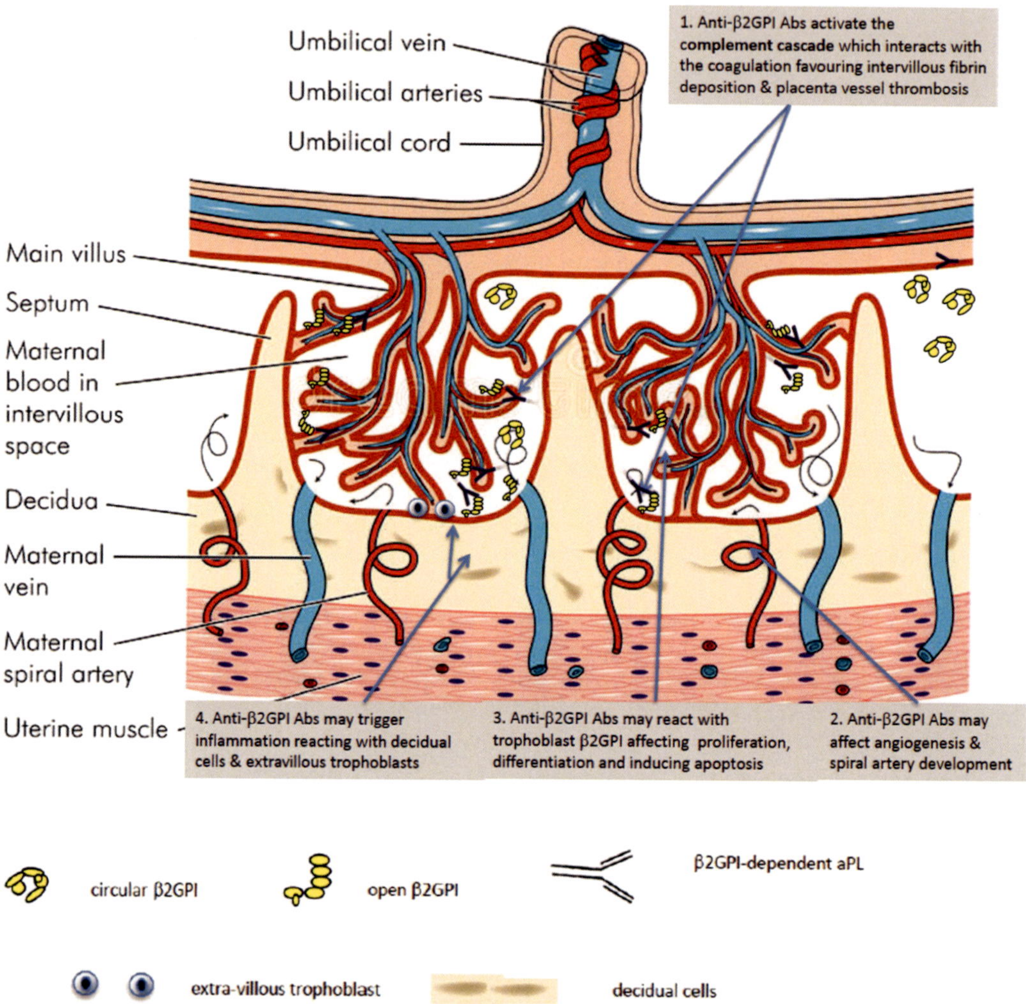

FIGURE 1 **Pathogenic mechanisms mediated by β2GPI-dependent aPL on placenta.** β2GPI-dependent aPL do not react with circulating β2GPI (closed form) but only with the open molecule present on the cell membranes. β2GPI-dependent aPL react mainly through the immunodominant D1 epitope. The open form of β2GPI is physiologically present on decidual endothelium and on villous and extravillous trophoblast and can be bound by β2GPI-dependent aPL. The antibody binding can be responsible for (1) complement activation, (2) antiangiogenic effect, (3) trophoblast inhibition and apoptosis, (4) inflammatory response on decidual and extravillous trophoblast cells.

hand, late obstetrical manifestations (e.g., preeclampsia, intrauterine growth restriction, and intrauterine fetal death) can be related to the failure of extravillous trophoblasts in remodeling spiral arteries, eventually resulting in hypoxic injury, inadequate delivery of nutrients to the fetus, and high-velocity and pressure placental blood flow.

2 Evaluations

2.1 Diagnosis

As stated in the Introduction section, the classification of APS requires the presence of at least one clinical and one laboratory criterion (Table 1). The formal classification laboratory tests for detecting antibodies against anionic PL-binding proteins are LA, β2GPI-dependent aCL, and anti-β2GPI assays [1]. As for other systemic autoimmune diseases, we do not have specific "diagnostic" criteria, and APS diagnosis is currently made using the classification criteria.

The term "lupus anticoagulant" refers to a panel of functional assays detecting a heterogeneous group of immunoglobulins that react with different PL-binding proteins behaving as acquired in-vitro inhibitors of the coagulation. There is evidence that antibodies against β2GPI and prothrombin (PT) are the most important ones, displaying the highest diagnostic specificity. LA is an intriguing paradox since it causes a PL-dependent prolongation of the clotting time but is associated with an increased risk of thrombosis and pregnancy morbidity. LA detection is based on a three-step laboratory procedure, including screening, mixing, and confirmation step. Two test systems, the activated partial thromboplastin time and the dilute Russell's viper venom time, are currently used [17,18]. Results of LA are expressed as positive or negative since there is no evidence whether the strength of the LA effect (the degree of prolongation of the clotting tests) can offer better information on the risk stratification for the clinical manifestations. So far, LA is regarded positive if the three steps, within at least one test system, display a normalized ratio above the local cut-off value. The interpretation of the results is challenging owing to interfering factors, such as concomitant anticoagulant drugs and high plasma levels of acute-phase proteins leading to false-positive results [17,18]. The International Society of Thrombosis and Hemostasis has recently provided updated guidelines for LA detection and interpretation [17]. Anti-β2GPI antibodies have been shown to prolong the PL-dependent coagulation tests, being responsible in part for the LA phenomenon [17–19]. On the other hand, antibodies against prothrombin (aPT), and in particular those reacting with the phosphatidylserine (PS)-PT complex (aPS/PT), have also been shown to mediate the LA phenomenon [17–19]. Finally, "isolated" LA, without any anti-β2GPI or aPS/PT antibodies, have been described. In these samples, the coagulation inhibitors (antibodies?) are still a matter of research [17,19,20]. Isolated positivity of LA, in the absence of aCL and anti-β2GPI, is regarded as a lower risk factor [21,22].

aPL does not recognize anionic PL alone but a complex between PL and PL-binding proteins. β2GPI was identified as the most relevant cofactor for aPL [19]. β2GPI-dependent antibodies bind their antigen: (1) when complexed with CL, for example, in the aCL assay in which β2GPI is added to CL-coated plates or (2) when the plates are directly coated with β2GPI. It has been suggested that once bound to CL, β2GPI displays conformational changes and/or increases its antigenic density, favoring antibody binding [19]. The direct coating of β2GPI to γ-irradiated polystyrene plates is thought to reproduce a comparable molecule presentation, ultimately offering the right antigen structure to the antibodies [19]. So, β2GPI-dependent antibodies are apparently responsible for positive results in the two solid-phase assays (i.e., aCL and anti-β2GPI assays) that are the formal laboratory classification criteria for APS. Since β2GPI-dependent aPL are also associated with the LA phenomenon, this finding supports the idea that β2GPI-dependent aPL can be responsible for the positivities in all three formal laboratory classification (and diagnostic) tests for APS.

At variance of LA, aCL, and anti-β2GPI assays are semiquantitative solid-phase analyses and are reported as low, medium, and high positive in comparison with a normal cutoff [19]. The higher the titer of the antibodies, the higher the diagnostic/prognostic value of the test, as discussed in Section 2.3. Transient aPL can be detected during infectious disorders of different origins and are not usually associated with the APS clinical manifestations [19]. Consistently, only aPL positive tests, which are persistently positive for at least 12 weeks apart from the first detection, are kept into consideration as classification and diagnostic tools (Table 1).

Although not yet formally included in the laboratory classification tests, there are additional assays that look promising in the diagnosis and risk stratification of both vascular and obstetric APS. Among these nonclassification criteria, anti-PT antibodies, particularly those detectable by PS/PT coated plates, seem to display a useful diagnostic and prognostic value [19,23]. However, their inclusion in the laboratory classification criteria for APS is not yet formalized. The anti-PS/PT test can help in detecting aPL potentially responsible for the LA phenomenon. So, anti-PS/PT and anti-β2GPI assays may represent a surrogate tool when variables interfering with the LA determination are present (e.g., when the patient is taking anticoagulant drugs or displaying high circulating levels of CRP) [19].

β2GPI-dependent aPL are directed against different epitopes of the molecule, but there is growing evidence that antibodies reacting with the conformational immunodominant region within domain 1 (DI) (spanning the amino acid sequence Arg39-Arg43) mediate pathogenic mechanisms in experimental models [24–26]. Moreover, a recent metanalysis showed that the presence of IgG anti-D1 β2GPI displays a higher specificity and predictive value than IgG against the whole molecule [27]. Accordingly, anti-D1 antibody detection has been suggested as a new laboratory criterion for APS. However, some patients result positive for antibodies against the whole β2GPI molecule but negative in the anti-D1 assays, suggesting that anti-D1 cannot replace the test for antibodies against the whole molecule. Nevertheless, anti-D1 antibodies are present more frequently in full-blown APS patients with double/triple aPL positivity but are not usually detected in aPL occurring in other unrelated APS conditions. Hence, the search for anti-D1 antibodies can be used as a sort of a confirmatory test (19). There is no sound evidence that the detection of other aPL (e.g., anti-PS, antiphosphatidylethanolamine) or antibodies against other coagulation proteins (e.g., anti-Annexin V, anti-ProteinS/C) may improve the diagnostic power of the classification laboratory assays [19].

2.2 When should we look for aPL?

Testing for aPL antibodies is indicated in all women with a history of unexplained pregnancy morbidity or arterial/venous thrombosis. In the clinical practice, aPL are often tested because of other reasons, including (1) the presence of so-called noncriteria manifestations (e.g., livedo reticularis, skin ulcers, thrombocytopenia, etc.), (2) a previous diagnosis of an autoimmune disease, and (3) past pregnancy complications not fulfilling the formal classification criteria (e.g., less than three early miscarriages or infertility) [28].

2.3 Antiphospholipid antibodies as a risk factor

Antiphospholipid antibodies are widely accepted as risk factors for APS-related clinical manifestations, including obstetric complications [12,19]. The levels of aPL are correlated with the risk of clinical manifestations, being medium/high titers much more predictive than low titers. Moreover, the positivity for two or three classification laboratory tests represents a further risk. However, it should be underlined that the persistent positivity in two assays even at a low titer is associated with a significant risk for miscarriages. The clinical significance of this finding is further supported by the response to treatment (i.e., low dose aspirin and low molecular weight heparin) of all the patients, while up to 20% of patients with medium/high aPL titers have recurrences [19]. While aPL of the IgG isotype displays a more predictive value for the vascular manifestations than IgM; the same association is still debated for the obstetric manifestations [9,19,29–31].

Among the three formal classification laboratory assays, LA is widely accepted as the most predictive one, even if isolated LA positive cases (with no anti-β2GPI or anti-PS/PT antibodies) can be found not associated with both vascular and obstetric APS [21,22]. The clinical significance of isolated aCL or anti-β2GPI positive results is doubtful. It should be critically evaluated from a clinical point of view. In particular, transient positivities, such as those associated with concurrent infectious processes, should always be ruled out. On the other hand, the use of tests for detecting antibodies against β2GPI domain 1 is promising as second-line assays for confirming a high-risk antibody profile (see paragraph 2.1) [27]. The whole risk profile for the obstetric APS is supported not only by the aPL profile (e.g., titer, isotype, type of the assay) but by aPL unrelated variables such as traditional risk factors for miscarriages and the presence of an associated underlying systemic autoimmune rheumatic disease. In addition, the presence of a thrombotic event in the past history is itself a risk for a recurrence further favored by the thrombophilic profile associated with the pregnancy (Fig. 2).

Obstetric APS

Higher risk depending on the aPL profile:

aPL assay type positivity
Triple/Double positivity
LA positivity
Isotype (IgG > IgM)
Titer (Medium/High titer)

Further aPL-unrelated risk factors:

Previous obstetric history
Traditional obstetric risk factors (age, diabetes, arterial hypertension, congenital thrombophilia, dyslipidemia, BMI, smoking, sedentary life style etc.)
Underlying autoimmune diseases
Thrombotic events in the past history

FIGURE 2 **Antiphospholipid antibodies as a risk factor.** The aPL profile, isotype, titer, and aPL-unrelated factors are ranking risks for obstetric antiphospholipid syndrome.

3 Clinical practice management

3.1 Risk stratification

Pregnant women with APS display different obstetrical risks depending on both the clinical and laboratory features. Risk stratification during preconceptional counseling is therefore mandatory for a tailored therapeutic approach. As stated in paragraph 2.3, the "aPL profile" as well as the clinical features identify the ultimate risk for APOs (Fig. 2). Consistent with the role of these variables, the patients must be optimized by preconception interventions, including cessation of smoking, reduction in BMI, and treatment of some of the comorbidities known to impact pregnancy outcomes.

3.1.1 Antibody profile

Regarding the antibody profile, the 2019 EULAR recommendations for the management of APS pregnant women defined a high-risk profile as the persistent presence of LA or double (any combination of LA, aCL or anti-β2GPI antibodies) or triple (all three subtypes) aPL positivity or the presence of persistently high aPL titers [32]. More recently, further effort has been made to better detail the risk profile of aPL positive women. The EUREKA study showed that also low aPL titers confer risk for pregnancy morbidity in addition to the widely accepted risk associated with medium/high aPL titers. In particular, the persistent double positivity for LA and low titers of anti-β2GPI IgG was found to display a higher risk than low aCL and IgG anti-β2GPI, either alone or associated [33]. There is a general agreement that aPL of the IgG isotype are more predictive for the clinical manifestations than IgM antibodies [29–31]. As stated in paragraph 2.1, isolated LA without anti-β2GPI or anti-PS/PT antibodies display a low-risk profile, while the demonstration of β2GPI-dependent aPL reacting with Domain 1 may support the presence of aPL associated risk for APO [32].

3.1.2 Patients not fulfilling the classification criteria

A more complex scenario is present in the routine clinical setting, such as in the case of asymptomatic aPL positive women (the so-called "aPL carriers") and women not fulfilling the formal clinical or laboratory classification criteria [28]. Indeed, women with a risk aPL profile but without any previous pregnancy complications (e.g., asymptomatic carriers) or women with pregnancy morbidity but isolated aPL positivity and low antibody titers are frequently sent for a consultation to tertiary referral centers. The optimal management of these women is still a matter of debate since there are no ad hoc clinical studies. However, the 2019 EULAR and ACR recommendations for the management of APS pregnancies suggest that aPL carriers with triple positivity or women with severe obstetric history and low aPL risk profile can be considered at high risk for AOP and ultimately treated with LDASA alone or even in combination with heparin in selected cases [33,34].

3.2 Treatment

3.2.1 Therapeutic algorithm

Fig. 3 shows the therapeutic algorithm for obstetric APS. Despite the heterogeneity of the studies, a recent Cochrane systematic review confirmed that the combination of

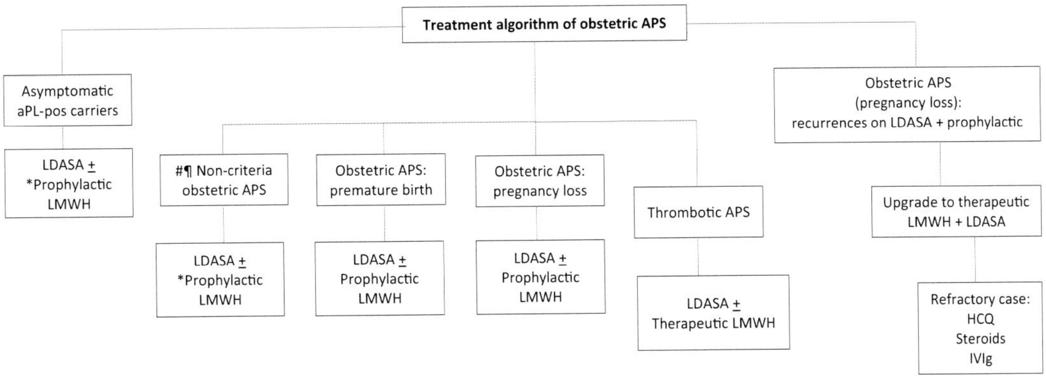

FIGURE 3 Therapeutic algorithm of obstetric APS. *The use of prophylactic LMWH in addition to LDASA is tailored to benefits/risks of the individual woman. ¶ Noncriteria APS: women that do not fulfill the clinical and/or the laboratory classification criteria. *aPL*, Antiphospholipid antibodies; *APS*, antiphospholipid syndrome; *LDASA*, low-dose aspirin; LMWH, low-molecular weight heparin; *HCQ*, hydroxychloroquine; *IVIg*, intravenous immunoglobulins.

heparin [unfractionated (UFH) or low molecular weight heparin (LMWH)] plus low dose aspirin (LDA) might increase the live birth rate in women with persistent aPL when compared with aspirin treatment alone [35]. There is evidence that the association is effective in protecting aPL positive pregnant women from recurrences of pregnancy complications regardless of the type of manifestations [3]. Ideally, LDA should be started before conception, while heparin should be prescribed as soon as the fetal heartbeat is detected. The dosage of ASA varies depending on the local rules, ranging from 75 to 150 mg/day, but there are no comparative studies to identify the best dosage [33,34]. Prophylactic heparinization is suggested for APS women with no previous thrombotic events, and therapeutic dosage is prescribed for patients with thrombosis in the past [33,34]. Data from the 2020 Cochrane review did not report significant adverse events (bleeding, heparin-induced thrombocytopenia, allergic reactions) in the mothers or the child (congenital malformations, neonatal bleeding) for the treatment with LDA and/or heparin [35].

Treatment tailoring is recommended for noncriteria APS pregnant women. For example, LDA only is suggested for asymptomatic aPL-positive carriers or women not fulfilling the classification criteria for obstetric APS; however, some authors are prescribing heparin as well, if the risk profile is high (e.g., triple positive and/or medium/high aPL titers) [33,34]. Moreover, the EUREKA monocentric study showed that LDA plus heparin improves the pregnancy outcome also in women with persistent low-titer of aPL [36]. The fact that nonthrombotic mechanisms play a major role in aPL-related defective placentation suggests that heparin and aspirin display their protective effect through mechanisms alternative to the interference in the coagulation cascade [5].

3.2.2 *Refractory cases*

Despite conventional treatment, up to 20%–30% of APS pregnant patients display a poor obstetric outcome [3]. Additional therapeutic approaches have been reported in cases refractory to the standard therapy with LDA and heparin (Fig. 3). Consistent with the evidence that thrombosis plays a marginal role in APO and placental inflammation may be more important (see paragraph 1.3), early (first trimester) low-dose prednisolone added to LDA and heparin was found beneficial in terms of live birth rate with scarce side-effects [37]. This finding was also supported by the beneficial effect of adding corticosteroids in a multicenter European registry on refractory obstetric APS [38]. Besides its antiplatelet activity and protective effect on the annexin-V placenta shield, hydroxychloroquine (HCQ) possesses some antiinflammatory properties at the trophoblast level and the ability to inhibit complement activation [39–45]. Together with the good safety profile during pregnancy, these pharmacological activities make HCQ an appealing add-on option for treating refractory cases. Although ad hoc randomized controlled trials are still ongoing, EULAR and ACR recommended its use in women nonresponsive to LDA and heparin [33,34].

Upgrading heparin (in particular LMWH) from prophylactic to therapeutic dosage in refractory cases has been suggested as rescue therapy, even though no ad hoc clinical trials have been carried out [3]. The use of intravenous immunoglobulins with or without previous apheresis in selected cases has also been reported in line with the idea that the infused Ig may counteract the pathogenic effect of the aPL [46–48]. Finally, two additional therapeutic tools are under active investigation: the use of TNF alpha inhibitors for damping placenta inflammation and the pharmacological effect of statins (in particular pravastatin) on both placental inflammation and vascular damage [3,4,49–51].

4 Monitoring

Pregnant APS patients should be considered at high risk and strictly monitored at monthly intervals or even more frequently if required. A multidisciplinary team of obstetricians and rheumatologists is suggested; in particular, if the mother is suffering from an associated systemic autoimmune disease, the disease activity and the related immunosuppressive therapy should be critically evaluated [33,34]. Given the central role of the complement cascade in obstetric APS pathogenesis, it is not surprising that serum complement assessment has been proposed as a supplementary predictor of obstetric outcome. Lower serum C3 and C4 levels were reported in some APS pregnant women, although an association with a negative pregnancy outcome is still a matter of research [14,52]. The huge physiological variation of complement levels and their increase even during normal pregnancy may make the measurement of C3/C4 components not sensitive enough. Consistently, the detection of soluble complement activation products looks like a more sensitive and promising biomarker for the negative outcome [53,54]. An imbalance between pro- and antiangiogenic factors like placental growth factor (PlGF) and soluble fms-like tyrosine kinase-1 (sflt-1) contributes to late pregnancy placenta-related complications in the general population. So far, very little evidence on their application in the APS population has been described, and further studies are needed to support their potential prognostic role [55,56]. Antepartum fetal monitoring is routinely performed in all pregnant women at 11–13 + 6 gestational weeks (gw) to calculate the gestational age and for prenatal diagnosis and at 19–21 gw for assessing fetal morphology. Additional fetal surveillance is carried out in APS women every month by US examination during the third trimester for evaluating fetal biometry, amniotic fluid volume, and Doppler flow velocimetry waveform analysis (e.g., absent end-diastolic flow in the umbilical artery) [33,34]. This approach was shown to be useful in identifying early or late intrauterine growth restriction and for better scheduling the time of delivery [57].

5 Delivery and puerperium

Time and mode of delivery should be tailored to the maternal and fetal status. In the absence of common medical or obstetric indications for early delivery, induction of labor or cesarean section is scheduled at 39 weeks gestation to control the cessation time for antithrombotic drugs and prevent the postpartum thrombotic risks. There are no randomized trials on the best delivery mode in APS. However, physicians should consider the significantly higher thromboembolic risk associated with cesarean section in comparison with vaginal delivery. The latest dose of heparin is usually administered 12 and 24 hours before the planned cesarean section or labor induction, respectively, for the prophylactic or therapeutic dosage [58]. Vaginal birth or epidural anesthesia can be safely performed in APS women 12 and 24 hours after the last dose of prophylactic or therapeutic heparin, respectively [58]. Even though ASA does not increase the risk of spinal hematoma, anesthesiologists might prefer stopping LDASA 3–7 days before performing a spinal tap [3,58]. Because of the high thromboembolic risk during the puerperium, both EULAR and ACR guidelines recommend continuing the ASA and heparin for 4–6 weeks after delivery [33,34]. Women with APS must be encouraged to breastfeed if they desire since both ASA and heparin are compatible with breastfeeding [33,34,59].

6 Fetal and neonatal outcome

Maternal IgG aPL can cross the placenta being detectable in newborns until 12 months after delivery [60]. Nevertheless, neonatal thrombosis was reported in very few cases despite the neonatal thrombophilic state [60–62]. It is speculated that the massive presence of β2GPI in the placental tissues may be responsible for absorbing the pathogenic aPL, eventually protecting the baby. In addition, aspirin may cross the placenta and display its protecting antiplatelet activity.

7 aPL and infertility

The theoretical role of aPL on infertility might be due to the negative interference with oocyte development, preimplantation embryo morphology, uterine receptivity, and proper decidualization. Besides some experimental models, a causative link between aPL and infertility is poorly supported [63]. For example, a recent systematic review did not find a relation between assisted reproductive technologies (ART) failure and the presence of aPL [64]. Consistent with this finding, aPL testing should not be routinely included in the diagnostic panel of infertile women. On the other hand, the thrombotic risk of ovarian stimulation regimens is a concern for ART in aPL-positive and APS patients. Currently, ART in APS and aPL-positive patients is regarded as a generally safe procedure, and recommendations have been published to avoid the thrombotic risk by prophylactic heparinization [33,34,59].

8 Summary and recommendations

1. *Definition of aPL.* aPL are auto-antibodies against PL-binding proteins. aPL react with two major auto-antigens: β2GPI & PT.
2. *Detection assays/classification laboratory criteria.* Three assays are used as classification laboratory criteria: (1) LA is a qualitative functional PL-dependent coagulation assay mediated mainly by anti-β2GPI and/or anti-PT antibodies. (2) aCL and anti-β2GPI tests are semiquantitative solid-phase assays detecting mainly β2GPI-dependent aPL.
3. *Epidemiology.* APS is among the most frequent acquired-risk factors for a treatable cause of recurrent pregnancy loss.
4. *aPL diagnostic value.* The persistent presence of the positivity for at least one of the three classification lab criteria is required for APS classification. The same criteria are currently used as diagnostic tools in association with the clinical manifestations. Although the aPL positivity at medium/high titer in the solid-phase assays is widely accepted as the most predictive and diagnostic, low aPL titers may be associated with obstetric complications as well.
5. *aPL as pathogenic antibodies.* There is good evidence from experimental models and epidemiological studies that β2GPI-dependent aPL are pathogenic for aPL-associated pregnancy complications.
6. *aPL as a risk factor.* Double/triple positivity for the aPL tests and medium/high aPL titers in the solid-phase assays display the highest risk for obstetric APS-related manifestations.
7. *When looking for aPL?* aPL testing should be required in all women with a previous unexplained history of pregnancy morbidity and in women with unexplained previous arterial or venous thrombosis or those suffering from systemic autoimmune diseases and planning a pregnancy.
8. *How to test aPL?*. All three formal classification laboratory assays should be performed, and the tests should be repeated 12 weeks after the first detection to rule out transient positivities.
9. *Standard therapy for obstetric APS.* LDASA started during conception, and prophylactic heparin (most frequently LMWH) when pregnancy is confirmed are the standard therapeutic approach. Therapeutic heparin is used in pregnant women on oral anticoagulation because of a previous thrombotic event. Upgrading to therapeutic heparin even in the absence of previous thrombotic history or on top addition of corticosteroids or IVIg and HCQ are used in cases refractory to the standard therapy. LDASA and heparin should be continued for 4–6 weeks after delivery.
10. *APS pregnancy monitoring.* A strict multidisciplinary clinical, laboratory, and ultrasonography monitoring, including fetal biometry assessment and uterine/umbilical artery velocimetry, is suggested.

References

[1] Miyakis S, Lockshin MD, Atsumi T, Branch DW, Brey RL, Cervera R, et al. International consensus statement on an update of the classification criteria for definite antiphospholipid syndrome (APS). J Thromb Haemost 2006;4:295–306.

[2] Barbhaiya M, Zuily S, Ahmadzadeh Y, Amigo MC, Avcin T, Bertolaccini ML. New APS classification criteria collaborators. development of a new international antiphospholipid syndrome classification criteria phase I/II report: generation and reduction of candidate criteria. Arthritis Care Res (Hoboken) 2021;73(10):1490–501. Available from: https://doi.org/10.1002/acr.24520.

[3] Beltagy A, Trespidi L, Gerosa M, Ossola MW, Meroni PL, Chighizola CB. Anti-phospholipid antibodies and reproductive failures. Am J Reprod Immunol 2021;85(4):e13258. Available from: https://doi.org/10.1111/aji.13258.

[4] de Jesús GR, Benson AE, Chighizola B, Sciascia S, Branch DW. 16th International congress on antiphospholipid antibodies task force report on obstetric antiphospholipid syndrome. Lupus 2020;29(12):1601–15. Available from: https://doi.org/10.1177/0961203320954520.

[5] Meroni PL, Borghi MO, Grossi C, Chighizola CB, Durigutto P, Tedesco F. Obstetric and vascular antiphospholipid syndrome: same antibodies but different diseases? Nat Rev Rheumatol 2018;14(7):433–40. Available from: https://doi.org/10.1038/s41584-018-0032-6.

[6] Duarte-García A, Pham MM, Crowson CS, Amin S, Moder KG, Pruthi RK, et al. The epidemiology of antiphospholipid syndrome: a population-based study. Arthritis Rheumatol 2019;71(9):1545–52. Available from: https://doi.org/10.1002/art.40901.

[7] Cervera R, Serrano R, Pons-Estel GJ, Ceberio-Hualde L, Shoenfeld Y, de Ramón E, et al. Morbidity and mortality in the antiphospholipid syndrome during a 10-year period: a multicentre prospective study of 1000 patients. Ann Rheum Dis 2015;74(6):1011–18. Available from: https://doi.org/10.1136/annrheumdis-2013-204838.

[8] Yelnik CM, Laskin CA, Porter TF, Branch DW, Buyon JP, Guerra MM, et al. Lupus anticoagulant is the main predictor of adverse pregnancy outcomes in aPL-positive patients: validation of PROMISSE study results. Lupus Sci Med 2016;3:e000131. Available from: https://doi.org/10.1136/lupus-2015-000131.

[9] Alijotas-Reig J, Esteve-Valverde E, Ferrer-Oliveras R, Sáez-Comet L, Lefkou E, Mekinian A, et al. The European registry on obstetric antiphospholipid syndrome (EUROAPS): a survey of 1000 consecutive cases. Autoimmun Rev 2019;18(4):406–14. Available from: https://doi.org/10.1016/j.autrev.2018.12.006.

[10] Andreoli L, Chighizola CB, Banzato A, Pons-Estel GJ, Ramires de Jesus G, Erkan D. Estimated frequency of antiphospholipid antibodies in patients with pregnancy morbidity, stroke, myocardial infarction, and deep vein thrombosis: a critical review of the literature. Arthritis Care Res 2013;65:1869e1873.

[11] Viall CA, Chamley LW. Histopathology in the placentae of women with antiphospholipid antibodies: a systematic review of the literature. Autoimmun Rev 2015;14(5):446–71. Available from: https://doi.org/10.1016/j.autrev.2015.01.008.

[12] Meroni PL, Borghi MO, Raschi E, Tedesco F. Pathogenesis of antiphospholipid syndrome: understanding the antibodies. Nat Rev Rheumatol 2011;7(6):330–9. Available from: https://doi.org/10.1038/nrrheum.2011.52.

[13] Shamonki JM, Salmon JE, Hyjek E, Baergen RN. Excessive complement activation is associated with placental injury in patients with antiphospholipid antibodies. Am J Obst Gynecol 2007;196(2). Available from: https://doi.org/10.1016/j.ajog.2006 167.e1–5.

[14] Tedesco F, Borghi MO, Gerosa M, Chighizola CB, Macor P, Lonati PA, et al. Pathogenic role of complement in antiphospholipid syndrome and therapeutic implications. Front Immunol 2018;9:1388. Available from: https://doi.org/10.3389/fimmu.2018.01388.

[15] Salmon JE, Girardi G. Antiphospholipid antibodies and pregnancy loss: a disorder of inflammation. J Reprod Immunol 2008;77(1):51–6. Available from: https://doi.org/10.1016/j.jri.2007.02.007.

[16] Martinez de la Torre Y, Buracchi C, Borroni EM, Dupor J, Bonecchi R, Nebuloni M, et al. Protection against inflammation- and autoantibody-caused fetal loss by the chemokine decoy receptor D6. Proc Natl Acad Sci U S A 2007;104(7):2319–24. Available from: https://doi.org/10.1073/pnas.0607514104.

[17] Devreese KMJ, de Groot G, de Laat B, Erkan D, Favaloro EJ, Mackie I, et al. Guidance from the Scientific and Standardization Committee for lupus anticoagulant/antiphospholipid antibodies of the International Society on Thrombosis and Haemostasis: Update of the guidelines for lupus anticoagulant detection and interpretation. J Thromb Haemost 2020;18(11):2828–39. Available from: https://doi.org/10.1111/jth.15047.

[18] Chighizola CB, Raschi E, Banzato A, Borghi MO, Pengo V, Meroni PL. The challenges of lupus anticoagulants. Expert Rev Hematol 2016;9(4):389–400. Available from: https://doi.org/10.1586/17474086.2016.1140034.

[19] Meroni PL, Borghi MO. Antiphospholipid antibody assays in 2021: looking for a predictive value in addition to a diagnostic one. Front Immunol 2021;12:726820. Available from: https://doi.org/10.3389/fimmu.2021.726820.

[20] Oosting JD, Derksen RH, Bobbink IW, Hackeng TM, Bouma BN, de Groot P, et al. Antiphospholipid antibodies directed against a combination of phospholipids with prothrombin, protein C, or protein S: an explanation for their pathogenic mechanism? Blood. 1993;81(10):2618–25.

[21] Yin, de Groot PG, Ninivaggi M, Devreese KMJ, de Laat B. Clinical relevance of isolated lupus anticoagulant positivity in patients with thrombotic antiphospholipid syndrome. Thromb Haemost 2021;121(9):1220–7. Available from: https://doi.org/10.1055/a-1344-4271.

[22] Pengo V, Testa S, Martinelli I, Ghirarduzzi A, Legnani C, Gresele P, et al. Incidence of a first thromboembolic event in carriers of isolated lupus anticoagulant. Thrombosis Res 2014;135(1):46–9. Available from: https://doi.org/10.1016/j.thromres.2014.10.013.

[23] Sciascia S, Sanna G, Murru V, Roccatello D, Khamashta MA, Bertolaccini ML. Anti-prothrombin (aPT) and anti-phosphatidylserine/prothrombin (aPS/PT) antibodies and the risk of thrombosis in the antiphospholipid syndrome. A systematic review. Thromb Haemost 2014;111(2):354–64. Available from: https://doi.org/10.1160/TH13-06-5410509.

[24] de Laat B, de Groot PG. Autoantibodies directed against domain I of 421 beta2-glycoprotein I. Curr Rheumatol Rep 2011;13(1):70–6. Available from: https://doi.org/10.1007/s11926-422010-0144-8.

[25] Durigutto P, Grossi C, Borghi MO, Macor P, Pregnolato F, Raschi E, et al. New insight into antiphospholipid syndrome: antibodies to β2glycoprotein I-domain 5 fail to induce thrombi in rats. Haematologica 2019;104 (4):819–26. Available from: https://doi.org/10.3324/haematol.2018.198119.

[26] Agostinis C, Durigutto P, Sblattero D, Borghi MO, Grossi C, Guida F, et al. A non-complement-fixing antibody to β2 glycoprotein I as a novel therapy for antiphospholipid syndrome. Blood 2014;123(22):3478–87. Available from: https://doi.org/10.1182/blood-2013-11-537704.

[27] Radin M, Cecchi I, Roccatello D, Meroni PL, Sciascia S. Prevalence and thrombotic risk assessment of anti-β2 glycoprotein i domain i antibodies: a systematic review. Semin Thromb Hemost 2018;44(5):466–74. Available from: https://doi.org/10.1055/s-0037-1603936.

[28] Pires da Rosa G, Bettencourt P, Rodríguez-Pintó I, Cervera R, Espinosa G. "Non-criteria" antiphospholipid syndrome: a nomenclature proposal. Autoimmun Rev 2020;19(12):102689. Available from: https://doi.org/10.1016/j.autrev.2020.102689.

[29] Chayoua W, Kelchtermans H, Gris JC, Moore GW, Musiał J, Wahl D, et al. The (non-)sense of detecting anticardiolipin and anti-b2glycoprotein I IgM antibodies in the antiphospholipid syndrome. J Thromb Haemost 2020;18(1):169–79. Available from: https://doi.org/10.1111/jth.14633.

[30] Bouvier S, Cochery-Nouvellon E, Lavigne-Lissalde G, Mercier E, Marchetti T, Balducchi JP, et al. Comparative incidence of pregnancy outcomes in treated obstetric antiphospholipid syndrome: the NOHAPS observational study. Blood 2014;123(3):404–13. Available from: https://doi.org/10.1182/blood-2013-08-522623.

[31] Kelchtermans H, Pelkmans L, de Laat B, Devreese KM. IgG/IgM antiphospholipid antibodies present in the classification criteria for the antiphospholipid syndrome: a critical review of their association with thrombosis. J Thromb Haemost 2016;14(8):1530–48. Available from: https://doi.org/10.1111/jth.13379.

[32] Chighizola CB, Pregnolato F, Andreoli L, Bodio C, Cesana L, Comerio C, et al. Beyond thrombosis: anti-β2GPI domain 1 antibodies identify late pregnancy morbidity in anti-phospholipid syndrome. J Autoimmun 2018;90:76–83. Available from: https://doi.org/10.1016/j.jaut.2018.02.002.

[33] Tektonidou MG, Andreoli L, Limper M, Amoura Z, Cervera R, Costedoat-Chalumeau N, et al. EULAR recommendations for the management of antiphospholipid syndrome in adults. Ann Rheum Dis 2019;78 (19):1296–304. Available from: https://doi.org/10.1136/annrheumdis-2019-215213.

[34] Sammaritano LR, Bermas BL, Chakravarty EE, Chambers C, Clowse MEB, Lockshin MD, et al. American College of rheumatology guideline for the management of reproductive health in rheumatic and musculoskeletal diseases. Arthritis Rheumatol2020 2020;72(4):529–56. Available from: https://doi.org/10.1002/art.41191.

[35] Hamulyák EN, Scheres LJ, Marijnen MC, Goddijn M, Middeldorp S. Aspirin or heparin or both for improving pregnancy outcomes in women with persistent antiphospholipid antibodies and recurrent pregnancy loss. Cochrane Database Syst Rev 2020;5(5):CD012852. Available from: https://doi.org/10.1002/14651858.CD012852.

[36] Pregnolato F, Gerosa M, Raimondo MG, Comerio C, Bartoli F, Lonati PA, et al. EUREKA algorithm predicts obstetric risk and response to treatment in women with different subsets of anti-phospholipid antibodies. Rheumatology (Oxf) 2021;60(3):1114–24. Available from: https://doi.org/10.1093/rheumatology/keaa203.

[37] Bramham K, Thomas M, Nelson-Piercy C, Khamashta M, Hunt BJ. First-trimester low-dose prednisolone in refractory antiphospholipid antibody–related pregnancy loss. Blood 2011;117(25):6948–51. Available from: https://doi.org/10.1182/blood-2011-02-339234.

[38] Mekinian A, Alijotas-Reig J, Carrat F, Costedoat-Chalumeau N, Ruffatti A, Lazzaroni MG, et al. Refractory obstetrical antiphospholipid syndrome: features, treatment and outcome in a European multicenter retrospective study. Autoimmun Rev 2017;16(7):730–4. Available from: https://doi.org/10.1016/j.autrev.2017.05.006.

[39] Edwards MH, Pierangeli S, Liu X, Barker JH, Anderson G, Harris EN. Hydroxychloroquine reverses thrombogenic properties of antiphospholipid antibodies in mice. Circulation 1997;96(12):4380–4. Available from: https://doi.org/10.1161/01.cir.96.12.4380.

[40] Rand JH, Wu X-X, Quinn AS, Ashton AW, Chen PP, Hathcock JJ, et al. Hydroxychloroquine protects the annexin A5 anticoagulant shield from disruption by antiphospholipid antibodies: evidence for a novel effect for an old antimalarial drug. Blood 2010;115(11):2292–9. Available from: https://doi.org/10.1182/blood-2009-04-213520.

[41] Meroni PL. Prevention & treatment of obstetrical complications in APS: is hydroxychloroquine the Holy Grail we are looking for? J Autoimmun 2016;75:1–5. Available from: https://doi.org/10.1016/j.jaut.2016.07.003.

[42] Marchetti T, Ruffatti A, Wuillemin C, de Moerloose P, Cohen M. Hydroxychloroquine restores trophoblast fusion affected by antiphospholipid antibodies. J Thromb Haemost 2014;12(6):910–20. Available from: https://doi.org/10.1111/jth.12570.

[43] Albert CR, Schlesinger WJ, Viall CA, Mulla MJ, Brosens JJ, Chamley LW, et al. Effect of hydroxychloroquine on antiphospholipid antibody-induced changes in first trimester trophoblast function. Am J Reprod Immunol 2014;71(2):154–64. Available from: https://doi.org/10.1111/aji.12184.

[44] Scott RE, Greenwood SL, Hayes DJL, Baker BC, Jones RL, Heazell AEP. Effects of hydroxychloroquine on the human placenta—findings from in vitro experimental data and a systematic review. Reprod Toxicol 2019;87:50–9. Available from: https://doi.org/10.1016/j.reprotox.2019.05.056.

[45] Bertolaccini ML, Content G, Lennen R, Sanna G, Blower PJ, Ma MT, et al. Complement inhibition by hydroxychloroquine prevents placental and fetal brain abnormalities in antiphospholipid syndrome. J Autoimmun 2016;75:30–8. Available from: https://doi.org/10.1016/j.jaut.2016.04.008.

[46] Hoxha A, Favaro M, Calligaro A, Del Ross T, Ruffatti AT, Infantolino C, et al. Upgrading therapy strategy improves pregnancy outcome in antiphospholipid syndrome: a cohort management study. Thromb Haemost 2020;120(1):36–43. Available from: https://doi.org/10.1055/s-0039-1697665.

[47] Urban ML, Bettiol A, Serena C, Comito C, Turrini I, et al. Intravenous immunoglobulin for the secondary prevention of stillbirth in obstetric antiphospholipid syndrome: a case series and systematic review of literature. Autoimmun Rev 2020;19(9):102620. Available from: https://doi.org/10.1016/j.autrev.2020.102620.

[48] Ruffatti A, Tonello, Hoxha A, Sciascia S, Cuadrado MJ, Latino JO, et al. Effect of additional treatments combined with conventional therapies in pregnant patients with high-risk antiphospholipid syndrome: a multicentre study. Thromb Haemost 2018;118(4):639–46. Available from: https://doi.org/10.1055/s-0038-1632388.

[49] Lefkou E, Mamopoulos A, Dagklis T, Vosnakis C, Rousso D, Girardi G. Pravastatin improves pregnancy outcomes in obstetric antiphospholipid syndrome refractory to antithrombotic therapy. J Clin Invest 2016;126(8):2933–40. Available from: https://doi.org/10.1172/JCI86957.

[50] Lefkou E, Varoudi K, Pombo J, Jurisic A, Jurisic Z, Contento G, et al. Triple therapy with pravastatin, low molecular weight heparin and low dose aspirin improves placental haemodynamics and pregnancy outcomes in obstetric antiphospholipid syndrome in mice and women through a nitric oxide-dependent mechanism. Biochem Pharmacol 2020;182:114217. Available from: https://doi.org/10.1016/j.bcp.2020.114217.

[51] Vahedian-Azimi A, Karimi L, Reiner Z, Makvandi S, Sahebkar A. Effects of statins on preeclampsia: a systematic review. Pregnancy Hypertension 2021;23:123–30. Available from: https://doi.org/10.1016/j.preghy.2020.11.014.

[52] Nalli C, Lini D, Andreoli L, Crisafulli F, Fredi M, Lazzaroni MG, et al. Low preconception complement levels are associated with adverse pregnancy outcomes in a multicenter study of 260 pregnancies in 197 women with antiphospholipid syndrome or carriers of antiphospholipid antibodies. Biomedicines 2021;9(6):671. Available from: https://doi.org/10.3390/biomedicines9060671.

[53] Kim MY, Guerra MM, Kaplowitz E, Laskin CA, Petri M, Branch DW, et al. Complement activation predicts adverse pregnancy outcome in patients with systemic lupus erythematosus and/or antiphospholipid antibodies. Ann Rheum Dis 2018;77(4):549–55. Available from: https://doi.org/10.1136/annrheumdis-2017-212224.

[54] Scambi C, Ugolini S, Tonello M, Bortolami O, De Franceschi L, Castagna A, et al. Complement activation in the plasma and placentas of women with different subsets of antiphospholipid syndrome. Am J Reprod Immunol 2019;82(6):e13185. Available from: https://doi.org/10.1111/aji.13185.
[55] Kim MY, Buyon JP, Guerra MM, Rana S, Zhang D, Laskin CA, et al. Angiogenic factor imbalance early in pregnancy predicts adverse outcomes in patients with lupus and antiphospholipid antibodies: results of the PROMISSE study. Am J Obstet Gynecol 2016;214(1):108.e101–14. Available from: https://doi.org/10.1016/j.ajog.2015.09.066.
[56] Cochery-Nouvellon É, Mercier É, Bouvier S, Balducchi JP, Quéré I, Perez-Martin A, et al. Obstetric antiphospholipid syndrome: early variations of angiogenic factors are associated with adverse outcomes. Haematologica 2017;102(5):835–42. Available from: https://doi.org/10.3324/haematol.2016.155184.
[57] Alfirevic Z, Stampalija T, Dowswell T. Fetal and umbilical Doppler ultrasound in high-risk pregnancies. Cochrane Database Syst Rev 2017;6(6):Cd007529. Available from: https://doi.org/10.1002/14651858.CD006066.pub3.
[58] Bates SM, Greer JA, Pabinger L, Sofaer SJH. Venous thromboembolism, thrombophilia, antithrombotic therapy, and pregnancy: American College of chest physicians evidence-based clinical practice guidelines (8th edition). Chest 2008;133(6 Suppl.):844Se886S. Available from: https://doi.org/10.1378/chest.08-0761.
[59] Andreoli L, Bertsias GK, Agmon-Levin N, Brown S, Cervera R, Costedoat-Chalumeau N, et al. EULAR recommendations for women's health and the management of family planning, assisted reproduction, pregnancy and menopause in patients with systemic lupus erythematosus and/or antiphospholipid syndrome. Ann Rheum Dis 2017;76(3):476–85. Available from: https://doi.org/10.1136/annrheumdis-2016-209770.
[60] Motta M, Chirico G, Rebaioli CB, Faden D, Lojacono A, Allegri F, et al. Anticardiolipin and anti-beta2 glycoprotein I antibodies in infants born to mothers with antiphospholipid antibody-positive autoimmune disease: a follow-up study. Am J Perinatol 2006;2384:247–51. Available from: https://doi.org/10.1055/s-2006-939533.
[61] Mekinian A, Lachassinne E, Nicaise-Roland P, Carbillon L, Motta M, Vicaut E, et al. European registry of babies born to mothers with antiphospholipid syndrome. Ann Rheum Dis 2013;72(2):217–22. Available from: https://doi.org/10.1136/annrheumdis-2011-201167.
[62] Boffa MC, Lachassinne E. Infant perinatal thrombosis and antiphospholipid antibodies: a review. Lupus 2007;16(8):634–41. Available from: https://doi.org/10.1177/0961203307079039.
[63] Chighizola CB, de Jesus GR, Branch DW. The hidden world of anti-phospholipid antibodies and female infertility: a literature appraisal. Autoimmun Rev 2016;15(6):493–500. Available from: https://doi.org/10.1016/j.autrev.2016.01.018.
[64] Simopoulou M, Sfakianoudis K, Maziotis E, Grigoriadis S, Giannelou P, Rapani A, et al. The impact of autoantibodies on IVF treatment and outcome: a systematic review. Int J Mol Sci 2019;20(4):892. Available from: https://doi.org/10.3390/ijms20040892.

CHAPTER

9

Antisperm antibodies and reproductive failure

Hiroaki Shibahara, Yuekun Chen, Ayano Yamaya, Yu Wakimoto, Atsushi Fukui and Akiko Hasegawa

Department of Obstetrics and Gynecology, School of Medicine, Hyogo Medical University, Nishinomiya, Hyogo, Japan

1 Introduction

The clinical definition of infertility is a disease of the reproductive system defined by the failure to achieve a clinical pregnancy after 12 or more months of regular unprotected sexual intercourse [1]. The incidence of infertility has been shown in approximately 10%–15% of couples. There are three major causes of infertility, including endocrine, tubal, and male factors. In addition, decreased ovarian function has manifested due to delayed marriage and the timing of conception.

Reproductive failures include infertility and recurrent pregnancy loss (RPL) [2]. A pregnancy loss is defined as the spontaneous demise of a pregnancy before the fetus reaches viability. The term includes all pregnancy losses from the time of conception until 22 weeks of gestation. The Guideline Development Group concluded that a diagnosis of RPL could be considered after the loss of two or more pregnancies [3]. This definition includes pregnancy losses both after spontaneous conception and assisted reproductive technology but excludes ectopic and molar pregnancies and implantation failure. The incidence of RPL has been shown up to 2% of childbearing women. There are various etiologies of RPL. However, approximately 60% of RPL is without any risk factors. Although the relative importance of immunological factors in human reproduction remains unclear, there may be immune-mediated reproductive failures in a portion of unexplained infertility and RPL [4].

As a cause of immunological factors, antisperm antibodies (ASA) are produced both in men and women. Previously, we have demonstrated the diagnosis and treatment of immunologically infertile men and women with ASA independently [5,6]. There have

Immunology of Recurrent Pregnancy Loss and Implantation Failure
DOI: https://doi.org/10.1016/B978-0-323-90805-4.00003-1

been reported several antigens in the surface of sperm that are especially foreign to women. Therefore alloimmunity can develop in women.

2 Definition

ASA are antibodies produced against sperm antigens which are immunoglobulins consisting of IgG, IgA, and/or IgM. In female immunological infertility, sperm-immobilizing antibodies are highly related to reproductive failures [7]. The presence of ASA in the sera of infertile women has been shown to inhibit sperm migration in the female genital tract, especially sperm-immobilizing antibodies detected by sperm-immobilization test (SIT) [8,9]. Such antibodies can also exert inhibitory effects on various stages of sperm−egg interaction and subsequent embryo development in vitro. Contrarily, the relationship between ASA and RPL has been reported to be controversial [7].

3 Prevalence

Several assay methods were developed to detect ASA. However, it is obvious that the most important consideration is the selection of the method. The presence of ASA in the fertile population suggests that not all ASA cause infertility. Therefore it has been recommended that clinically specific tests for female infertility should be chosen for detecting circulating ASA. ASA with proven bioactivities, including sperm-immobilizing antibodies, sperm-agglutinating antibodies, and fertilization-blocking antibodies, are suitable initial testing for infertile women. For example, the incidence of infertile women having sperm-immobilizing antibodies in their sera is approximately 3%, among whom visited infertility outpatient clinics for the first time [10].

4 Underlying immunopathologies related to reproduction

The presence of ASA, especially sperm-immobilizing antibodies, in the sera of infertile women has been shown to inhibit sperm migration in the female genital tract. Such antibodies can also exert inhibitory effects on various stages of sperm−egg interaction and subsequent embryo development in vitro.

4.1 Inhibition of sperm transport by antisperm antibodies at the level of cervical mucus

If any adverse effects of ASA on the process of fertilization are present, ASA should be present in the secretory fluids of the reproductive tracts [11]. ASA could cause infertility in women at different stages of reproduction, including postcoital sperm survival in cervical mucus (CM). Koyama et al. [12] reported that the frequency of poor postcoital test (PCT) results was significantly higher in patients with sperm-immobilizing antibodies in their sera than in those without the antibodies. Other investigators also arrived at similar conclusions [13–15]. These studies indicated the inhibitory effect of sperm-immobilizing antibodies on sperm penetration through the CM.

We investigated whether the antibody titers (50% sperm immobilization unit: SI_{50} unit), assessed by the quantitative SIT, [16,17] in infertile women with sperm-immobilizing antibodies, correlate with the results of PCT [16]. When patients with sperm-immobilizing antibodies were divided into two groups according to the SI_{50} titers, the abnormal results of PCT were obtained in all 100% (10/10) of patients with high SI_{50} titers (>10), and 66.7% (14/21) of patients with low ($\leqq 10$) SI_{50} titers ($P < 0.05$). These findings suggested that the SI_{50} titer in the serum can predict inhibitory effects on sperm migration through CM in infertile women with sperm-immobilizing antibodies [18]. To overcome such an inhibitory effect on sperm transport in CM, intrauterine insemination (IUI) has been selected for the treatment of infertile women with sperm-immobilizing antibodies [19,20]. However, its effectiveness has been limited because there are still other inhibitory effects such as on sperm transport from the uterine cavity to Fallopian tubes and sperm–egg interaction.

4.2 Inhibition of sperm transport by antisperm antibodies from uterine cavity to Fallopian tubes

It was unclear whether sperm transport from the uterine cavity to and through the Fallopian tubes could be impaired by the antibodies. Peritoneal sperm recovery test (PSRT) [21] examines the possible impairment of sperm migration in the Fallopian tubes via laparoscopic examinations and investigation of the presence of motile sperm in the peritoneal fluid following IUI [22].

The peritoneal fluid was aspirated from the Douglas pouch and centrifuged immediately. The sediments were microscopically examined for the presence of motile sperm. As a result, sperm recovery in the peritoneal fluid was not observed in 88.9% (24/27) of patients with sperm-immobilizing antibodies and 66.0% (140/212) of patients without the antibodies, which was significantly different ($P < 0.05$) [22]. When the 27 patients with sperm-immobilizing antibodies were divided into two groups according to the SI_{50} titers, the abnormal result of PSRT was obtained in 95.7% (22/23) of patients with high (>10) SI_{50} titers and 50.0% (2/4) of patients with low ($\leqq 10$) SI_{50} titers ($P < 0.05$). These findings suggested that the SI_{50} titer in the serum can predict inhibitory effects on sperm migration from the uterine cavity to and through the Fallopian tubes in infertile women with sperm immobilizing antibodies.

4.3 Inhibition of sperm–egg interaction by antisperm antibodies

It has been shown that sperm-immobilizing antibodies can exert inhibitory effects on sperm–egg interaction at the level of the acrosome reaction, [23–25] zona pellucida recognition and penetration [26–32]. We have reported a significant correlation in the SI_{50} titers between the patients' sera and their follicular fluid ($P < 0.0001$) [33]. In this study, patients' sera and their follicular fluid were collected from infertile women with sperm-immobilizing antibodies in their sera when they were treated by in vitro fertilization-embryo transfer (IVF-ET). Therefore thorough washing of collected oocytes is important prior to allocating the oocytes in the culture medium to prevent contamination of the culture medium, which can prohibit fertilization in IVF and obtain better results.

As an example, we retrospectively investigated the clinical effects of sperm-immobilizing antibodies in the sera of infertile women on fertilization rates [34]. Before the introduction of SIT as a routine test for female infertility, 85 oocytes were collected in nine IVF cycles from four infertile women who were found to have sperm-immobilizing antibodies in their sera. The oocytes were subsequently inseminated with swim-up sperm in a medium containing the patient's serum. As controls, 50 oocytes were collected in five IVF cycles from five infertile women possessing the antibodies in their sera, and the oocytes were inseminated with swim-up sperm in a medium supplemented with human serum albumin (HSA). In the former group, 41 (48.2%) of 85 oocytes were fertilized, and in the latter group, 43 (86.0%) of 50 oocytes were fertilized. There was a significant difference between the two groups ($P < 0.0001$). These findings indicated that the sperm-immobilizing antibodies in the sera of infertile women might cause a low fertilization rate [34]. However, with a thorough washing of the collected eggs and cultured in a medium without the patient's serum before IVF, there was no difference in the fertilization rate in the patients with high (≧10) and low (<10) SI_{50} titers in their follicular fluid ($P = 0.62$) [33].

4.4 Impairment of early embryo development by antisperm antibodies

As the sperm surface antigens were proven to be present in developing embryos, embryo development could be impaired by ASA [35,36]. By using immunofluorescent staining, Koyama et al. [36] showed that the sperm-specific antigens were transferred to the fertilized eggs in rats. They were located on the cell membrane in the various developmental stages from one cell to the blastocyst stage. ASA against rat sperm impaired the in vitro development of fertilized eggs in the presence of complement [36].

Naz [37] found that early cleavage of zygotes was abnormal after oocytes were fertilized with sperm cells from one of 64 infertile couples studied in their IVF and embryo development procedure. The abnormal cleavage, which did not affect fertilization, was not due to polyspermy or parthenogenetic activation of oocytes but to ASA present on the sperm cells that reacted with the acrosomal (strongly) and tail (weakly) regions of human sperm. The antibodies recognized a double band comprising two glycoproteins of 14 ± 3 and 18 ± 3 kDa and a protein band of 22 ± 3 kDa. Antisera were raised in rabbits against the double band and the protein. Passive transfer of affinity-purified human or rabbit immunoglobulins directed against the double band, but not those directed against the protein, caused a significant inhibition of early cleavage of oocytes without affecting pronuclear formation in mice. It was suggested that the sperm surface antigens of 14 ± 3 and 18 ± 3 kDa might provide an extra-nuclear signal for oocyte division in mice and humans.

In humans, we found that the sperm-immobilizing antibodies in the sera of infertile women cause poor embryo quality following in vitro fertilization [34]. As mentioned above, the sperm-immobilizing antibody in the sera of infertile women might cause a low fertilization rate. Following fertilization in a medium containing the patient's serum with sperm-immobilizing antibody, an evaluation of embryo cleavage was performed on the second or the third day after oocyte retrieval according to Veeck's classification [38]. The grade 1 and 2 embryos were considered good quality. Using this classification, 16 (39.0%) of 41 embryos incubated in the medium containing the patient's serum were of good quality, while 34 (79.1%) of 43 embryos incubated in the medium supplemented with HSA were good quality. There was a significant difference between the two groups ($P = 0.0003$). These findings might indicate that sperm-immobilizing antibodies in the sera of infertile women cause poor embryo quality in vitro.

Because human embryos are not easily available due to ethical reasons, studies were limited to exploring a possible mechanism of inducing embryo fragmentation by sperm-immobilizing antibodies. Hence, we investigated if patients' sera with antihuman sperm antibodies show inhibitory effects on IVF and embryo development in mice [39] Patients' sera were collected from 8 infertile women having sperm-immobilizing antibodies and 17 infertile women without antibodies. Male ICR mice and female F1 mice (BALB/c X C57BL/6 J) were used. In mice IVF, preincubated sperm were cultured in the medium containing the patient's serum with or without sperm-immobilizing antibodies or bovine serum albumin (BSA) as a control. The fertilization rates and the incidences of blastocyst formation were compared. A mouse SIT which was established prior to the experiment was measured. Five (62.5%) of eight serum samples with sperm-immobilizing antibodies and nine (52.9%) of 17 serum samples without the antibodies showed sperm-immobilizing activities in mice. There was no significant difference between the two groups. Five sera with sperm-immobilizing activities in humans and mice and five sera without sperm-immobilizing activities in humans or mice were used for the further experiments. The fertilization rates in BSA, patient's serum with sperm immobilizing antibodies, and that without the antibodies were 82.5% (746/904), 43.6% (508/1165), and 64.5% (669/1037), respectively. There were significant differences between the groups ($P < 0.001$). The incidences of blastocyst formation, 59.9% (447/746), 31.7% (161/508), and 47.7% (319/669), respectively, were also significant differences between the groups ($P < 0.001$).

4.5 Antisperm antibodies and recurrent pregnancy loss

The controversial relationship between ASA and RPL has been reported. Increased miscarriage rates in women with ASA were demonstrated by some authors (Table 1) [40–43]. In contrast, a lack of association between ASA and RPL was reported (Table 2) [44–47].

4.6 Reports showing a positive association between antisperm antibodies and recurrent pregnancy loss

Haas et al. [40] reported that no pregnancies were subsequently gestated to term in women who were ASA-positive unless they were inoculated with their husband's leukocytes as a treatment for an immune basis (not related to ASA) for their recurrent abortions. In this paper, they used a radiolabeled antiglobulin assay (RAA), a modified enzyme-linked

TABLE 1 Summary of the manuscripts showing positive association between.

Authors	Subjects	No. of patients	Assay method for antisperm antibodies	Association	Reference
Haas GG, et al.	RPL (>3)	173	RAA/ELISA/ TAT/SIT	Yes	[40]
Mathur S, et al.	RPL (>2) versus control	26 (cont. 53)	cytotoxic/ hemagglutination	Yes	[41]
Witkin SS, et al.	infertile	109	ELISA	Yes	[42]
Witkin SS, et al.	RPL versus control	44 (cont. 616)	IBT or MAR	Yes	[43]

Abbreviations: *ELISA*: enzyme-linked immunosorbent assay; *IBT*: immunobead test; *MAR*: mixed antiglobulin reaction; *RAA*: radiolabeled antiglobulin assay; *SIT*: sperm-immobilization test; *TAT*: tray agglutination test.

TABLE 2 Summary of the manuscripts showing negative association between antisperm antibodies and recurrent pregnancy loss.

Authors	Subjects	No. of patients	Assay method for ASA	Association	Reference
Maier DB, et al.	RPL (<3) vs control	34 (cont. 24)	IBT	No	[44]
Clarke GN, et al.	RPL (>3) vs control	70 (cont. 1133)	IBT	No	[45]
Christiansen OB, et al.	RPL (<3)	123	SAT/MAR	No	[46]
Al-Hussein K, et al.	RPL (>3) vs control	24 (cont. 6)	FCA	No	[47]

Abbreviations: *FCA*: flow cytometric assay; *IBT*: immunobead test; *MAR*: mixed antiglobulin reaction; *SAT*: sperm-agglutination test.

immunosorbent assay (ELISA), a tray agglutination test (TAT), and SIT to detect ASA for 173 women with a history of three or more recurrent consecutive abortions (Table 1). Mathur et al. [41] investigated the types for human leukocyte antigens (HLA) of A and B loci and screened for the presence of cytotoxic and hemagglutinating antibodies to sperm in 26 couples with two or more early spontaneous abortions and 53 couples with recently proven fertility (control group). They suggested that the presence of ASA is associated with early pregnancy loss. Histocompatibility antigens B7 and B35 may play a role through their association with ASA and early spontaneous abortions. Witkin et al. [42] examined the relationship between ASA, conception, and miscarriage in 109 infertile couples. ASA present on the surface of husbands' ejaculated sperm and antibodies in husbands' or wives' sera that reacted with a purified population of the husbands' motile sperm was detected by an ELISA. They concluded that ASA in female sera and on ejaculated sperm was associated with a failure to conceive and first-trimester miscarriage. Later, they also reported a statistically significant association ($P < 0.005$) between the presence of IgG tail-directed ASA and a history of unexplained RPL [43]. ASA may be a marker for defective immunosuppression in women with RPL. Alternatively, sperm exposure to sperm-sensitized pregnant women may activate the maternal immune system to respond to paternal antigens present on the embryo.

4.7 Reports showing a negative association between antisperm antibodies and recurrent pregnancy loss

In contrast, several authors have reported the lack of association between ASA and RPL (Table 2). Maier et al. [44] evaluated 34 women with RPL for the presence of lupus-associated autoantibodies, ASA detected by immunobead test (IBT), and evidence of complement abnormalities. In their study, no woman had positive ASA. They suggested that interpretation of the study by Haas et al. [40] is difficult because of the numerous ASA tests used, the high positive rates in both groups, and the absence of a control group. In addition, most of the positive results were from an ELISA assay. These assays can detect antibodies to internal sperm antigens, which are of questionable importance. They concluded that it was unable to duplicate their results using the IBT. Clarke et al. [45] reported that there is a lack of association between ASA and RPL. They tested 70 women with a history of unexplained RPL and 1133 infertile women as a control. IBT was used for the detection of ASA in their sera. Patients with recurrent abortion have a lower incidence (1.4% vs 7.4%) of circulating ASA. They concluded that women with a history of RPL have a significantly lower incidence of circulating ASA. Christiansen et al. [46] studied the associations between human leukocyte antigen (HLA) class II and antiphospholipid (APL) antibodies in 123 women with RPL. In the same study, antinuclear antibodies (ANA), antizona pellucida antibodies, and ASA were also investigated using their sera. While 19 (15.4%) patients were positive for antizona pellucida antibodies, only six (4.9%) patients were positive for ASA. Of these, two were carrying the HLA-Dr3 allele, which seemed to predispose to the formation of ACL antibodies and ANA. It was concluded that the numbers were too small for meaningful statistical comparisons. Al-Hussein et al. [47] used the flow cytometric assay (FCA) for the detection of ASA in the sera obtained from 24 women with RPL (Group 1), and lymphocytes and semen samples collected from their husbands at the same time. Sera, lymphocytes and semen samples were also collected from six couples with no history of abortion who served as controls (Group 2). Using a sensitive FCA to analyze the samples, without knowledge of clinical status, elevated levels of both IgG and IgM were detected in Group 1. However, no significant association was found when compared with normal females who had healthy pregnancies. They concluded that maternal antipaternal antibodies with respect to IgG and IgM classes do not play a major role for RPL, despite the presence of increased levels of antibodies against three different sources of paternal antigens.

4.8 Pregnancy outcome in infertile women with sperm-immobilizing antibodies

In our previous study, the effects of sperm-immobilizing antibodies on pregnancy outcomes were examined in infertile women treated with IVF-ET (Fig. 1) [48]. 143 ET cycles

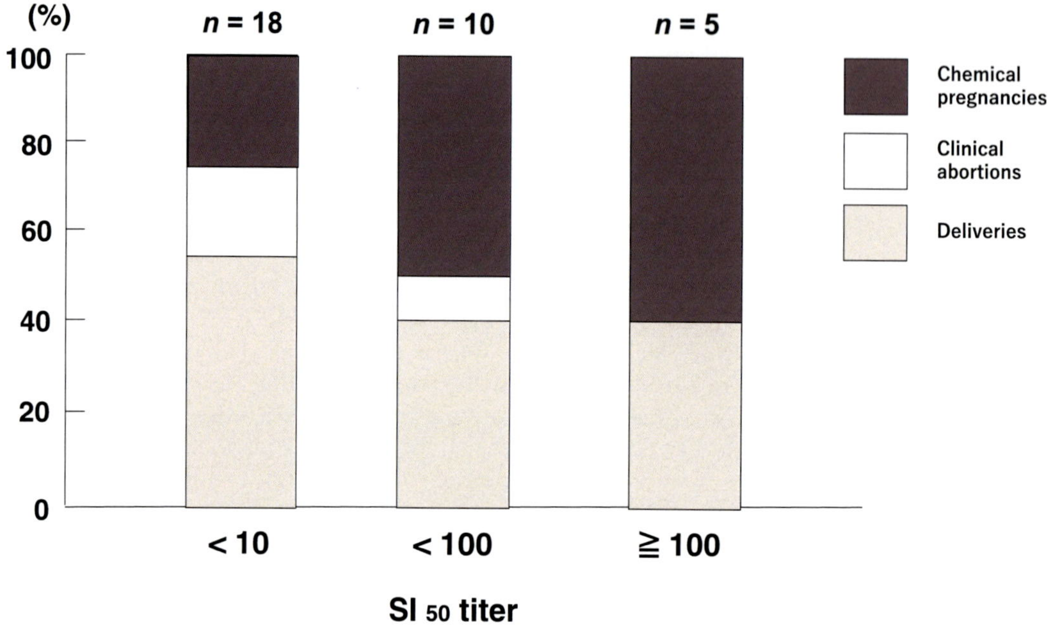

FIGURE 1 Relationship between pregnancy outcome and sperm-immobilizing antibody titers (SI_{50}) in patients' sera at the time of conception. The outcomes of 33 cycles of pregnancy were compared by dividing the patients into three groups, according to the SI_{50} titers in their sera at the time of conception. The chemical pregnancy rates were 22.2% in patients with SI_{50} titers below 10 units, 50.0% in patients with SI_{50} titers above 10 units and 60.0% in patients with SI_{50} titers above 100 units, respectively.

in 58 infertile women with sperm-immobilizing antibodies and 363 ET cycles in patients with tubal infertility as control were compared. Diagnosis of chemical pregnancy was made when the urinary hCG level had risen over 50 IU/L, but a gestational sac could not be demonstrated. 33 (23.1%) of 143 cycles in the patients with sperm-immobilizing antibodies and 56 (15.4%) of 363 cycles in the control patients were diagnosed as pregnant. There was a significant difference between the two groups ($P < 0.05$). In the patients with sperm-immobilizing antibodies, 12 (36.4%) chemical pregnancies, 5 (15.2%) abortions, and 16 (48.5%) deliveries were reported. In the control group, 18 (32.1%) chemical pregnancies, 10 (17.9%) clinical abortions, including ectopic pregnancies, and 28 (50.0%) deliveries were reported. There were no differences in each category. When the SI_{50} titers at the time of conception were considered, chemical pregnancy rates were 22.2% (4/18) in patients with SI_{50} titers below 10 units, 50.0% (5/10) when the SI_{50} titers were above 10 units, and 60.0% (3/5) when the titers were above 100 units ($P > 0.05$). In four of five patients who had both chemical and clinical pregnancies, the SI_{50} titers at the time of conception were higher in the chemical pregnancy cycles than in the clinical pregnancy cycles. Though the pregnancy rates were significantly higher in the patients with sperm-immobilizing antibodies as compared to those with tubal infertility, chemical pregnancy rates were also higher in the patients with higher sperm-immobilizing antibody titers. These results suggest that sperm-immobilizing antibodies may cause the damage of early development of human embryos in vivo in patients with a high titer of the antibodies.

In our later study that demonstrated the adverse effects of sperm-immobilizing antibodies in the sera of infertile women on fertilization and embryo quality, there were no patients who aborted after successfully established pregnancies following IVF-ET among the study population, although the number of patients was low [34].

5 Evaluation

5.1 Test for antisperm antibodies

Detection of ASA has evolved considerably since the first serum agglutination assay was reported in the 1950s [49]. Franklin and Dukes found that 20.1 % of 214 women undergoing infertility investigations had detectable sperm agglutinating activity in their serum [50]. Women with unexplained infertility had a much higher incidence (72.1%) than women with organic causes for their infertility (8.4%) or fertile women (5.7%). It should be noted that this study found a very high incidence of ASA, and the results are not supported by recent reports using such as the IBT. However, this report was notable from a historical perspective since it stimulated significant interest in the idea that female immunological responses to sperm could be involved in the development of otherwise unexplained infertility and the concept of an antisperm contraceptive vaccine [51].

Later, ASA has been studied as one of the important causes of infertility since the 1970s [8,9]. Several assays have been developed for detecting ASA [52–56]. Among them, Isojima et al. developed a complement-dependent SIT that detects sperm immobilizing antibodies in the sera of infertile women [8,9]. Of the several assay methods detecting ASA, SIT is the most reliable assay for determining female infertility because of its specificity.

SIT, Kibrick's, and Franklin-Dukes' sperm-agglutination tests [49,50] were compared as methods for detecting ASA in the sera of sterile, pregnant, and unmarried women [8,9]. Positive reactions in the SIT were detected in the sera of 17.2% of the patients with infertility of unexplained cause but none in normal pregnant and unmarried women. In contrast, Kibrick's and Franklin-Dukes' tests gave rather higher percentages of positive sperm-agglutination reactions in normal pregnant women than in cases of infertility of unexplained infertility but a very low percentage of positive responses in unmarried women.

5.2 Application of computer-aided sperm analysis for the detection of sperm-immobilizing antibodies

Recently, our group applied computer-aided sperm analysis (CASA) for detecting sperm-immobilizing antibodies. The SIT has been carried out by counting the number of motile sperm under a microscope which may yield varying results due to the subjective judgment of the operator, as seen with routine semen examination [57]. We developed a novel method using CASA, and the results were compared with those obtained by the traditional method. The results were identical, and 25 of 78 samples tested were positive and 53 samples were negative for sperm-immobilizing antibodies based on both methods.

For SI-positive samples, the SI_{50} values obtained using the two methods correlated closely with high coefficiency. We concluded that the novel method using CASA would make it possible to objectively evaluate SIT and SI_{50} data at clinical facilities and increase the convenience of using the SIT as a clinical indicator. Furthermore, we recently developed a modified method of SIT using CASA by substituting fresh sperm to cryopreserved sperm when fresh sperms are unavailable [58].

6 Treatment

6.1 Treatment modalities

Women with ASA, especially sperm-immobilizing antibodies, are refractory to traditional treatments for infertility except for IVF-ET. Before the introduction of IVF-ET, there were essentially two strategies for the treatments of infertile women with ASA. One is to reduce ASA production, and the other is to increase the rate of sperm and egg encounters. For the former, consistent use of condoms or abstinence to avoid sperm sensitization and immunosuppressive therapy such as corticosteroid therapy to suppress antibody production have been tried. However, these methods are no longer used due to the lack of efficacy and possible side effects, respectively [59]. For the latter, IUI and IVF-ET have been successfully applied to eliminate the impairment of sperm passage through the female reproductive tract by ASA [19,48,60].

If infertile women with ASA successfully conceived with methods such as IVF-ET but developed RPL, further investigation is required. When no other reason for RPL is identified, corticosteroid treatment with careful observation might be considered to expect decreasing the level of ASA.

6.2 Strategy for the treatment of infertile women with sperm-immobilizing antibodies

Kobayashi et al. [19] proposed a strategy for the treatment of infertile women with sperm-immobilizing antibodies according to the undulation patterns of individual patient's SI_{50} titers. They categorized patients with sperm-immobilizing antibodies into three groups according to their follow-up SI_{50} titers. Group A consists of patients with continuously high SI_{50} titers (>10 units) who did not conceive by ordinary or repeated IUI; however, a satisfactory pregnancy rate was obtained by IVF-ET. Group B consists of patients who had intermediate SI_{50} titer patterns around 10 and showed a low rate of success with IUI. Group C are patients with continuously low SI_{50} titers (<10 units) who conceived by repeated or ordinary IUI, although the success rates were lower than using IVF-ET. Taken together, a strategy for the treatment of infertile women with sperm-immobilizing antibodies was suggested in Fig. 2 [6]. The therapeutic strategy emphasizes the importance of assessing SI_{50} titers to select a remedy for infertile women with sperm-immobilizing antibodies.

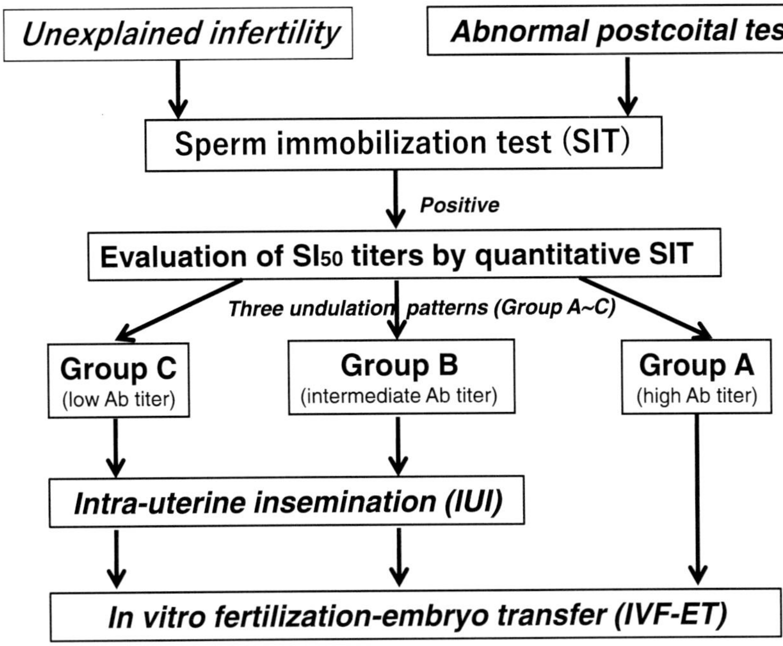

FIGURE 2 Strategy for the treatment of infertile women with sperm-immobilizing antibodies. It is important to assess the SI_{50} titers for selecting treatments for infertile women with sperm-immobilizing antibodies. A strategy for the treatments of infertile women with sperm-immobilizing antibodies is shown according to the undulation patterns of individual patient's SI_{50} titers. Patients with sperm-immobilizing antibodies are divided into three groups according to their follow-up SI_{50} titers; Group A: the patients with continuously high SI_{50} titers (>10 units) are recommended to be treated by in vitro fertilization-embryo transfer (IVF-ET). Group B: the patients have intermediate SI_{50} titer patterns around 10, and Group C; the patients with continuously low SI_{50} titers (<10 units) are initially recommended to be treated by repeated or ordinary intrauterine insemination (IUI). Because the pregnancy rate by the repeated IUI in Group C was reported to be better than that in Group B (25.0% vs 17.6%), earlier step up from IUI to IVF-ET is better to be considered for Group B. If they are not able to conceive by IUI, then the treatment by IVF-ET should be recommended.

7 Reproductive outcomes

It is well known that infertile women with sperm-immobilizing antibodies are refractory to conventional infertility treatments such as timed intercourse or IUI because the antibodies secreted in the female genital tract inhibit sperm migration within CM [12,18] and the Fallopian tubes [22]. To overcome this cause of infertility in vivo, IVF-ET has been applied, and satisfactory outcomes have been obtained resulting from suitable culture conditions for gametes and embryos [19,48,60]. It is suggested that SIT in the sera of infertile women should be performed prior to proceeding with IVF. The manipulation of gametes and embryos from patients with sperm immobilizing antibodies should be carefully carried out, especially to avoid contamination of the culture media by the patient's serum and follicular fluid. These steps should be performed to overcome the immunological causes of female infertility by ASA and obtain satisfactory results under suitable conditions for gametes and embryos [19,20,48,60–62].

8 Summary and recommendations

Association between ASA and reproductive failures has been described. ASA are heterogeneous, and there are two kinds of ASA, those related to and those unrelated to infertility. The former could be the cause of reproductive failures. ASA inhibit sperm migration through entire female reproductive tracts. They show the inhibitory effects on various stages of sperm–egg interaction. Furthermore, some ASA show the harmful effect on embryo development. However, ASA might have little impact on RPL. The assessment of SIT for infertile women is the most important before proceeding with IVF.

1. ASA are heterogeneous, and there are two kinds of ASA, including antibodies that are related to infertility and not related to infertility.
2. ASA with bioactivities, including sperm-immobilizing antibodies, sperm-agglutinating antibodies, and fertilization-blocking antibodies, are suitable initial testing for infertile women.
3. The presence of ASA, especially sperm-immobilizing antibodies, in the sera of infertile women has been shown to inhibit sperm migration through entire female reproductive tracts.
4. The ASA inhibitory effects were demonstrated on various stages of sperm–egg interaction. Some ASA show the harmful impact on embryo development.
5. To select an appropriate treatment, it is important to assess SI_{50} titers for infertile women with sperm-immobilizing antibodies.
6. To overcome infertility with ASA, IVF-ET can be considered for satisfactory outcomes resulting from suitable culture conditions for gametes and embryos.
7. In regards to RPL and ASA, a controversial relationship has been reported.
8. If infertile women with ASA successfully conceived with methods such as IVF-ET but developed RPL, further investigation is required.
9. In infertile women with ASA, corticosteroid treatment with careful observation might be considered to expect decreasing the amount of ASA when no other reason for RPL is identified.

References

[1] Zegers-Hochschild F, Adamson GD, de Mouzon J, et al. International committee for monitoring assisted reproductive technology (ICMART) and the World Health Organization (WHO) revised glossary of ART terminology, 2009. Fertil Steril, 92. 2009. p. 1520–4.

[2] Kokcu A, Yavuz E, Celik H, et al. A panoramic view to relationships between reproductive failure and immunological factors. Arch Gynecol Obstet 2012. Available from: https://doi.org/10.1007/s00404-012-2480-6.

[3] Atik RB, Christiansen OB, Elson J, et al. ESHRE guideline: recurrent pregnancy loss. Hum Reprod Open 2018;1–12.

[4] Choudhury SR, Knapp LA. Human reproductive failure I: immunological factors. Hum Reprod Update 2000;7:113–34.

[5] Shibahara H, Shiraishi Y, Suzuki M. Diagnosis and treatment of immunologically infertile males with anti-sperm antibodies. Reprod Med Biol 2005;4:133–41.
[6] Shibahara H, Koriyama J, Shiraishi Y, et al. Diagnosis and treatment of immunologically infertile women with sperm-immobilizing antibodies in their sera. J Reprod Immunol 2009;83:139–44.
[7] Shibahara H, Wakimoto Y, Fukui A, et al. Anti-sperm antibodies and reproductive failures. Am J Reprod Immunol 2021;85(4):e13337. Available from: https://doi.org/10.1111/aji.13337.
[8] Isojima S, Li TS, Ashitaka Y. Immunologic analysis of sperm immobilizing factor found in sera of women with unexplained sterility. Am J Obstet Gynecol 1968;101:677–83.
[9] Isojima S, Tsuchiya K, Koyama K, et al. Further studies on sperm-immobilizing antibody found in sera of unexplained cases of sterility in women. Am J Obstet Gynecol 1972;112:199–207.
[10] Shibahara H, Koriyama J. Methods for direct and indirect antisperm antibody testing. Methods Mol Biol 2013;927:51–60.
[11] Isojima S, Koyama K. Techniques for sperm immobilization test. Arch Androl 1989;23:185–99.
[12] Koyama K, Ikuma K, Kubota K, et al. Effects of antisperm antibodies on sperm migration through cervical mucus. Excerpta Med Int Congr Ser 1980;512:705–8.
[13] Cantuaria AA. Sperm immobilizing antibodies in the serum and cervicovaginal secretions of infertile and normal women. Br J Obstet Gynaecol 1977;84:865–8.
[14] Chen C, Jones WR. Application of a sperm micro-immobilization test to cervical mucus in the investigation of immunologic infertility. Fertil Steril 1981;35:542–5.
[15] Jager S, Kremer J, De Wilde-Janssen IW. Are sperm immobilizing antibodies in cervical mucus an explanation for a poor post-coital test? Am J Reprod Immunol 1984;5:56–60.
[16] Isojima S, Koyama K. Quantitative estimation of sperm immobilizing antibody in the sera of women with sterility of unknown etiology: the 50% sperm immobilization unit (SI_{50}). Excerpta Med Int Congr Ser 1974;370:10–15.
[17] Koyama K, Kubota K, Ikuma K, et al. Correlation between quantitative antibody titers of sperm immobilizing antibodies and pregnancy rates by treatments. Fertil Steril 1988;33:201–6.
[18] Shibahara H, Shiraishi Y, Hirano Y, et al. Relationship between level of serum sperm immobilizing antibody and its inhibitory effect on sperm migration through cervical mucus in immunologically infertile women. Am J Reprod Immunol 2007;57:142–6.
[19] Kobayashi S, Bessho T, Shigeta M, et al. Correlation between quantitative antibody titers of sperm immobilizing antibodies and pregnancy rates by treatments. Fertil Steril 1990;54:1107–13.
[20] Hasegawa A, Shigeta M, Shibahara H. Sperm immobilizing antibody and its target antigen. In: Krause WKH, Naz RK, editors. Immune infertility. Switzerland: Springer; 2016. p. 173–84.
[21] Templeton AA, Mortimer D. Laparoscopic sperm recovery in infertile women. Br J Obstet Gynecol 1980;87:1128–31.
[22] Shibahara H, Shigeta M, Toji H, et al. Sperm immobilizing antibodies interfere with sperm migration from the uterine cavity through the Fallopian tubes. Am J Reprod Immunol 1995;34:120–4.
[23] Bandoh R, Yamano S, Kamada M, et al. Effect of sperm-immobilizing antibodies on the acrosome reaction of human spermatozoa. Fertil Steril 1992;57:387–92.
[24] Cheng GY, Shi JL, Wang M, et al. Inhibition of mouse acrosome reaction and sperm-zona pellucida binding by anti-human sperm membrane protein 1 antibody. Asian J Androl 2007;9:23–9.
[25] Zhang J, Ding X, Bian Z, et al. The effect of anti-eppin antibodies on ionophore A23187-induced calcium influx and acrosome reaction of human spermatozoa. Hum Reprod 2010;25:29–36.
[26] Bronson RA, Cooper GW, Rosenfeld DL. Sperm-specific isoantibodies and autoantibodies inhibit the binding of human sperm to the human zona pellucida. Fertil Steril 1982;38:724–9.
[27] Alexander NJ. Antibodies to human spermatozoa impede sperm penetration of cervical mucus or hamster eggs. Fertil Steril 1984;41:433–9.
[28] Kamada M, Daitoh T, Hasebe H, et al. Blocking of human fertilization in vitro by sera with sperm-immobilizing antibodies. Am J Obstet Gynecol 1985;153:328–31.
[29] Mahony MC, Blackmore PF, Bronson RA, et al. Inhibition of human sperm-zona pellucida tight binding in the presence of antisperm antibody positive polyclonal patient sera. J Reprod Immunol 1991;19:287–301.
[30] Tsukui S, Noda Y, Fukuda A, et al. Blocking effect of sperm immobilizing antibodies on sperm penetration of human zonae pellucida. Vitro Fert Embryo Transf 1988;5:123–8.

[31] Shibahara H, Shigeta M, Koyama K, et al. Inhibition of sperm-zona pellucida tight binding by sperm immobilizing antibodies as assessed by the hemizona assay (HZA). Acta Obstet Gynaecol Jpn 1991;43:237–8.
[32] Shibahara H, Burkman LJ, Isojima S, et al. Effects of sperm-immobilizing antibodies on sperm-zona pellucida tight binding. Fertil Steril 1993;60:533–9.
[33] Shibahara H, Hirano Y, Shiraishi Y, et al. Effects of in vivo exposure to eggs with sperm-immobilizing antibodies in follicular fluid on subsequent fertilization and embryo development in vitro. Reprod Med Biol 2006;5:137–43.
[34] Taneichi A, Shibahara H, Hirano Y, et al. Sperm immobilizing antibodies in the sera of infertile women cause low fertilization rates and poor embryo quality in vitro. Am J Reprod Immunol 2002;47:46–51.
[35] Menge AC. Effect of isoimmunization and isoantisera against seminal antigens on fertility processes in female rabbits. Biol Reprod 1971;4:137–44.
[36] Koyama K, Hasegawa A, Isojima S. Effects of antisperm antibody on the rat embryos. Gamete Res 1984;10:143–52.
[37] Naz RK. Effects of antisperm antibodies on early cleavage of fertilized ova. Biol Reprod 1992;46:130–9.
[38] Veeck LL. Atlas of the human oocytes and early conceptus. 1991, 2. Baltimore, Williams & Willkins Co.
[39] Taneichi A, Shibahara H, Takahashi K, et al. Effects of sera from infertile women with sperm immobilizing antibodies on fertilization and embryo development in vitro in mice. Am J Reprod Immunol 2003;50:146–51.
[40] Haas GG, Kubota K, Quebbeman JF, et al. Circulating antisperm antibodies in recurrently aborting women. Fertil Steril 1986;45:209–15.
[41] Mathur S, Neff MR, Williamson HO, et al. Sperm antibodies and human leukocyte antigens in couples with early spontaneous abortions. Int J Fertil 1987;32:59–65.
[42] Witkin SS, David SS. Effect of sperm antibodies on pregnancy outcome in a subfertile population. Am J Obstet Gynecol 1988;158:59–62.
[43] Witkin SS, Chaudhry A. Association between recurrent spontaneous abortions and circulating IgG antibodies to sperm tails in women. J Reprod Immunol 1989;15:151–8.
[44] Maier DB, Parke A. Subclinical autoimmunity in recurrent aborters. Fertil Steril 1989;51:280–5.
[45] Clarke GN, Baker HW. Lack of association between sperm antibodies and recurrent spontaneous abortion. Fertil Steril 1993;59:463–4.
[46] Christiansen OB, Ulcova-Gallova Z, Mohapeloa H, et al. Studies on associations between human leukocyte antigen (HLA) class II alleles and antiphospholipid antibodies in Danish and Czech women with recurrent miscarriages. Hum Reprod 1998;13:3326–31.
[47] Al-Hussein K, Al-Mukhalafi Z, Bertilsson P-A, et al. Value of flow cytometric assay for the detection of antisperm antibodies in women with a history of recurrent abortion. Am J Reprod Immunol 2002;31–7.
[48] Shibahara H, Mitsuo M, Ikeda Y, et al. Effects of sperm immobilizing antibodies on pregnancy outcome in infertile women treated with IVF-ET. Am J Reprod Immunol 1996;36:96–100.
[49] Kibrick S, Belding DL, Merrill B. Methods for the detection of antibodies against mammalian spermatozoa. I. A modified macroscopic agglutination test. Fertil Steril 1952;3:419–29.
[50] Franklin RR, Dukes CD. Further studies on sperm-agglutinating antibody and unexplained infertility. JAMA 1964;190:682–3.
[51] Clarke GN. ASA in the female. In Immune infertility. Impact of immune reactions on human fertility, 2nd ed., (Eds) Krause WKH, Naz RK, Springer, 2017: 161–172.
[52] Bronson R, Cooper G, Rosenfeld D. Ability of antibody-bound human sperm to penetrate zona- free hamster ova in vitro. Fertil Steril 1981;36:778–83.
[53] Shibahara H, Hirano Y, Takamizawa S, et al. Effects of sperm-immobilizing antibodies bound to the surface of ejaculated human spermatozoa on sperm motility in immunologically infertile men. Fertil Steril 2003;79:641–2.
[54] Jager S, Kremer J, van Slochteren-Draaisma T. A simple method of screening for antisperm antibodies in the human male. Detection of spermatozoal surface IgG with the direct mixed antiglobulin re-action carried out on untreated fresh human semen. Int J Fertil 1978;23:12–21.
[55] Clarke GN, Elliott PJ, Smaila C. Detection of sperm antibodies in semen using the immunobead test: a survey of 813 consecutive patients. Am J Reprod Immunol Microbiol 1985;7:118–23.
[56] Centola GM, Andolina E, Deutsch A. Comparison of the immunobead binding test (IBT) and immunospheres (IS) assay for detecting serum antisperm antibodies. Am J Reprod Immunol 1997;37:300–3.

[57] Wakimoto Y, Fukui A, Kojima T. Application of computer-aided sperm analysis (CASA) for detecting sperm-immobilizing antibody. Am J Reprod Immunol 2018;79(3). Available from: https://doi.org/10.1111/aji.12814.

[58] Wakimoto Y, Fukui A, Kojima T, et al. Sperm immobilization test and quantitative sperm immobilization test using frozen-thawed sperm preparation applied with computer-aided sperm analysis. Reprod Med Biol 2021;20:321–6.

[59] Koyama K. Gamete immunology: infertility and contraception. Reprod Immunol Biol 2009;24:1–17.

[60] Sugimoto Y, Hasegawa A, Yokoyama K, et al. Successful application of in vitro fertilization and embryo replacement in the treatment of infertile women with sperm immobilizing antibody. Acta Obstet Gynaecol Jpn 1987;38:1135–6.

[61] Yovich J, Stranger JD, Kay D, et al. In vitro fertilization of oocytes from women with serum antisperm antibodies. Lancet 1984;1:369–70.

[62] Daitoh T, Kamada M, Yamano S, et al. High implantation rate and consequently high pregnancy rate by in vitro fertilization-embryo transfer in infertile women with antisperm antibody. Fertil Steril 1995;63:87–91.

CHAPTER

10

Antithyroid antibodies and reproductive function

Elena Borodina[1,2], *Alexander M. Gzgzyan*[1,3], *Lyailya Kh. Dzhemlikhanova*[1,3], *Dariko A. Niauri*[1,2,3] *and Yehuda Shoenfeld*[2,4,5]

[1]Department of Obstetrics, Gynecology and Reproductive Sciences, Saint Petersburg State University, Saint Petersburg, Russia [2]Laboratory of the Mosaic of Autoimmunity, Saint Petersburg State University, Saint Petersburg, Russia [3]FSBSI "The Research Institute of Obstetrics, Gynecology and Reproductology named after D.O. Ott," Saint Petersburg, Russia [4]Ariel University, Ariel, Israel [5]Zabludowicz Center for Autoimmune Diseases; Sheba Medical Center, Tel-Aviv, Israel

1 Introduction

Today, the negative impact of overt hypothyroidism on reproductive function is beyond doubt. Hypothyroidism can affect reproductive outcomes at all stages of human reproduction, starting with ovulatory dysfunction, early pregnancy losses, and ending with miscarriage, premature birth, intrauterine growth restriction [1–3], and even neurocognitive disorders in offspring [4,5]. The most common cause of hypothyroidism in developed countries is autoimmune thyroiditis (AIT) [6]; however, the direct role of autoantibodies to the thyroid tissue in euthyroid women is still being discussed.

2 Definition of autoimmune thyroiditis

AIT is a chronic progressive disease in which an autoreactive process develops against the thyroid gland. This process involves all subpopulations of lymphocytes, manifested by mononuclear infiltration of the thyroid gland and signs of inflammation, leading to the gradual extinction of thyroid function with several complications.

Immunology of Recurrent Pregnancy Loss and Implantation Failure
DOI: https://doi.org/10.1016/B978-0-323-90805-4.00023-7

3 Prevalence

AIT is 5–10 times more common among women of reproductive age than men and the most frequent endocrine pathology [7]. Nowadays, researchers suggest three hypotheses for the underlying mechanism of AIT without overt hypothyroidism on reproductive function.

4 Underlying immunopathology

4.1 Antithyroid antibodies as an epiphenomenon

Some researchers assume that antithyroid antibodies (ATA) (like other organ-specific and especially nonorgan-specific antibodies) are markers of generalized T-cell dysfunction. This assumption was put forward more than 20 years ago when scientists questioned, "How can such different antibodies lead to the same result?" According to this hypothesis, ATA and other autoantibodies are an epiphenomenon that occurs with the background of generalized T-cell dysfunction [8,9]. It is known that shifted Th1/Th2 ratio in favor of Th1 immunity is associated with reproductive failures [10–12] and proinflammatory Th1 cytokines (IFN-γ, TNF-α, IL-2) contribute to the production of cytotoxic antibodies and activation of NK and T-cell effectors, leading to the damage of trophoblasts [11]. Stewart-Akers et al. reported a significant increase in IFN-γ and a decrease in antiinflammatory cytokines, such as IL-4 and IL-10, in the endometrial samples from women with AIT [13]. Kim et al. had similar results that women with AIT had increased TNF-α, cytotoxic NK cell counts, Th1/Th2 ratios in favor of Th1 immunity, and more common organ-specific autoantibodies in the peripheral blood [14]. Other studies also demonstrated the frequent combination of AIT with antiphospholipid antibodies [15], organ-specific autoantibodies, other autoimmune diseases, and increased NK cytotoxicity [16,17]. Additionally, there is an association of AIT with gynecological diseases, leading to infertility [18,19].

4.2 Antithyroid antibodies as a reason for the compromised function of the thyroid gland

Another point of view explains the role of AIT in infertility and recurrent pregnancy losses (RPLs) by developing thyroid disorders. During pregnancy, the need for thyroid hormones increases by an average of 47%, starting from 5 weeks of gestation to 16–20 weeks, reaching the plateau level and remaining approximately at the same level until delivery [19,20]. In 40% of women with ATA, overt or subclinical hypothyroidism is detected [21], and women with subclinical hypothyroidism enter pregnancy with the euthyroid state. However, many experts believe that these women have compromised thyroid function, which is not suitable to fully satisfy the body's increasing demand.

Ashoor et al. [22], after analyzing the levels of TSH, T4, T3, and ATA, concluded that the altered thyroid function affects pregnancy outcomes but not the presence of ATA. In this study, hormone levels did not go beyond the euthyroid state. Median TSH and free T4 of women with pregnancy losses and favorable pregnancy outcomes were 1.133 versus

1.007 multiple of median (MoM) and 0.958 versus 0.992 MoM, respectively. It was suggested that the thyroid function of women with pregnancy losses was reduced compared to women with favorable pregnancy outcomes while remaining within the euthyroid status. Similar data were obtained in a retrospective analysis by Negro et al. [23], examining data from women with euthyroid status who underwent assisted reproductive technology (ART) program. The presence of ATA alone did not affect either pregnancy rate or miscarriage. However, TSH level before pregnancy was significant for pregnancy outcomes. In women with antithyroperoxidase antibodies (anti-TPO) and unfavorable pregnancy outcomes, the TSH level was 2.8 mIU/L, while 1.6 mIU/L in women with anti-TPO and favorable pregnancy outcomes, and 1.1 mIU/L in women without anti-TPO. The results of the *meta*-analysis by He et al. supported the previous studies that ATA alone did not play a role if a woman does not have subclinical hypothyroidism [24]. Besides, Lepoutre et al. [25]. reported that among women with anti-TPO, adverse pregnancy outcomes were observed only in those who had not received levothyroxine therapy since the first prenatal visit. The subsequent *meta*-analysis, including almost 8000 women, came to a similar conclusion that levothyroxine treatment reduced the risk of pregnancy loss and premature birth in anti-TPO positive patients [26]. Similar results were obtained by others [27–29].

4.3 Mechanisms by which antithyroid antibodies can cause reproductive failures

4.3.1 *The theory of cross-reacting antibodies*

Currently, possible mechanisms of ATA that directly affect the tissues of the reproductive system negatively have been studied. It was shown that anti-TPO and antithyroglobulin antibodies (anti-TG) were present in the follicular fluid of women with AIT [30,31]. Concentrations of these antibodies were about half of the serum concentrations after passing the blood-ovarian barrier. In women with ATA, decreases in the percentage of fertilization (63% vs 72%, $p < 0.05$) and the proportion (%) of high-quality embryos (25% vs 48%, $p < 0.05$) were shown as compared to negative controls. Interestingly, in ATA-positive women, pregnancy occurred only in those for whom the in vitro fertilization/intracytoplasmic sperm injection procedure (IVF/ICSI) was performed, while in the control group, pregnancy occurred both in the cases of classic IVF and ICSI [30]. In general, the best results were observed in those with lower antibody concentrations, although the sample size was small. Weghofer et al. also investigated the quality of embryos obtained from women with and without ATA. The study included older reproductive-age women with a reduced ovarian reserve and euthyroid status. Patients were classified according to their TSH levels into those with TSH <2.5 mU/mL and those with 2.5–4.5 mU/mL. The qualities of embryos in women with TSH ≤2.5 or TSH >2.5 mU/mL were comparable. The presence of anti-TPO significantly worsened the quality of obtained embryos in the group with TSH ≤2.5 mU/mL ($p = 0.045$). In the group with higher levels of TSH and anti-TPO, the quality of embryos was worse, probably due to both components ($p = 0.057$) [32]. Medenica et al. did not show a decrease in the quality and number of embryos but noted a reduced rate of pregnancy in women with ATA for each induced cycle (30.8% vs 61.5%, $p = 0.026$) and embryo transfer (34.8% vs 66.7%, $p = 0.029$). Multivariate analysis showed that ATA-positive women have a lower chance to achieve pregnancy ($p = 0.004$, OR = 0.036, 95% CI 0.004–0.347) than age and body mass index-matched ATA-negative controls [31].

These data correspond well with the theory of cross-reacting antibodies to the thyroid tissue and the ovarian antigen, zona pellucida. This observation was made by Kelkar et al., who studied the cross-reactivity of antibodies from the sera of women suffering from autoimmune oophoritis and found that antibodies against zona pellucida can bind to the thyroid tissue in the mice model [33]. Thus this finding suggested that ATA could bind to zona pellucida. Damages in zona pellucida may lead to a disturbance in the early stages of embryogenesis, including fertilization, hatching, and implantation, as well as impaired immune recognition of the embryo. This assumption implies that IVF/ICSI procedure could prevent possible ATA-related complications [30]. Chen et al. evaluated the reproductive outcomes of IVF/ICSI in 235 women with ATA and 214 women with anti-TPO. There was no association between IVF/ICSI failures and the presence of ATA [34], suggesting that the use of the IVF/ICSI procedure probably compensated for the negative impact of ATA. The study by Tan et al. also found no negative effect of ATAs on pregnancy outcomes achieved with IVF/ICSI [35].

Besides, the impact of AIT with euthyroid status on the results of IVF/ICSI was evaluated by Lukaszuk et al. The presence of ATA did not affect fertilization, implantation, pregnancy, live birth, or miscarriage rates in women who received ICSI procedures [36]. The recent study performed by He et al. [37] revealed the influence of anti-TPO in euthyroid women. The outcome analysis of the first IVF/ICSI in 1170 women revealed that the fertilization rate in anti-TPO positive women with $2.5 \leq$ TSH < 4.2 mIU/L was significantly lower than in anti-TPO negative women. However, this result might be due to the synergistic effect of anti-TPO and TSH levels because the fertilization rate in this group was significantly lower than in other groups with TSH < 2.5 mIU/L. In anti-TPO positive women with $2.5 \leq$ TSH < 4.2 mIU/L, the fertilization rate was 73% compared to 82% in patients with $2.5 \leq$ TSH < 4.2 mIU/L and anti-TPO negative, 83% in women with $0.3 \leq$ TSH < 2.5 mIU/L and anti-TPO negative, and 84% in patients with $0.3 \leq$ TSH < 2.5 mIU/L and anti-TPO positive.

In women with premature ovarian depletion (which is often associated with autoimmune damage of ovaries), the incidence of ATA was higher (15.3% vs 6%, $p = 0.009$) than women with age-related depletion of ovarian reserve [38]. The presence of ATA was recognized as a risk factor for an adverse reproductive outcome (RR 2.8, 95% CI 1.2–6.3). The pregnancy rate was higher in women with age-related depletion of ovarian reserve (16.7% vs 8.7% per cycle, $p = 0.037$) than women with premature ovarian depletion [38]. It is suggested that a decrease in the pregnancy rate in women with premature ovarian depletion could be associated with disturbed endometrial receptivity because of autoimmune background in the reproductive organs [38]. Another study reported that the pregnancy rate through IVF/ICSI was decreased in ATA-positive women with euthyroid status. None of the euthyroid women with ATA (24 cases) got pregnant, while 23.9% of ATA-negative women achieved pregnancy. The two groups of women did not differ in the number of oocytes received and embryos transferred [39]. Contrarily, other studies indicated that either ATA or TSH levels did not affect the pregnancy rates, the percentages of miscarriage, or premature birth, regardless of the ART method [40–42].

4.3.2 *Animal studies to support the theory of cross-reacting antibodies*

In addition to the potential cross-reacting antigens from zona pellucida, there may be other targets for anti-TG and anti-TPO in reproductive organs. To determine the role of anti-TG in the development of reproductive failures and the mechanism of their action, Matalon et al. immunized mice with thyroglobulin [43]. The mice developed thyroid antibodies; however, hypothyroidism was not observed. Thyroid function was comparable to controls. The immunized group demonstrated an increased percentage of fetal resorption ($p = 0.04$) and a reduced average weight of offspring (194 ± 4 mg vs 240 ± 6 mg; $p < 0.001$) and placentas (105 ± 2 mg vs 130 ± 3 mg; $p < 0.001$). The binding of anti-TG on the placenta explained the negative effect of anti-TG on reproductive outcomes. To determine the target for anti-TG in the mice model, Moravej et al. evaluated the presence of thyroglobulin in the ovaries, placenta, and decidual tissue of mice in the early, middle, and late stages of pregnancy [44]. Thyroglobulin was not detected in the tissues by either the IHC method or the dot-blot method after homogenization of the placenta, ovary, and decidual tissue. In addition, the thyroglobulin gene was not detected in the reproductive tissues by the reverse transcriptase-polymerase chain reaction (RT-PCR). This study suggested that the negative effect of anti-TG was caused by either an indirect or a direct mechanism, but possibly due to cross-reaction of ATA with other antigens of reproductive organs with similar epitopes to thyroglobulin. In light of this study, it can be assumed that the anti-TG in the placentas of immunized mice were bound to cross-reacting antigens with a similar molecular structure [43].

4.3.3 *Possible influence of antithyroid antibodies on implantation*

Another target for ATA, thyroperoxidase (TPO), was found in human reproductive organs, such as the endometrium and placenta [45]. TPO was found in the cytoplasm of glandular and luminal epithelial cells in the proliferative stage endometrium. The highest expression density was observed on the luminal surface of endometrial epithelial cells. In addition, the presence of TPO in the first trimester villous trophoblast was also confirmed by the immunohistochemistry study [45]. These findings suggested that anti-TPO can directly interfere implantation process, leading to early pregnancy losses. Furthermore, biochemical pregnancy rates were reported to be 5.3 times higher in anti-TPO positive women than anti-TPO-negative women regardless of the TSH level ($p = 0.002$, OR = 5.311, 95% CI 1.859–15.169) [37]. This finding suggests that the anti-TPO impact the implantation process.

To investigate the underlying mechanisms of anti-TPO, a euthyroid-AIT animal model was developed by immunizing mice with recombinant mouse TPO [46]. The production of anti-TPO did not lead to changing the T4 level; however, the TSH level in the immunized group was significantly higher than in the control group while remaining within the reference ranges. In this euthyroid AIT animal model, an increase in the percentage of fetal resorption (14.75% vs 0%; $p < 0.05$) and a decrease in the average weight of the offspring were reported. Placental pathology was speculated for these changes since there were no changes in the implantation rate. When the uterine cavity was irrigated with anti-TPO sera on the eve of implantation, the implantation rate of the study group was not different from that of the controls. Thus it seems likely that anti-TPO has an effect after the implantation process is over [46].

In addition, an IHC study revealed that anti-TPO bind to the surface of 3–4 cells to morula stage embryos in the mice study [45]. Although a slower development was noted in 4-cell stage embryos, the further stages of embryo development, such as blastulation, and hatching, were not different from controls. Therefore the embryo could adapt to such conditions since the detrimental effect of anti-TPO probably did not occur in preimplantation embryos. Contrary to mice embryos, human preimplantation embryos at the blastocyst stage did not express TPO [45]. This discrepancy may be related to the embryo age; 3–4 cells to the morula stage embryo in mice model versus blastocyst stage in human embryos.

4.3.4 *Placenta is a possible target for antithyroid antibodies*

Rahnama et al. evaluated full-term placentas obtained during cesarean section to detect targets for anti-TPO. With the IHC study, TPO was detected in the syncytiotrophoblast and on the surface of the invasive trophoblast. Further, the presence of the TPO gene in these tissues was confirmed by RT-PCR. This finding suggests that anti-TPO can directly bind to placental tissue.

The presence of TPO on the surface of invasive trophoblasts allows the binding of anti-TPO to their target, damaging the membrane of the invasive trophoblast. The decreased invasive capacity of the trophoblast results in placental insufficiency often observed in patients with AIT. Previously, we demonstrated a placental insufficiency in mice passively immunized with TPO [47]. We believe that placental insufficiency in our murine model might explain the placental pathology in patients with AIT. The proximity to the invasive trophoblast in the murine placenta is a giant cell layer [48], which directly contacts the uterus. These cells are functionally comparable to multinucleated trophoblastic giant cells, syncytiotrophoblasts in the human decidua, representing the final stage of differentiation of the invasive trophoblast [49]. It is unclear if TPO is present in the murine placental tissue. However, in the group of mice immunized with TPO, the giant trophoblast cell layer and their cellular nuclei are decreased [47]. These findings support the study by Rahnama et al., which reported a possible TPO presence in the placenta.

The presence of anti-TPO targets in human syncytiotrophoblast may suggest that AIT may contribute to placental insufficiency, which leads to decreased placental hormone production and lessened trafficking of nutrients and oxygen to a fetus. This is because the syncytiotrophoblast plays a major role in transferring nutrients and oxygen to the human placenta and becomes the main source of placental hormones and growth factors.

In the murine placenta, syncytiotrophoblast is involved in the transport of nutrients and oxygen to the fetus since it is located at the intervascular barrier in the labyrinth part of the placenta. The intervascular barrier is consists of the following layers: the endothelium of the fetal capillaries, the basement membrane of the endothelium, the two-layer syncytiotrophoblasts, and the cytotrophoblast, lining the lumen of the sinuses filled with maternal blood [50]. In our experiment, thickening of the intervascular barriers in the labyrinth part of the placenta and deterioration of blood flow were predominantly detected in animals after anti-TPO sera injection [47]. In consideration of these findings, the barrier thickening is probably due to the syncytiotrophoblast damage. According to this fact, it seems that a direct target for these antibodies, TPO, might be present on the murine syncytiotrophoblast cells. Further research is needed.

5 Evaluation

Evaluation of thyroid function is an important part of prenatal care. Therefore it is often regulated by national guidelines. All over the world, the assessment of thyroid function is quite similar, and many national guidelines follow the comprehensive guidelines of the American Thyroid Association [51], which recommends assessing the level of TSH in all patients seeking for pregnancy or newly pregnant, and with risk factors of thyroid dysfunction, such as:

1. Known thyroid antibody positivity or the presence of a goiter;
2. History of the head or neck radiation or prior thyroid surgery;
3. Age >30 years;
4. Type 1 diabetes or other autoimmune disorders;
5. History of pregnancy loss, preterm delivery, or infertility;
6. Multiple prior pregnancies;
7. Family history of autoimmune thyroid disease or thyroid dysfunction;
8. Morbid obesity;
9. Use of amiodarone or lithium, or recent administration of iodinated radiologic contrast;
10. Residing in an area of known moderate to severe iodine insufficiency.

Moreover, pregnant women at risk for hypothyroidism, especially ATA-positive women, should be monitored with a serum TSH level every month until midgestation and at least once near 30 weeks gestation.

At the same time, ATA should be assessed in cases of TSH concentrations >2.5 mU/L. For example, routine screening for serum TSH level is recommended to all women planning a pregnancy or newly pregnant in the Russian Federation. The Russian Association of Endocrinologists oversees this practice [52], considering that almost the entire territory of the Russian Federation has an iodine deficiency. For ATA screening, TSH concentration >2.5 mU/L is an indication for ATA assessment.

6 Treatment

The guidelines of the American Thyroid Association [51] for ATA-positive euthyroid women are the followings: levothyroxine therapy can be considered for ATA-positive women with TSH concentrations above 2.5 mU/L and below 4 mU/L (in the majority of cases, TSH concentration 4 mU/L is considered as an upper limit of the pregnancy-specific reference range). This approach can be applied to women planning pregnancy, pregnant, and those who are undergoing ART cycles. In all cases, it is reasonable to target maternal TSH concentrations below 2.5 mU/L using levothyroxine therapy. In addition, women with subclinical hypothyroidism and especially overt hypothyroidism should be treated with levothyroxine. Glucocorticoid and intravenous immunoglobulin therapy are not recommended for ATA-positive euthyroid women [51].

7 Reproductive outcomes

At the moment, there is enough information about the independent role of ATA in the pathogenesis of reproductive failures. Studies that ATA can negatively affect reproductive function even in conditions of euthyroidism have been published [53–67]. In a study involving 90 ATA-positive and 676 ATA-negative women, the frequency of fertilization, implantation, and pregnancy rates in ATA-positive women were significantly lower (64.3% vs 74.6%, $p<0.001$, 17.8% vs 27.1%, $p<0.001$, 33.3% vs 46.7%, $p=0.002$) and the miscarriage rate was significantly higher (26.9% vs 11.8%, $p=0.002$) than those of ATA negative women. However, in this study, not all patients had known TSH. Hence, the data should be carefully interpreted [53].

In the group of euthyroid women with ATA, increased miscarriage and premature birth rates have been reported [54–59]. In addition, there was an increased risk of perinatal mortality in this group of women [60]. Contrarily, the study of Seungdamrong et al. evaluated 1,468 patients with a history of miscarriage using logistic regression analysis. The presence of anti-TPO before conception increased the risk of pregnancy loss (43.9% vs 25.3%, OR 2.17, 95% CI 1.12–4.22) and reduced the probability of live birth (17.1% vs 25.4%, OR 0.58, 95% CI 0.35–0.96) compared to patients without anti-TPO. However, the percentage of pregnancy in the groups did not differ significantly (33.3% vs 36.3%), and there was no effect of TSH levels on pregnancy outcomes in this study [61]. Meena et al. also published data that confirmed a negative impact of ATAs on pregnancy outcomes. Euthyroid women with ATA had increased percentages of miscarriage (13.33 vs 2.34%, $p<0.001$), the birth of small fetuses (25 vs 5.12%, $p<0.001$), and early preterm birth (<34 weeks) (5 vs 1.80%, $p>0.05$) compared to those without ATAs [62]. In a study performed by Poppe et al. [63], which included euthyroid women with anti-TPO ($n=32$) and without anti-TPO ($n=202$) who underwent IVF programs, the risk of miscarriage in the first trimester was significantly higher in women with anti-TPO (53% vs 23%, OR 3.77, 95% CI (1.29–11.05); $p=0.016$), but the percentage of pregnancies did not statistically differ from the control group.

Dendrinos et al. showed a significantly increased frequency of ATA in the group of women with RPL (37% vs 13%, $p<0.05$) [64], and in the study of Bellver et al., an increased frequency of ATA was described in women with implantation failures [65]. *Meta*-analyses by Dhillon-Smith et al. [66] and Chen [67] supported the negative impact of ATAs on pregnancy: an increase in the proportion of miscarriage and preterm birth in euthyroid women with anti-TPO; however, no improvement in reproductive outcomes was recorded in women who took levothyroxine [66].

8 Summary and recommendation

In summary, despite the numerous observed impacts of ATA on reproductive function, the underlying mechanism of ATA remains unclear. Most likely, human reproductive function suffers from the synergistic effect of several mechanisms described in this review. Therefore a complete and timely examination is crucial.

1. At the stage of pregnancy planning (regardless of whether it is natural or assisted), all women should be examined for serum TSH levels, and the screening for ATA should be considered.
2. If serum TSH is above 2.5 mU/L, a woman should be examined for ATA.
3. If a woman did not receive the evaluation before pregnancy, TSH should be tested during the first prenatal visit.
4. In a woman with positive ATA and TSH level below 2.5 mU/L, periodic TSH check-up is recommended up to mid-gestation and at least once near 30 weeks gestation (due to growing strain on the thyroid gland until the term).
5. In case of positive ATA and TSH higher than 2.5 mU/L, levothyroxine therapy should be prescribed (25–50 μg/day as a starting dose).
6. The goal of therapy in pregnant patients is to reduce TSH below 2.5 mU/L.

References

[1] Zhang Y, Wang H, Pan X, Teng W, Shan Z. Patients with subclinical hypothyroidism before 20 weeks of pregnancy have a higher risk of miscarriage: a systematic review and meta-analysis. PLoS One 2017;12(4). Available from: https://doi.org/10.1371/journal.pone.0175708.

[2] Maraka S, Ospina NMS, O'Keeffe DT, et al. Subclinical hypothyroidism in pregnancy: a systematic review and meta-analysis. Thyroid 2016;26(4):580–90. Available from: https://doi.org/10.1089/thy.2015.0418.

[3] Shinohara DR, Santos TDS, De Carvalho HC, et al. Pregnancy complications associated with maternal hypothyroidism: a systematic review. Obstet Gynecol Surv 2018;73(4):219–30. Available from: https://doi.org/10.1097/OGX.0000000000000547.

[4] Drover SSM, Villanger GD, Aase H, et al. Maternal thyroid function during pregnancy or neonatal thyroid function and attention deficit hyperactivity disorder: a systematic review. Epidemiology 2019;30(1):130–44. Available from: https://doi.org/10.1097/EDE.0000000000000937.

[5] Liu Y, Chen H, Jing C, Li F. The association between maternal subclinical hypothyroidism and growth, development, and childhood intelligence: a meta-analysis. JCRPE J Clin Res Pediatr Endocrinol 2018;10(2):153–61. Available from: https://doi.org/10.4274/jcrpe.4931.

[6] Petnehazy E, Buchinger W. Hashimoto thyroiditis, therapeutic options and extrathyroidal options–an up-to-date overview. Wiener Medizinische Wochenschrift 2020;170(1–2):26–34. Available from: https://doi.org/10.1007/s10354-019-0691-1.

[7] Ralli M, Angeletti D, Fiore M, et al. Hashimoto's thyroiditis: an update on pathogenic mechanisms, diagnostic protocols, therapeutic strategies, and potential malignant transformation. Autoimmun Rev 2020;19(10):102649. Available from: https://doi.org/10.1016/j.autrev.2020.102649.

[8] Gleicher N. Autoimmunity and reproductive failure. Ann N Y Acad Sci 1991;626(1 Frontiers in):537–44. Available from: https://doi.org/10.1111/j.1749-6632.1991.tb37945.x.

[9] Gleicher N. Some thoughts on the reproductive autoimmune failure syndrome (RAFS) and Th-1 vs Th-2 immune responses. Am J Reprod Immunol 2002;48(4):252–4. Available from: https://doi.org/10.1034/j.1600-0897.2002.01111.x.

[10] Kwak-Kim JYH, Chung-Bang HS, Ng SC, et al. Increased T helper 1 cytokine responses by circulating T cells are present in women with recurrent pregnancy losses and in infertile women with multiple implantation failures after IVF. Hum Reprod 2003;18(4):767–73. Available from: https://doi.org/10.1093/humrep/deg156.

[11] Raghupathy R, Makhseed M, Azizieh R, Omu A, Gupta M, Farhat R. Cytokine production by maternal lymphocytes during normal human pregnancy and in unexplained recurrent spontaneous abortion. Hum Reprod 2000;15(3):713–18. Available from: https://doi.org/10.1093/humrep/15.3.713.

[12] Raghupathy R. Th1-type immunity is incompatible with successful pregnancy. Immunol Today 1997;18(10):478–82. Available from: https://doi.org/10.1016/S0167-5699(97)01127-4.

[13] Stewart-Akers AM, Krasnow JS, Brekosky J, DeLoia JA. Endometrial leukocytes are altered numerically and functionally in women with implantation defects. Am J Reprod Immunol 1998;39(1):1–11. Available from: https://doi.org/10.1111/j.1600-0897.1998.tb00326.x.

[14] Kim NY, Cho HJ, Kim HY, et al. Thyroid autoimmunity and its association with cellular and humoral immunity in women with reproductive failures. Am J Reprod Immunol 2011;65(1):78–87. Available from: https://doi.org/10.1111/j.1600-0897.2010.00911.x.

[15] Promberger R, Walch K, Seemann R, Pils S, Ott J. A retrospective study on the association between thyroid autoantibodies with β2-glycoprotein and cardiolipin antibodies in recurrent miscarriage. 2017;16 <http://ijaai.tums.ac.ir> [accessed 12.03.21].

[16] Kaider AS, Kaider BD, Janowicz PB, Roussev RG. Immunodiagnostic evaluation in women with reproductive failure. Am J Reprod Immunol 1999;42(6):335–46. Available from: https://doi.org/10.1111/j.1600-0897.1999.tb00110.x.

[17] Conigliaro P, D'Antonio A, Pinto S, et al. Autoimmune thyroid disorders and rheumatoid arthritis: a bidirectional interplay. Autoimmun Rev 2020;19(6):102529. Available from: https://doi.org/10.1016/j.autrev.2020.102529.

[18] Ho CW, Chen HH, Hsieh MC, et al. Increased risk of polycystic ovary syndrome and it's comorbidities in women with autoimmune thyroid disease. Int J Environ Res Public Health 2020;17(7). Available from: https://doi.org/10.3390/ijerph17072422.

[19] Twig G, Shina A, Amital H, Shoenfeld Y. Pathogenesis of infertility and recurrent pregnancy loss in thyroid autoimmunity. J Autoimmun 2012;38(2–3). Available from: https://doi.org/10.1016/j.jaut.2011.11.014.

[20] Alexander EK, Marqusee E, Lawrence J, Jarolim P, Fischer GA, Larsen PR. Timing and magnitude of increases in levothyroxine requirements during pregnancy in women with hypothyroidism. N Engl J Med 2004;351(3):241–9. Available from: https://doi.org/10.1056/nejmoa040079.

[21] Dhanwal D, Bajaj S, Rajput R, et al. Prevalence of hypothyroidism in pregnancy: an epidemiological study from 11 cities in 9 states of India. Indian J Endocrinol Metab 2016;20(3):387–90. Available from: https://doi.org/10.4103/2230-8210.179992.

[22] Ashoor G, Maiz N, Rotas M, Jawdat F, Nicolaides KH. Maternal thyroid function at 11 to 13 weeks of gestation and subsequent fetal death. Thyroid 2010;20(9):989–93. Available from: https://doi.org/10.1089/thy.2010.0058.

[23] Negro R, Formosa G, Coppola L, et al. Euthyroid women with autoimmune disease undergoing assisted reproduction technologies: the role of autoimmunity and thyroid function. J Endocrinol Invest 2007;30 (1):3–8. Available from: https://doi.org/10.1007/BF03347388.

[24] He H, Jing S, Gong F, Tan YQ, Lu GX, Lin G. Effect of thyroid autoimmunity per se on assisted reproduction treatment outcomes: a meta-analysis. Taiwan J Obstet Gynecol 2016;55(2):159–65. Available from: https://doi.org/10.1016/j.tjog.2015.09.003.

[25] Lepoutre T, Debiève F, Gruson D, Daumerie C. Reduction of miscarriages through universal screening and treatment of thyroid autoimmune diseases. Gynecol Obstet Invest 2012;74(4):265–73. Available from: https://doi.org/10.1159/000343759.

[26] Rao M, Zeng Z, Zhou F, et al. Effect of levothyroxine supplementation on pregnancy loss and preterm birth in women with subclinical hypothyroidism and thyroid autoimmunity: a systematic review and meta-analysis. Hum Reprod Update 2019;25(3):344–61. Available from: https://doi.org/10.1093/humupd/dmz003.

[27] Velkeniers B, Van Meerhaeghe A, Poppe K, Unuane D, Tournaye H, Haentjens P. Levothyroxine treatment and pregnancy outcome in women with subclinical hypothyroidism undergoing assisted reproduction technologies: systematic review and meta-analysis of RCTs. Hum Reprod Update 2013;19(3):251–8. Available from: https://doi.org/10.1093/humupd/dms052.

[28] Thangaratinam S, Tan A, Knox E, Kilby MD, Franklyn J, Coomarasamy A. Association between thyroid autoantibodies and miscarriage and preterm birth: meta-analysis of evidence. BMJ 2011;342(7806). Available from: https://doi.org/10.1136/bmj.d2616.

[29] Bartáková J, Potluková E, Rogalewicz V, et al. Screening for autoimmune thyroid disorders after spontaneous abortion is cost-saving and it improves the subsequent pregnancy rate. BMC Pregnancy Childbirth 2013;13. Available from: https://doi.org/10.1186/1471-2393-13-217.

[30] Monteleone P, Parrini D, Faviana P, et al. Female infertility related to thyroid autoimmunity: the ovarian follicle hypothesis. Am J Reprod Immunol 2011;66(2):108–14. Available from: https://doi.org/10.1111/j.1600-0897.2010.00961.x.

[31] Medenica S, Garalejic E, Arsic B, et al. Follicular fluid thyroid autoantibodies, thyrotropin, free thyroxine levels and assisted reproductive technology outcome. PLoS One 2018;13(10). Available from: https://doi.org/10.1371/journal.pone.0206652.

[32] Weghofer A, Himaya E, Kushnir VA, Barad DH, Gleicher N. The impact of thyroid function and thyroid autoimmunity on embryo quality in women with low functional ovarian reserve: a case-control study. Reprod Biol Endocrinol 2015;13(1). Available from: https://doi.org/10.1186/s12958-015-0041-0.

[33] Kelkar RL, Meherji PK, Kadam SS, Gupta SK, Nandedkar TD. Circulating auto-antibodies against the zona pellucida and thyroid microsomal antigen in women with premature ovarian failure. J Reprod Immunol 2005;66(1):53–67. Available from: https://doi.org/10.1016/j.jri.2005.02.003.

[34] Chen X, Mo ML, Huang CY, et al. Association of serum autoantibodies with pregnancy outcome of patients undergoing first IVF/ICSI treatment: a prospective cohort study. J Reprod Immunol 2017;122:14–20. Available from: https://doi.org/10.1016/j.jri.2017.08.002.

[35] Tan S, Dieterle S, Pechlavanis S, Janssen OE, Fuhrer D. Thyroid autoantibodies per se do not impair intracytoplasmic sperm injection outcome in euthyroid healthy women. Eur J Endocrinol 2014;170(4):495–500. Available from: https://doi.org/10.1530/EJE-13-0790.

[36] Łukaszuk K, Kunicki M, Kulwikowska P, et al. The impact of the presence of antithyroid antibodies on pregnancy outcome following intracytoplasmatic sperm injection-ICSI and embryo transfer in women with normal thyreotropine levels. J Endocrinol Invest 2015;38(12):1335–43. Available from: https://doi.org/10.1007/s40618-015-0377-5.

[37] He Q, Zhang Y, Qiu W, Fan J, Zhang C, Kwak-Kim J. Does thyroid autoimmunity affect the reproductive outcome in women with thyroid autoimmunity undergoing assisted reproductive technology? Am J Reprod Immunol 2020;84(6):e13321. Available from: https://doi.org/10.1111/aji.13321 Epub 2020 Sep 9. PMID:33119203.

[38] Beydilli Nacak G, Ozkaya E, Yayla Abide C, Bilgic BE, Devranoglu B, Gokcen Iscan R. The impact of autoimmunity-related early ovarian aging on ICSI cycle outcome. Gynecol Endocrinol 2018;34(11):940–3. Available from: https://doi.org/10.1080/09513590.2018.1469612.

[39] Fumarola A, Grani G, Romanzi D, et al. Thyroid function in infertile patients undergoing assisted reproduction. Am J Reprod Immunol 2013;70(4):336–41. Available from: https://doi.org/10.1111/aji.12113.

[40] Unuane D, Velkeniers B, Bravenboer B, et al. Impact of thyroid autoimmunity in euthyroid women on live birth rate after IUI. Hum Reprod 2017;32(4):915–22. Available from: https://doi.org/10.1093/humrep/dex033.

[41] Unuane D, Velkeniers B, Deridder S, Bravenboer B, Tournaye H, De Brucker M. Impact of thyroid autoimmunity on cumulative delivery rates in in vitro fertilization/intracytoplasmic sperm injection patients. Fertil Steril 2016;106(1):144–50. Available from: https://doi.org/10.1016/j.fertnstert.2016.03.011.

[42] Chai J, Yeung WYT, Lee CYV, Li HWR, Ho PC, Ng HYE. Live birth rates following in vitro fertilization in women with thyroid autoimmunity and/or subclinical hypothyroidism. Clin Endocrinol (Oxf) 2014;80(1):122–7. Available from: https://doi.org/10.1111/cen.12220.

[43] Matalon ST, Blank M, Levy Y, et al. The pathogenic role of anti-thyroglobulin antibody on pregnancy: evidence from an active immunization model in mice. Hum Reprod 2003;18(5):1094–9. Available from: https://doi.org/10.1093/humrep/deg210.

[44] Moravej A, Jeddi-Tehrani M, Salek-Moghaddam AR, et al. Evaluation of thyroglobulin expression in murine reproductive organs during pregnancy. Am J Reprod Immunol 2010;64(2):97–103. Available from: https://doi.org/10.1111/j.1600-0897.2010.00827.x.

[45] Rahnama R, Mahmoudi AR, Kazemnejad S, et al. Thyroid peroxidase in human endometrium and placenta: a potential target for anti-TPO antibodies. Clin Exp Med 2021;21(1):79–88. Available from: https://doi.org/10.1007/s10238-020-00663-y.

[46] Lee YL, Ng HP, Lau KS, et al. Increased fetal abortion rate in autoimmune thyroid disease is related to circulating TPO autoantibodies in an autoimmune thyroiditis animal model. Fertil Steril 2009;91(5 SUPPL):2104–9. Available from: https://doi.org/10.1016/j.fertnstert.2008.07.1704.

[47] Borodina E, Katz I, Antonelli A, et al. The pathogenic role of circulating Hashimoto's thyroiditis-derived TPO-positive IgG on fetal loss in naïve mice. Am J Reprod Immunol 2021;85(1). Available from: https://doi.org/10.1111/aji.13331.

[48] Boltovskaya MN, Artemyeva KA, Nazimova SV, Stepanova II, Starosvetskaya NA, Kalujin OV. Structural and functional comparison of human and mouse placenta as a justification for modeling miscarriage. Clin Exp Morphol 2014;1(9):73–8 Russian.

[49] Aĭlamazian E, Stepanova O, Selkov S, Sokolov D. Cells of immune system of mother and trophoblast cells: constructive cooperation for the sake of achievement of the joint purpose. Vestnik Rossiĭ skoĭ Akademii Meditsinskikh Nauk 2013;01(1):12–2149 Russian.
[50] Watson ED, Cross JC. Development of structures and transport functions in the mouse placenta. Physiology 2005;20(3):180–93. Available from: https://doi.org/10.1152/physiol.00001.2005.
[51] Alexander EK, Pearce EN, Brent GA, Brown RS, Chen H, Dosiou C, et al. Guidelines of the American thyroid association for the diagnosis and management of thyroid disease during pregnancy and the postpartum. Thyroid 2017;27(3):315–89. Available from: https://doi.org/10.1089/thy.2016.0457.
[52] Russian Association of Endocrinologists. Clinical Recommendations: Hypothyroidism, 2019. <https://rae-org.ru/system/files/documents/pdf/568_gipotireoz_vzroslye.finalnaya.versiya.pdf> Russian.
[53] Zhong YP, Ying Y, Wu HT, et al. Relationship between antithyroid antibody and pregnancy outcome following in vitro fertilization and embryo transfer. Int J Med Sci 2012;9(2):121–5. Available from: https://doi.org/10.7150/ijms.3467.
[54] Rajput R, Yadav T, Seth S, Nanda S. Prevalence of thyroid peroxidase antibody and pregnancy outcome in euthyroid autoimmune positive pregnant women from a tertiary care center in Haryana. Ind J Endocrinol Metab 2017;21(4):577–80. Available from: https://doi.org/10.4103/ijem.IJEM_397_16.
[55] Meena M, Chopra S, Jain V, Aggarwal N. The effect of antithyroid peroxidase antibodies on pregnancy outcomes in euthyroid women. J Clin Diagnostic Res 2016;10(9):QC04–7. Available from: https://doi.org/10.7860/JCDR/2016/19009.8403.
[56] Han Y, Mao LJ, Ge X, et al. Impact of maternal thyroid autoantibodies positivity on the risk of early term birth: Ma'anshan birth cohort study. Endocrine 2018;60(2):329–38. Available from: https://doi.org/10.1007/s12020-018-1576-6.
[57] Negro R. Thyroid autoimmunity and preterm delivery: brief review and meta-analysis. J Endocrinol Invest 2011;34(2):155–8. Available from: https://doi.org/10.1007/bf03347047.
[58] Negro R, Schwartz A, Gismondi R, Tinelli A, Mangieri T, Stagnaro-Green A. Thyroid antibody positivity in the first trimester of pregnancy is associated with negative pregnancy outcomes. J Clin Endocrinol Metab 2011;96(6). Available from: https://doi.org/10.1210/jc.2011-0026.
[59] Stagnaro-Green A, Roman SH, Cobin RH, Harazy E, Alvarez Marfany M, Davies TF. Detection of at-risk pregnancy by means of highly sensitive assays for thyroid autoantibodies. JAMA J Am Med Assoc 1990;264(11):1422–5. Available from: https://doi.org/10.1001/jama.1990.03450110068029.
[60] Männistö T, Vääräsmäki M, Pouta A, et al. Perinatal outcome of children born to mothers with thyroid dysfunction or antibodies: a prospective population-based cohort study. J Clin Endocrinol Metab 2009;94(3):772–9. Available from: https://doi.org/10.1210/jc.2008-1520.
[61] Seungdamrong A, Steiner AZ, Gracia CR, et al. Preconceptional antithyroid peroxidase antibodies, but not thyroid-stimulating hormone, are associated with decreased live birth rates in infertile women. Fertil Steril 2017;108(5):843–50. Available from: https://doi.org/10.1016/j.fertnstert.2017.08.026.
[62] Meena A, Nagar P. Pregnancy outcome in euthyroid women with antithyroid peroxidase antibodies. J Obstet Gynecol India 2016;66(3):160–5. Available from: https://doi.org/10.1007/s13224-014-0657-6.
[63] Poppe K, Glinoer D, Tournaye H, et al. Assisted reproduction and thyroid autoimmunity: an unfortunate combination? J Clin Endocrinol Metab 2003;88(9):4149–52. Available from: https://doi.org/10.1210/jc.2003-030268.
[64] Dendrinos S, Papasteriades C, Tarassi K, Christodoulakos G, Prasinos G, Creatsas G. Thyroid autoimmunity in patients with recurrent spontaneous miscarriages. Gynecol Endocrinol 2000;14(4):270–4. Available from: https://doi.org/10.3109/09513590009167693.
[65] Bellver J, Soares SR, Álvarez C, et al. The role of thrombophilia and thyroid autoimmunity in unexplained infertility, implantation failure and recurrent spontaneous abortion. Hum Reprod 2008;23(2):278–84. Available from: https://doi.org/10.1093/humrep/dem383.
[66] Dhillon-Smith RK, Coomarasamy A. TPO antibody positivity and adverse pregnancy outcomes. Best Pract Res Clin Endocrinol Metab 2020;34(4). Available from: https://doi.org/10.1016/j.beem.2020.101433.
[67] Chen L, Hu R. Thyroid autoimmunity and miscarriage: a meta-analysis. Clin Endocrinol (Oxf) 2011;74(4):513–19. Available from: https://doi.org/10.1111/j.1365-2265.2010.03974.x.

CHAPTER

11

Fetal/neonatal alloimmune-mediated thrombocytopenia and recurrent pregnancy loss

Si won Lee[1], *Tiffany Alexis Clinton*[2] *and Sun Kwon Kim*[3]

[1]Department of Obstetrics and Gynecology, Mount Sinai Medical Center, Miami Beach, FL, United States [2]Department of Obstetrics and Gynecology, Henry Ford Health System, Detroit, MI, United States [3]Department of Obstetrics and Gynecology, Division of Maternal-Fetal Medicine, Henry Ford Health System, Detroit, MI, United States

1 Introduction

Fetal and neonatal immune-mediated thrombocytopenia (FNAIT) is a life-threatening disease characterized by severe thrombocytopenia caused by a maternal immune response against fetal platelet antigens, resulting in the destruction of fetal platelets [1]. Clinical manifestations can vary, from asymptomatic disease to severe intracranial hemorrhage (ICH) of the fetus and neonate, recurrent miscarriages, fetal growth restriction (FGR), or fetal and neonatal death [1]. Incompatibility between maternal and fetal human platelet antigens (HPAs) can lead to maternal alloantibody formation against the fetal platelets that results in fetal platelet destruction and thrombocytopenia.

2 Definition

FNAIT is defined and categorized based on the history of fetal thrombocytopenia, the presence of maternal alloantibodies to HPAs, and a history of fetal and neonatal ICH in prior pregnancies (Table 1) [2,3].

- *Unknown risk*: A history of a previous infant with thrombocytopenia with or without ICH of unknown etiology and negative maternal HPA antibody.

Immunology of Recurrent Pregnancy Loss and Implantation Failure
DOI: https://doi.org/10.1016/B978-0-323-90805-4.00014-6

TABLE 1 Definition, monitoring, and management of fetal/neonatal alloimmune thrombocytopenia by history-based risk stratification.

Risk	Definition	Monitoring	Medical management	Delivery
Unknown	Previous pregnancy with fetal/neonatal thrombocytopenia or ICH of unknown etiology. No detected HPA antibody	Serial testing with anti-HPA antibodies and serological crossmatching with paternal platelets every 4 weeks, starting at 8 weeks gestation	Empiric treatment not recommended	
Standard	Previous infant with thrombocytopenia but without ICH. Serologically confirmed FNAIT	Monitor hemoglobin and hematocrit monthly Detailed anatomy ultrasound scan at 16–20 weeks Serial ultrasound every 4–6 weeks until delivery to detect ICH	1. Start therapy at 20 weeks • IVIG 2 g/kg/week **OR** IVIG 1 g/kg/week plus prednisone 0.5 mg/kg/day 2. At 32 weeks of GA, switch all patients • -IVIG 2 g/kg/week plus prednisone 0.5 mg/kg/day	CS at 37–38 weeks
High	Previous infant with thrombocytopenia and ICH that developed after 28 weeks of GA or during the neonatal period. Serologically confirmed FNAIT.		1. Start at 12 weeks • IVIG 1 g/kg/week 2. At 20 weeks • IVIG 1 g/kg/week plus prednisone 0.5 mg/kg/day **OR** increase IVIG to 2 g/kg/week 3. At 28 weeks • IVIG 2 g/kg/week plus prednisone 0.5 mg/kg/day	CS at 37–38 weeks
Extremely high	Previous infant with thrombocytopenia and ICH that developed at less than 28 weeks of GA. Serologically confirmed FNAIT		1. Start at 12 weeks • IVIG 2 g/kg/week 2. At 20 weeks • IVIG 2 g/kg/week plus prednisone 1 mg/kg/day until delivery	CS at 36–37 weeks
Modified IVIG protocol	To provide earlier suppression of platelet antibodies.	Monitor CBC, anti-HPA antibody and crossmatching with paternal platelets monthly. Detailed anatomy ultrasound scan at 16–20 weeks. Serial ultrasound every 4 weeks until delivery to detect ICH.	1. IVIG 400 mg/kg/week is initiated before conception or at 4–6 weeks of gestation with prednisone 20 mg per day. 2. At 16 weeks, IVIG is increased to 1 g/kg/week. 3. At 32 weeks, IVIG 2 g/kg/week with prednisone 20 mg per day.	—

FNAIT, Fetal/neonatal alloimmune thrombocytopenia; *ICH*, intracranial hemorrhage; *GA*, gestational age; *IVIG*, intravenous immunoglobulin; *HPA*, human platelet antigen; *CS*, cesarean section.

- *Standard risk*: A history of a previous infant with thrombocytopenia without ICH and serologically confirmed FNAIT.
- *High-risk*: A history of a previous infant with thrombocytopenia and ICH developed after 28 weeks of gestation or during the neonatal period and serologically confirmed FNAIT.
- *Extremely high-risk*: A history of the previous infant with thrombocytopenia and ICH that developed at or before 28 weeks of gestation and serologically confirmed FNAIT.

The definition of neonatal thrombocytopenia is a platelet count below 150,000/μL. This can be further classified into mild (platelet 100,000–150,000/μL), moderate (platelet 50,000–99,000/μL) and severe thrombocytopenia (platelet <50,000/μL) [4].

3 Prevalence

Severe thrombocytopenia in FNAIT has a frequency of 1 in 1000–1200 per live birth [5]. ICH related to FNAIT occurs in 0.02 to 0.1 in 1000 live births that can be detected as early as 14–16 weeks of gestation by routine ultrasound. FNAIT frequently develops before 28 weeks with fatal outcomes [6–8]. Postnatal ICH is reported in about 10%–20% of FNAIT neonates and is one of the major causes of long-term disabilities or neonatal death in 5% of cases [9]. Several large prospective studies have shown that the most common trigger for FNAIT is HPA-1 histo-incompatibility between the maternal and fetal HPA antigens. The prevalence of HPA-1b/1b (1a negative) is about 2.5% in Caucasian women [3].

4 Underlying immunopathology related to reproduction

4.1 Maternal allosensitization by fetal human platelet antigen

Presently, 37 types of alloantigens are described to be related to FNAIT. The majority of FNAIT cases are due to HPA-1 antigen incompatibility, accounting for almost 85% of the cases in families of Caucasian and African ancestry [1]. About half of these alloantigens are located on the domains of integrin β3 subunit, which are expressed on platelets, endothelial cells, and trophoblast/placental cells [1,10]. Integrins are early contributors to blastocyst implantation and spiral artery remodeling, hence playing an essential role in placental development [1,10].

An HPA-1a negative (1b/1b) mother who carries an HPA-1a baby produces immunoglobulin G alloantibodies specific to HPA-1a antigens from the fetal circulation. These alloantibodies then cross the placenta via Fc receptor transport and target the fetal platelets, resulting in fetal platelet opsonization and destruction [3,10–12]. Thus, the determination of paternal zygosity plays a major role in the prediction of FNAIT. If a mother is negative for HPA-1a (1b/1b), 100% of fetuses with an HPA-1a homozygous father and 50% of fetuses with a heterozygous father will be histoincompatible to the mother's HPA-1a antigen. The second most commonly involved HPA for FNAIT is HPA-5b antigen in Caucasians and HPA-4b in Asians [11]. In a study of 569 FNAIT, anti-HPA-1a antibody was present in 75.2% of cases, followed by anti-HPA-5b antibody (17.8%), anti-HPA-1a + HPA-5b antibodies (2.3%), anti-HPA 3a antibody (1.8%) and antibodies against low-frequency alloantigens (HPA-8bW, -11bW, -12bW, -13bW, and -14bW, 1.1% total) [13].

4.2 Thrombosis and cellular immune activation

A few studies using murine FNAIT models suggested that the immune complex formed with fetal platelet antigen can initiate thrombosis leading to placental pathology which contributes to miscarriages, FGR, and fetal deaths [1,9]. These studies showed that anti-β3 integrin antibodies target fetal semiallogeneic trophoblasts and provide binding sites for natural killer (NK) cell Fc-δ receptors [9]. In normal pregnancy, uterine NK cells participate in the spiral artery remodeling process while protecting the placenta from pathogens and consequential miscarriages. In FNAIT cases, increased immune complexes deposit into the maternal-fetal interface and activate uterine NK cells. This results in chronic inflammation at the maternal-fetal junction, leading to trophoblast apoptosis, impaired spiral artery remodeling, and placental dysfunction [1,9,14]. Contrariwise, NK cell-mediated placental inflammation and antibody-dependent cell-mediated cytotoxicity of invasive trophoblasts play a crucial role in developing FNAIT, pregnancy failure, FGR, ICH, intrauterine fetal death, or neonatal death [1,14]. Expansion of $CD3^{-}CD56^{+}$ NK and $CD3^{+}CD56^{+}$ NKT cells subsets has been reported in patients with active ITP, which has a temporary negative correlation with periods of disease progression or remission [15,16].

In a case series of 19 women with FNAIT, significantly higher TNF-α/IL-10 (33.58 ± 7.56 vs 23.93 ± 5.27, $P = .005$) and IFN-γ/IL-10 producing Th cell ratios (18.36 ± 4.51 vs 9.21 ± 4.22, $P = .000$) were reported compared to those of controls [6]. NK cell levels ($10.68 \pm 5.01\%$ vs $9.80 \pm 2.01\%$, $P =$ NS) and cytotoxicity (23.09 ± 6.21 vs 20.88 ± 3.39, $P =$ NS) were higher in the FNAIT group than those of normal controls but not statistically significant [6]. Hence, FNAIT is significantly associated with increased Th1/Th2 cell ratios and increased NK cell [6] numbers and cytotoxicities, which have been reported to be the risk factors for pregnancy losses and preeclampsia [6,17].

5 Evaluation

5.1 Diagnostic criteria

Suspicion for FNAIT is raised based on the presence of thrombocytopenia or ICH in the perinatal period. However, diagnosis can only be made based on confirmatory lab values.

Clinical criteria:

1. Presence of thrombocytopenia in a neonate at birth or within seven days after birth below 100,000/microL.
2. Presence of fetal ICH [18,19].

Laboratory diagnostic criteria

1. Incompatibility of maternal and paternal HPA antigens.
2. Maternal anti-HPA antibodies are specific to the incompatible antigen [18].

5.2 Clinical features

The mothers of the affected infants are usually asymptomatic. Patients can present with a personal or a family history of FNAIT. Infants with mild thrombocytopenia generally do not show any signs or symptoms and are usually diagnosed during postnatal workup. In infants with severe thrombocytopenia, ICH is the most common and severe complication detected during prenatal ultrasounds as early as 14–16 weeks of gestation [6,7]. FNAIT may also present with recurrent miscarriages, FGR, and fetal death [1]. Unlike Rh alloimmunization, FNAIT is detected in more than 50% of primary pregnancies of at-risk women. The risk and severity of FNAIT are increased with each subsequent pregnancy [7,20–22], and particularly, the severity of FNAIT is dependent on the maternal HPA-1a antibody level [18]. The recurrence rate is almost 100% in subsequent pregnancies with the added complication of earlier onset and more severe symptoms of FNAIT [9].

5.3 Testing modalities

5.3.1 *Candidates*

The routine screening of FNAIT is not performed in current practices due to the concern of the low prevalence and the cost of the testing [2]. However, the prevalence is more frequent than previously reported, and the cost of PCR testing has become reasonable enough to apply screening test to high-risk populations. Most FNAIT cases are diagnosed antenatally during the first pregnancy or at the time of delivery. When detected during prenatal care or the neonatal period, evaluation for bacterial and viral infections, disseminated intravascular coagulation, and other congenital conditions associated with thrombocytopenia should be undertaken. After these causes are excluded, diagnostic modalities, including maternal and paternal HPA typing and maternal platelet HPA antibody testing using sensitive assays, should be performed.

5.3.2 *Assays*

Currently available third-generation assays are monoclonal antibody immobilization of platelet antigens, modified antigen-capture enzyme-linked immunosorbent assay, and simultaneous analysis of specific platelet antibodies [23]. In addition to these in-house assays, commercial glycoprotein (GP)-specific assays are available, such as PakPlus ELISA (Immucor, Dreieich, Germany) and the bead-based PakLx immunoassay (Immucor, Dreieich, Germany). Free platelet auto- and alloantibodies in plasma can be detected using direct and indirect platelet antibody tests using intact platelets or with GP-specific tests. GP-specific immunoassays are state-of-the-art in platelet antibody testing. Additional techniques are also available, including eluates, immunological crossmatch, and anti-HPA-1a quantitation.

Antiplatelet antibody screening and identification are often performed by serological/immunological platelet crossmatch with a platelet test panel. The platelet test panel is comprised of platelet collections or platelet GP preparations representing all relevant platelet GPs. HPA histocompatibility between the couple can be confirmed by HPA typing by DNA-based methods. When incompatibility at HPA loci (1, 2, 3, 4, 5, 6, 9, or 15) and specific maternal anti-HPA antibodies are identified, a diagnosis of FNAIT is established [4].

5.4 Screening and diagnosis

Universal screening of FNAIT is not currently recommended [2]. However, screening of couples at risk is considered reasonable. Fig. 1 is the recommended algorithm for FNAIT diagnosis and management. At-risk couples include a personal or family history of FNAIT or ICH in previous pregnancies. Screening is done using maternal platelet counts, maternal and paternal HPA typing for incompatibility screen, and maternal HPA antibody testing [24].

If there is incompatibility at HPA loci between the couple, but negative maternal anti-HPA antibody, maternal anti-HPA antibody testing and/or crossmatch testing with paternal platelets should be repeated every 4 weeks starting at 8 weeks gestation to allow for early detection of sensitization and treatment. Previously, screening at 12, 24, and 32 weeks gestation has also been recommended; however, this approach may miss early detection of alloimmunity during pregnancy [4,6]. If all of the results are negative, then FNAIT can be excluded [4].

When FNAIT is suspected during pregnancy, fetal platelet antigen type can be determined by obtaining fetal DNA from chorionic villus sampling (CVS) or amniocentesis. Theoretically, noninvasive cell-free DNA (cfDNA) testing can be another option, but this is not yet wildly available in the U.S. [25,26]. If serological tests take time to result, treatment should be initiated based on the clinical suspicion and fetal platelet count (Table 1).

If HPA alloantibodies are present, serial titers may be helpful to determine the risk of developing FNAIT, and these women should be referred to comprehensive care centers with experience in FNAIT management. In addition, patients at risk should be monitored every 4 weeks beginning at 8–12 weeks of gestation until delivery with serial antibody testing and ultrasonography to screen for fetal ICH [6].

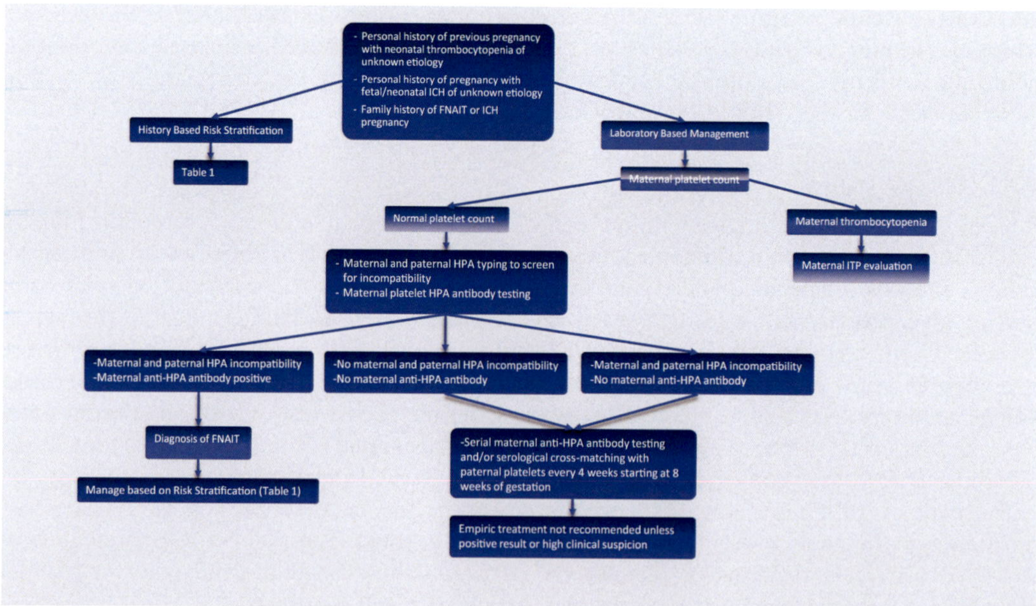

FIGURE 1 The recommended algorithm for fetal/neonatal alloimmune thrombocytopenia diagnosis and management.

6 Treatment

6.1 Medications

Management of FNAIT is controversial, and there is no consensus or professional committee guidelines [27]. Currently, the most acceptable approach is made by risk-based stratification involving intravenous immunoglobulin (IVIG) with or without corticosteroids.

- Unknown risk populations can be followed by serial testing with anti-HPA antibodies and serological crossmatching with paternal platelets every 4 weeks starting at 8 weeks of gestation. Empiric treatment is not recommended unless an HPA antibody is detected or high clinical suspicion for FNAIT is present [2,3].
- In a standard risk population, IVIG 2 g/kg/week or IVIG 1 g/kg/week plus prednisone 0.5 mg/kg/day is initiated at 20 weeks of gestation [2,3,28]. Data suggest both regimens are practical with similar efficacy in preventing severe fetal thrombocytopenia in standard risk pregnancies [24]. With either regimen, it is suggested to switch all patients to IVIG 2 g/kg/week plus prednisone 0.5 mg/kg/day at 32 weeks of gestation [2,3,28,29]. However, this protocol has a limitation since it cannot prevent ICH before 20 weeks of gestation.
- In a high-risk population, IVIG 1 g/kg/week is initiated at 12 weeks of gestation, and at 20 weeks, prednisone 0.5 mg/kg/day is added, or IVIG is increased to 2 g/kg/week. At 28 weeks, IVIG is increased to 2 g/kg/week and prednisone to 0.5 mg/kg/day [2].
- In an extremely high-risk population, IVIG 2 g/kg/week is initiated at 12 weeks gestation followed by additional prednisone 1 mg/kg/day at 20 weeks until delivery [30].

  Modified IVIG protocol has been utilized in patients with HPA antibody positivity, which provides earlier suppression of platelet antibodies during pregnancy and minimize the side effect of IVIG and prednisone.
- Low dose IVIG 400 mg/kg/week is initiated before conception or at 4–6 weeks of gestation with prednisone 20 mg/day [4]. Patients are monitored every 4 weeks for HPA antibody testing and serological crossmatching with paternal platelets, and the prednisone dose is adjusted based on the test results. IVIG dose is increased to 1 g/kg/week at 16 weeks gestation and further increased to 2 g/kg/week with prednisone 20 mg/day at 32 weeks of gestation.

The primary purpose of this risk strategy management based on the previous pregnancy with an affected fetus is to avoid fetal blood sampling to decrease complications while minimizing the risk of ICH in the current pregnancy [29,30]. However, possible side effects related to high dose IVIG and prednisone, particularly recommended to high and extremely high-risk populations, should be carefully monitored. When ICH is detected despite treatment, a case-by-case approach is suggested.

6.2 Timing and mode of delivery

There is no consensus on the mode of delivery for this condition. However, cesarean section is favored in patients with standard-, high-, and extremely high-risk FNAITs due to the increased risk of ICH in the fetus or neonates. Recommended timing of delivery is between 37 and 38 weeks of gestation for standard- and high-risk pregnancies and at 36–37 weeks of gestation for extremely high-risk pregnancies [3,31]. Further investigation is necessary to find a more flexible approach allowing for vaginal delivery, especially in multiparous women with standard risk who had IVIG and/or steroid treatment during pregnancy [32]. With unknown fetal platelet count, operative vaginal deliveries or invasive procedures involving the fetus (e.g., fetal scalp electrodes) should be avoided. Additionally, platelet counts of the cord blood should be sent out immediately after the delivery.

6.3 Intrapartum and postpartum management

Most patients at risk of developing FNAIT receive a prolonged period of steroids during prenatal care. The effects of long-term corticosteroid during pregnancy is unclear. However, it is theoretically possible to suppress the hypothalamic-pituitary-adrenal (HPA) axis resulting in an insufficient response to endogenous stress during labor or cesarean section [33]. In usual circumstances of moderate surgical stress (e.g., lower extremity revascularization, total joint replacement), the scheduled morning steroid dose is followed by a stress dose of intravenous hydrocortisone 50 mg prior to the procedure, followed by hydrocortisone 25 mg every 8 hours for the next 24 hours. Following completion of the stress dose regimen, normal steroid dosing is continued. However, there are no randomized controlled studies on whether stress dose steroids are required in pregnant women undergoing labor or cesarean section. During the postpartum period, a rapid decrease in endogenous corticosteroid levels takes place after placental delivery. Therefore, a controlled withdrawal of steroids and individualized use of stress dose hydrocortisone is recommended [33]. In current practice, a steroid stress dose is given to most patients who received more than 20 mg/day of prednisone for over three weeks due to the high likelihood of HPA axis suppression. Nonetheless, a recent study proposed that stress dose steroid is not indicated in patients with an intact HPA axis [34]. Therefore, a controlled prospective randomized study is necessary to elucidate this finding.

The need for additional fluorinated corticosteroids for fetal lung maturity in women receiving long-term prednisone is unclear. However, additional betamethasone or dexamethasone is recommended to decrease the risk for neonatal morbidity and mortality for patients who are between 23 weeks 0 day and 33 weeks and 6 days of gestation and at increased risk of preterm delivery within the next seven days [35].

6.4 Possible side effects: risks and monitoring

Long-term use of systemic corticosteroids can cause maternal adverse effects, including osteoporosis, impaired glucose tolerance, gestational diabetes, immune suppression, and HPA axis suppression [18]. Accordingly, screening for diabetes is recommended after one month of treatment. Previously, multiple *meta*-analyses have demonstrated an increased risk of preterm delivery, preterm premature rupture of membranes, and FGR [33].

Possible side effects of maternal IVIG treatment include local reaction at the injection site, fever, rash, immune alteration, aseptic meningitis, infection, and anemia in women with blood type A, B, or AB [36]. Therefore, an iron supplement and screening for anemia with hemoglobin and hematocrit every month are recommended for women receiving IVIG [36].

7 Reproductive outcomes

7.1 Postpartum management and evaluation

There is no indication for immediate maternal intervention in the postpartum period other than a steroid taper to prevent HPA suppression. Nevertheless, extensive counseling regarding the importance of screening and testing for the parents in future pregnancy should be pursued.

7.2 Future reproductive options

To avoid the conception of FNAIT affected fetuses in screen-positive couples, the use of assisted reproductive technologies can be considered. Using HPA compatible donor sperm might be the best option in patients with high-risk and extremely high-risk pregnancies who cannot tolerate or have contraindications for IVIG and prednisone treatment. If the baby's father is heterozygous for the HPA gene, IVF with a preimplantation genetic diagnosis and transferring the unaffected embryo should also be considered.

7.3 Preconceptional counseling

Couples with a history of suspected FNAIT or previous pregnancy with fetal/neonatal ICH of uncertain etiology should receive preconception counseling for subsequent pregnancies. Evaluation of these couples starts with the maternal platelet count to exclude maternal thrombocytopenia. If maternal platelet counts are normal, maternal anti-HPA antibody and both maternal and paternal platelet antigen genotyping should be performed to predict future risk of FNAIT. If HPA incompatibility is detected and maternal serum HPA alloantibodies specific to a foreign paternal platelet antigen is present, FNAIT is diagnosed [3,18,24]. If the patient has a family history without a personal history of an affected pregnancy, the same testing strategy is recommended without HPA antibody testing [18].

7.4 Prenatal care in the subsequent pregnancy

FNAIT develops more severely and at an earlier gestational age in subsequent pregnancies. Therefore, patients at risk should be referred to centers for high-risk pregnancies with experience in FNAIT management as soon as pregnancy is confirmed. If paternity is unknown, fetal HPA genotyping using CVS at 10–13 weeks or amniocentesis at 16–20 weeks should be performed to determine whether the fetus has an incompatible platelet antigen with the maternal HPA genotype [8]. If paternity is known and HPA incompatibility is detected, the next step is to test for paternal HPA genotype to estimate whether the infant has a 100% (homozygous) or 50% (heterozygous) chance of inheriting the specific platelet antigen. If the fetus is determined to be at risk, the patient should be counseled on therapy during pregnancy based on appropriate risk stratification (Table 1).

8 Summary and recommendations

FNAIT is caused by incompatibility between maternal and fetal HPA, leading to maternal alloantibody formation against fetal platelets, resulting in fetal thrombocytopenia and/or ICH.

- FNAIT is associated with fetal/neonatal ICH, miscarriage, FGR, and fetal/neonatal death.
- Unlike red blood cell alloimmunization, FNAIT can develop in the first pregnancy with a recurrence rate of almost 100% in the subsequent pregnancy with earlier onset and more severe symptoms in affected fetuses.
- Screening for couples at risk includes maternal and paternal HPA typing, as well as maternal antiplatelet antibody evaluation. At present, routine screening for all pregnant women is not recommended due to lack of cost-effectiveness.
- Initiation of therapy using IVIG and/or prednisone, repeating labs every 4-weeks and serial ultrasound follow-up is recommended based on risk-stratification.

References

[1] Yougbaré I, Tai W-S, Zdravic D, et al. Activated NK cells cause placental dysfunction and miscarriages in fetal alloimmune thrombocytopenia. Nat Commun 2017;8(1):1–13.

[2] Pacheco LD, Berkowitz RL, Moise Jr KJ, Bussel JB, McFarland JG, Saade GR. Fetal and neonatal alloimmune thrombocytopenia: a management algorithm based on risk stratification. Obstet Gynecol 2011;118(5):1157–63.

[3] Peterson JA, McFarland JG, Curtis BR, Aster RH. Neonatal alloimmune thrombocytopenia: pathogenesis, diagnosis and management. Br J Haematol 2013;161(1):3–14.

[4] Fernandes CJ, Mahoney Jr DH. Neonatal immune-mediated thrombocytopenia.

[5] Williamson LM, Hackett G, Rennie J, et al. The natural history of fetomaternal alloimmunization to the platelet-specific antigen HPA-1a (PlA1, Zwa) as determined by antenatal screening. Blood J Am Soc Hematol 1998;92(7):2280–7.

[6] Skariah A, Sung N, Salazar Garcia MD, et al. Low-dose prednisone and immunoglobulin G treatment for woman at risk for neonatal alloimmune thrombocytopenia and T helper 1 immunity. Am J Reprod Immunol 2017;77(6):e12649.

[7] Tiller H, Kamphuis MM, Flodmark O, et al. Fetal intracranial haemorrhages caused by fetal and neonatal alloimmune thrombocytopenia: an observational cohort study of 43 cases from an international multicentre registry. BMJ Open 2013;3(3).

[8] Lieberman L, Greinacher A, Murphy MF, et al. Fetal and neonatal alloimmune thrombocytopenia: recommendations for evidence-based practice, an international approach. Br J Haematol 2019;185(3):549–62.

[9] Yougbaré I, Zdravic D, Ni H. Fetal and neonatal alloimmune thrombocytopenia: Novel mechanisms of miscarriage learned from placental pathology in animal models. J Pediatrics Pediatric Med 2018;2(1).

[10] Curtis BR, Bussel JB, Manco-Johnson MJ, Aster RH, McFarland JG. Fetal and neonatal alloimmune thrombocytopenia in pregnancies involving in vitro fertilization: a report of four cases. Am J Obstet Gynecol 2005;192(2):543–7.

[11] Davoren A, Curtis BR, Aster RH, McFarland JG. Human platelet antigen-specific alloantibodies implicated in 1162 cases of neonatal alloimmune thrombocytopenia. Transfusion 2004;44(8):1220–5.

[12] Yougbare I, Wei-She T, Zdravic D, et al. Natural killer cells contribute to pathophysiology of placenta leading to miscarriage in fetal and neonatal alloimmune thrombocytopenia. In: American Society of Hematology, Washington, DC; 2015.

[13] Kroll H, Yates J, Santoso S. Immunization against a low-frequency human platelet alloantigen in fetal alloimmune thrombocytopenia is not a single event: characterization by the combined use of reference DNA and novel allele-specific cell lines expressing recombinant antigens. Transfusion 2005;45(3):353–8.

[14] Dubruc E, Lebreton F, Giannoli C, et al. Placental histological lesions in fetal and neonatal alloimmune thrombocytopenia: a retrospective cohort study of 21 cases. Placenta 2016;48:104–9.

[15] Talaat R, Elmaghraby A, Barakat S, El-Shahat M. Alterations in immune cell subsets and their cytokine secretion profile in childhood idiopathic thrombocytopenic purpura (ITP). Clin Exp Immunol 2014;176 (2):291–300.

[16] Xu R, Zheng Z, Ma Y, et al. Elevated NKT cell levels in adults with severe chronic immune thrombocytopenia. Exp Ther Med 2014;7(1):149–54.

[17] Garcia MS, Mobley Y, Henson J, et al. Early pregnancy immune biomarkers in peripheral blood may predict preeclampsia. J Reprod Immunol 2018;125:25–31.

[18] Paidas MJ, Wilkins-Haug L, Barss V. Fetal and neonatal alloimmune thrombocytopenia: parental evaluation and pregnancy management. In: UpToDate Inc. Waltham, MA; 2020.

[19] Petermann R, Bakchoul T, Curtis BR, et al. Investigations for fetal and neonatal alloimmune thrombocytopenia: communication from the SSC of the ISTH. J Thrombosis Haemost 2018;16(12):2526–9.

[20] Kamphuis MM, Paridaans NP, Porcelijn L, Lopriore E, Oepkes D. Incidence and consequences of neonatal alloimmune thrombocytopenia: a systematic review. Pediatrics 2014;133(4):715–21.

[21] Delbos F, Bertrand G, Croisille L, Ansart-Pirenne H, Bierling P, Kaplan C. Fetal and neonatal alloimmune thrombocytopenia: predictive factors of intracranial hemorrhage. Transfusion 2016;56(1):59–66.

[22] Bussel JB, Sola-Visner M. Current approaches to the evaluation and management of the fetus and neonate with immune thrombocytopenia. Paper presented at Seminars in Perinatology; 2009.

[23] Kiefel V. Platelet antibodies in immune thrombocytopenia and related conditions. J Laboratory Med 2020;44 (5):273–84.

[24] Lipitz S, Ryan G, Murphy M, et al. Neonatal alloimmune thrombocytopenia due to anti-P1A1 (anti-hpa-1a): Importance of paternal and fetal platelet typing for assessment of fetal risk. Prenat Diagnosis 1992;12 (11):955–8.

[25] Nogués N. Recent advances in non-invasive fetal HPA-1a typing. Transfus Apheresis Sci 2020;59(1):102708.

[26] Ohto H, Kato K, Tohyama Y, et al. Prenatal determination of human platelet antigen type 4 by DNA amplification of amniotic fluid cells. Transfus Sci 1997;18(1):85–9.

[27] Winkelhorst D, Oepkes D, Lopriore E. Fetal and neonatal alloimmune thrombocytopenia: evidence based antenatal and postnatal management strategies. Expert Rev Hematol 2017;10(8):729–37.

[28] Lakkaraja M, Berkowitz RL, Vinograd CA, et al. Omission of fetal sampling in treatment of subsequent pregnancies in fetal-neonatal alloimmune thrombocytopenia. Am J Obstet Gynecol 2016;215(4):471 e471–471. e479.

[29] Berkowitz RL, Lesser ML, McFarland JG, et al. Antepartum treatment without early cordocentesis for standard-risk alloimmune thrombocytopenia: a randomized controlled trial. Obstet Gynecol 2007;110(2 Part 1):249–55.

[30] Bussel JB, Berkowitz RL, Hung C, et al. Intracranial hemorrhage in alloimmune thrombocytopenia: stratified management to prevent recurrence in the subsequent affected fetus. Am J Obstet Gynecol 2010;203(2):135 e131–135. e114.

[31] Van den Akker E, Oepkes D, Lopriore E, Brand A, Kanhai H. Noninvasive antenatal management of fetal and neonatal alloimmune thrombocytopenia: safe and effective. BJOG: An Int J Obstet Gynaecol 2007;114 (4):469–73.

[32] Ronzoni S, Keunen J, Shah PS, et al. Management and neonatal outcomes of pregnancies with fetal/neonatal alloimmune thrombocytopenia: a single-center retrospective cohort study. Fetal Diagnosis Ther 2019;45 (2):85–93.

[33] AlSaad D, Lindow S, Lee BH, Tarannum A, Abdulrouf PV. Maternal, fetal, and neonatal outcomes associated with long-term use of corticosteroids during pregnancy. Obstet Gynaecol 2019;21(2):117–25.

[34] Sylvester-Armstrong KR, Duff P, Genç MR. Are peripartum stress-dose steroids necessary? Obstet Gynecol 2020;135(3):522–5.

[35] Lee M-J, Guinn D, Martin R. Antenatal corticosteroid therapy for reduction of neonatal respiratory morbidity and mortality from preterm delivery. UpToDate Waltham, MA, UpToDate; 2018.

[36] Berger M. Adverse effects of IgG therapy. J Allergy Clin Immunol: Pract 2013;1(6):558–66.

CHAPTER

12

Infectious and noninfectious endometritis and recurrent pregnancy loss

Fuminori Kimura[1], Aina Morimune[2], Akiko Nakamura[2], Jun Kitazawa[2], Tetsuro Hanada[2] and Takashi Murakami[2]

[1]Department of Obstetrics and Gynecology, Nara Medical University, Kashihara, Nara, Japan

[2]Deparment of Obstetrics and Gynecology, Shiga University of Medical Science, Otsu, Shiga, Japan

1 Introduction

The adverse effects of endometrial inflammation on the establishment and subsequent maintenance of pregnancy have attracted attention. This inflammation does not refer to severely or rapidly progressing inflammation, such as an acute infection, but to persistent and minute inflammation. Endometrial infections with microorganisms are the typical cause of persistent and subtle inflammation in the endometrium. However, the effects of other diseases that influence endometrial immunity should also be considered as potential causes. Recent studies have shown that persistent inflammation is deeply involved in recurrent pregnancy loss (RPL), regardless of its causes. In this chapter, the mechanism of RPL in association to infection and noninfectious endometritis are discussed.

2 Definition

Inflammation has both detrimental and beneficial effects on the body that is dependent on its severity and duration. The beneficial effects on tissues damaged are stemmed from mechanical insult, chemical toxins, microbial invasion, and hypersensitivity reactions [1,2]. Inflammation can be divided into acute and chronic stages. Acute inflammation is a response to pathogens or changes induced by physical or chemical damage that works to eliminate the cause of the damage and restore homeostasis of the affected tissue [1,2]. In contrast, chronic inflammation indicates a persistent inflammatory response that may

Immunology of Recurrent Pregnancy Loss and Implantation Failure
DOI: https://doi.org/10.1016/B978-0-323-90805-4.00002-X

involve progressive changes, causing long-term dysfunction and progressive destruction. Although the chronic inflammation is usually weak that manifests with or without subtle symptoms, the progressive destruction can be partially repaired [3]. Chronic inflammation is thus characterized by the simultaneous destruction and repair of tissue in response to the inflammatory process [3,4]. It may follow acute inflammation or exist as a prolonged low-grade reaction. Chronic infections, autoimmune diseases, and benign gynecological diseases, such as endometriosis and adenomyosis, are the conditions often related to chronic endometritis (CE) and inflammatory responses [3,4].

When inflammation occurs in the endometrium, it is referred to as endometritis, but the term "endometritis" is generally accepted as an infectious etiology. For this reason, the expression describing the condition of mild inflammation that affects the establishment and maintenance of pregnancy in the uterus may need to be changed. In this sense, "impaired inflammatory state of the endometrium" might be a more appropriate term than "endometritis," as it directly indicates "inflammation of the endometrium," regardless of the cause [5].

Although further discussion of these terms may be needed, inflammation in the endometrium is referred to as endometritis in this chapter. Inflammation in the endometrium that is thought to be caused by a microorganism is referred to as infectious endometritis, and that induced by other causes is referred to as noninfectious endometritis. The effects of each condition on RPL are described.

2.1 Infectious endometritis

2.1.1 *Etiology of infectious endometritis*

The endometrium sheds during menstruation in women of reproductive age and then regenerates in the next cycle. Therefore whether inflammation is persistently present in the endometrium is questionable. A state of persistent inflammation in the endometrium that exists across menstrual cycles is known as CE. A definitive diagnosis of CE can only be made histologically and is noted by the presence of plasma cells in the stroma of the endometrium [6,7]. In addition to plasma cells, high stromal cell proliferation, dissociated maturation between the epithelium and stroma, and a pronounced predecidual reaction may be present [6,7]. If taking the role of plasma cells (i.e., secreting large concentrations of antibodies) into account, CE may describe a condition in which immune cells monitor aberrant pathogens that reside in the uterine cavity for an extended period and regulate them to prevent progression to intense inflammation [8]. CE may thus be a state of obsolete inflammation after acute endometritis. However, the relationship between acute endometritis and CE remains to be determined. CE has been shown to have adverse effects on implantation. Microorganisms are now considered to be the leading cause of CE, as most of them are cured by antibiotics. Women with CE and infertility often have improved implantation rates when they are treated by antibiotic treatments [7,9]. Therefore in this chapter, CE is described as synonymous with infectious endometritis.

The inside of the uterine cavity in a normal woman has long been considered sterile. However, with the development of next-generation sequencing (NGS), small numbers of bacterial species can be detected. This has revealed that the female upper reproductive organs, including the ovaries, fallopian tubes, and uterus, are not sterile but have specific bacterial flora [10,11].

Bacteria in the vagina represented by *Lactobacillus* spp. are estimated to account for approximately 9% of all bacteria in women, indicating the substantial presence of bacterial populations in the vagina [12,13]. Conversely, according to Chen et al., the upper genitalia only has 1/10,000 of the number of vaginal bacteria [14]. These bacteria have also been proposed to regulate the function of both endometrial cells and the local immune system, compete with pathogenic microorganisms, and produce protective molecules to prevent uterine infections [15,16]. This suggests the presence of a symbiotic relationship in the uterus.

CE may be a condition in which this mechanism is disrupted due to the existence of pathogenic microorganisms in the uterus and a subtle immunological response to them [7]. In CE, *Streptococcus* spp., *Escherichia coli, Enterococcus faecalis, Gardnerella vaginalis, Klebsiella pneumoniae, Pseudomonas aeruginosa, Proteus* spp., *Staphylococcus* spp., and *mycoplasma/urea plasma* are frequently detected in microbial cultures, polymerase chain reaction, and NGS. These bacteria are currently considered the main causative pathogens of CE [10,17].

Microparticles existing in the vagina can be moved to the uterus through the cervix within a matter of minutes by the uterine peristaltic pump [18]. Thus vaginal microorganisms might invade the uterine cavity through the uterine cervix. Certain bacterial species, such as *Fusobacterium*, have been reported to colonize both mouse and bovine uteri [19]. Colonization of this particular microorganism in mice can occur via blood circulation (blood flow). It is also possible that oral or intestinal microorganisms penetrate the vascular floor by breaking the defense barrier of tissues due to local infection, such as gingivitis and enteritis, and form colonies inside the uterus. Furthermore, recent research has demonstrated the existence of microorganisms in the peritoneum [14]. Peritoneal microorganisms from the gastrointestinal tract may reach the uterus via the fallopian tube. Further investigation is necessary to elucidate the origin and pathway of colonized microorganisms causing CE. The factors behind the formation of the endometrial flora also need to be investigated in the future.

2.1.2 Prevalence and clinical features of chronic endometritis

The prevalence of CE ranges from 9.3% to 67.6% in women with RPL [20,21]. This large variance among studies seems to be due to the differences in diagnostic criteria employed and patients' backgrounds. Thus epidemiological studies have revealed that the rate of CE diagnosis is high in RPL patients. Studies have been conducted to investigate the risk factors for CE. Bacterial vaginosis, endometrial polyps, and endometriosis are gynecological diseases that have been reported to be associated with CE [22]. Noninfectious endometritis has been associated with endometrial polyps and endometriosis is detailed in the later section.

2.1.3 Diagnosis of chronic endometritis

CE is diagnosed based on the pathology evaluation of the endometrial tissue section. Although the excessive proliferation of stromal cells, differences in the maturation status between the stroma and epithelia, and marked decidual changes (edema) are found in the CE, the diagnosis is made by identifying the distinctive distribution of immunocompetent cells, especially the presence of plasma cells in the endometrial stroma in routine clinical practice [6,23].

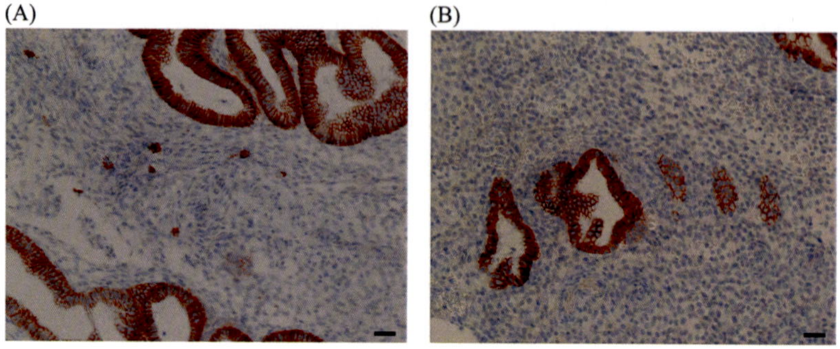

FIGURE 1 Immunohistochemical staining with anti-CD138 of the endometrium with and without chronic endometritis (CE). (A) CE endometrium. $CD138^+$ plasma cells are seen in the endometrial stroma. Epithelial cells are also stained for CD138. (B) Non-CE endometrium. CD138 is stained only in the epithelial cells. Bar = 100 μm.

Plasma cells are absent from normal endometrial stroma, except during and just before menstruation; on HE staining, plasma cells show a characteristic morphology described as axles. However, the presence of monocytic infiltration, stromal division, and plasmacytoid stromal cells in the endometrium makes it difficult for even experienced pathologists to detect plasma cells in the endometrium because they are morphologically similar. Therefore plasma cells are generally identified by immunohistochemistry for their specific marker, CD138 (syndecan-1). It is typical for the periphery of plasma cells to be stained because CD138 is found in the membrane [5,7,21,24], but they often form clusters in the tissue (Fig. 1A and B).

Diagnostic criteria of CE at fluid hysteroscopy were established with the use of the "Delphi method" in the International Working Group for Standardization of CE. They suggest the following criteria [25].

1. *Strawberry aspect*: A large area of hyperemic endometrium flushed with white central points (Fig. 2A).
2. *Focal hyperemia*: Small areas of hyperemic endometrium (Fig. 2B).
3. *Hemorrhagic spots*: Localized red areas with sharp and irregular borders that may be continuous with capillaries (Fig. 2C).
4. *Micropolyps*: Small intrauterine new growths, less than 1 mm, with a distinct connective vascular axis, distributed on focal areas (Fig. 2D), or over the entire endometrial surface (Fig. 2E).
5. *Thick and pale endometrium in follicular phase*: Evidence of stromal edema (A normal finding in the secretory phase) (Fig. 2F).

The diagnosis is made by the presence of at least one of these features.

Currently, these criteria are considered to be the most reliable diagnostic criteria for CE for fluid media hysteroscopy, but they are subject to further validation.

2.1.4 Pathophysiology of chronic endometritis

Progesterone not only has an immunosuppressive effect in the uterus [26] but also suppresses inflammation caused by *E. coli* and lipopolysaccharide demonstrated in animal

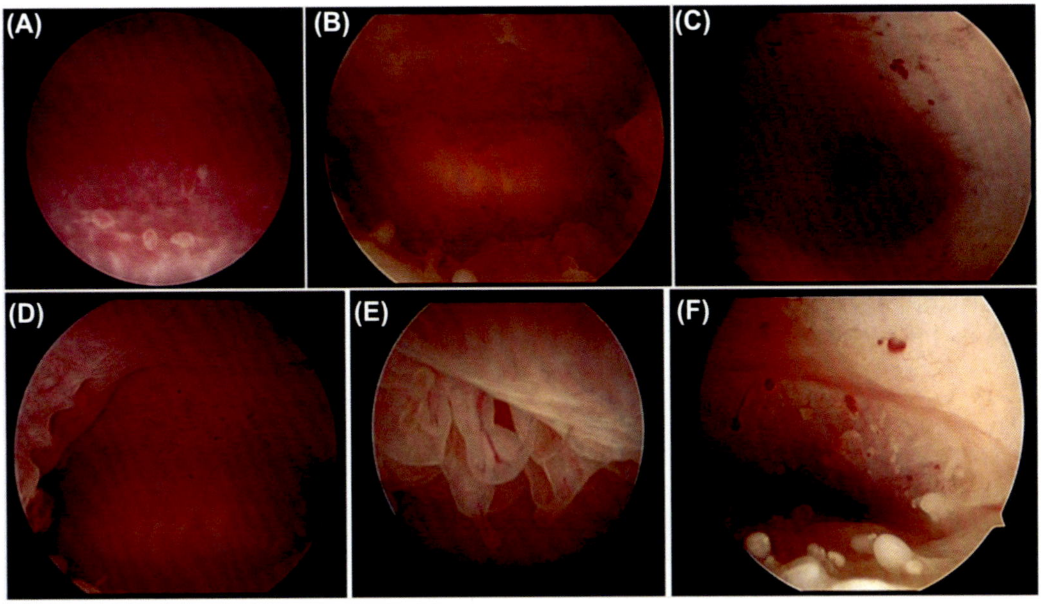

FIGURE 2 Diagnostic criteria for chronic endometritis at fluid hysteroscopy. (A) Strawberry aspect. (B) Focal hyperemia. (C) Hemorrhagic spots. (D) Focal micropolyps. (E) Diffuse micropolyps. (F) Endometrial edema. *Source: Taken from Cicinelli E., Vitagliano A., Kumar A., et al. Unified diagnostic criteria for chronic endometritis at fluid hysteroscopy: proposal and reliability evaluation through an international randomized-controlled observer study. Fertil Steril 2019;112(1):162–173.e162.*

species [27]. Therefore when considering the effects of CE on implantation and the maintenance of early pregnancy, it is unclear whether inflammation continues in the postpartum uterus. An analysis of miscarriage specimens following the histological diagnosis of CE or non-CE showed that chronic deciduitis (CD) with multiple plasma cells were present in more than half of cases in CE patients who miscarried again [28]. CD, like CE, is thought to be a persistent inflammation that occurs within the decidua. This result suggests that inflammation might persist even after the establishment of pregnancy and causes miscarriage in some CE patients.

CE is biochemically and morphologically inflamed. It has been clearly shown that the levels of proinflammatory cytokines, for example, interleukin-6, interleukin-1β, and tumor necrosis factor α, in menstrual blood are higher in CE patients than in non-CE patients [29]. The increased inflammatory substances have generally been shown to affect cell proliferation, apoptosis, and cellular function. It has also been revealed that the distribution and composition of immunocompetent cells differ due to the effects of such inflammation. The number of B cells is low in the normal endometrium. However, in CE, a large number of B cells infiltrate the stroma and glandular epithelia of the functional layer and push forward to the glandular lumina [24]. Although this B cell infiltration may be associated with the emergence of plasma cells in the stromal area, how these cells affect the maintenance of pregnancy is unclear. T cells are scattered in the stroma and epithelial area. The majority of endometrial T cells are $CD8^+$ cells, $CD4^+$ helper T cells (Th cells), and the subpopulation composition of Th cells, which are known to be important for the establishment and maintenance of pregnancy. In CE patients, the ratio of Th1 is increased while that of Th2 is decreased according to an analysis of fluorescence-assisted cell sorting [30]. The numbers of $CD8^+$ cytotoxic T cells and $Foxp3^+$ regulatory T cells are significantly elevated according to an analysis of immunohistochemistry [31].

Natural killer (NK) cells are also thought to play an important role in the establishment and maintenance of pregnancy. The major phenotype of NK cells, $CD56^{bright}CD16^-$ NK cells, with low cytotoxicity, increases up to 30%–40% of cells in the stromal area in the normal endometrium during the late secretory phase. A recent study revealed that the subpopulations of $CD56^{bright}CD16^-$ or $CD56^+CD16^-$ NK cells are decreased with increasing T cells in the uterine endometrium of CE patients [32].

For the successful maintenance of pregnancy, the regulation of sex steroid hormones influencing endometrial proliferation and differentiation is needed. In CE, the effect of progesterone on the endometrium might be aberrant. The expression of Ki-67 (a nuclear marker of cell proliferation) and BCL2 and BAX (regulators of apoptosis) is upregulated in the endometrium during the implantation phase [33]. CE modifies decidualization via the aberrant expression of estrogen and progesterone receptors (PGRs) [34]. These results suggest that CE disturbs the decidualization process and consequently influences the establishment and maintenance of pregnancy.

2.1.5 *Effect of chronic endometritis on maintenance of pregnancy and efficacy of treatment for chronic endometritis*

The effect of CE on the outcomes of subsequent pregnancies has been investigated in patients with RPL. A case-control observational study showed that the miscarriage rates in patients with untreated CE (32.3%; 11/34) tended to be higher than that in patients without CE who had RPL due to other causes (12.9%; 4/31) [35]. This suggests that CE may cause PRL.

McQueen et al. reported the treatment effect on CE patients with early RPL. Ofloxacin (800 mg) and metronidazole (1000 mg) for 2 weeks were given to 26 of 35 CE patients, and an alternative antibiotic—either doxycycline alone, doxycycline and metronidazole, or ciprofloxacin and metronidazole—was given to the remaining 9 patients [36]. Around 31 of 35 patients underwent a rebiopsy to determine the therapeutic effect, and 7 of these 31 patients were diagnosed with persistent CE. All 7 persistent CE patients were cured by a second course of antibiotics (2/7) or with natural resolution (5/7) without the need for additional antibiotic treatment. Therefore the cure rates are 94% (29/31) and 100% (31/31) after a single or two courses of antibiotics treatment, respectively. When patients with RPL are diagnosed with CE, antibiotics have been shown to cure most cases. In this way, the efficacy of antibiotics treatment for CE appeared to be high in RPL patients [36].

The effects of antibiotics treatment for CE on subsequent pregnancies have also been studied. A significant difference ($P = 0.032$) in the miscarriage rate was noted among three groups of patients: those without CE (60%: 12/20), treated CE (15%: 2/13), and untreated CE (56%: 5/9). The miscarriage rate tended to be higher in patients with treated CE when comparing those with untreated CE [37]. However, the efficacy of antibiotics on the cure rate of CE and the maintenance of the next pregnancy for the patients with RPL could not be concluded as the number of study participants was small. However, in infertile patients, the effect of CE on implantation and the effects of pregnancy maintenance have been well evaluated. The effect of CE on the miscarriage rate should therefore be investigated.

The miscarriage rates in CE and non-CE patients have been retrospectively examined in patients undergoing in vitro fertilization-embryo transfer (IVF-ET) who were not diagnosed with implantation failure, RPL, adenomyosis, submucosal myoma, or uterine anomaly, at the time of pregnancy. The study revealed a significant difference ($P < 0.03$) in miscarriage rates per pregnancy between the CE group (40.0%: 14/35) and the non-CE group (12.8%: 5/39) [38]. Another study reported that the miscarriage rate did not differ markedly between CE and non-CE patients, possibly due to different diagnostic criteria [39]. However, the pooling of these results yielded a significant difference in the miscarriage rate between the CE and non-CE groups (Table 1).

The effect of antibiotics on the cure rate of CE in infertile patients treated with IVF-ET has been well studied. A systematic review and meta-analysis revealed that patients with cured CE showed a higher ongoing pregnancy rate/live birth rate [odds ratio (OR): 6.81], clinical pregnancy rate (OR 4.02), and implantation rate (OR 3.24) than those with persistent CE [9]. The miscarriage rates of patients undergoing IVF-ET have been compared between cured and persistent CE patients in several studies concerning fertility treatment (Table 2) [40–43]. In those studies, the numbers of patients in the persistent CE group were small because the cure rate of CE by antibiotics treatment is high, and patients with persistent CE rarely conceive. These reasons might explain the lack of any significant difference in the miscarriage rate in these previous studies. When the results were pooled, the miscarriage rate was shown to be increased in cases with persistent CE, as shown in Table 2.

TABLE 1 Miscarriage rates in patients with and without chronic endometritis.

Objects	CE miscarriage /total	Non-CE miscarriage /total	*p* Value	Reference number
Infertility treated with IVF	14/35	5/39	$p < 0.03$	37
Infertility treated with IVF	1/17	53/431	NS	38
Total	15/52	58/470	$p < 0.003$	

CE: chronic endometritis, IVF: in vitro fertilization, NS: not significance.

TABLE 2 Miscarriage rates in patients with persistent and cured chronic endometritis.

Objects	Persistent miscarriage /total	Cured miscarriage /total	*P* value	Reference number
Infertility treated with IVF	3/11	15/135	NS	39
RIF	2/5	2/30	NS	40
RIF	0/0	8/53	-	41
RIF	1/3	3/9	NS	42
Total	6/19	28/227	$p < 0.05$	

CE, chronic endometritis; IVF, in vitro fertilization; NS, not significance; RIF. recurrent implantation failure.

In summary, patients undergoing IVF-ET may have an increased miscarriage rate if they become pregnant while still having untreated or persistent CE. Antibiotics treatments have a high cure rate for CE, and curing CE seems to reduce miscarriage rates as well as improve implantation rates.

2.2 Noninfectious (immune) endometritis

Etiology of noninfectious (immune) endometritis: Multiple conditions and diseases can cause noninfectious endometritis, including Asherman's syndrome, endometrial polyp, adenomyosis, and endometriosis.

2.2.1 *Asherman's syndrome*

Asherman's syndrome is morphological changes in which fibrosis and adhesions are formed between the entire or partial endometrial basal layer and/or myometrium of the uterine cavity or cervix [44,45]. Although the detailed mechanism by which Asherman's syndrome develops remains poorly understood, these alterations are encountered after intrauterine infection or manipulation, such as uterine curettage, intrauterine surgery, and obstetrical procedures for retained fragments of the placenta. Under such conditions, acute inflammation is considered to destroy the endometrial functional layers, inducing persistent chronic inflammation and subsequently leading to adhesions and fibrosis, causing an altered endometrial function.

Asherman's syndrome is detected in 1.3%–9.6% of RPL cases [44,45]. Although the prevalence of Asherman's syndrome is high in RPL, this may be secondary to the increased risk of receiving dilatation and curettage in this patient population [44,45]. In addition, the presence of Asherman's syndrome may affect obstetrical complications, as intrauterine adhesions cause pregnancy loss due to a lack of sufficient endometrium to support fetal and placental development and a sufficient blood supply caused by defective vascularization in the fibrotic endometrium [46]. The relationship between RPL and Asherman's syndrome—which one is the cause and the result—is unclear. A previous study reported that 40% (66/165) and 23% (38/165) of subsequent pregnancies resulted in spontaneous miscarriages and preterm deliveries, respectively. However, according to an analysis of women with untreated Asherman's syndrome [47], there is insufficient evidence to show that Asherman's syndrome causes RPL, and it has been suggested that the link between the two conditions is still controversial due to the difficulty conducting randomized controlled trials [45]. There is also no clear treatment strategy for RPL with Asherman's syndrome. In addition, insufficient evidence is available concerning the benefits of surgical removal of intrauterine adhesions for the pregnancy outcome, as the surgery itself can promote more adhesion formation in the uterine cavity. Precaution is thus important for preventing the formation of Asherman's syndrome when intrauterine manipulation is first performed [45].

2.2.2 *Endometrial polyps*

Endometrial polyps are the localized hyperplastic overgrowth of the endometrial gland and stroma that form a protruded or pedunculated lesion from the surface of the endometrium [48]. Endometrial polyps, which are thought to be caused by inflammation, produce proinflammatory mediators by themselves and are considered responsible for chronic inflammation [45]. The prevalence of endometrial polyps related to RPL ranges from 6.0% to 8.5% [49,50]. The presence of endometrial polyps is thought to adversely impact the maintenance of pregnancy as well as fertility by forming a physical obstacle and altering the release of substances from the endometrium, such as insulin-like growth factor-1 and glycodelin [51]. Although theoretically, endometrial polyps may affect the maintenance of pregnancy, no evidence has shown its direct association with RPL due to a lack of studies assessing the efficacy of polypectomy on the miscarriage rate [52]. Evidence is thus insufficient to recommend hysteroscopic removal of polyps to reduce the risk of RPL [45,53,54]. However, it should be noted that endometrial polyps are frequently associated with CE, and this relationship should be explored in further studies.

2.2.3 *Endometriosis*

Endometriosis is defined as the presence of endometrial or endometrial-like tissue outside the uterus [55]. Although the fundamental cause and natural history of the disease remain unclear, excessive inflammation and altered immune responses are present in the eutopic endometrium and are considered to be related to the etiology and pathophysiology of the symptoms, including pain, infertility, and adverse pregnancy outcomes [56]. It has been shown that the eutopic endometrium of endometriosis patients produces elevated levels of Cox-2, prostaglandin, and proinflammatory cytokines, such as IL6, IL8, and IL17, suggesting the presence of chronic inflammation in the eutopic endometrium. The endometrium of endometriosis is thus considered to be affected by chronic inflammation [57].

The final diagnosis of endometriosis is made by the intraperitoneal observation of laparoscopy. Since not all patients with RPL undergo laparoscopy, the prevalence of endometriosis among patients with RPL is still unknown. For this reason, it has not been possible to evaluate whether endometriosis causes RPL. The presence of endometriosis has not been diagnosed in patients with RPL in our daily clinical practice. While endometriosis is recognized as a cause of infertility worldwide, it is not recognized as a cause of RPL even according to the American Society of Reproductive Medicine (ASRM) practical committee and European Society of Human Reproduction and Embryology (ESHRE) guidelines [53,54]. However, in patients with endometriosis diagnosed with laparoscopy, differences in the pelvic environment and characteristics of ovum and eutopic endometrium between endometriosis and nonendometriosis groups have been well reported [57–61]. In addition, the effects of endometriosis on clinical pregnancy outcomes have been investigated [62,63]. Recently, endometriosis has received attention as a factor that adversely affects the outcomes of pregnancy and as a factor of RPL.

The effects of endometriosis on miscarriage are assumed to be due to multiple mechanisms. First, inflammation caused by endometriosis may affect the environment surrounding gametes/embryos and alter the quality of embryos when they move from the pelvis to the uterus. This may increase the risk of embryonic aneuploidy. IVF-ET can salvage gametes/embryos from these inflamed environments. However, according to a recent analysis of embryos with preimplantation genetic screening, there was no statistically significant difference in the aneuploidy rate of patients with or without endometriosis [64]. Second, the endometrial function can be altered due to the inflammation present in the endometrium. This may impair the endometrial epithelial function and decidualization required for implantation. Molecular biological analysis of endometrial tissue in patients with endometriosis showed the altered endometrial expression of substances related to implantation in the epithelium and stroma. Such changes that occur in endometriosis are thought to be associated with estradiol dominance and progesterone resistance due to inflammation [65].

Aromatase is an enzyme that converts androgens to estrogens, appears in endometriotic tissues. Additionally, there is reduced expression of 17bHSD1, which converts estradiol to weak estrone. This allows for a predominance of estrogen in endometriosis, creating an environment that promotes the proliferation of endometriotic tissue and inflammation [56,65]. Estradiol exerts its effect by binding to its receptors: estrogen receptor (ESR)1 and ESR2. The expression of ESR2 is increased, and ESR1 is decreased in endometriotic tissues [66]. Cyclooxygenase-2 (COX2) and prostaglandins produced by the activity of COX2 are promoted by the de novo synthesis of estrogen, and altered estrogen receptors for estrogen in the endometriotic tissue cause inflammation [67]. These inflammatory substances might affect endometrial function. The gene expression of FOXA2 in the endometrial epithelium affects the gene expression of leukemia inhibitory factor is reduced by inflammation;[68] this is seen in patients with endometriosis [69]. In contrast, it has been reported that the expression of ESR2 is unchanged while the expression of ESR1 is increased in eutopic endometrium during the implantation window in women with endometriosis [66]. The involvement of ESR expression in the pathogenesis and development of endometriosis requires further investigation.

The effect of progesterone is well known to be diminished in endometriotic tissues and in its eutopic endometrium. This phenomenon is referred to as progesterone resistance. A decreased reactivity of progesterone impairs decidualization and promotes cell proliferation due to reduced counteracting estrogen activity. The diminished progesterone effect may be due to an altered PGR expression. PGR is reduced in endometriotic tissues and the eutopic endometrium of these patients [70]. As a result, there is reduced expression of mediators of PGR stimulation, including Wnt Family Member 4 (WNT4) [71,72]. Additionally, the expression of progesterone activity enhancers, including Notch homolog 1 (NOTCH1) and FKBP52, are decreased in endometriosis and its eutopic endometrium. Therefore these decreased gene expressions are considered to be a potential mechanism to acquire progesterone resistance [73,74].

The endometrium of endometriosis patients is thought to have a reduced receptivity to embryos due to the impaired gene expression, as mentioned before. In addition, progesterone resistance causes delayed decidualization, resulting in a delayed opening of the implantation window. In recent years, it has been shown that women with an altered implantation window carry a risk of implantation failure, and those who undergo implantation outside the normal implantation window had an increased risk of miscarriage [75]. Given this theoretical background, increased miscarriage is expected in patients with endometriosis.

A recent meta-analysis revealed that women with endometriosis have a significantly higher risk of miscarriage than women without endometriosis. For example, a study analyzing 29 studies including 697,984 women in total revealed that the miscarriage risk was increased in women with endometriosis compared with those without endometriosis (OR: 1.81, 95% confidence interval [CI]: 1.44–2.28, $I^2 = 96\%$) [76]. Another analysis of 24 studies, including 1,924,114 women, showed that women with endometriosis had a significantly higher risk of miscarriage (OR 1.75; 95% CI, 1.29–2.37) than healthy controls [77]. However, these results were based on the clinical data of a combination of naturally conceived patients and patients treated with IVF-ET. When comparing endometriosis and nonendometriosis groups among patients who were only conceived by IVF-ET, the miscarriage rate was not markedly different [78]. In IVF-ET, an embryo develops outside the body without inflammation after fertilization before embryo transfer. This, in combination with a sufficient amount of progesterone replacement before embryo transfer, might overcome progesterone resistance. These factors might contribute to the decreased miscarriage rate in patients with endometriosis.

2.2.4 *Adenomyosis*

Adenomyosis is defined by the presence of endometrial or endometrial-like glands and stroma inside the uterine myometrium. Although a pathological evaluation makes a definite adenomyosis diagnosis of resected specimens, patients who wish to have children undergo diagnostic imaging, such as ultrasonography and magnetic resonance imaging, and are treated for their symptoms of subfertility and obstetrical complications. Excessive inflammation and an altered immune system in the adenomyotic tissue and eutopic endometrium may cause these conditions and the physical impact on the uterus.

The production of IFNγ, TNFα, and IL-1β is significantly increased in eutopic endometrial mononuclear cells with adenomyosis [79]. Furthermore, higher amounts of IL-6, IL-8, and MCP-1 are secreted in the eutopic endometrium of adenomyosis patients than in the endometrium of controls [80]. Thus the presence of inflammation within the endometrium can alter the endometrial function. Adenomyosis causes the alteration of the normal response to progesterone in the eutopic endometrium, which is also seen in patients with endometriosis. The expression of HOXA10, which plays an important role in uterine receptivity and implantation as well as decidualization, is decreased in eutopic endometrium of adenomyosis patients [81]. The secretion of IGFBP-1 and PRL was shown to be decreased in the culture supernatant of endometrial stromal cells obtained from adenomyotic patients [80]. This means that progesterone resistance is present in the eutopic endometrium of adenomyosis patients.

Another cause of miscarriage may be the physiological effect of adenomyosis on the uterus, which induces abnormal peristalsis in the endometrium and uterine muscles [82]. Adenomyosis causes a high frequency of peristalsis in the uterus during implantation [83]. Aromatase is present in adenomyotic as well as endometriotic tissue, and an aberrantly high level of estrogen in myometrium induced by the local production of this enzyme is assumed to cause abnormal peristalsis of the uterus [82]. Through this mechanism, adenomyosis adversely affects the establishment and maintenance of pregnancy.

The prevalence of adenomyosis in women with RPL was reported to be 13.6% [84]. The pregnancy and miscarriage rates with and without uterine adenomyosis have been mainly investigated in patients treated with IVF. The presence of adenomyosis has also been shown to increase miscarriage rates in meta-analyses. One meta-analysis showed that the miscarriage rate was higher in women with adenomyosis (31.9%) than in those without adenomyosis (14.1%), with a relative risk of 2.12 (95% CI, 1.20–3.75) [85]. This finding was confirmed by another analysis which revealed an increased risk of miscarriage in women with adenomyosis (OR 3.49, 95% CI 1.41–8.65) [78]. In addition, while the miscarriage rate in the second trimester was difficult to determine in these analyses, a recent retrospective study showed that the miscarriage rate in the second-trimester miscarriage was higher in pregnant women with adenomyosis than in those without adenomyosis (12.2% vs 1.2%, OR 11.2, 95% CI, 2.2–71.2) [86].

Fertility-sparing surgery for adenomyosis is now being performed, but its preventive effect on miscarriage has not yet been clarified. In addition, pretreatment or the long-term use of a GnRH agonist in IVF reportedly reduces the miscarriage rate in the patients with adenomyosis. Suppression of inflammation before implantation might also prevent miscarriage [87].

3 Summary and recommendations

1. When encountering patients with RPL, the presence of CE needs to be evaluated.
2. The effect of antibiotic treatment before pregnancy on RPL patients with CE needs to be verified in the future, although a positive effect has been documented. be expected.
3. Endometriosis and adenomyosis may cause miscarriage due to the presence of inflammation and an altered gene expression related to estrogen dominance and progesterone resistance in the ectopic and eutopic endometrium.
4. When patients with endometriosis become pregnant using IVF-ET, the miscarriage rate is not increased, possibly due to the escape of germ cells from the high-inflammation environment and strong supplementation of progesterone.
5. Long-term suppression of inflammation through medication, such as GnRH agonists or anti-inflammatory drugs, may help to prevent miscarriage in adenomyosis patients.

References

[1] Chen L, Deng H, Cui H, et al. Inflammatory responses and inflammation-associated diseases in organs. Oncotarget 2018;9(6):7204–18.

[2] Pahwa R, Goyal A, Bansal P, Jialal I. Chronic inflammation. StatPearls. FL: Treasure Island; 2021.

[3] Kunnumakkara AB, Sailo BL, Banik K, et al. Chronic diseases, inflammation, and spices: how are they linked? J Transl Med 2018;16(1):14.

[4] Katsuyama T, Tsokos GC, Moulton VR. Aberrant T cell signaling and subsets in systemic lupus erythematosus. Front Immunol 2018;9:1088.

[5] Puente E, Alonso L, Lagana AS, Ghezzi F, Casarin J, Carugno J. Chronic endometritis: old problem, novel insights and future challenges. Int J Fertil Steril 2020;13(4):250–6.

[6] Greenwood SM, Moran JJ. Chronic endometritis: morphologic and clinical observations. Obstet Gynecol 1981;58(2):176–84.

[7] Kimura F, Takebayashi A, Ishida M, et al. Review: chronic endometritis and its effect on reproduction. J Obstet Gynaecol Res 2019;45(5):951–60.

[8] Inoue T, Moran I, Shinnakasu R, Phan TG, Kurosaki T. Generation of memory B cells and their reactivation. Immunol Rev 2018;283(1):138–49.

[9] Vitagliano A, Saccardi C, Noventa M, et al. Effects of chronic endometritis therapy on in vitro fertilization outcome in women with repeated implantation failure: a systematic review and meta-analysis. Fertil Steril 2018;110(1):103–12 e101.

[10] Moreno I, Codoner FM, Vilella F, et al. Evidence that the endometrial microbiota has an effect on implantation success or failure. Am J Obstet Gynecol 2016;215(6):684–703.

[11] Kyono K, Hashimoto T, Nagai Y, Sakuraba Y. Analysis of endometrial microbiota by 16S ribosomal RNA gene sequencing among infertile patients: a single-center pilot study. Reprod Med Biol 2018;17(3):297–306.
[12] Gonzalez A, Vazquez-Baeza Y, Knight R. SnapShot: the human microbiome. Cell. 2014;158(3):690 690 e691.
[13] Human Microbiome Project C. Structure, function and diversity of the healthy human microbiome. Nature 2012;486(7402):207–14.
[14] Chen C, Song X, Wei W, et al. The microbiota continuum along the female reproductive tract and its relation to uterine-related diseases. Nat Commun 2017;8(1):875.
[15] Benner M, Ferwerda G, Joosten I, van der Molen RG. How uterine microbiota might be responsible for a receptive, fertile endometrium. Hum Reprod Update 2018;24(4):393–415.
[16] Crha I, Ventruba P, Zakova J, et al. Uterine microbiome and endometrial receptivity. Ceska Gynekol 2019;84 (1):49–54.
[17] Cicinelli E, De Ziegler D, Nicoletti R, et al. Chronic endometritis: correlation among hysteroscopic, histologic, and bacteriologic findings in a prospective trial with 2190 consecutive office hysteroscopies. Fertil Steril 2008;89(3):677–84.
[18] Quayle AJ. The innate and early immune response to pathogen challenge in the female genital tract and the pivotal role of epithelial cells. J Reprod Immunol 2002;57(1–2):61–79.
[19] Santos TM, Bicalho RC. Diversity and succession of bacterial communities in the uterine fluid of postpartum metritic, endometritic and healthy dairy cows. PLoS One 2012;7(12):e53048.
[20] Cicinelli E, Matteo M, Tinelli R, et al. Chronic endometritis due to common bacteria is prevalent in women with recurrent miscarriage as confirmed by improved pregnancy outcome after antibiotic treatment. Reprod Sci 2014;21(5):640–7.
[21] Kitaya K. Prevalence of chronic endometritis in recurrent miscarriages. Fertil Steril 2011;95(3):1156–8.
[22] Takebayashi A, Kimura F, Kishi Y, et al. The association between endometriosis and chronic endometritis. PLoS One 2014;9(2):e88354.
[23] Cicinelli E, Matteo M, Trojano G, et al. Chronic endometritis in patients with unexplained infertility: prevalence and effects of antibiotic treatment on spontaneous conception. Am J Reprod Immunol 2018;79(1).
[24] Kitaya K, Yasuo T. Aberrant expression of selectin E, CXCL1, and CXCL13 in chronic endometritis. Mod Pathol 2010;23(8):1136–46.
[25] Cicinelli E, Vitagliano A, Kumar A, et al. Unified diagnostic criteria for chronic endometritis at fluid hysteroscopy: proposal and reliability evaluation through an international randomized-controlled observer study. Fertil Steril 2019;112(1):162–73 e162.
[26] van der Burg B, van der Saag PT. Nuclear factor-kappa-B/steroid hormone receptor interactions as a functional basis of anti-inflammatory action of steroids in reproductive organs. Mol Hum Reprod 1996;2 (6):433–8.
[27] Cui L, Wang H, Lin J, et al. Progesterone inhibits inflammatory response in E.coli- or LPS-Stimulated bovine endometrial epithelial cells by NF-kappaB and MAPK pathways. Dev Comp Immunol 2020;105:103568.
[28] Kaku S, Kubo T, Kimura F, et al. Relationship of chronic endometritis with chronic deciduitis in cases of miscarriage. BMC Womens Health 2020;20(1):114.
[29] Tortorella C, Piazzolla G, Matteo M, et al. Interleukin-6, interleukin-1beta, and tumor necrosis factor alpha in menstrual effluents as biomarkers of chronic endometritis. Fertil Steril 2014;101(1):242–7.
[30] Kitazawa J, Kimura F, Nakamura A, et al. Alteration in endometrial helper T-cell subgroups in chronic endometritis. Am J Reprod Immunol 2021;85(3):e13372.
[31] Li Y, Yu S, Huang C, et al. Evaluation of peripheral and uterine immune status of chronic endometritis in patients with recurrent reproductive failure. Fertil Steril 2020;113(1):187–96 e181.
[32] Matteo M, Cicinelli E, Greco P, et al. Abnormal pattern of lymphocyte subpopulations in the endometrium of infertile women with chronic endometritis. Am J Reprod Immunol 2009;61(5):322–9.
[33] Di Pietro C, Cicinelli E, Guglielmino MR, et al. Altered transcriptional regulation of cytokines, growth factors, and apoptotic proteins in the endometrium of infertile women with chronic endometritis. Am J Reprod Immunol 2013;69(5):509–17.
[34] Wu D, Kimura F, Zheng L, et al. Chronic endometritis modifies decidualization in human endometrial stromal cells. Reprod Biol Endocrinol 2017;15(1):16.
[35] McQueen DB, Perfetto CO, Hazard FK, Lathi RB. Pregnancy outcomes in women with chronic endometritis and recurrent pregnancy loss. Fertil Steril 2015;104(4):927–31.

[36] McQueen DB, Bernardi LA, Stephenson MD. Chronic endometritis in women with recurrent early pregnancy loss and/or fetal demise. Fertil Steril 2014;101(4):1026–30.
[37] Gay C, Hamdaoui N, Pauly V, et al. Impact of antibiotic treatment for chronic endometritis on unexplained recurrent pregnancy loss. J Gynecol Obstet Hum Reprod 2021;50(5):102034.
[38] Morimune A, Kimura F, Nakamura A, et al. The effects of chronic endometritis on the pregnancy outcomes. Am J Reprod Immunol 2021;85(3):e13357.
[39] Li Y, Xu S, Yu S, et al. Diagnosis of chronic endometritis: how many CD138(+) cells/HPF in endometrial stroma affect pregnancy outcome of infertile women? Am J Reprod Immunol 2021;85(5):e13369.
[40] Cicinelli E, Matteo M, Tinelli R, et al. Prevalence of chronic endometritis in repeated unexplained implantation failure and the IVF success rate after antibiotic therapy. Hum Reprod 2015;30(2):323–30.
[41] Kitaya K, Matsubayashi H, Takaya Y, et al. Live birth rate following oral antibiotic treatment for chronic endometritis in infertile women with repeated implantation failure. Am J Reprod Immunol 2017;78(5).
[42] Tersoglio AE, Salatino DR, Reinchisi G, Gonzalez A, Tersoglio S, Marlia C. Repeated implantation failure in oocyte donation. What to do to improve the endometrial receptivity? JBRA Assist Reprod 2015;19(2):44–52.
[43] Xiong Y, Chen Q, Chen C, et al. Impact of oral antibiotic treatment for chronic endometritis on pregnancy outcomes in the following frozen-thawed embryo transfer cycles of infertile women: a cohort study of 640 embryo transfer cycles. Fertil Steril 2021;116(2):413–21.
[44] Hooker AB, Lemmers M, Thurkow AL, et al. Systematic review and meta-analysis of intrauterine adhesions after miscarriage: prevalence, risk factors and long-term reproductive outcome. Hum Reprod Update 2014;20 (2):262–78.
[45] Carbonnel M, Pirtea P, de Ziegler D, Ayoubi JM. Uterine factors in recurrent pregnancy losses. Fertil Steril 2021;115(3):538–45.
[46] Yu D, Wong YM, Cheong Y, Xia E, Li TC. Asherman syndrome—one century later. Fertil Steril 2008;89 (4):759–79.
[47] Schenker JG, Margalioth EJ. Intrauterine adhesions: an updated appraisal. Fertil Steril 1982;37(5):593–610.
[48] Kim KR, Peng R, Ro JY, Robboy SJ. A diagnostically useful histopathologic feature of endometrial polyp: the long axis of endometrial glands arranged parallel to surface epithelium. Am J Surg Pathol 2004;28 (8):1057–62.
[49] Elsokkary M, Elshourbagy M, Labib K, et al. Assessment of hysteroscopic role in management of women with recurrent pregnancy loss. J Matern Fetal Neonatal Med 2018;31(11):1494–504.
[50] Seckin B, Sarikaya E, Oruc AS, Celen S, Cicek N. Office hysteroscopic findings in patients with two, three, and four or more, consecutive miscarriages. Eur J Contracept Reprod Health Care 2012;17(5):393–8.
[51] Elbehery MM, Nouh AA, Mohamed ML, Alanwar AA, Abd-Allah SH, Shalaby SM. Insulin-like growth factor binding protein-1 and glycodelin levels in uterine flushing before and after hysteroscopic polypectomy. Clin Lab 2011;57(11–12):953–7.
[52] Lieng M, Istre O, Qvigstad E. Treatment of endometrial polyps: a systematic review. Acta Obstet Gynecol Scand 2010;89(8):992–1002.
[53] Practice Committee of the American Society for Reproductive M. Evaluation and treatment of recurrent pregnancy loss: a committee opinion. Fertil Steril 2012;98(5):1103–11.
[54] EGGo RPL, Bender Atik R, Christiansen OB, et al. ESHRE guideline: recurrent pregnancy loss. Hum Reprod Open 2018;2018(2):hoy004.
[55] Saunders PTK, Horne AW. Endometriosis: etiology, pathobiology, and therapeutic prospects. Cell 2021;184 (11):2807–24.
[56] Pirtea P, Cicinelli E, De Nola R, de Ziegler D, Ayoubi JM. Endometrial causes of recurrent pregnancy losses: endometriosis, adenomyosis, and chronic endometritis. Fertil Steril 2021;115(3):546–60.
[57] Lessey BA, Kim JJ. Endometrial receptivity in the eutopic endometrium of women with endometriosis: it is affected, and let me show you why. Fertil Steril 2017;108(1):19–27.
[58] Vallve-Juanico J, Houshdaran S, Giudice LC. The endometrial immune environment of women with endometriosis. Hum Reprod Update 2019;25(5):564–91.
[59] Sanchez AM, Vanni VS, Bartiromo L, et al. Is the oocyte quality affected by endometriosis? A review of the literature. J Ovarian Res 2017;10(1):43.

[60] Patel BG, Rudnicki M, Yu J, Shu Y, Taylor RN. Progesterone resistance in endometriosis: origins, consequences and interventions. Acta Obstet Gynecol Scand 2017;96(6):623–32.

[61] Zondervan KT, Becker CM, Missmer SA. Endometriosis. N Engl J Med 2020;382(13):1244–56.

[62] Borisova AV, Konnon SRD, Tosto V, Gerli S, Radzinsky VE. Obstetrical complications and outcome in patients with endometriosis. J Matern Fetal Neonatal Med 2020;1–15.

[63] Lalani S, Choudhry AJ, Firth B, et al. Endometriosis and adverse maternal, fetal and neonatal outcomes, a systematic review and meta-analysis. Hum Reprod 2018;33(10):1854–65.

[64] Juneau C, Kraus E, Werner M, et al. Patients with endometriosis have aneuploidy rates equivalent to their age-matched peers in the in vitro fertilization population. Fertil Steril 2017;108(2):284–8.

[65] Marquardt RM, Kim TH, Shin JH, Jeong JW. Progesterone and estrogen signaling in the endometrium: what goes wrong in endometriosis? Int J Mol Sci 2019;20(15).

[66] Osinski M, Wirstlein P, Wender-Ozegowska E, Mikolajczyk M, Jagodzinski PP, Szczepanska M. HSD3B2, HSD17B1, HSD17B2, ESR1, ESR2 and AR expression in infertile women with endometriosis. Ginekol Pol 2018;89(3):125–34.

[67] Tamura M, Deb S, Sebastian S, Okamura K, Bulun SE. Estrogen up-regulates cyclooxygenase-2 via estrogen receptor in human uterine microvascular endothelial cells. Fertil Steril 2004;81(5):1351–6.

[68] Kelleher AM, Peng W, Pru JK, Pru CA, DeMayo FJ, Spencer TE. Forkhead box a2 (FOXA2) is essential for uterine function and fertility. Proc Natl Acad Sci USA 2017;114(6):E1018–26.

[69] Lin A, Yin J, Cheng C, Yang Z, Yang H. Decreased expression of FOXA2 promotes eutopic endometrial cell proliferation and migration in patients with endometriosis. Reprod Biomed Online 2018;36(2):181–7.

[70] Igarashi TM, Bruner-Tran KL, Yeaman GR, et al. Reduced expression of progesterone receptor-B in the endometrium of women with endometriosis and in cocultures of endometrial cells exposed to 2,3,7,8-tetrachlorodibenzo-p-dioxin. Fertil Steril 2005;84(1):67–74.

[71] Burney RO, Talbi S, Hamilton AE, et al. Gene expression analysis of endometrium reveals progesterone resistance and candidate susceptibility genes in women with endometriosis. Endocrinology 2007;148(8):3814–26.

[72] Liang Y, Li Y, Liu K, Chen P, Wang D. Expression and significance of WNT4 in ectopic and eutopic endometrium of human endometriosis. Reprod Sci 2016;23(3):379–85.

[73] Yang H, Zhou Y, Edelshain B, Schatz F, Lockwood CJ, Taylor HS. FKBP4 is regulated by HOXA10 during decidualization and in endometriosis. Reproduction 2012;143(4):531–8.

[74] Brown DM, Lee HC, Liu S, et al. Notch-1 signaling activation and progesterone receptor expression in ectopic lesions of women with endometriosis. J Endocr Soc 2018;2(7):765–78.

[75] Wilcox AJ, Baird DD, Weinberg CR. Time of implantation of the conceptus and loss of pregnancy. N Engl J Med 1999;340(23):1796–9.

[76] Huang Y, Zhao X, Chen Y, Wang J, Zheng W, Cao L. Miscarriage on Endometriosis and adenomyosis in women by assisted reproductive technology or with spontaneous conception: a systematic review and meta-analysis. Biomed Res Int 2020;2020:4381346.

[77] Zullo F, Spagnolo E, Saccone G, et al. Endometriosis and obstetrics complications: a systematic review and meta-analysis. Fertil Steril 2017;108(4):667–72 e665.

[78] Horton J, Sterrenburg M, Lane S, Maheshwari A, Li TC, Cheong Y. Reproductive, obstetric, and perinatal outcomes of women with adenomyosis and endometriosis: a systematic review and meta-analysis. Hum Reprod Update 2019;25(5):592–632.

[79] Sotnikova N, Antsiferova I, Malyshkina A. Cytokine network of eutopic and ectopic endometrium in women with adenomyosis. Am J Reprod Immunol 2002;47(4):251–5.

[80] Peng Y, Jin Z, Liu H, Xu C. Impaired decidualization of human endometrial stromal cells from women with adenomyosisdagger. Biol Reprod 2021;104(5):1034–44.

[81] Fischer CP, Kayisili U, Taylor HS. HOXA10 expression is decreased in endometrium of women with adenomyosis. Fertil Steril 2011;95(3):1133–6.

[82] Munro MG. Uterine polyps, adenomyosis, leiomyomas, and endometrial receptivity. Fertil Steril 2019;111(4):629–40.

[83] Zhu L, Che HS, Xiao L, Li YP. Uterine peristalsis before embryo transfer affects the chance of clinical pregnancy in fresh and frozen-thawed embryo transfer cycles. Hum Reprod 2014;29(6):1238–43.

[84] Atabekoglu CS, Sukur YE, Kalafat E, et al. The association between adenomyosis and recurrent miscarriage. Eur J Obstet Gynecol Reprod Biol 2020;250:107–11.
[85] Vercellini P, Consonni D, Dridi D, Bracco B, Frattaruolo MP, Somigliana E. Uterine adenomyosis and in vitro fertilization outcome: a systematic review and meta-analysis. Hum Reprod 2014;29(5):964–77.
[86] Hashimoto A, Iriyama T, Sayama S, et al. Adenomyosis and adverse perinatal outcomes: increased risk of second trimester miscarriage, preeclampsia, and placental malposition. J Matern Fetal Neonatal Med 2018;31(3):364–9.
[87] Stanekova V, Woodman RJ, Tremellen K. The rate of euploid miscarriage is increased in the setting of adenomyosis. Hum Reprod Open 2018;2018(3):hoy011.

CHAPTER

13

Thrombophilic pathologies in recurrent pregnancy losses

Ae Ra Han[1] *and Sung Ki Lee*[2]

[1]Cha University, Fertility Center, Daegu, Republic of Korea [2]Department of Obstetrics and Gynecology, Konyang University College of Medicine, Daejeon, Republic of Korea

1 Introduction

Although there is no unified definition of recurrent pregnancy losses (RPL) yet, the American Society for Reproductive Medicine and the European Society of Human Reproduction and Embryology (ESHRE), two of the most influential academic societies in reproductive medicine, define RPL as two or more pregnancy losses [1,2]. Approximately 15% of all clinical pregnancies are lost sporadically, 2% of pregnant women experience two consecutive miscarriages, and 1% or less have more than three consecutive losses [3].

Placentation is established under a low O_2 environment with active vascular remodeling and proper coagulation. In the early stages of pregnancy, the intrauterine oxygen saturation is relatively low when implantation and placentation are progressing, while blood vessels in the decidua are transformed to prepare for a sufficient blood supply to the placenta. Aberrant placentation occurs when the blood supply to the placenta is abnormally reduced during pregnancy, leading to diverse obstetric complications such as miscarriage, preeclampsia (PE), preterm birth, intrauterine growth restriction (IUGR), and stillbirth [4]. Although the underlying pathology of aberrant placentation is unclear, abnormal expression of angiogenic markers and hypercoagulation have been suggested as possible etiologies [5–8].

Thrombophilia is an inherited or acquired predisposition to form clots inappropriately, and it increases the risk for venous thromboembolism (VTE) and thromboembolic diseases [9,10]. A typically acquired thrombophilia, antiphospholipid syndrome (APS), is characterized by the increased risk of venous or arterial thrombosis and/or placenta-mediated pregnancy complications [11]. Inherited thrombophilia is a condition in which a genetic mutation affects the amount or function of a protein in the coagulation system [12]. Here, we focused on the pathophysiology and management of inherited and acquired thrombophilia, except APS in relation to RPL.

Immunology of Recurrent Pregnancy Loss and Implantation Failure
DOI: https://doi.org/10.1016/B978-0-323-90805-4.00011-0

2 Thrombophilia

2.1 Inherited thrombophilia

The most popular forms of inherited thrombophilia are factor V Leiden G1691A (FVL) and prothrombin G20210A polymorphisms (PT), and deficiencies of protein C and S, and antithrombin (AT). Other inherited thrombophilia, such as factor V H1299R (R2) polymorphism, a missense mutation (Y1702C) in factor V, factor XIII V34L polymorphism, β-fibrinogen-455G > A polymorphism, and HPA1 a/b (L33P) polymorphisms, hyperhomocysteinemia (Hcy) caused by MethyleneTetraHydrofolate Reductase (MTHFR) C677T and A1298C polymorphisms, and PAI-1 4G/5G polymorphism were also investigated for their association with obstetric complications, including RPL [5,13–17].

These conditions can be divided by the VTE risk: (1) high-risk types—AT deficiency, FVL homozygosity, PT homozygosity, and heterozygosity for both FVL and PT, and (2) low-risk types—FVL heterozygosity, PT heterozygosity, and protein C and S deficiency. Although several reports have described the effect of inherited thrombophilia on the development of various obstetric complications [18,19], there is no consensus on the established risk of each inherited thrombophilia for RPL or other obstetric complications (Table 1).

2.2 Acquired thrombophilia

APS is the most common form of acquired thrombophilia, which is reviewed in a separate chapter. Defective folate metabolism, which is often caused by folate and vitamin B12 deficiencies,

TABLE 1 Association between inherited thrombophilia and recurrent pregnancy loss [3,18,20–28].

Thrombophilia	RPL[a]	Late obstetrical complications[b]	The prevalence of polymorphism in the general population or controls
FVL G1691A	Yes	Yes	13%–15% in Greeks and Middle Eastern descendants 5%–9% in European whites, Absent in Asians.
PT G20210A	Yes	Yes	2%–4% in Caucasians. Absent in Asian and African populations.
MTHFR C677T, A1298C	Yes/No	Yes (PE)	10%–16% and 4%–6% in Europeans, respectively. 20.6% and 3.8% in Korean women, respectively. 19.7% (C677T) in Japanese women.
Antithrombin	Yes	No	0.02%–1.15% in European whites. 0.15% in Japanese.
Protein C	No	No	0.2%–0.3% in Europeans. 0.13% in Japanese.
Protein S	Yes	Yes (late pregnancy loss)	0.03%–0.13% in European white. 1.12% in males and 1.60% in Japanese women.

[a] *Recurrent pregnancy loss.*
[b] *The second trimester and late pregnancy losses, preeclampsia (PE), placental abruption.*
Modified table from inherited thrombophilia and anticoagulation therapy from women with reproductive failure. Am J Reprod Immunol 2021;85(4):e13378.

obesity, and diabetes, has been found to be a reversible thrombophilia and a common risk factor for cerebral venous thrombosis, congenital anomalies, and pregnancy losses [29].

2.3 Prevalence of inherited thrombophilia

The prevalence of inherited thrombophilia varies by race and geographic region [30]. Compared to Asians, Caucasians had a higher incidence of FVL and PT polymorphisms but a lower incidence of MTHFR polymorphism and protein S deficiency (Table 1) [3,20–24]. In a study of race-specific reference ranges, protein C and S levels were significantly lower in black subjects than in whites [31]. Another study showed the differences in the prevalence of each inherited thrombophilia between Asians and Middle Easterners [30]. Other factors such as sex, age, and hormonal levels also influence the plasma levels of natural anticoagulants. In a study regarding normal reference ranges of AT, protein C, and protein S, reproductive-age women had lower protein S levels than same-age men [32]. The plasma protein C level increased with age, but AT and protein S did not. These findings were confirmed by Sakata et al. [20]. Hormones may regulate these markers, too. Menopause increased the concentration of AT and protein S [32–34]. Oral-contraceptive (OC) users have lower AT and protein S levels but higher protein C levels than non-OC users [32]. Since appropriate ranges of thrombophilic markers for each obstetrical condition are not well defined, diagnosis of inherited thrombophilia should be made carefully.

3 Physiologic hypercoagulability in pregnancy

The placenta allows gas exchange and nutritional supply to the fetus and control the immune balance and coagulation system during pregnancy. Decidual stromal cells produce thrombin and form a hemostatic envelope against hemorrhage during the capillary breaching, subsequent extravascular trophoblast invasion, and remodeling of the spiral arteries during early placentation [35].

In the first and second trimesters, clotting factors I (fibrinogen), II (prothrombin, PT), VII, IX, and XII are increased, and free protein S is decreased [36]. Resistance in activated protein C is also reduced [37] but protein C activity and AT level change little or remain constant [38]. This prothrombotic condition lasts up to 12 weeks postpartum [39]. Low pressure and low velocity of uteroplacental circulation are susceptible to thrombotic complications [35]. The prothrombotic condition during pregnancy helps to decrease intrapartum blood loss [40–42]; however, if the physiologic hypercoagulability gets abnormally exaggerated, thromboembolism and/or various placenta-mediated obstetrical complications, such as RPL, PE, IUGR, placenta abruption, and stillbirth can develop. Indeed, placental pathologies from miscarriage and stillbirth often exhibit thrombotic changes, such as perivillous fibrin deposition, fibrinoid necrosis of the decidual bed, and thrombi in intervillous spaces and fetal vessels [43]. Inadequate decidual and chorionic circulation in early pregnancy is followed by secondary impairment of trophoblast invasion, excessive thrombosis of placental vessels, and secondary uteroplacental insufficiency in the second and third trimester, contributing to mid-late obstetric complications [44–47]. To counter-balance, these thrombophilic conditions in placenta circulation, antithrombotic and fibrinolytic activities of blood must be well maintained before conception and activated appropriately during pregnancy.

Since pregnant women have a physiologic hypercoagulable state, either inherited or acquired thrombophilia can quickly induce pathologic coagulation. In addition, vascular endothelial injuries from the immune complex or complements in autoimmune diseases and pro-inflammatory cytokines and chemokines from activated immune effectors act as direct stimulants for thrombosis. Indeed, several studies showed a significant relationship between thrombophilia or circulatory parameters and several immune markers in RPL [48,49]. In addition, well-known anticoagulants such as aspirin and heparin have various immune-modulatory functions [50,51]. In summary, inflammation and coagulation are linked to each other, and the abnormal activation of one or the other can lead to various pregnancy complications.

4 Inherited thrombophilia and recurrent pregnancy loss

It is evident that inherited thrombophilia is a risk factor for VTE in both women and men [9,52], but the role of each inherited thrombophilia in women with RPL has not been elucidated well. A recent *meta*-analysis described the prevalence of various thrombophilias in RPL women [53]. Among the studies investigating various etiologies of RPL, seven studies exploring inherited thrombophilia were included ($n = 1713$). The prevalence of FVL ranges from 0% to 17.0%, and PT gene polymorphism was reported in 0%–13.6% of women with RPL. The protein S and protein C deficiencies were detected in 0%–18.2% and 1.1%–13.6% of women with RPL, respectively. Hence, the prevalence of inherited thrombophilia in RPL varies significantly according to the study and patient population.

The results have been inconsistent in terms of the prevalence of inherited thrombophilia in women with RPL and fertile controls. In a case-control study done by Carp et al., the incidences of FVL, PT, and MTHFR C677T mutations were not different between 108 women with a history of more than three miscarriages and 82 healthy controls [52]. However, Coulam et al. observed that the number of whom have more than three mutations among 10 thrombophilia genes were higher in RPL women than in fertile controls (68% vs 21%, respectively, $P = .02$) [17]. According to the serial reports of thrombophilia gene polymorphism investigated in Korean unexplained RPL women, the genotypes of MTHFR C677T and A1298C, PAI-1 4G/5G, angiotensin-converting enzyme I/D, p53 codon 72, and prolactin receptor gene C/T mutations were not different between unexplained RPL women and fertile controls [16,22,54,55]. On the other hand, in a systemic review published in 2006, the odds ratios of FVL, PT, and MTHFR gene mutations, and protein S deficiency were 1.40–3.55 for early pregnancy loss, 1.91–2.70 for recurrent first-trimester loss, 4.12–8.60 for second-trimester loss, and 1.31–20.09 for late pregnancy loss [18].

5 Evaluation

5.1 Who needs to be tested?

Routine screening for inherited thrombophilia in unselected populations is not recommended, and there is no strong evidence regarding whom to test or the optimal panel tests yet [56]. The most recent ESHRE guideline for RPL is against routine screening for inherited thrombophilia unless in the context of research or in women with additional risk factors for thrombophilia [2]. The American College of Obstetricians and Gynecologists suggests screening for inherited thrombophilia only in the following subjects who are planning to conceive with a personal history of VTE or a first-degree relative of a high-risk thrombophilia [42]. However, in RPL women who have no explained etiology, inherited thrombophilia and immune abnormalities have been reported. Hence, thrombophilia workups may help women with unexplained RPL.

5.2 Which tests should be done?

Frequently tested inherited thrombophilias are FVL and PT G20210A polymorphism, functional tests for protein C and S, and AT deficiency. Additionally, serum homocysteine level, MTHFR C667T and PAI-1 4G/5G polymorphisms, and factor VIII level are often tested. Inherited protein C and S, and AT deficiency may result from genetic mutations, but the mutated points vary. Therefore, functional tests of them are used to define those abnormalities. Protein S deficiency is the most challenging inherited thrombophilia to document with certainty due to the broader variation of its results affected by testing modalities and the condition of each individual. The most reliable approach to screening protein S deficiency is free protein S level using an immunoassay [57]. The enzymatic assay using amidolytic cleavage for protein C deficiency is preferred because of its superior performance to clotting assays [58]. For AT deficiency, a chromogenic assay is preferred [59].

5.3 What criteria can be applied?

The controversies in the role of inherited thrombophilia in RPL may come from the various factors affecting the cutoff values of each thrombophilic marker for specific populations. According to our unpublished data, known reference ranges of each thrombophilic marker were significantly different from the mean ± 2 standard deviation (SD) value of the fertile women. In detail, the levels of ATIII and protein S activity in RPL women had wider dispersion than established reference ranges. In addition, the prevalence of abnormal thrombophilia in RPL women based on the known reference range was significantly different from that of the fertile controls (mean ± 2 SD). Current reference ranges for each thrombophilic marker were initially developed to evaluate VTE risk and came from the central 95% of apparently healthy men and women between 18 and 65 years old [60]. It suggests that the current reference range of thrombophilic markers may not be appropriate to apply RPL evaluation. Thus, it is necessary to verify whether these thrombophilic markers with the current reference range are adequate to evaluate RPL women. We need to establish the new reference ranges with healthy reproductive-age women according to the clinical and laboratory standards institute.

6 Treatment of inherited thrombophilia

RPL women with inherited thrombophilia without current thromboembolism, like APS, are generally treated with a low dose of aspirin and a prophylactic dose of low-molecular-weight heparins (LMWH).

6.1 Anticoagulants

Aspirin, an acetylsalicylic acid, is classified as a nonsteroidal anti-inflammatory drug, mainly used as an antithrombotic agent [61]. Aspirin shows an antithrombotic effect by binding to a long hydrophobic channel in COX-1 and inhibiting the conversion of arachidonic acid to prostaglandin (PG) H2, a precursor of PGD2, PGE2, PGF2a, PGI2, and thromboxane A2 (TXA2) [62]. Reduced production of PG and TXA2 leads to suppression of platelet activation and aggregation. Aspirin is also involved in the formation of the fibrin network and the fibrinolytic process [61]. This broad antithrombotic action of aspirin can lead to severe internal bleeding in careless use.

Heparin is a glycosaminoglycan structurally similar to heparan sulfate, an endogenous macromolecule that binds to various protein ligands associated with blood coagulation. This resemblance attributes to various features of heparin and LMWH. The main anticoagulation effect of heparin is resulted from binding to AT. This heparin-AT complex primarily inactivates factors Xa and IIa, but also IXa, XIa, and XIIa [63]. The high-affinity pentasaccharide unit of heparin contributes to the activation of AT that accelerates the interaction with thrombin (FIIa) and activated factor X (factor Xa), which inactivates factor Xa. Other than factor Xa inhibition, catalysis of AT-mediated inactivation of FIIa is processed by the formation of a ternary heparin-antithrombin-thrombin complex. LMWH is a mixture of polymers with molecular weights that vary from 1000 to 10,000 Da, with a mean molecular weight between 4000 and 5000 Da, approximately one-third the size of unfractionated heparin (UFH) [64]. LMWH mainly inhibits factor Xa but less thrombin. Thus, it exhibits a lower bleeding tendency than UFH [65]. LMWH is almost 100% bioavailable when administered subcutaneously, and it has a dose-independent clearance mechanism, which makes it predictable in anticoagulation [66]. LMWH reduces the frequency of injection and injection-related pain more than UFH because its average half-life is longer than UFH [10]. The usual dosage of LMWH is documented in Table 2.

6.2 Management of inherited thrombophilia in recurrent pregnancy loss women

Because there is no consistent and robust evidence for managing inherited thrombophilia in RPL women, almost all clinical guidelines from major societies recommend LMWH prophylaxis only in women with family history (first-degree relatives) or a history of VTE

TABLE 2 Dosage of low-molecular-weight heparin for anticoagulation treatment.

Definition of regimen	Dosage of commonly used LMWH
Prophylactic LMWH	Enoxaparin, 40 mg SC q.d. Dalteparrin, 5000 IU SC q.d.
Intermediate-dose LMWH	Enoxaparin, 40 mg SC b.i.d. Dalteparin, 5000 IU SC b.i.d.
Therapeutic LMWH	Enoxaparin, 1 mg/kg SC b.i.d. Dalteparin, 200 IU/kg q.d.

LMWH, Low-molecular-weight heparin; *SC*, subcutaneously. It may require modifying the dose at extremes of body weight.

regardless of inherited thrombophilia or RPL. Brenner et al. first published supportive results for anticoagulation therapy in women with a history of miscarriage and inherited thrombophilia [67]. In this prospective cohort study, 50 unexplained RPL women with inherited thrombophilias were treated with enoxaparin 40 mg a day for a single thrombophilia and 80 mg/day for the combined thrombophilic defect. The enoxaparin-treated group had a significantly higher live birth rate (LBR) than untreated women (75% vs 20%). This study was followed by multiple consistent studies, suggesting that LMWH with or without aspirin improves pregnancy outcomes in women with RPL and inherited thrombophilia [68–70]. However, these studies had limitations such as small sample sizes and the study design (many of them were not randomized). The first open-label randomized trial (ALIFE1) of aspirin plus nadroparin, aspirin alone, and placebo treatment in uRPL women, including women with inherited thrombophilia (FVL, PT, PC, PS, and AT), was published in 2010 [71]. There were no differences in miscarriage rate among the whole study groups. In this study, only 13 of 123 women with aspirin and nadroparin treatment, 17 of 120 with aspirin, and 17 of 121 with placebo treatment had inherited thrombophilia. Hence, this study was insufficiently powered to assess subgroup effects for inherited thrombophilia [71]. It was similar to the HABENOX I trial [72], which was prematurely ended due to low recruitment.

Discrepancies between several observational studies and randomized controlled trials (RCTs) on the effect of anticoagulation treatment for RPL with inherited thrombophilia may result from the lack of studies revealing appropriate thrombophilia markers for RPL as described above. In addition, even with the same inherited thrombophilia, the risk of thrombus formation and decreased blood flow to the placenta may be different in cases with abnormal immune activation, such as chronic inflammatory conditions or autoimmune diseases. Indeed, the LBR of unexplained RPL women with thrombophilia with abnormal cellular immunity was different from those without abnormal cellular immunity, even with anticoagulation and immunomodulation treatment [73]. Thus, the evaluation of other factors which may influence thrombophilic conditions, such as immune abnormalities, should be considered for women with inherited thrombophilia.

Another obstacle is the heterogeneity of previous studies. Prior RCTs used diverse research protocols. There was inconsistency in selected inherited thrombophilic markers, inclusion and exclusion criteria of the subjects, the timing of initiation and duration of heparin therapy, as well as the types and doses of heparin. Besides, all RCTs about RPL women with inherited thrombophilia were done only in Western countries up to date [71,72]. Because there is a difference in the incidence of inherited thrombophilia between different races, studies in non-Western countries may show different results. To conclude the role of inherited thrombophilia in RPL, further studies are needed to find appropriate inherited thrombophilias and their reference ranges. Depending on the results, anticoagulation treatment can be performed in the appropriate subgroups.

7 Reproductive outcomes

Studies on the reproductive outcomes focusing on RPL women with inherited thrombophilia are insufficient. Our previous reports demonstrated an increased LBR and a decreased incidence of obstetric complications in RPL women with inherited thrombophilia after anticoagulation therapy [73,74]. In these retrospective cohort studies, RPL women with immune etiologies and/or thrombophilia were treated based on their underlying etiologies, while controls were RPL women without treatment or with empirical treatment without proper evaluation of immunologic and thrombophilic markers. Although it is not an RCT, a recent multicenter cohort study in unexplained RPL women with thrombophilia, OTTILTA registry, showed significant higher pregnancy loss [OR, 2.9 (95% CI, 1.4–6.1)] in the absence of anticoagulation, and the administration of LMWH (with or without ASA) was associated with higher LBR (OR, 10.6; 95% CI, 5.0–22.3) [75]. In addition, it also showed that LBR in RPL women without thrombophilia was independent of LMWH prophylaxis (alone or in association with ASA). Together with our previous studies [73,74], it could support "etiology-based treatment" for inherited thrombophilia in RPL women.

8 Summary and recommendations

Unlike established treatment for RPL women with APS, there is yet insufficient data for anticoagulation therapy in RPL women with inherited thrombophilia [76]. However, recent studies with etiology-based anticoagulation treatment showed clearly favorable effects of anticoagulation treatment on pregnancy outcomes in RPL women with inherited thrombophilia. Further studies are needed to find appropriate inherited thrombophilic markers and their reference ranges and the clinical efficacy of anticoagulation for RPL women with thrombophilia and not for all RPL women.

- When physiologic hypercoagulability, which prevents intrapartum and peripartum overt blood loss, gets abnormally exaggerated, maternal thromboembolism and various placenta-mediated obstetric complications including RPL may develop.
- The establishment of reference ranges for fertile reproductive-age women may help investigate inherited thrombophilia in RPL women.
- Current clinical guidelines recommend prophylactic anticoagulation treatment for pregnant women with family history (first-degree relatives) or personal history of VTE regardless of thrombophilia.
- Although clinical guidelines do not recommend anticoagulation in RPL women with inherited thrombophilia due to the lack of clear evidence, recent studies show favorable outcomes of prophylactic anticoagulation with LMWH for inherited thrombophilia in RPL women.
- LMWH and aspirin, well-known prophylaxis for VTE, are relatively safe to use before and during pregnancy.

References

[1] Practice Committee of the American Society for Reproductive M. Evaluation and treatment of recurrent pregnancy loss: a committee opinion. Fertil Steril 2012;98(5):1103–11.

[2] ESHRE Guideline Group on RPL, et al. ESHRE guideline: recurrent pregnancy loss. Hum Reprod Open 2018;2018(2):hoy004.

[3] Pritchard AM, Hendrix PW, Paidas MJ. Hereditary thrombophilia and recurrent pregnancy loss. Clin Obstet Gynecol 2016;59(3):487–97.
[4] Adu-Gyamfi EA, Ding YB, Wang YX. Regulation of placentation by the transforming growth factor beta superfamily†. Biol Reprod 2020;102(1):18–26.
[5] Bogdanova N, et al. A common haplotype of the annexin A5 (ANXA5) gene promoter is associated with recurrent pregnancy loss. Hum Mol Genet 2007;16(5):573–8.
[6] Tiscia G, et al. Haplotype M2 in the annexin A5 (ANXA5) gene and the occurrence of obstetric complications. Thromb Haemost 2009;102(2):309–13.
[7] Alfaidy N, et al. The emerging role of the prokineticins and homeobox genes in the vascularization of the placenta: physiological and pathological aspects. Front Physiol 2020;11:591850.
[8] Sugimura M, et al. Intraplacental coagulation in intrauterine growth restriction: cause or result? Semin Thromb Hemost 2001;27(2):107–13.
[9] Khan S, Dickerman JD. Hereditary thrombophilia. Thromb J 2006;4:15.
[10] Arachchillage DRJ, Makris M. Inherited thrombophilia and pregnancy complications: should we test? Semin Thromb Hemost 2019;45(1):50–60.
[11] Alijotas-Reig J, et al. Inherited thrombophilia in women with poor aPL-related obstetric history: prevalence and outcomes. Survey of 208 cases from the European registry on obstetric antiphospholipid syndrome cohort. Am J Reprod Immunol 2016;76(2):164–71.
[12] Stevens SM, et al. Guidance for the evaluation and treatment of hereditary and acquired thrombophilia. J Thromb Thrombolysis 2016;41(1):154–64.
[13] Zhao L, et al. Association between the SERPINE1 (PAI-1) 4G/5G insertion/deletion promoter polymorphism (rs1799889) and pre-eclampsia: a systematic review and meta-analysis. Mol Hum Reprod 2013;19(3):136–43.
[14] Wu X, et al. Folate metabolism gene polymorphisms MTHFR C677T and A1298C and risk for preeclampsia: a meta-analysis. J Assist Reprod Genet 2015;32(5):797–805.
[15] Park CW, et al. The role of methylenetetrahydrofolate reductase C677T polymorphism on the peripheral blood natural killer cell proportion in women with unexplained recurrent miscarriages. Clin Exp Reprod Med 2011;38(3):168–73.
[16] Kim JJ, et al. The PAI-1 4G/5G and ACE I/D polymorphisms and risk of recurrent pregnancy loss: a case-control study. Am J Reprod Immunol 2014;72(6):571–6.
[17] Coulam CB, et al. Multiple thrombophilic gene mutations rather than specific gene mutations are risk factors for recurrent miscarriage. Am J Reprod Immunol 2006;55(5):360–8.
[18] Robertson L, et al. Thrombophilia in pregnancy: a systematic review. Br J Haematol 2006;132(2):171–96.
[19] Rey E, et al. Thrombophilic disorders and fetal loss: a meta-analysis. Lancet 2003;361(9361):901–8.
[20] Sakata T, et al. Prevalence of protein S deficiency in the Japanese general population: the Suita Study. J Thromb Haemost 2004;2(6):1012–13.
[21] Ivanov P, et al. Inherited thrombophilia and IVF failure: the impact of coagulation disorders on implantation process. Am J Reprod Immunol 2012;68(3):189–98.
[22] Hwang KR, et al. Methylenetetrahydrofolate reductase polymorphisms and risk of recurrent pregnancy loss: a case-control study. J Korean Med Sci 2017;32(12):2029–34.
[23] Makino A, et al. No association of C677T methylenetetrahydrofolate reductase and an endothelial nitric oxide synthase polymorphism with recurrent pregnancy loss. Am J Reprod Immunol 2004;52(1):60–6.
[24] Tait RC, et al. Prevalence of protein C deficiency in the healthy population. Thromb Haemost 1995;73 (1):87–93.
[25] Bennett SA, Bagot CN, Arya R. Pregnancy loss and thrombophilia: the elusive link. Br J Haematol 2012;157 (5):529–42.
[26] Di Nisio M, et al. Thrombophilia and outcomes of assisted reproduction technologies: a systematic review and meta-analysis. Blood 2011;118(10):2670–8.
[27] Wu X, et al. Association between the MTHFR C677T polymorphism and recurrent pregnancy loss: a meta-analysis. Genet Test Mol Biomarkers 2012;16(7):806–11.
[28] Hansda J, Roychowdhury J. Study of thrombophilia in recurrent pregnancy loss. J Obstet Gynaecol India 2012;62(5):536–40.
[29] Gogineni S, et al. Deep cerebral venous thrombosis-a clinicoradiological study. J Neurosci Rural Pract 2021;12(3):560–5.

[30] Dugalic S, et al. Perinatal complications related to inherited thrombophilia: review of evidence in different regions of the world. J Matern Fetal Neonatal Med 2021;34(15):2567–76.
[31] Jerrard-Dunne P, et al. Ethnic differences in markers of thrombophilia: implications for the investigation of ischemic stroke in multiethnic populations: the South London ethnicity and stroke study. Stroke 2003;34(8):1821–6.
[32] Franchi F, et al. Normal reference ranges of antithrombin, protein C and protein S: effect of sex, age and hormonal status. Thromb Res 2013;132(2):e152–7.
[33] Rodeghiero F, Tosetto A. The VITA project: population-based distributions of protein C, antithrombin III, heparin-cofactor II and plasminogen–relationship with physiological variables and establishment of reference ranges. Thromb Haemost 1996;76(2):226–33.
[34] Gari M, et al. The influence of low protein S plasma levels in young women, on the definition of normal range. Thromb Res 1994;73(2):149–52.
[35] Lanir N, Aharon A, Brenner B. Haemostatic mechanisms in human placenta. Best Pract Res Clin Haematol 2003;16(2):183–95.
[36] Brenner B. Haemostatic changes in pregnancy. Thromb Res 2004;114(5–6):409–14.
[37] Bremme KA. Haemostatic changes in pregnancy. Best Pract Res Clin Haematol 2003;16(2):153–68.
[38] Leaf RK, Connors JM. The role of anticoagulants in the prevention of pregnancy complications. Clin Appl Thromb Hemost 2017;23(2):116–23.
[39] Kamel H, et al. Risk of a thrombotic event after the 6-week postpartum period. N Engl J Med 2014;370(14):1307–15.
[40] Bennett SA, et al. Women with unexplained recurrent pregnancy loss do not have evidence of an underlying prothrombotic state: experience with calibrated automated thrombography and rotational thromboelastometry. Thromb Res 2014;133(5):892–9.
[41] Papadakis E, et al. Low molecular weight heparins use in pregnancy: a practice survey from Greece and a review of the literature. Thromb J 2019;17:23.
[42] American College of, O. and B.-O. Gynecologists' Committee on Practice. ACOG practice bulletin no. 197: inherited thrombophilias in pregnancy. Obstet Gynecol 2018;132(1):e18–34.
[43] Salafia C, et al. Placental and decidual histology in spontaneous abortion: detailed description and correlations with chromosome number. Obstet Gynecol 1993;82(2):295–303.
[44] Arias F, et al. Thrombophilia: a mechanism of disease in women with adverse pregnancy outcome and thrombotic lesions in the placenta. J Matern Fetal Med 1998;7(6):277–86.
[45] Mousa HA, Alfirevic Z. Do placental lesions reflect thrombophilia state in women with adverse pregnancy outcome? Hum Reprod 2000;15(8):1830–3.
[46] Rand JH, et al. Antiphospholipid immunoglobulin G antibodies reduce annexin-V levels on syncytiotrophoblast apical membranes and in culture media of placental villi. Am J Obstet Gynecol 1997;177(4):918–23.
[47] Kuperman A, Di Micco P, Brenner B, et al. Fertility, infertility and thrombophilia. Womens Health (Lond) 2011;7(5):545–53.
[48] Ota K, et al. Effects of MTHFR C677T polymorphism on vitamin D, homocysteine and natural killer cell cytotoxicity in women with recurrent pregnancy losses. Hum Reprod 2020;35(6):1276–87.
[49] Koo HS, et al. Resistance of uterine radial artery blood flow was correlated with peripheral blood NK cell fraction and improved with low molecular weight heparin therapy in women with unexplained recurrent pregnancy loss. Am J Reprod Immunol 2015;73(2):175–84.
[50] Köller M, et al. Influence of low molecular weight heparin (certoparin) and unfractionated heparin on the release of cytokines from human leukocytes. Inflammation 2001;25(5):331–7.
[51] Kharbanda RK, et al. Prevention of inflammation-induced endothelial dysfunction: a novel vasculo-protective action of aspirin. Circulation 2002;105(22):2600–4.
[52] Carp H, et al. Prevalence of genetic markers for thrombophilia in recurrent pregnancy loss. Hum Reprod 2002;17(6):1633–7.
[53] van Dijk MM, et al. Recurrent pregnancy loss: diagnostic workup after two or three pregnancy losses? A systematic review of the literature and meta-analysis. Hum Reprod Update 2020;26(3):356–67.
[54] Yoon SH, et al. No association of p53 codon 72 polymorphism with idiopathic recurrent pregnancy loss in Korean population. Eur J Obstet Gynecol Reprod Biol 2015;192:6–9.
[55] Kim JJ, et al. Prolactin receptor gene polymorphism and the risk of recurrent pregnancy loss: a case-control study. J Obstet Gynaecol 2018;38(2):261–4.

[56] Haemostasis, B.C.f.S.i.H. Thrombosis Task Force, Investigation and management of heritable thrombophilia. Br J Haematol 2001;114(3):512–28.
[57] Smock KJ, et al. Protein S testing in patients with protein S deficiency, factor V Leiden, and rivaroxaban by North American Specialized Coagulation Laboratories. Thromb Haemost 2016;116(1):50–7.
[58] Sala N, Owen WG, Collen D. A functional assay of protein C in human plasma. Blood 1984;63(3):671–5.
[59] Van Cott EM, et al. Recommendations for clinical laboratory testing for antithrombin deficiency; communication from the SSC of the ISTH. J Thromb Haemost 2020;18(1):17–22.
[60] CLSI. Defining, establishing, and verifying reference intervals in the clinical laboratory; approved guideline -. CLSI document EP28-A3c. 3rd ed. Wayne, PA: Clinical and Laboratory Standards Institute; 2008.
[61] Undas A, Brummel-Ziedins KE, Mann KG. Antithrombotic properties of aspirin and resistance to aspirin: beyond strictly antiplatelet actions. Blood 2007;109(6):2285–92.
[62] Picot D, Loll PJ, Garavito RM. The X-ray crystal structure of the membrane protein prostaglandin H2 synthase-1. Nature 1994;367(6460):243–9.
[63] Hirsh J, et al. Guide to anticoagulant therapy: Heparin: a statement for healthcare professionals from the American Heart Association. Circulation 2001;103(24):2994–3018.
[64] Nelson SM, Greer IA. The potential role of heparin in assisted conception. Hum Reprod Update 2008;14 (6):623–45.
[65] Kim JH, Lim KM, Gwak HS. New anticoagulants for the prevention and treatment of venous thromboembolism. Biomol Ther (Seoul) 2017;25(5):461–70.
[66] Weitz JI. Low-molecular-weight heparins. N Engl J Med 1997;337(10):688–98.
[67] Brenner B, et al. Gestational outcome in thrombophilic women with recurrent pregnancy loss treated by enoxaparin. Thromb Haemost 2000;83(5):693–7.
[68] Carp H, Dolitzky M, Inbal A. Thromboprophylaxis improves the live birth rate in women with consecutive recurrent miscarriages and hereditary thrombophilia. J Thromb Haemost 2003;1(3):433–8.
[69] Brenner B, et al. Efficacy and safety of two doses of enoxaparin in women with thrombophilia and recurrent pregnancy loss: the LIVE-ENOX study. J Thromb Haemost 2005;3(2):227–9.
[70] Folkeringa N, et al. Reduction of high fetal loss rate by anticoagulant treatment during pregnancy in antithrombin, protein C or protein S deficient women. Br J Haematol 2007;136(4):656–61.
[71] Kaandorp SP, et al. Aspirin plus heparin or aspirin alone in women with recurrent miscarriage. N Engl J Med 2010;362(17):1586–96.
[72] Visser J, et al. Thromboprophylaxis for recurrent miscarriage in women with or without thrombophilia. HABENOX: a randomised multicentre trial. Thromb Haemost 2011;105(2):295–301.
[73] Lee SK, et al. Intravenous immunoglobulin G improves pregnancy outcome in women with recurrent pregnancy losses with cellular immune abnormalities. Am J Reprod Immunol 2016;75(1):59–68.
[74] Han AR, et al. Obstetrical outcome of anti-inflammatory and anticoagulation therapy in women with recurrent pregnancy loss or unexplained infertility. Am J Reprod Immunol 2012;68(5):418–27.
[75] Grandone E, et al. Findings from a multicentre, observational study on reproductive outcomes in women with unexplained recurrent pregnancy loss: the OTTILIA registry. Hum Reprod 2021;36(8):2083–90.
[76] Royal College of Obstetricians and Gynaecologists. The Investigation and Treatment of Couples with Recurrent Firsttrimester and Second-trimester Miscarriage (Green-top Guideline No. 17). [cited 2020]. Available from: <https://www.rcog.org.uk/globalassets/documents/guidelines/gtg_17.pdf>; 2011.

CHAPTER

14

Rheumatic diseases and reproductive outcomes

Brooke Mills and Bonnie L. Bermas

Division of Rheumatic Diseases, University of Texas Southwestern Medical Center, Dallas, TX, United States

1 Rheumatic conditions and pregnancy outcome

Rheumatologic disorders preferentially affect women during their reproductive years. Given that rheumatic diseases are autoimmune diseases, and the immune system changes during pregnancy, the impact of rheumatic disease on reproduction and vice versa is inevitable. In this chapter, we will discuss the more common rheumatologic conditions, including rheumatoid arthritis (RA), systemic lupus erythematosus (SLE), systemic sclerosis, and antiphospholipid syndrome, and their impact on reproductive outcomes. We will also discuss the role of anti-Ro (SSA) and anti-La (SSB) antibodies in neonatal lupus (NL).

Pregnancy is a natural experiment in immunologic tolerance. The mother must be tolerant of paternal antigens expressed by the fetus. This tolerance is induced through the anatomic separation of the fetus from the mother via the placenta, the evasion of fetal cell recognition by the maternal immune system, and pregnancy-induced shifts in the maternal immune system. Nobel laureate Peter Medawar first posited that the fetus, as a semiallograft, induces maternal immune tolerance to circumvent pregnancy failure [1]. HLA- mismatch is an important factor in inducing immune tolerance and indeed, the greater the HLA- disparity between the mother and the fetus, the more likely that the pregnancy will be successful [2]. Over the past several decades, research in reproductive immunology has incorporated knowledge garnered from our understanding of the immune system. For example, in the 1990s, scientists described two types of T-helper cells: T-helper 1 (TH1) and T-helper 2 (TH2) cells. It was hypothesized that cell-mediated immune responses were mediated by TH1 cells, and humoral immune responses were mediated by TH2 cells [3]. Wegmann and Mosman applied this construct to understanding reproduction and suggested that pregnancy was a TH2 phenomenon in which TH2 type cytokines would mediate fetal protection while TH1 cytokines would mediate fetal loss [4]. Recent data

Immunology of Recurrent Pregnancy Loss and Implantation Failure
DOI: https://doi.org/10.1016/B978-0-323-90805-4.00006-7

suggest that this model is an over-simplification and that TH1, TH2, T-helper 17 (TH17), and T regulatory (Treg) cells are all involved [5]. Moreover, pregnancy success is less likely the result of a shift to a TH2 profile but rather results from soluble antagonists to proinflammatory cytokines [6]. Concurrent to the growth in research regarding the role of cytokines in pregnancy success, cytokines were found to be instrumental in the mediation of disease activity in autoimmune conditions [7]. These findings have led to biological drug development that has transformed rheumatic disease management [8]. The burgeoning information on the role of the immune system in pregnancy success, as well as rheumatic disease activity, has revealed important insights into how rheumatic diseases activity is impacted by pregnancy and how rheumatic diseases impact pregnancy outcome. These insights are supported by what has been observed clinically for many decades.

2 Rheumatoid arthritis

2.1 Definition and prevalence

RA is a prototypic chronic inflammatory disorder in which persons make proinflammatory cytokines that cause inflammation both systemically and locally in the joints. The systemic inflammation can cause symptoms such as malaise and fatigue, as well as lead to vasculitis and inflammatory lung disease. The local inflammation leads to hypertrophy of the joint synovial lining causing joint swelling that can lead to joint destruction [9]. Inflammatory cytokines such as IL-6 and IL-1 are thought to drive this process [10]. RA is more common in women and has a prevalence of approximately 1% [11].

2.2 Underlying immune mechanism contributing to reproductive outcomes in rheumatoid arthritis

During the fetal growth stage of pregnancy, the relative downregulation of proinflammatory cytokines IL-6 and IL-1 could translate into the improvement of RA [12]. This is concordant with what the literature reports. Hench first observed the amelioration of disease activity during pregnancy in up to 70% of patients with RA [13]. A systemic review and *meta*-analysis found that disease activity improves in 60% of pregnant women with RA and flares postpartum in 46.7% [14]. Factors related to disease improvement include the absence of a rheumatoid factor or anticitrullinated protein antibody [15]. Studies have also shown that HLA disparity between the mother and fetus is associated with diminished disease activity during pregnancy [16]. In addition, data suggest that HLA-DRB1 alleles, which are associated with the risk of developing RA, can be acquired through microchimerism resulting from fetal-maternal cell exchange [17].

2.3 Evaluation

Individuals who present with one or more swollen joints for more than 6 weeks that are not explained by another disease process should be evaluated for RA. Testing for a rheumatoid factor and an anticyclic citrullinated peptide (anti-ccp) antibody should be performed as one or both of these tests are positive in the majority of patients with RA, although the absence of these autoantibodies does not preclude the diagnosis of this disorder. The inflammatory markers, ESR (sedimentation rate) and c-reactive protein should be checked. Persons are classified as having RA if they have synovitis in at least one joint and a score of 6 from scores in four domains that include the number of joints involved, seropositivity (presence of RF and/or anti-ccp), elevated inflammatory markers, and duration of symptoms for at least 6 weeks (2010 ACR/EULAR criteria) [18].

2.4 Treatment

The treatment of RA during pregnancy is focused on keeping the disease under good control, as active disease can increase the risk of adverse pregnancy outcomes. Many, but not all, of the medications commonly used to treat RA can be continued [19]. These include hydroxychloroquine, sulfasalazine, and TNF-α blockers; the latter are generally discontinued in the third trimester. Nonsteroidal antiinflammatory drugs (NSAIDs) can be used up until 20 weeks gestation; at which time there is a risk of kidney issues in the developing fetus that can lead to low amniotic fluid-oligohydramnios [20]. After 30 weeks, NSAIDs increase the risk of premature closure of the ductus arteriosus. Low-dose aspirin does not have these restrictions and may be used throughout pregnancy. There are conflicting data on whether NSAID drugs can interfere with embryonic implantation causing early pregnancy losses [21,22]. Other biologics such as abatacept, tocilizumab, and rituximab may be continued through conception [19]. There are insufficient data to conclude whether small molecules such as JAK kinase inhibitors are compatible with pregnancy. Methotrexate and leflunomide are not compatible with pregnancy, the former being teratogenic and the latter having an unclear safety profile. Methotrexate should be discontinued 1–3 months prior to conception, and leflunomide should be either discontinued 2 years before conception or washed out of the mother's system with a cholestyramine wash-out until levels are undetectable. For disease flares, nonfluorinated glucocorticoids can be used in as low a dose as possible as these medications can contribute to gestational hypertension and diabetes as well as small for gestational aged infants and preterm premature rupture of the membranes [19].

2.5 Reproductive outcomes

Women with RA report having smaller families when RA is diagnosed prior to childbearing compared to women who are diagnosed after completing childbearing [23]. The reasons for this are multifactorial and include delays in attempting conception due to disease activity, concerns regarding the ability to care for a child, and a longer time to achieve pregnancy [24]. Data regarding the risk of infertility in RA patients are conflicting. While 42% of RA patients report infertility, the reasons are not clear as studies have shown normal ovarian reserve as measured by anti-Mullerian hormone [25,26]. Women with RA have a higher risk of other adverse pregnancy outcomes, including preeclampsia, preterm delivery, small for gestational age infants, and higher cesarean section rate [27,28]. The increased incidence of preeclampsia, especially in women with high disease activity, is particularly interesting as tumor necrosis factor-α and IL-6 levels are increased in both RA disease flare and preeclampsia [29].

3 Systemic lupus erythematosus

3.1 Definition and prevalence

SLE is a multiorgan systemic disease that predominantly impacts women during their reproductive years. The disease affects roughly 1 in 1000 persons worldwide, with an increased burden on women of African ancestry, Asian race, and Hispanic ethnicity. The disease can range from mild, in which patients may suffer from mild skin rashes, joint pain, and benign cytopenias, to severe, in which patients may have extensive organ involvement of their kidneys, lungs, nervous and cardiovascular systems that is associated with high morbidity and mortality [30].

3.2 Underlying immune mechanisms contributing to reproductive outcomes in systemic lupus erythematosus

The shift from a proinflammatory state to an antiinflammatory antibody-mediated state during pregnancy can potentially increase disease activity in antibody-mediated diseases such as SLE. Moreover, hormonal changes in pregnancy can also trigger SLE flares [31]. Indeed, some women with lupus will have disease flares during pregnancy [32]. In addition, women with lupus are more predisposed to preeclampsia. The mechanisms underlying this are unclear but may be related to complement deposition in the placenta as well as local placental inflammation [33]. Differentiating between SLE flare and preeclampsia can be challenging, and often, the ultimate diagnosis is unclear.

3.3 Evaluation

Reproductive age women who present with multiorgan system symptoms including rashes, cytopenias, arthritis, serositis, renal disease, and nervous system should be evaluated for SLE. An antinuclear antibody (ANA) test should be done, with levels of 1:80 or greater considered positive. A positive ANA test in itself is not diagnostic of SLE, as ANAs can be present in upwards of 10% of the population. Individuals are classified by having SLE when they have a positive ANA and accumulate $>$ 10 points of weighted criteria that include seven clinical and three immunologic domains. A combination of both clinical and immunologic domains is necessary [34].

3.4 Treatment during pregnancy

The primary tenet of pregnancy management in SLE patients is disease control. Patients should be in remission or have low disease activity for several months prior to attempting pregnancy as an active disease at the time of conception portends a worse pregnancy outcome. In addition, patients who have ever had SLE renal disease are more likely to flare during pregnancy [35]. All patients with SLE should be on hydroxychloroquine during pregnancy as this medication has been shown to both decrease the risk of preeclampsia and improve pregnancy outcomes [36]. For flares during pregnancy, nonfluorinated glucocorticoids can be used; however, clinicians should aim to prescribe the lowest dose possible as glucocorticoids can cause pregnancy-induced hypertension and gestational diabetes as well as small for gestational age infants and preterm premature rupture of the membranes. Immunosuppressive agents such as azathioprine, cyclosporine, and tacrolimus can be used to manage both flares and as a transition agent prior to attempting pregnancy. Mycophenolate mofetil and cyclophosphamide are teratogenic and not compatible with pregnancy [37].

3.5 Reproductive outcome

Primary fertility does not appear to be impaired in SLE women. Women who have been previously treated with cyclophosphamide may develop infertility due to medication-induced premature ovarian failure. Secondary infertility is due to the coexistence of antiphospholipid antibodies (see below). These antibodies are common amongst SLE patients. There is a higher risk of preeclampsia in women with SLE as well as a higher fetal death rate. Small for gestational age infants and preterm premature rupture of membranes is commonly seen in SLE patients; however, dissecting out the contribution of medication use from disease activity is challenging.

4 Systemic sclerosis

4.1 Definition and prevalence

Systemic Sclerosis (SSc) is a rare heterogeneous autoimmune disease characterized by vasculopathy and progressive fibrosis of the skin and internal organs [38]. SSc has an annual incidence ranging from 10 to 20 new cases per million adults. The prevalence of SSc varies based on gender, geographic region, and race. It is over four times more prevalent in women than men. It can occur at any age, with a peak age of onset being between ages 35 and 65 [39].

4.2 Underlying immune mechanism contributing to reproductive outcomes in systemic Sclerosis

There is an increased rate of pregnancy loss in women with SSC prior to systemic disease onset. This is hypothesized to be partially due to maternal–fetal microchimerism [40]. Microchimerism is defined as the presence of cells from one individual in a genetically distinct individual. The most common source of microchimerism is maternal–fetal cell exchange during pregnancy, including failed and terminated pregnancies. Viable immunologic stem cells of fetal origin have been found in healthy women-years following pregnancy, and women with SSc have been found to have more circulating fetal cells than healthy controls [41]. Some reviews attribute the development of SSc to a lack of immune tolerance to these persistent fetal cells, similar to what is seen in graft versus host disease [42]. This could partially explain the increased prevalence of SSc, specifically in women and in later reproductive years. However, nulliparous women can also develop SSc. Thus there are likely genetic and environmental factors that play a role in the future development of SSc.

4.3 Evaluation

Individuals who present with skin thickening of the fingers should be evaluated for systemic sclerosis by either a rheumatologist or a dermatologist. The diagnosis is clinical and is supported by the presence of associated features of the Raynaud phenomenon, characteristic nail fold changes, calcinosis, telangiectasias, and heartburn. Basic renal function and pulmonary function should be performed. Anticentromere antibodies and anti-scl-70 antibodies are suggestive of limited systemic sclerosis and diffuse systemic sclerosis, respectively, although the presence of these antibodies is not necessary to make the diagnosis nor in themselves diagnostic of this condition in the absence of clinical features [43].

4.4 Treatment modalities

There are currently no formal treatment guidelines regarding SSc and pregnancy. Medications often used for the skin tightening of SSc, such as methotrexate and mycophenolate mofetil, are contra-indicated during pregnancy. SSc often overlaps with APS and Sjogren's Syndrome; therefore antiphospholipid and SSA antibodies should be checked in patients with SSc. Patients should start hydroxychloroquine and/or low dose aspirin, respectively, if these tests are positive [19].

4.5 Reproductive outcomes

There is a paucity of data regarding SSc and pregnancy, as this is a rare disease with a peak age of onset toward the end of reproductive years. A 2020 *meta*-analysis found that patients with systemic sclerosis have a significantly higher rate of miscarriage than healthy controls [44]. However, prior small reviews in the past have not seen increased pregnancy loss in SSc patients. Studies have also shown an increased risk of pregnancy loss in patients prior to disease onset [45]. Patients with SSc are predisposed to pulmonary artery hypertension (PAH) and need to be monitored for this entity prior to becoming pregnant. While PAH is a relative contraindication to pregnancy, recent data suggests better pregnancy outcomes in women with PAH than previously thought [46]. However, patients with the stable disease can have uneventful pregnancies with good maternal and fetal outcomes. The limited literature on pregnancy and systemic sclerosis have found an increased risk of miscarriages (1.6 times the amount of healthy controls), intrauterine growth restriction, and preterm delivery in these patients [44].

5 Antiphospholipid antibody syndrome

5.1 Definition and prevalence

Antiphospholipid antibody syndrome (APS) is an autoimmune disease characterized by thrombosis and/or pregnancy morbidity in the presence of persistently positive antiphospholipid antibodies (APLs), which include the lupus anticoagulant, anti-β2-Glycoprotein 1 antibodies, and anticardiolipin antibodies. Obstetric APS is defined as the presence of positive APL found on two occasions 12 weeks apart in addition to one of the following: three or more recurrent pregnancy losses before 10 weeks gestation; one or more pregnancy loss beyond 10 weeks gestation with normal anatomy by prenatal ultrasound examination; one or more preterm delivery of a morphologically normal infant before 34 weeks gestation due to severe preeclampsia, eclampsia or features consistent with placental insufficiency [47]. APS can occur as an isolated disease or as a manifestation of other systemic diseases, with SLE being the most common. The prevalence of APS is reported to be between two and 50 per 100,000 people [48]. Approximately 35% of APS cases have concomitant SLE [49]. Positive APL are three times more prevalent in patients with recurrent pregnancy loss than those with healthy pregnancies [50].

5.2 Underlying immunopathology related to reproduction

The mechanism of recurrent pregnancy loss in APS is not well understood and felt to be multifactorial. Placenta histopathology in women with APL shows evidence of placental infarction, impaired spiral artery remodeling, decidual inflammation, increased syncytial knots, decreased vasculo-syncytial membranes, and complement deposition. Placentae rarely showed evidence of intravascular clotting [51]. It is felt that inflammation plays more of a role in recurrent pregnancy loss from APS than thrombosis. Antiphospholipid antibodies directly target the trophoblast and decidual cells, resulting in complement activation and trophoblast apoptosis [52]. Trophoblasts are unable to adequately remodel the spiral arteries, which can result in preeclampsia, intrauterine growth restriction, and stillbirth [53]. Recurrent pregnancy loss is felt to be related to defective placentation due to the aforementioned factors. Manifestations of placental insufficiency include abnormal fetal surveillance tests, abnormal doppler wave-form, oligohydramnios, and birth weight of less than the 10th percentile.

Excessive neutrophil extracellular traps (NETs) are seen in APS and play a role in its pathogenesis [50,54]. The presence of NETs results in a proinflammatory state, subsequent endothelial dysfunction, and NETs in the placental intervillous spaces. NETs have been associated with decidual necrosis, maternal vasculitis, hemorrhage at the maternal–fetal interface, and thrombotic vasculopathy [55].

5.3 Evaluation

Women who have a history of unexplained or provoked thrombotic events, recurrent first-trimester pregnancy losses, second-trimester loss, or stillborn should be evaluated for APL. The lupus anticoagulant, anti-β2-Glycoprotein 1 antibodies, and anticardiolipin antibodies should all be checked. If any of these are positive, they should be repeated in 12 weeks to confirm the diagnosis.

5.4 Treatment of antiphospholipid antibody syndrome during pregnancy

Three situations need to be taken into account when considering the need for prophylactic anticoagulation or antiplatelet therapy during pregnancy: patients with positive APL without a history of obstetric APS or thrombosis, patients with obstetric APS without a history of thrombosis, and patients with APS and history of thrombosis [56]. Primary thromboprophylaxis and treatment of APS vary between these three groups of patients, as outlined as follows.

5.4.1 *Asymptomatically positive antiphospholipid antibodies*

Asymptomatic patients with positive APL are not usually treated with prophylactic anticoagulation to prevent thrombosis. A randomized, double-blind placebo-controlled trial in 2007 showed no difference in thrombosis rates between asymptomatic APL-positive patients treated with low-dose aspirin and those who were given a placebo [57]. However, the presence of APL is an independent risk factor for preeclampsia. There is significant evidence that low-dose aspirin is beneficial for the prevention of preeclampsia [58]. Therefore it is conditionally recommended by the American College of Rheumatology to start this group of patients on low dose aspirin for preeclampsia prophylaxis [56].

Observational studies indicate that hydroxychloroquine may be protective against thrombosis in patients with SLE and coexisting APL [59,60]. However, it is unclear whether hydroxychloroquine is effective in preventing thrombosis in patients with positive APL in the absence of an associated SLE diagnosis. The only randomized controlled trial conducted to study this hypothesis was terminated early due to low patient recruitment and hydroxychloroquine price increase and shortages [54,61]. Currently, it is not recommended to use hydroxychloroquine as primary prophylaxis against thrombosis in this patient population [56].

5.4.2 *Obstetric antiphospholipid antibody syndrome*

For patients who meet obstetric APS criteria, the use of combination low-dose aspirin and prophylactic low molecular weight heparin (LMWH) until 6–12 weeks postpartum is strongly recommended by the American College of Rheumatology [56]. This combination, when compared to low-dose aspirin alone, has been shown to improve the number of live births in patients with recurrent pregnancy loss [62].

Intravenous immunoglobulin (IVIG), hydroxychloroquine, pravastatin, and prednisone have been used in refractory APS in addition to standard therapy [63]. Prednisone may improve live birth rates in patients with obstetric APS. However, its use has been associated with increased rates of preterm delivery and preeclampsia, and therefore not recommended for treatment of obstetric APS [64]. There are no adequate data to support the use of IVIG and pravastatin in refractory obstetric APS, and the use of these treatments is not currently recommended [56,63]. Two retrospective studies, including one that excluded patients with SLE, have shown that hydroxychloroquine could be beneficial in improving live birth rates in refractory obstetric APS, and its use is currently recommended for patients with obstetric APS [56,65,66].

5.4.3 *Thrombotic antiphospholipid antibody syndrome*

Therapeutic anticoagulation with heparin, in addition to low-dose aspirin, is recommended for the treatment of pregnant patients with a history of thrombosis due to APS [56]. In addition to its role in anticoagulation, heparin has antiinflammatory and anticomplement properties that make it effective for the treatment of APS during pregnancy. For the reasons stated in the paragraph above, hydroxychloroquine therapy is also recommended during pregnancy for this subgroup of patients.

5.4.4 *Other considerations*

While there are general guidelines for the treatment of APS in pregnancy, many patients have unique circumstances that need to be taken into consideration. The number and level of APL titers carry different risks. Patients with triple-positive APLs at high titers have worse clinical outcomes than those with single or double APL positivity [63]. The decision to use prophylactic heparin therapy in patients with triple-positive APL that do not quite fit the criteria for obstetric APS should be made on an individualized basis.

Women who are attempting to conceive via assisted reproductive technologies (ARTs) will be exposed to hormonal treatments, such as estrogen, that affect thrombosis risk. Prophylactic LMWH is recommended for patients with positive APL undergoing ART, regardless of whether they fit obstetric APS criteria [56].

5.5 Reproductive outcomes

APS pregnancy complications include recurrent early pregnancy loss, fetal death, and placental insufficiency that could result in preeclampsia, eclampsia, and intrauterine growth restriction. The presence of lupus anticoagulant and triple positive APL are associated with poor pregnancy outcomes [67,68]. In a multicenter, retrospective cohort study called the PREGNANTS study, only 30% of triple APL positive patients had pregnancies that resulted in a live birth, despite treatment with prophylactic heparin and low-dose aspirin [68]. Appropriate medical therapy improves reproductive outcomes in patients with APS. However, this diagnosis is still associated with significant pregnancy morbidity. In addition, treatment failure occurs in 20%–30% of patients [69]. More studies are needed to identify additional treatment modalities.

6 Neonatal lupus

6.1 Definition and prevalence

NL occurs with the passive transfer of antibodies from the mother to the developing fetus. Anti-Ro/SSA and anti-La/SSB are involved in the majority of cases, but NL has been caused by anti-RNP antibodies as well [70]. These antibodies occur in a variety of rheumatologic conditions, including Sjogren's syndrome, RA, and SLE. They can also exist without a known diagnosis of rheumatic disorders in 0.2%–0.86% of the population [71,72]. They are often discovered in the mothers whose child was affected by NL without a prior rheumatologic diagnosis; however, half of these asymptomatic mothers will ultimately develop a rheumatologic disorder [73]. Manifestations of NL include cutaneous and/or cardiac manifestations. Cutaneous NL occurs in approximately 10% of children born from anti-Ro and/or anti-La positive mothers. Infants present with a photosensitive rash that can be accompanied by elevated liver enzymes and thrombocytopenia. These symptoms resolve by 6 months of age [74]. Cardiac NL occurs in approximately 1:15,000 live births and 1%–2% of pregnancies in which the mother has Ro/La antibodies [75]. Recurrence rates for congenital complete heart block (CCHB) in subsequent pregnancies is roughly 18% [76]. Fetuses can develop first, second, or third-degree heart block between weeks 18 and 25 of gestation. In third-degree heart block, there is no conduction between the atria and the ventricles. Infants born with this condition will likely need a pacemaker at birth [77].

6.2 Underlying immune mechanisms related to reproductive outcome

In cutaneous NL, exposure to UV light releases antigens in the neonate's skin that react with circulating anti-Ro/SSA and anti-La/SSB antibodies to cause a rash [74]. In the case of CCHB, anti-Ro, and anti-La antibodies cross the placenta and bind to SSA/Ro and SSB/La antigens expressed on fetal cardiocytes. This can result in proinflammatory cytokines that cause tissue damage in the developing fetal heart and fibrosis of the atrioventricular node [78].

6.3 Evaluation

Women with SLE, RA, and Sjogren's syndrome should have anti-Ro/SSA and anti-La/SSB antibodies checked prior to conception if possible. Women without a known diagnosis of rheumatologic disorders who have a fetus or child with CCHB should have these antibodies checked and, if positive, be sent for rheumatologic evaluation. Patients with positive antibodies will need a screening fetal echocardiogram between 16 and 26 weeks to evaluate for CCHB.

6.4 Treatment modalities

Unfortunately, treatment to reverse evolving heart block has been disappointing. While case reports of using high-dose dexamethasone and IVIG have shown reversal of second-degree heart block, randomized controlled studies have not been positive [79,80]. Recent data suggest that hydroxychloroquine can reduce the incidence of this disorder, especially as secondary prevention [81].

6.5 Reproductive outcomes

Pregnancies in which maternal anti-Ro and anti-La antibodies are present have a 10% chance of resulting in neonatal cutaneous lupus and a 1%–2% chance of resulting in CCHB.

When CCHB occurs, there is a 17% chance of fetal death [82]. In women who have had a child with CCHB, they have an 18% risk of a subsequent child being born with CCHB.

7 Conclusion

Rheumatologic disorders are common in reproductive-aged women. Disease activity can be impacted by pregnancy, with RA potentially improving and SLE potentially worsening. Overall, primary fertility does not seem to be impacted by rheumatologic disorders, although secondary infertility as manifested by recurrent pregnancy losses can be seen when APL coexist. Adverse pregnancy outcomes are more common in rheumatic disease patients and include a higher risk of preeclampsia, preterm premature rupture of membranes, and small for gestational age infants. Conceiving when the disease is quiescent is crucial to improving pregnancy outcomes. Women who are anti-Ro and anti-La positive have a 1%–2% chance of having a fetus with CCHB and a 5%–10% chance of having a child with NL.

8 Summary and recommendations

1. RA tends to improve during pregnancy.
2. SLE tends to worsen during pregnancy.
3. Increased rate of miscarriage is seen prior to and following systemic sclerosis diagnosis; however, the literature regarding systemic sclerosis and pregnancy is limited.
4. Regardless of underlying diagnosis, rheumatic disease activity is associated with adverse pregnancy outcomes, including preeclampsia, small for gestational age infants, and preterm premature rupture of membranes.
5. All patients with positive antiphospholipid antibodies, regardless of a formal APS diagnosis, should be treated with aspirin throughout pregnancy.
6. The treatment for APS during pregnancy differs for patients with obstetric versus thrombotic APS.
7. Patients with triple-positive APL have worse pregnancy outcomes than patients with only one or two positive APL.
8. Anti-Ro (SSA) and anti-La (SSB) antibodies are associated with NL.
9. NL can either present as a self-limited disease manifested mainly by skin rash or less frequently as CCHB.
10. Rheumatic disease should be under good control with pregnancy-compatible medications prior to conception.

References

[1] Deshmukh H, Way SS. Immunological basis for recurrent fetal loss and pregnancy complications. Annu Rev Pathol Mech Dis 2019;14(1):185–210. Available from: https://doi.org/10.1146/annurev-pathmechdis-012418-012743.

[2] Ober C. Current topic: HLA and reproduction: lessons from studies in the Hutterites. Placenta 1995;16 (7):569–77. Available from: https://doi.org/10.1016/0143-4004(95)90026-8.
[3] Mosmann TR, Cherwinski H, Bond MW, Giedlin MA, Coffman RL. Two types of murine helper T cell clone. I. Definition according to profiles of lymphokine activities and secreted proteins. J Immunol 1986;136 (7):2348–57. Available from: http://www.ncbi.nlm.nih.gov/pubmed/2419430.
[4] Wegmann TG, Lin H, Guilbert L, Mosmann TR. Bidirectional cytokine interactions in the maternal-fetal relationship: is successful pregnancy a TH2 phenomenon? Immunol Today 1993;14(7):353–6. Available from: https://doi.org/10.1016/0167-5699(93)90235-D.
[5] Saito S, Nakashima A, Shima T, Ito M. Th1/Th2/Th17 and regulatory T-Cell paradigm in pregnancy. Am J Reprod Immunol 2010;63(6):601–10. Available from: https://doi.org/10.1111/j.1600-0897.2010.00852.x.
[6] Ostensen M. Cytokines and pregnancy in rheumatic disease. Ann N Y Acad Sci 2006;1069(1):353–63. Available from: https://doi.org/10.1196/annals.1351.033.
[7] Brennan FM, Feldmann M. Cytokines in autoimmunity. Curr Opin Immunol 1992;4(6):754–9. Available from: https://doi.org/10.1016/0952-7915(92)90057-L.
[8] Maini RN, Brennan FM, Williams R, et al. TNF-alpha in rheumatoid arthritis and prospects of anti-TNF therapy. Clin Exp Rheumatol 1993;11(Suppl 8):S173–5. Available from: http://www.ncbi.nlm.nih.gov/pubmed/8391952.
[9] Lee DM, Weinblatt ME. Rheumatoid arthritis. Lancet 2001;358(9285):903–11. Available from: https://doi.org/10.1016/S0140-6736(01)06075-5.
[10] McInnes IB, Schett G. Cytokines in the pathogenesis of rheumatoid arthritis. Nat Rev Immunol 2007;7 (6):429–42. Available from: https://doi.org/10.1038/nri2094.
[11] Alamanos Y, Voulgari PV, Drosos AA. Incidence and prevalence of rheumatoid arthritis, based on the 1987 American College of Rheumatology criteria: a systematic review. Semin Arthritis Rheum 2006;36(3):182–8. Available from: https://doi.org/10.1016/j.semarthrit.2006.08.006.
[12] Förger F, Villiger PM. Immunological adaptations in pregnancy that modulate rheumatoid arthritis disease activity. Nat Rev Rheumatol 2020;16(2):113–22. Available from: https://doi.org/10.1038/s41584-019-0351-2.
[13] Hench PS. The amelorating effect of pregnancy on chronic atrophic (infectious, rheumatoid) arthritis, fibrositis, and intermittent hydrarthrosis. Mayo Clin Proc 1938;13:161–7.
[14] Jethwa H, Lam S, Smith C, Giles I. Does rheumatoid arthritis really improve during pregnancy? A systematic review and metaanalysis. J Rheumatol 2019;46(3):245–50. Available from: https://doi.org/10.3899/jrheum.180226.
[15] de Man YA, Bakker-Jonges LE, Goorbergh CMDd, et al. Women with rheumatoid arthritis negative for anti-cyclic citrullinated peptide and rheumatoid factor are more likely to improve during pregnancy, whereas in autoantibody-positive women autoantibody levels are not influenced by pregnancy. Ann Rheum Dis 2010;69 (2):420–3. Available from: https://doi.org/10.1136/ard.2008.104331.
[16] Nelson JL, Hughes KA, Smith AG, Nisperos BB, Branchaud AM, Hansen JA. Remission of rheumatoid arthritis during pregnancy and maternal-fetal class II alloantigen disparity. Am J Reprod Immunol 1992;28(3-4):226–7. Available from: https://doi.org/10.1111/j.1600-0897.1992.tb00798.x.
[17] Yan Z, Aydelotte T, Gadi VK, Guthrie KA, Nelson JL. Acquisition of the rheumatoid arthritis HLA shared epitope through microchimerism. Arthritis Rheum 2011;63(3):640–4. Available from: https://doi.org/10.1002/art.30160.
[18] Aletaha D, Neogi T, Silman AJ, et al. 2010 rheumatoid arthritis classification criteria: an American College of Rheumatology/European League Against Rheumatism collaborative initiative. Ann Rheum Dis 2010;69 (9):1580–8. Available from: https://doi.org/10.1136/ard.2010.138461.
[19] Sammaritano LR, Bermas BL, Chakravarty EE, et al. 2020 American College of Rheumatology guideline for the management of reproductive health in rheumatic and musculoskeletal diseases. Arthritis Rheumatol (Hoboken, NJ) 2020;72(4):529–56. Available from: https://doi.org/10.1002/art.41191.
[20] US FDA Drug Safety Communication. FDA recommends avoiding use of NSAIDs in pregnancy at 20 weeks or later because they can result in low amniotic fluid. <https://www.fda.gov/media/142967/download>.
[21] Li D-K, Ferber JR, Odouli R, Quesenberry C. Use of nonsteroidal anti-inflammatory drugs during pregnancy and the risk of miscarriage. Am J Obstet Gynecol 2018;219(3):275.e1–8. Available from: https://doi.org/10.1016/j.ajog.2018.06.002.
[22] Edwards DRV, Aldridge T, Baird DD, Funk MJ, Savitz DA, Hartmann KE. Periconceptional over-the-counter nonsteroidal anti-inflammatory drug exposure and risk for spontaneous abortion. Obstet Gynecol 2012;120 (1):113–22. Available from: https://doi.org/10.1097/AOG.0b013e3182595671.

[23] Katz PP. Childbearing decisions and family size among women with rheumatoid arthritis. Arthritis Rheum 2006;55(2):217–23. Available from: https://doi.org/10.1002/art.21859.
[24] Jawaheer D, Zhu JL, Nohr EA, Olsen J. Time to pregnancy among women with rheumatoid arthritis. Arthritis Rheum 2011;63(6):1517–21. Available from: https://doi.org/10.1002/art.30327.
[25] Clowse MEB, Chakravarty E, Costenbader KH, Chambers C, Michaud K. Effects of infertility, pregnancy loss, and patient concerns on family size of women with rheumatoid arthritis and systemic lupus erythematosus. Arthritis Care Res (Hoboken) 2012;64(5):668–74. Available from: https://doi.org/10.1002/acr.21593.
[26] Brouwer J, Laven JSE, Hazes JMW, Schipper I, Dolhain RJEM. Levels of serum anti-Müllerian hormone, a marker for ovarian reserve, in women with rheumatoid arthritis. Arthritis Care Res (Hoboken) 2013;65(9):1534–8. Available from: https://doi.org/10.1002/acr.22013.
[27] Aljary H, Czuzoj-Shulman N, Spence AR, Abenhaim HA. Pregnancy outcomes in women with rheumatoid arthritis: a retrospective population-based cohort study. J Matern Fetal Neonatal Med 2020;33(4):618–24. Available from: https://doi.org/10.1080/14767058.2018.1498835.
[28] Chakravarty EF, Nelson L, Krishnan E. Obstetric hospitalizations in the United States for women with systemic lupus erythematosus and rheumatoid arthritis. Arthritis Rheum 2006;54(3):899–907. Available from: https://doi.org/10.1002/art.21663.
[29] Pinheiro MB, Carvalho MG, Martins-Filho OA, et al. Severe preeclampsia: are hemostatic and inflammatory parameters associated? Clin Chim Acta 2014;427:65–70. Available from: https://doi.org/10.1016/j.cca.2013.09.050.
[30] Kiriakidou M, Ching CL. Systemic lupus erythematosus. Ann Intern Med 2020;172(11):ITC81–96. Available from: https://doi.org/10.7326/AITC202006020.
[31] Oktem O, Yagmur H, Bengisu H, Urman B. Reproductive aspects of systemic lupus erythematosus. J Reprod Immunol 2016;117:57–65. Available from: https://doi.org/10.1016/j.jri.2016.07.001.
[32] Ruiz-Irastorza G, Lima F, Alves J, et al. Increased rate of lupus flare during pregnancy and the puerperium: a prospective study of 78 pregnancies. Br J Rheumatol 1996;35(2):133–8. Available from: http://www.ncbi.nlm.nih.gov/pubmed/8612024.
[33] Abramson SB, Buyon JP. Activation of the complement pathway: comparison of normal pregnancy, preeclampsia, and systemic lupus erythematosus during pregnancy. Am J Reprod Immunol 1992;28(3-4):183–7. Available from: https://doi.org/10.1111/j.1600-0897.1992.tb00787.x.
[34] Aringer M, Costenbader K, Daikh D, et al. 2019 European League against rheumatism/American College of Rheumatology classification criteria for systemic lupus erythematosus. Arthritis Rheumatol (Hoboken, NJ) 2019;71(9):1400–12. Available from: https://doi.org/10.1002/art.40930.
[35] Jara LJ, Medina G, Cruz-Dominguez P, Navarro C, Vera-Lastra O, Saavedra MA. Risk factors of systemic lupus erythematosus flares during pregnancy. Immunol Res 2014;60(2-3):184–92. Available from: https://doi.org/10.1007/s12026-014-8577-1.
[36] Clowse MEB, Magder L, Witter F, Petri M. Hydroxychloroquine in lupus pregnancy. Arthritis Rheum 2006;54(11):3640–7. Available from: https://doi.org/10.1002/art.22159.
[37] Andreoli L, Bertsias GK, Agmon-Levin N, et al. EULAR recommendations for women's health and the management of family planning, assisted reproduction, pregnancy and menopause in patients with systemic lupus erythematosus and/or antiphospholipid syndrome. Ann Rheum Dis 2017;76(3):476–85. Available from: https://doi.org/10.1136/annrheumdis-2016-209770.
[38] van den Hoogen F, Khanna D, Fransen J, et al. 2013 classification criteria for systemic sclerosis: an American College of Rheumatology/European League against rheumatism collaborative initiative. Ann Rheum Dis 2013;72(11):1747–55. Available from: https://doi.org/10.1136/annrheumdis-2013-204424.
[39] Mayes MD, Lacey JV, Beebe-Dimmer J, et al. Prevalence, incidence, survival, and disease characteristics of systemic sclerosis in a large US population. Arthritis Rheum 2003;48(8):2246–55. Available from: https://doi.org/10.1002/art.11073.
[40] Abbot S, Bossingham D, Proudman S, de Costa C, Ho-Huynh A. Risk factors for the development of systemic sclerosis: a systematic review of the literature. Rheumatol Adv Pract 2018;2(2). Available from: https://doi.org/10.1093/rap/rky041.
[41] Di Cristofaro J, Karlmark KR, Kanaan SB, et al. Soluble HLA-G expression inversely correlates with fetal microchimerism levels in peripheral blood from women with scleroderma. Front Immunol 2018;9. Available from: https://doi.org/10.3389/fimmu.2018.01685.

[42] Nelson JL. Maternal–fetal immunology and autoimmune disease. Is some autoimmune disease auto-alloimmune or allo-autoimmune? Arthritis Rheum 1996;39(2):191–4. Available from: https://doi.org/10.1002/art.1780390203.

[43] Poormoghim H, Lucas M, Fertig N, Medsger TA. Systemic sclerosis sine scleroderma: demographic, clinical, and serologic features and survival in forty-eight patients. Arthritis Rheum 2000;43(2):444–51 10.1002/1529-0131(200002)43:2 < 444::AID-ANR27 > 3.0.CO;2-G.

[44] Blagojevic J, AlOdhaibi KA, Aly AM, et al. Pregnancy in systemic sclerosis: results of a systematic review and metaanalysis. J Rheumatol 2020;47(6):881–7. Available from: https://doi.org/10.3899/jrheum.181460.

[45] Silman AJ, Black C. Increased incidence of spontaneous abortion and infertility in women with scleroderma before disease onset: a controlled study. 1988:441–444.

[46] Ekici H, Imamoglu M, Okmen F, Ogultarhan R, Yeniel AO. Pulmonary hypertension in pregnancy: experience from 45 cases at a tertiary care center. J Matern Fetal Neonatal Med 2020;1–6. Available from: https://doi.org/10.1080/14767058.2020.1770216.

[47] Miyakis S, Lockshin MD, Atsumi T, et al. International consensus statement on an update of the classification criteria for definite antiphospholipid syndrome (APS). J Thromb Haemost 2006;4(2):295–306. Available from: https://doi.org/10.1111/j.1538-7836.2006.01753.x.

[48] Duarte-García A, Pham MM, Crowson CS, et al. The epidemiology of antiphospholipid syndrome: a population-based study. Arthritis Rheumatol (Hoboken, NJ) 2019;71(9):1545–52. Available from: https://doi.org/10.1002/art.40901.

[49] Cervera R, Serrano R, Pons-Estel GJ, et al. Morbidity and mortality in the antiphospholipid syndrome during a 10-year period: a multicentre prospective study of 1000 patients. Ann Rheum Dis 2015;74(6):1011–18. Available from: https://doi.org/10.1136/annrheumdis-2013-204838.

[50] D'Ippolito S, Ticconi C, Tersigni C, et al. The pathogenic role of autoantibodies in recurrent pregnancy loss. Am J Reprod Immunol 2020;83(1). Available from: https://doi.org/10.1111/aji.13200.

[51] Viall CA, Chamley LW. Histopathology in the placentae of women with antiphospholipid antibodies: a systematic review of the literature. Autoimmun Rev 2015;14(5):446–71. Available from: https://doi.org/10.1016/j.autrev.2015.01.008.

[52] Garcia D, Erkan D. Diagnosis and management of the antiphospholipid syndrome. Longo DL, (ed.) N Engl J Med. 2018; 378(21):2010-2021. doi:10.1056/NEJMra1705454

[53] Abrahams VM, Chamley LW, Salmon JE. Emerging treatment models in rheumatology: antiphospholipid syndrome and pregnancy: pathogenesis to translation. Arthritis Rheumatol 2017;69(9):1710–21. Available from: https://doi.org/10.1002/art.40136.

[54] Sammaritano LR. Anti-phospholipid syndrome. Best Pract Res Clin Rheumatol 2020;34(1):101463. Available from: https://doi.org/10.1016/j.berh.2019.101463.

[55] Marder W, Knight JS, Kaplan MJ, et al. Placental histology and neutrophil extracellular traps in lupus and preeclampsia pregnancies. Lupus Sci Med 2016;3(1):e000134. Available from: https://doi.org/10.1136/lupus-2015-000134.

[56] Sammaritano LR, Bermas BL, Chakravarty EE, et al. 2020 American College of Rheumatology guideline for the management of reproductive health in rheumatic and musculoskeletal diseases. Arthritis Care Res (Hoboken) 2020;72(4):461–88. Available from: https://doi.org/10.1002/acr.24130.

[57] Erkan D, Harrison MJ, Levy R, et al. Aspirin for primary thrombosis prevention in the antiphospholipid syndrome: a randomized, double-blind, placebo-controlled trial in asymptomatic antiphospholipid antibody–positive individuals. Arthritis Rheum 2007;56(7):2382–91. Available from: https://doi.org/10.1002/art.22663.

[58] LeFevre ML. Low-dose aspirin use for the prevention of morbidity and mortality from preeclampsia: U.S. preventive services task force recommendation statement. Ann Intern Med 2014;161(11):819. Available from: https://doi.org/10.7326/M14-1884.

[59] Kaiser R, Cleveland CM, Criswell LA. Risk and protective factors for thrombosis in systemic lupus erythematosus: results from a large, multi-ethnic cohort. Ann Rheum Dis 2009;68(2):238–41. Available from: https://doi.org/10.1136/ard.2008.093013.

[60] Jung H, Bobba R, Su J, et al. The protective effect of antimalarial drugs on thrombovascular events in systemic lupus erythematosus. Arthritis Rheum 2010;62(3):863–8. Available from: https://doi.org/10.1002/art.27289.

[61] Erkan D, Unlu O, Sciascia S, et al. Hydroxychloroquine in the primary thrombosis prophylaxis of antiphospholipid antibody positive patients without systemic autoimmune disease. Lupus 2018;27(3):399–406. Available from: https://doi.org/10.1177/0961203317724219.

[62] Mak A, Cheung MW-L, Cheak AA, Ho RC-M. Combination of heparin and aspirin is superior to aspirin alone in enhancing live births in patients with recurrent pregnancy loss and positive antiphospholipid antibodies: a meta-analysis of randomized controlled trials and meta-regression. Rheumatology (Oxf) 2010;49(2):281–8. Available from: https://doi.org/10.1093/rheumatology/kep373.

[63] Arslan E, Fellow V, Medicine M, Branch DW, James R, Endowed JS. Best practice & research clinical obstetrics and gynaecology antiphospholipid syndrome: diagnosis and management in the obstetric patient. Best Pract Res Clin Obstet Gynaecol 2020;64:31–40. Available from: https://doi.org/10.1016/j.bpobgyn.2019.10.001.

[64] Cowchock FS, Reece EA, Balaban D, Branch DW, Plouffe L. Repeated fetal losses associated with antiphospholipid antibodies: a collaborative randomized trial comparing prednisone with low-dose heparin treatment. Am J Obstet Gynecol 1992;166(5):1318–23. Available from: http://www.ncbi.nlm.nih.gov/pubmed/1595785.

[65] Sciascia S, Hunt BJ, Talavera-Garcia E, Lliso G, Khamashta MA, Cuadrado MJ. The impact of hydroxychloroquine treatment on pregnancy outcome in women with antiphospholipid antibodies. Am J Obstet Gynecol 2016;214(2):273.e1–8. Available from: https://doi.org/10.1016/j.ajog.2015.09.078.

[66] Mekinian A, Lazzaroni MG, Kuzenko A, et al. The efficacy of hydroxychloroquine for obstetrical outcome in antiphospholipid syndrome: Data from a European multicenter retrospective study. Autoimmun Rev 2015;14(6):498–502. Available from: https://doi.org/10.1016/j.autrev.2015.01.012.

[67] Lockshin MD, Kim M, Laskin CA, et al. Prediction of adverse pregnancy outcome by the presence of lupus anticoagulant, but not anticardiolipin antibody, in patients with antiphospholipid antibodies. Arthritis Rheum 2012;64(7):2311–18. Available from: https://doi.org/10.1002/art.34402.

[68] Saccone G, Berghella V, Maruotti GM, et al. Anti-phospholipid antibody profile based obstetric outcomes of primary anti-phospholipid syndrome: the PREGNANTS study. Am J Obstet Gynecol 2017;216(5):525.e1–525.e12. Available from: https://doi.org/10.1016/j.ajog.2017.01.026.

[69] Schreiber K, Hunt BJ. Managing antiphospholipid syndrome in pregnancy. Thromb Res 2019;181:S41–6. Available from: https://doi.org/10.1016/S0049-3848(19)30366-4.

[70] Izmirly PM, Rivera TL, Buyon JP. Neonatal lupus syndromes. Rheum Dis Clin North Am 2007;33(2):267–85. Available from: https://doi.org/10.1016/j.rdc.2007.02.005.

[71] Fritzler MJ, Pauls JD, Kinsella TD, Bowen TJ. Antinuclear, anticytoplasmic, and anti-Sjogren's syndrome antigen A (SS-A/Ro) antibodies in female blood donors. Clin Immunol Immunopathol 1985;36(1):120–8. Available from: https://doi.org/10.1016/0090-1229(85)90045-5.

[72] Satoh M, Chan EKL, Ho LA, et al. Prevalence and sociodemographic correlates of anti-nuclear antibodies in the United States. Arthritis Rheum 2012;64(7):2319–27. Available from: https://doi.org/10.1002/art.34380.

[73] Rivera TL, Izmirly PM, Birnbaum BK, et al. Disease progression in mothers of children enrolled in the research registry for neonatal lupus. Ann Rheum Dis 2009;68(6):828–35. Available from: https://doi.org/10.1136/ard.2008.088054.

[74] Neiman AR, Lee LA, Weston WL, Buyon JP. Cutaneous manifestations of neonatal lupus without heart block: characteristics of mothers and children enrolled in a national registry. J Pediatr 2000;137(5):674–80. Available from: https://doi.org/10.1067/mpd.2000.109108.

[75] Brucato A, Frassi M, Franceschini F, et al. Risk of congenital complete heart block in newborns of mothers with anti-Ro/SSA antibodies detected by counterimmunoelectrophoresis: a prospective study of 100 women. Arthritis Rheum 2001;44(8):1832–5. Available from: https://doi.org/10.1002/1529-0131(200108)44:8<1832::AID-ART320>3.0.CO;2-C.

[76] Llanos C, Izmirly PM, Katholi M, et al. Recurrence rates of cardiac manifestations associated with neonatal lupus and maternal/fetal risk factors. Arthritis Rheum 2009;60(10):3091–7. Available from: https://doi.org/10.1002/art.24768.

[77] Waltuck J, Buyon JP. Autoantibody-associated congenital heart block: outcome in mothers and children. Ann Intern Med 1994;120(7):544–51. Available from: https://doi.org/10.7326/0003-4819-120-7-199404010-00003.

[78] Miranda-Carús ME, Askanase AD, Clancy RM, et al. Anti-SSA/Ro and anti-SSB/La autoantibodies bind the surface of apoptotic fetal cardiocytes and promote secretion of TNF-alpha by macrophages. J Immunol 2000;165(9):5345–51. Available from: https://doi.org/10.4049/jimmunol.165.9.5345.

[79] Jaeggi ET, Fouron J-C, Silverman ED, Ryan G, Smallhorn J, Hornberger LK. Transplacental fetal treatment improves the outcome of prenatally diagnosed complete atrioventricular block without structural heart disease. Circulation 2004;110(12):1542–8. Available from: https://doi.org/10.1161/01.CIR.0000142046.58632.3A.

[80] Eliasson H, Sonesson S-E, Sharland G, et al. Isolated atrioventricular block in the fetus: a retrospective, multinational, multicenter study of 175 patients. Circulation 2011;124(18):1919–26. Available from: https://doi.org/10.1161/CIRCULATIONAHA.111.041970.

[81] Barsalou J, Costedoat-Chalumeau N, Berhanu A, et al. Effect of in utero hydroxychloroquine exposure on the development of cutaneous neonatal lupus erythematosus. Ann Rheum Dis 2018;77(12):1742–9. Available from: https://doi.org/10.1136/annrheumdis-2018-213718.

[82] Izmirly PM, Saxena A, Kim MY, et al. Maternal and fetal factors associated with mortality and morbidity in a multi-racial/ethnic registry of anti-SSA/Ro-associated cardiac neonatal lupus. Circulation 2011;124(18):1927–35. Available from: https://doi.org/10.1161/CIRCULATIONAHA.111.033894.

SECTION III

Repeated implantation failures

CHAPTER

15

The role of reproductive immunology in recurrent pregnancy loss and repeated implantation failure

Jenny S. George, Roisin Mortimer and Raymond M. Anchan

Center for Infertility and Reproductive Surgery, Brigham and Women's Hospital, Harvard Medical School, Boston, MA, United States

1 Introduction

Although clinically recognized pregnancy loss occurs in 15%–25% of pregnancies, only 1%–5% of losses can be attributed to recurrent pregnancy loss (RPL), a unique disorder defined as two or more failed clinical pregnancies. The vast majority of sporadic losses can be attributed to errors in chromosomal segregation, resulting in trisomy, monosomy, and polyploidy [1]. Despite a comprehensive evaluation, the etiology of 50% of RPL losses will remain unknown. With the advent of preimplantation genetic testing for aneuploidy (PGT-A) and preimplantation genetic testing for monogenic disorders (PGT-M), we now have the opportunity to screen and select euploid embryos lacking specific mutations. Despite the preferential transfer of unaffected euploid embryos, a subset of patients will experience recurrent implantation failure (RIF), defined as three failed transfers of high-quality embryos [2]. Managing these patients is clinically challenging as the underlying etiology of these losses remains unknown. More recently, studies by several reproductive immunology groups have demonstrated a central therapeutic role for immune modulation, resulting in successful implantation and pregnancy outcomes. In this chapter, we review the role of a comprehensive reproductive immunologic evaluation in patients with unexplained RPL and RIF following euploid embryo transfers. Based on the current literature, we propose collaborative management of patients with RPL and RIF by reproductive endocrinologists and reproductive immunologists to help achieve successful pregnancy outcomes.

Immunology of Recurrent Pregnancy Loss and Implantation Failure
DOI: https://doi.org/10.1016/B978-0-323-90805-4.00005-5

2 Definitions and terminology

A pregnancy loss is defined as the spontaneous demise of a pregnancy prior to 20 weeks gestation [1]. Sporadic pregnancy loss is common, occurring in 15%–25% of pregnancies [1]. A large proportion of sporadic pregnancy losses are secondary to chromosomal errors, resulting in trisomy, monosomy, and polyploidy. These chromosomal errors are closely associated with increasing maternal age; consequently, a single clinical loss does not necessarily confer a higher risk of loss in subsequent pregnancies, nor does it imply a recurrent etiology for subsequent losses. In contrast, RPL is a disorder epidemiologically defined by three or more failed clinical pregnancies [1], verified by first-trimester ultrasound or the presence of chorionic villi on endometrial biopsy. Consequently, this definition excludes ectopic pregnancies, molar gestations, and implantation failure [3]. Furthermore, it is important to note that a threshold of three or more losses should only be employed for epidemiological studies, as clinical evaluation is warranted following two first-trimester pregnancy losses [3,4]. RPL can be further classified into primary and secondary RPL: primary RPL characterizes patients who have never carried a pregnancy to viability, whereas secondary RPL refers to patients who have had at least one live birth prior to the recurrent losses.

The definition of RIF, in contrast, remains controversial. In a systematic review analyzing the various definitions of RIF utilized in the literature, Polanski et al. defined RIF as an iatrogenic condition resulting from repetitive unsuccessful cycles of in vitro fertilization (IVF) or intracytoplasmic sperm injection (ICSI). They proposed "three previous IVF-embryo transfer attempts" as the most common definition of RIF. A minority of clinicians argued that the number and quality of transferred embryos must be considered [5], while others proposed rigid diagnostic criteria, defining RIF as a failure after the transfer of at least four good-quality embryos within a minimum of three fresh or frozen cycles under 40 years of age [6]. The absence of shared classification criteria to characterize "good quality" embryos only further complicated attempts to define RIF. In a novel study attempting to present a definition of RIF, which accounts for the effects of female age and anticipated blastocyst euploidy rates on cumulative implantation rates, Ata et al. argued that RIF should be a functional term that guides further management [7]. The authors demonstrated that when the euploidy status of the transferred embryo is unknown, at least seven blastocysts must be transferred to attain a 95% cumulative probability for the implantation of at least one embryo. Consequently, the authors proposed that RIF should be a term reserved for implantation failure, which is unlikely to be caused by embryo aneuploidy, the leading cause of implantation failure.

3 Prevalence

The prevalence of RPL is defined as the number of patients in a population, who at a specific time point, have experienced three or more clinical pregnancy losses. The prevalence of RPL is low, with fewer than 5% of patients experiencing two consecutive miscarriages and fewer than 1% of patients experiencing three or more losses [8,9]. In an intriguing retrospective study of women with RPL utilizing data provided by the Swedish National Patient Register, Rasmark, et al. demonstrated the average incidence of RPL in 2003 as 42 per 100,000 (0.042%) in women aged 18–42 years; notably, the average incidence rose to 73 per 100,000 (0.073%) by 2012. The authors postulated that this dramatic rise in the incidence of RPL could be due to a variety of factors, including advancing maternal age, environmental elements, and immunologic influences [10].

The incidence and prevalence of RIF have been challenging to elucidate, as RIF remains an important clinical entity without a universally established definition. In a retrospective cohort study aiming to study the true prevalence of RIF, Pirtea et al. assessed women ($n = 4429$) with an anatomically normal uterus who underwent up to three consecutive frozen euploid single embryo transfers. The authors demonstrated that the rate of true RIF is low, as less than 5% of patients failed to achieve a clinical pregnancy if three euploid embryos were transferred [11].

4 Underlying immunopathology related to reproduction

The female reproductive tract, consisting of the ovaries, fallopian tubes, uterus, cervix, and vagina, has unique requirements for the regulation of immune processes during both pregnant and nonpregnant states [12]. In the nonpregnant state, epithelial cells of the reproductive tract modulate mucosal immunity, protecting tissue from microbial infection and resulting inflammation. In contrast to epithelial cells within the reproductive tract, the ovary utilizes macrophages to orchestrate menstrual cycles: ovarian macrophages promote microscopic inflammatory reactions, which prompt ovulation and luteolysis. Additionally, macrophages produce growth factors that promote angiogenesis and cell proliferation and phagocytize apoptotic granulosa and luteal cells following follicular atresia [13]. It is apparent that macrophages are essential to ovarian immunity, as the disruption and dysregulation of ovarian macrophages have been implicated in polycystic ovarian syndrome [14], premature ovarian insufficiencies [13], and endometriosis [15].

During pregnancy, adaptive and innate immune systems must undergo extensive modification to ensure that the fetus and placental tissue are not rejected as "foreign" entities. Utilizing cross-signaling with trophoblast cells, uterine natural killer (uNK) cells downregulate their own cytotoxic activity, effectively creating an immunosuppressive environment for the fetus and placenta. uNK cells peak in volume within the decidua during the first trimester and continue to decline throughout gestation so that only small numbers are present at term [16]. A principal immune cell within the placenta as well, uNK cells represent 75% of the inflammatory cells of the placenta and are essential to promote trophoblast invasion of the decidua and spiral arteries, angiogenesis, and suppression of macrophages and T cells [17].

Decidual macrophages represent 15% of decidual leukocytes and can be classified as M2 (associated with immune-modulating and tissue remodeling) or M1 (proinflammatory) [18]. The majority of decidual macrophages are of the M2 subtype and are designed to decrease inflammation by producing anti-inflammatory cytokines, performing phagocytosis, and secreting angiogenic factors [17]. Decidual macrophages are also critical to normal reproductive function and pregnancy, as they function to phagocytose apoptotic trophoblasts [12].

Although these apoptotic cells normally occur in a developing placenta, apoptotic trophoblasts may be identified as foreign tissue, thereby eliciting a robust maternal immune response [19]. Decidual macrophages prevent this response by consuming apoptotic cells and preventing the release of proinflammatory debris.

5 Evaluation

The primary challenge providers face is to distinguish sporadic miscarriage from RPL. When considering whom to evaluate, it is important to note that a threshold of three or more losses should only be employed for epidemiological studies assessing RPL, as clinical evaluation is warranted following two first-trimester pregnancy losses [3,4]. The evaluation of RPL can be categorized based on the underlying etiology: chromosomal abnormalities, anatomic anomalies, antiphospholipid antibody syndrome (APLAS), hormonal disturbances, alloimmune abnormalities, environmental factors, and infection, summarized in Table 1. Unfortunately, a comprehensive evaluation fails to yield a diagnosis in up to 50% of patients with RPL.

TABLE 1 Suspected causes of recurrent pregnancy loss.

Cause	Contribution to recurrent pregnancy loss (%)	Evaluation
Cytogenetic	2–5	Balanced reciprocal translocations
		Karyotype, microarray analysis of POCs
Antiphospholipid syndrome	8–42	Lupus anticoagulant, anticardiolipin IgG or IgM antibody, anti-β2 glycoprotein I IgG or IgM
Anatomic	1.8–37.6	HSG, HyCoSy, MRI, hysteroscopy
Hormonal or metabolic	17–20	TSH, prolactin, HbA1c, ovarian reserve testing
Alloimmune and cellular immune	20	Peripheral and uterine NK cell proportions, NK cell cytotoxicity, cytokine profiles, T helper 1/T helper 2 ratios, blocking antibodies, HLA typing, antipaternal leukocyte antibodies
Environmental		History
Infection	0.5–5	Vaginal culture for bacterial vaginosis, endocervical culture, endometrial microbial gene studies by PCR
Thrombophilia	40–50	Factor V Leiden, factor II (prothrombin) gene mutation, hyperhomocysteinemia activated protein C resistance and protein S deficiency
Unexplained	Up to 50	

Adapted from Practice Committee of the American Society for Reproductive Medicine. Evaluation and treatment of recurrent pregnancy loss: a committee opinion. Fertil Steril. 2012; 98 (5):1103–11.

5.1 Cytogenetic abnormalities

Genetic abnormalities are the most frequent cause of early pregnancy loss, with up to 60% of products of conception examined following pregnancy loss demonstrating sporadic karyotypic abnormalities. Specifically, in an RPL cohort, 29%–46% of miscarriage is estimated to be due to aneuploidy [20], and this risk decreases with an increasing number of miscarriages [21]. The risk of aneuploidy increases with increasing maternal age [1], and it is accepted as the largest single risk factor for future miscarriage amongst the whole population and women with RPL [22]. Furthermore, as the number of miscarriages increases, the risk of euploid pregnancy loss also increases. Thus the cytogenetic analysis should be performed on products of conception in cases of RPL. If there is a chromosomal abnormality detected, the prognosis for a future pregnancy is improved [23].

Parental karyotypic abnormalities are much less frequent but must be considered when investigating the causes of RPL. This may be evaluated by performing parental karyotype analysis, specifically assessing for the presence of balanced translocations or Robertsonian translocations, which are estimated to comprise 2%–5% cases of RPL [24].

5.2 Anatomic anomalies

The prevalence of uterine anatomic abnormalities has been estimated at 6.7% amongst the general population, 7.3% amongst people with infertility, and between 12.6% and 16.7% amongst people with RPL [25,26]. Uterine anatomic abnormalities are strongly associated with second-trimester loss, although the association with first-trimester pregnancy loss is more circumspect than that with second or third trimester [1]. Congenital Mullerian anomalies which could contribute to RPL include septate and bicornuate uteri. Correction of septate uteri, in particular, has been associated with improved pregnancy outcomes in patients with RPL [26]. Imaging modalities that can be used to investigate this include three-dimensional ultrasound, MRI, hysterosalpingo contrast sonography (HyCoSy), and HSG. Definitive diagnosis and simultaneous treatment may be made by diagnostic and operative hysteroscopy with or without assisted diagnostic laparoscopy.

5.3 Antiphospholipid antibody syndrome

The international consensus of the classification criteria for APLAS has been published [27]. Between 8% and 42% reported ranges of patients with RPL test positive for antiphospholipid antibodies, generally agreed to 5%–20% ranges [28]. There are multiple antiphospholipid antibodies, but only lupus anticoagulant, anticardiolipin antibody (IgM or IgG), or Anti-β2-glycoprotein I (IgM or IgG) have standardized clinical assays and enough evidence to warrant routine screening in the RPL population.

5.4 Hormonal disturbances

Multiple hormones have been associated with RPL. An evaluation of thyroid hormone should begin with a TSH level. If TSH is <2.5, then a reflex T4 should be ordered [29]. Poorly controlled diabetes is associated with pregnancy loss and can be diagnosed with a hemoglobin A1c level [30]. Hyperprolactinemia has been associated with RPL. Alterations in the hypothalamic–pituitary–ovarian axis, resulting in impaired folliculogenesis and/or an insufficient luteal phase associated with hyperprolactinemia. In a study in which women with RPL and hyperprolactinemia were randomized to treatment with bromocriptine or none in a subsequent pregnancy, there was a significantly higher live birth rate (87% vs 52%) amongst the treated cohort [31].

5.5 Thrombophilias

A *meta*-analysis of 31 studies investigating thrombophilic factors and their association with pregnancy loss found that factor V Leiden, factor II (prothrombin) gene mutation, activated protein C resistance and protein S deficiency were associated with a significantly increased risk of both early and late RPL [32]. However, prospective studies have not confirmed this increased risk [33,34]. Screening for inherited thrombophilia is recommended in patients with a personal history of VTE or significant family history.

6 Treatment

6.1 Treatment of recurrent pregnancy loss

The clinical management of patients with RPL focuses on addressing the underlying etiology. Balanced reciprocal and Robertsonian translocations are identified in 2%–5% of patients with RPL [35]. If a balanced translocation is identified via peripheral karyotyping of the patient or her partner, genetic counseling is warranted, as the likelihood of live birth is dependent on the type of rearrangement and the specific chromosome(s) involved. Preimplantation genetic testing (PGT) is a form of genetic testing requiring the formation of embryos via IVF and blastocyst trophectoderm biopsy. PGT can also be employed to identify structural rearrangements (PGT-SR) in couples affected by a translocation. Over the years, PGT-SR has evolved from fluorescence in situ hybridization (FISH), quantitative real-time PCR (qPCR) to comprehensive methods, including array-based comparative genomic hybridization (CGH) and next-generation sequencing (NGS). The most dramatic change has been the ability to perform trophectoderm biopsies on day 5 instead of testing blastomere cells on day 3. In a retrospective study of 575 couples with chromosomal translocations, including 169 couples treated by SNP-PGD and 406 couples treated by FISH-PGD, Tan et al. demonstrated a higher pregnancy rate and a lower miscarriage rate in patients who underwent SNP-PGD compared to patients who underwent FISH-PGD [36]. In a novel study assessing pregnancy outcomes of reciprocal translocation carriers before and after PGT-SR, Huang et al. [37] found that after PGT-SR, the reciprocal translocation carriers who had previously experienced two or more RPL had a lower risk of miscarriage and a higher rate of live birth. Based on the literature, we recommend IVF with PGT-SR for patients affected by RPL due to a balanced translocation.

Addressing underlying uterine abnormalities is essential in the treatment of RPL. Congenital Mullerian tract anomalies, including unicornuate, didelphic, bicornuate, and septate uteri, can be identified via hysterosalpingography, HyCoSy, or three-dimensional ultrasound: suspicious findings can be confirmed with the gold standard of imaging, pelvic MRI. Previous studies have confirmed the role of uterine anomalies in reproductive outcomes: the septate uterus is associated with reduced fertility, increased miscarriage rates, and preterm delivery [38]. The threshold at which resection of the uterine septum remains controversial: the American Society of Reproductive Medicine (ASRM) defines an arcuate uterus as having a depth from the interstitial line to the apex of the indentation <1 cm, while a septate uterus has a depth >1.5 cm. Consequently, there is a lack of clinical guidance for scenarios in which the depth ranges from 1.0–1.5 cm. In the randomized uterine septum transection trial (TRUST), Rikken et al. randomized women to hysteroscopic resection or expectant management; the primary outcomes of the study were live birth, ongoing pregnancy, clinical pregnancy, and miscarriage [39]. 80 women were randomized, and live birth occurred in 12 of 39 women allocated to septum resection (31%) and in 14 of 40 women (35%) randomized to expectant management. The authors conclude that hysteroscopic resection does not improve reproductive outcomes in women with a septate uterus. However, it is essential to note that of the 80 women randomized, 71 patients had a partial uterine septum with depths <1.5 cm. Thus hysteroscopic resection of a uterine septum may be of significant reproductive benefit if the depth is >1.5 cm.

The standard treatment for documented criteria defined as APLAS consists of low-dose aspirin and twice-daily unfractionated heparin [40] as this treatment regimen has been found to reduce pregnancy loss in patients with APLAS by 54%. Although some providers may prescribe once-daily low molecular weight heparin in lieu of unfractionated heparin, comparable efficacy has yet to be established. In a novel double-blinded, placebo-controlled trial, Pasquier et al. [41] demonstrated that enoxaparin 40 mg daily did not improve the chance of live birth in nonthrombophilic women with unexplained RPL when compared to placebo. Similarly, the HepASA, a randomized controlled trial (RCT) comparing the efficacy of low dose aspirin to combination therapy of low molecular weight heparin and aspirin, found that combination therapy did not confer additional benefit compared to aspirin alone with regard to the live birth rates [42].

6.2 Treatment of recurrent implantation failure

There is some overlap in the treatment of RPL and RIF, with an evaluation of the uterine cavity and correction of uterine pathologies seen via hysteroscopy; hormonal evaluation (TSH, HbA1c, prolactin); and thrombophilia screen also performed in the workup of RIF. However, treatment options for RIF are currently still limited, and many are considered experimental. A recent *meta*-analysis divided proposed therapies and interventions for RIF into four categories (Table 2).

6.3 Uterine interventions

Uterine interventions include treatments that aim to regulate uterine inflammatory factors and stimulate immunomodulatory factors through various mechanisms.

6.3.1 Intentional endometrial injury

Intentional endometrial injury or "endometrial scratch" is theorized to cause an increase in pregnancy and live birth rate through inducing immunomodulatory factors. However,

TABLE 2 Proposed therapies for recurrent implantation failure.

Uterine interventions	Intentional endometrial injury
	Endometrial sampling, treatment of chronic endometritis
	Atosiban
	Copper IUD
	Stent placement
Laboratory and procedural technologies and interventions	Sequential ET
	Intracytoplasmic morphologically selected Sperm injection
	Blastocyst stage ET
	Zygote intrafallopian tube transfer (ZIFT)
	Assisted hatching
	PGT-A
	Timed lapse imaging
Immunomodulatory therapies	Low molecular weight heparin
	Intravenous immunoglobulin G
	Intrauterine peripheral blood mononuclear cell infusion
	Subcutaneous granulocyte colony-stimulating factor
	Intrauterine granulocyte colony-stimulating factor
	Intrauterine autologous platelet-rich plasma (PRP) infusion
	Lymphocyte immunization therapy (LIT)
	Intralipid
Endometrial therapies	Estradiol
	Progesterone
	Aspirin
	Pentoifylline +/− Vitamin E
	Intramuscular growth hormone
	Sildenafil
	Endometrial receptivity array
	Hysteroscopy
	Bone marrow stem cells endometrial mobilization
Others	Uterine transplant
	Gestational carrier

Adapted from Busnelli, Andrea et al. "Efficacy of therapies and interventions for repeated embryo implantation failure: a systematic review and meta-analysis." Scientific Reports vol. 11,1 1747. 18 Jan. 2021, doi:10.1038/s41598−021−81439−6.

the studies that have been performed are heterogeneous, with factors such as the instrument used (Pipelle biopsy, Karman biopsy, hysteroscopic cannula), the timing of the scratch (phase in the cycle, whether it is in the same cycle as transfer or the preceding cycle) and study outcomes all varying. Pooling of results from two observational studies did show a beneficial effect on implantation rate and pregnancy rate [43,44], but a *meta*-analysis of three RCTs did not show either increased chances of pregnancy or live birth and or any difference in miscarriage rate [45–47].

6.3.2 Chronic endometritis

Chronic endometritis has long been postulated to be a cause of RIF due to the inflammatory endometrial milieu it creates. Definitions vary, but it is commonly defined as the elevated presence of plasma cells (>5 per high power field) in a uterine sample [48]. A recent retrospective study evaluated peripheral blood and uterine samples of patients with RIF and RPL. Uterine samples were used to diagnose patients with chronic endometritis. Patients with chronic endometritis had similar blood immune cell levels but higher endometrial cytokine and lymphocyte levels, significantly reduced following antibiotics treatment [49]. However, the clinical benefit of treatment has not been firmly established. One comparative study found that patients with RIF diagnosed with chronic endometritis who were subsequently treated with antibiotics had lower implantation rates but similar ongoing clinical pregnancy rates to patients who did not have endometritis on biopsy or patients who did not undergo endometrial sampling [50].

6.3.3 Atosiban

Atosiban is a combined oxytocin/vasopressin V1A antagonist and has been postulated to be useful in RIF as uterine contractions in stimulated IVF cycles are increased compared to natural cycles and are associated with significantly lower pregnancy success. There is one prospective cohort study that has examined the effect of atosiban on RIF [51]. This study showed an increased chance of embryo implantation (OR 3.14, 95%CI 1.54–6.28) in women with two or more prior failed transfers without decreasing miscarriage risk (OR 1.66, 95% CI 0.43–6.35, $p = 0.46$). RCTs to replicate results and evaluate further are ongoing.

6.3.4 Copper IUD

One nonrandomized study evaluated patients with RIF who underwent hysteroscopy. Patients could opt to have a copper IUD placed at the completion of the procedure, with the hypothesis being that short-term placement of the copper IUD would alter the cytokine profile of the endometrium, which has been demonstrated to occur in mice. In this study, women who had an IUD placed had higher implantation and clinical pregnancy rates [52].

6.4 Laboratory and procedural technologies and interventions

6.4.1 Blastocyst transfer, sequential embryo transfer, and zygote intrafallopian tube transfer

Sequential embryo refers to the initial transfer of a cleavage stage embryo on day 2, followed by a blastocyst embryo transfer on day 5 of the same cycle. An RCT did not show an increase in chemical or clinical pregnancy rate in patients randomized to sequential transfer compared to a single embryo blastocyst transfer [53]. However, a *meta*-analysis of observational studies did show a beneficial effect of sequential transfer, with an increased chance of clinical pregnancy (OR 2.64, 95% CI1.56–4.47, $I^2 = 0$) [5]. Blastocyst stage embryo transfer has also been suggested to have better success outcomes in women with RIF, compared to D2–3 stage transfer, and one study showed a significantly increased chance of implantation with this strategy but no difference in clinical pregnancy rate or live birth rate. Zygote intrafallopian tube transfer (ZIFT) [54] has been investigated as a treatment for RIF, as it was theorized that the natural intrafallopian milieu results in better synchronization between embryonic and endometrial development, but has not been shown to increase clinical pregnancy or live birth rates, and appeared to have worse outcomes compared to blastocyst transfer in observational studies [54].

6.4.2 *Preimplantation genetic testing for aneuploidy*

PGT-A has been proposed as a technique to reduce the risk of RIF, as cytogenetically abnormal embryos are less likely to implant successfully. However, to date, there is no evidence to support this finding, with two RCTs and multiple observational studies finding no difference in clinical pregnancy rate or live birth rate in women with RIF who underwent PGT-A [55,56].

6.4.3 *Assisted hatching*

Prior to implantation, the blastocyst must "hatch" from the zona pellucida. Assisted hatching involves artificial thinning or breaching the zona pellucida, and many early reports suggested that this technique could increase implantation rates and clinical pregnancy rates [57]. There has been one randomized trial performed, which did not demonstrate an increased chance of clinical pregnancy in women who underwent assisted hatching [58].

6.4.4 *Intracytoplasmic morphologically selected sperm injection*

Specifically for the subgroup of couples with a male factor component to infertility and associated RIF, intracytoplasmic morphologically selected sperm injection (IMSI) has been proposed as a solution. Compared to ICSI, IMSI utilizes magnification of up to 6000 × to detect potential abnormalities such as an abnormal number of vacuoles in the sperm nucleus. This "ultramorphology testing" has been shown in observational studies to yield improved clinical pregnancy and live birth rates compared to ICSI in couples with RIF, but these studies are small, and retrospective in nature. further research is needed [59–61].

6.5 Immunomodulatory therapies

The role of immunomodulatory therapies provides a novel and exciting forefront of clinical medicine for the management of RPL and RIF. Ongoing RCT trials investigate immunomodulation using corticosteroids while modes of therapy are reviewed by Sung et al. [62]. A number of these therapies are discussed in this section.

6.5.1 *Intravenous immunoglobulin*

One study has evaluated the effect of IVIG treatment on women with RIF [63]. The proposed mechanism of action of IVIG is both a direct antibody effect and an immunomodulatory effect. In this study, Ho et al. first measured peripheral blood mononuclear cell levels (including natural killer cell levels) in patients with RIF. Patients with low levels of NK cells during the follicular phase of their cycle had lower implantation rates compared to patients with higher NK cell levels. Furthermore, amongst women with low levels of NK cells, those treated with IVIG had higher clinical pregnancy rates and live birth rates.

6.5.2 *Intrauterine peripheral blood mononuclear cell infusion*

Intrauterine PBMC holds promise as a therapy for women with RIF. In general, this involves intrauterine administration of PBMCs 1–3 days prior to scheduled embryo transfer. Multiple observational studies and three RCT have been performed, all of which show an increase in chances of implantation rate and clinical pregnancy rate [63–69]. The magnitude of effect in these studies is large; although only one study reported a live birth rate [64] and safety data on this therapy is still sparse.

6.5.3 *Intrauterine autologous platelet-rich plasma infusion*

Platelet-rich plasma (PRP) has also been proposed as a treatment for RIF, where this plasma is injected during the luteal phase of the preceding cycle. A recent study compared subendometrial and intrauterine PRP to standard FET in women with RIF and found a significantly higher ongoing pregnancy rate and live birth rate in women who had been treated with PRP [70], but no difference between subendometrial or intrauterine administration. Notably, the PRP protocol also involved subcutaneous G-CSF injections for 3 days after PRP infusion.

6.5.4 *Subcutaneous or intrauterine granulocyte colony-stimulating factor*

Granulocyte colony-stimulating factor (G-CSF) plays a role in suppressing the immune response in early pregnancy by affecting lymphocytes, macrophages, and T-helper cells [71]. Both subcutaneous and intrauterine routes of administration have been investigated in the treatment of RIF, and six RCTs have been performed. Overall, a fixed-effects *meta*-analysis of the RCTs showed an increased chance of clinical pregnancy in patients treated with G-CSF, although only one study looked at the live birth rate and found no difference [72]. In subgroup analysis, subcutaneous G-CSF administration was still associated with increased implantation rate and clinical pregnancy rate, with a large magnitude of effect, but intrauterine G-CSF administration was not.

6.5.5 *Low molecular weight heparin*

As thrombophilia has been associated with RPL particularly, antiplatelet and anticoagulant therapies have been investigated as potential treatments for RIF. Low molecular weight heparin has been studied in multiple RCT [73–77], which did not show a beneficial effect on clinical pregnancy rate or live birth rate. A recent *meta*-analysis included five RCTs and did not find a beneficial effect of LMWH. Since the RCTs included a small number of study participants, multicenter trials are required to address the question [78].

6.5.6 *Intrauterine hCG injection*

Meta-analysis of two observational studies showed that intrauterine hCG injections resulted in significantly increased chances of clinical pregnancy and live birth [79,80].

6.6 Endometrial therapies

Estradiol;
Progesterone support;
Aspirin;
Pentoifylline +/− vitamin E;
Hysteroscopy;
Bone marrow stem cells endometrial mobilization.

6.6.1 *Intramuscular growth hormone*

Vaginal sildenafil

One small RCT evaluated the effect of vaginal sildenafil on women with RIF attributed to poor endometrial thickness [81]. Sildenafil is a PDE inhibitor that increases functional levels of nitric oxide, which has vasodilatory effects. This trial of 105 women demonstrated significantly higher implantation and ongoing rates amongst women treated with sildenafil. However, the trial did not control for chronic endometritis, which was significantly different between groups.

6.6.2 *Endometrial sampling*

Evaluating the endometrium may be performed in a variety of ways—including via hysteroscopy and endometrial biopsy [82]. Currently, there are three commercially available arrays to gage endometrial receptivity, which determines whether or not an embryo will implant. It has been shown that the "implantation date" for humans ranges from cycle day 19–23 (with cycle day14 being ovulation). Therefore the endometrial receptivity analysis test aims to accurately date the endometrium to optimize transfer timing, with mixed results [83]. The endometrial function test is a second test that is comprised of two parts: a histologic assessment of the endometrium and dating of the endometrium (via a cycle day 15 and 24 biopsies) [84]. Finally, the ReceptivaDx takes a different approach, measuring an inflammatory marker (BCL6) associated with endometriosis, as endometriosis is associated with RIF [85]. These tests are expensive, with sparse data supporting their use as yet. In the future, these tests likely be combined with components of the immunomodulatory testing described above to create a clear picture of the endometrium in all its complexity.

6.7 Combination therapy

A combination of factors likely overlap, coalesce and lead to RIF and RPL, including cellular, humoral and thrombophilic factors [86], suggesting that multiple therapies may be combined for a successful outcome. One retrospective cohort study compared women with a history of RPL and RIF who received both immunomodulatory (prednisone and IVIG) and anticoagulation (LMWH and ASA) treatment whilst undergoing IVF treatment to historical controls. PR and LBR were significantly improved for patients with either RIF and RPL, or both, compared to control [62].

7 Reproductive outcomes

7.1 Recurrent pregnancy loss

Prior studies have consistently demonstrated that the risk of future pregnancy loss increases with the number of prior pregnancy losses: after three consecutive miscarriages, the risk of miscarriage with the subsequent pregnancy is approximately 29% and increases to approximately 53% after six or more previous losses [22]. The reason for this association, however, remains unclear. In an attempt to determine whether there is a different prognostic impact for consecutive and nonconsecutive early pregnancy losses in women with secondary RPL, Egerup et al. conducted a retrospective cohort study of pregnancy outcomes in patients with unexplained secondary RPL, who were included in three previously published, Danish double-blinded placebo-controlled trials of intravenous immunoglobulin (IVIg) conducted from 1991 to 2014 [87]. Of the 168 patients included in the trials, 127 had secondary RPL and experienced a subsequent live birth or unexplained pregnancy loss in the index pregnancy. In patients with secondary RPL, both a late and early loss before the last birth did not significantly influence the risk of a new pregnancy loss in the index pregnancy. The incidence rate ratio (IRR) was 1.31 (95% CI 0.62–2.77) for the late loss and 0.88 (95% CI 0.70–1.11) for the early loss. Thus the authors argue that birth in women with secondary RPL eradicates the negative prognostic impact of previous pregnancy losses.

7.2 Recurrent implantation failure

Although emerging treatments for RIF are beginning to show promise, the prognosis for achieving a live birth after a diagnosis of RIF remains unclear. Consequently, providers must counsel their patients on the utility of continuing to pursue ET. In a novel study assessing the cumulative live birth rate and mean time to pregnancy (via IVF/ICSI or natural conception) in women experiencing unexplained RIF following IVF/ICSI treatment, Koot et al. performed a retrospective cohort study involving 223 women under 39 years of age who had experienced RIF between January 2008 and December 2012 [88]. The authors demonstrated a cumulative live birth rate of 49% (95% CI 39%–59%) over 5.5 years; among women who gave birth, the calculated median time to pregnancy was 9 months after experiencing RIF, where 18% arose from natural conceptions.

8 Summary and recommendations

1. RPL is a disorder epidemiologically defined by three or more failed clinical pregnancies (<24 weeks gestation), verified by first-trimester ultrasound or the presence of chorionic villi on endometrial biopsy. Clinical evaluation is warranted following two first-trimester pregnancy losses.
2. The definition of RIF remains controversial; if defined as a failure to achieve pregnancy following three consecutive frozen euploid single embryo transfers, the rate of RIF is <5%.
3. Pregnancy requires extensive modification of adaptive and innate immune responses involving ovarian macrophages, uNK cells, and placental and decidual macrophages.

4. The evaluation of RPL can be categorized based on the underlying etiology: chromosomal abnormalities, anatomic anomalies, APLAS, hormonal disturbances, alloimmune abnormalities, environmental factors, and infection.
5. A comprehensive evaluation will fail to yield a diagnosis in up to 50% of patients with RPL.
6. Treatment of RPL involves PGT-SR for patients affected by a balanced translocation, correction of uterine anomalies, and low dose aspirin and twice-daily unfractionated heparin for patients with APLAS.
7. Treatment of RIF remains controversial, although intrauterine autologous PRP infusion and subcutaneous/intrauterine injection of G-CSF have been found to improve clinical pregnancy rates.
8. The risk of future pregnancy loss increases with the number of prior pregnancy losses. However, new data suggest a birth in women with secondary RPL eradicates the negative prognostic impact of previous pregnancy losses.
9. The prognosis for achieving a live birth after a diagnosis of RIF remains unclear, and counseling regarding pursuing additional embryo transfers remains the mainstay of treatment.

References

[1] Practice Committee of the American Society for Reproductive Medicine. Evaluation and treatment of recurrent pregnancy loss: a committee opinion. Fertil Steril 2012 Nov; 98(5):1103–11.
[2] Polanski LT, Baumgarten MN, Quenby S, Brosens J, Campbell BK, Raine-Fenning NJ. What exactly do we mean by "recurrent implantation failure"? A systematic review and opinion. Reprod Biomed Online 2014;28 (4):409–23.
[3] Practice Committee of the American Society for Reproductive Medicine. Electronic address: asrm@asrm.org. Definitions of infertility and recurrent pregnancy loss: a committee opinion. Fertil Steril 2020 Mar;113(3): 533–5.
[4] ESHRE Guideline: Recurrent Pregnancy Loss. Human reproduction [Open]. Oxford Academic [Internet]. [cited 2021 Sep 13]. <https://academic.oup.com/hropen/article/2018/2/hoy004/4963604>.
[5] Busnelli A, Somigliana E, Cirillo F, Baggiani A, Levi-Setti PE. Efficacy of therapies and interventions for repeated embryo implantation failure: a systematic review and *meta*-analysis. Sci Rep 2021;11(1):1747.
[6] Coughlan C. What to do when good-quality embryos repeatedly fail to implant. Best Pract Res Clin Obstet Gynaecol 2018;53:48–59.
[7] Ata B, Kalafat E, Somigliana E. A new definition of recurrent implantation failure on the basis of anticipated blastocyst aneuploidy rates across female age. Fertil Steril 2021;S0015–0282(21):00558-6.
[8] Stirrat GM. Recurrent miscarriage. Lancet Lond Engl 1990;336(8716):673–5.
[9] Jauniaux E, Farquharson RG, Christiansen OB, Exalto N. Evidence-based guidelines for the investigation and medical treatment of recurrent miscarriage. Hum Reprod Oxf Engl 2006;21(9):2216–22.
[10] Is the incidence of recurrent pregnancy loss increasing? A retrospective register-based study in Sweden, Rasmark Roepke, 2017. Acta Obstetricia et Gynecologica Scandinavica. Wiley Online Library [Internet]. [cited 2021 Sep 13]. <https://obgyn.onlinelibrary.wiley.com/doi/full/10.1111/aogs.13210>.
[11] Pirtea P, De Ziegler D, Tao X, Sun L, Zhan Y, Ayoubi JM, et al. Rate of true recurrent implantation failure is low: results of three successive frozen euploid single embryo transfers. Fertil Steril 2021;115(1):45–53.
[12] Picut CA, Dixon D, de Rijk EPCT. Immunopathology of the female reproductive tract and mammary gland In: Parker GA, editor. Immunopathology in toxicology and drug development, Volume 2. Cham: Springer International Publishing; 2017. p. 541–614[cited 2021 Sep 13] (Molecular and Integrative Toxicology). Available from. Available from: https://doi.org/10.1007/978-3-319-47385-7_11.

[13] Wu R, Van der Hoek KH, Ryan NK, Norman RJ, Robker RL. Macrophage contributions to ovarian function. Hum Reprod Update 2004;10(2):119–33.
[14] Lima PDA, Nivet A-L, Wang Q, Chen Y-A, Leader A, Cheung A, et al. Polycystic ovary syndrome: possible involvement of androgen-induced, chemerin-mediated ovarian recruitment of monocytes/macrophages†. Biol Reprod 2018;99(4):838–52.
[15] Li Q, Yuan M, Jiao X, Huang Y, Li J, Li D, et al. M1 Macrophage-derived nanovesicles repolarize M2 macrophages for inhibiting the development of endometriosis. Front Immunol 2021;12:707784.
[16] Gaynor LM, Colucci F. Uterine natural killer cells: functional distinctions and influence on pregnancy in humans and mice. Front Immunol 2017;8:467.
[17] Picut CA, Swanson CL, Parker RF, Scully KL, Parker GA. The metrial gland in the rat and its similarities to granular cell tumors. Toxicol Pathol 2009;37(4):474–80.
[18] Pinhal-Enfield G, Vasan NS, Leibovich SJ. The role of macrophages in the placenta [Internet]. Embryology—updates and highlights on classic topics. IntechOpen; 2012 [cited 2021 Sep 13]. <https://www.intechopen.com/chapters/34561>.
[19] Cao L, Tang Y, Niu X, Guo Q, Huang L. Mifepristone regulates macrophage-mediated natural killer cells function in decidua. Reprod Biol 2021;21(3):100541.
[20] Stephenson MD, Awartani KA, Robinson WP. Cytogenetic analysis of miscarriages from couples with recurrent miscarriage: a case-control study. Hum Reprod Oxf Engl 2002;17(2):446–51.
[21] Carp H, Toder V, Aviram A, Daniely M, Mashiach S, Barkai G. Karyotype of the abortus in recurrent miscarriage. Fertil Steril 2001;75(4):678–82.
[22] Clifford K, Rai R, Regan L. Future pregnancy outcome in unexplained recurrent first trimester miscarriage. Hum Reprod Oxf Engl 1997;12(2):387–9.
[23] Recurrent Miscarriage, Investigation and Treatment of Couples (Green-top Guideline No. 17) [Internet]. Royal College of Obstetricians and Gynaecologists. [cited 2021 Sep 13]. <https://www.rcog.org.uk/en/guidelines-research-services/guidelines/gtg17/>
[24] Stephenson MD, Sierra S. Reproductive outcomes in recurrent pregnancy loss associated with a parental carrier of a structural chromosome rearrangement. Hum Reprod Oxf Engl 2006;21(4):1076–82.
[25] Saravelos SH, Cocksedge KA, Li T-C. Prevalence and diagnosis of congenital uterine anomalies in women with reproductive failure: a critical appraisal. Hum Reprod Update 2008;14(5):415–29.
[26] Grimbizis GF, Camus M, Tarlatzis BC, Bontis JN, Devroey P. Clinical implications of uterine malformations and hysteroscopic treatment results. Hum Reprod Update 2001;7(2):161–74.
[27] Miyakis S, Lockshin MD, Atsumi T, Branch DW, Brey RL, Cervera R, et al. International consensus statement on an update of the classification criteria for definite antiphospholipid syndrome (APS). J Thromb Haemost JTH 2006;4(2):295–306.
[28] ACOG Practice Bulletin No. 118: antiphospholipid syndrome. Obstet Gynecol. 2011 Jan;117(1):192–9.
[29] Abalovich M, Gutierrez S, Alcaraz G, Maccallini G, Garcia A, Levalle O. Overt and subclinical hypothyroidism complicating pregnancy. Thyroid J Am Thyroid Assoc 2002;12(1):63–8.
[30] Jovanovic L, Knopp RH, Kim H, Cefalu WT, Zhu X-D, Lee YJ, et al. Elevated pregnancy losses at high and low extremes of maternal glucose in early normal and diabetic pregnancy: evidence for a protective adaptation in diabetes. Diabetes Care 2005;28(5):1113–17.
[31] Hirahara F, Andoh N, Sawai K, Hirabuki T, Uemura T, Minaguchi H. Hyperprolactinemic recurrent miscarriage and results of randomized bromocriptine treatment trials. Fertil Steril 1998;70(2):246–52.
[32] Rey E, Kahn SR, David M, Shrier I. Thrombophilic disorders and fetal loss: a *meta*-analysis. Lancet Lond Engl 2003;361(9361):901–8.
[33] Silver RM, Zhao Y, Spong CY, Sibai B, Wendel G, Wenstrom K, et al. Prothrombin gene G20210A mutation and obstetric complications. Obstet Gynecol 2010;115(1):14–20.
[34] Dizon-Townson D, Miller C, Sibai B, Spong CY, Thom E, Wendel G, et al. The relationship of the factor V Leiden mutation and pregnancy outcomes for mother and fetus. Obstet Gynecol 2005;106(3):517–24.
[35] Verdoni A, Hu J, Surti U, Babcock M, Sheehan E, Clemens M, et al. Reproductive outcomes in individuals with chromosomal reciprocal translocations. Genet Med J Am Coll Med Genet 2021; May 10.
[36] Tan Y-Q, Tan K, Zhang S-P, Gong F, Cheng D-H, Xiong B, et al. Single-nucleotide polymorphism microarray-based pre-implantation genetic diagnosis is likely to improve the clinical outcome for translocation carriers. Hum Reprod 2013;28(9):2581–92.

[37] Huang C, Jiang W, Zhu Y, Li H, Lu J, Yan J, et al. Pregnancy outcomes of reciprocal translocation carriers with two or more unfavorable pregnancy histories: before and after pre-implantation genetic testing. J Assist Reprod Genet 2019;36(11):2325–31.

[38] Chan YY, Jayaprakasan K, Tan A, Thornton JG, Coomarasamy A, Raine-Fenning NJ. Reproductive outcomes in women with congenital uterine anomalies: a systematic review. Ultrasound Obstet Gynecol J Int Soc Ultrasound Obstet Gynecol 2011;38(4):371–82.

[39] Rikken JFW, Kowalik CR, Emanuel MH, Bongers MY, Spinder T, Jansen FW, et al. Septum resection vs expectant management in women with a septate uterus: an international multicentre open-label randomized controlled trial. Hum Reprod Oxf Engl 2021;36(5):1260–7.

[40] Empson M, Lassere M, Craig JC, Scott JR. Recurrent pregnancy loss with antiphospholipid antibody: a systematic review of therapeutic trials. Obstet Gynecol 2002;99(1):135–44.

[41] Pasquier E, de Saint Martin L, Bohec C, Chauleur C, Bretelle F, Marhic G, et al. Enoxaparin for prevention of unexplained recurrent miscarriage: a multicenter randomized double-blind placebo-controlled trial. Blood 2015;125(14):2200–5.

[42] Laskin CA, Spitzer KA, Clark CA, Crowther MR, Ginsberg JS, Hawker GA, et al. Low molecular weight heparin and aspirin for recurrent pregnancy loss: results from the randomized, controlled HepASA Trial. J Rheumatol 2009;36(2):279–87.

[43] Matsumoto Y, Kokeguchi S, Shiotani M. Effects of endometrial injury on frozen-thawed blastocyst transfer in hormone replacement cycles. Reprod Med Biol 2017;16(2):196–9.

[44] Raziel A, Schachter M, Strassburger D, Bern O, Ron-El R, Friedler S. Favorable influence of local injury to the endometrium in intracytoplasmic sperm injection patients with high-order implantation failure. Fertil Steril 2007;87(1):198–201.

[45] Gürgan T, Kalem Z, Kalem MN, Ruso H, Benkhalifa M, Makrigiannakis A. Systematic and standardized hysteroscopic endometrial injury for treatment of recurrent implantation failure. Reprod Biomed Online 2019;39(3):477–83.

[46] Olesen MS, Hauge B, Ohrt L, Olesen TN, Roskær J, Bæk V, et al. Therapeutic endometrial scratching and implantation after in vitro fertilization: a multicenter randomized controlled trial. Fertil Steril 2019;112(6):1015–21.

[47] Baum M, Yerushalmi GM, Maman E, Kedem A, Machtinger R, Hourvitz A, et al. Does local injury to the endometrium before IVF cycle really affect treatment outcome? Results of a randomized placebo controlled trial. Gynecol Endocrinol J Int Soc Gynecol Endocrinol 2012;28(12):933–6.

[48] Zargar M, Ghafourian M, Nikbakht R, Mir Hosseini V, Moradi Choghakabodi P. Evaluating chronic endometritis in women with recurrent implantation failure and recurrent pregnancy loss by hysteroscopy and immunohistochemistry. J Minim Invasive Gynecol 2020;27(1):116–21.

[49] Li Y, Yu S, Huang C, Lian R, Chen C, Liu S, et al. Evaluation of peripheral and uterine immune status of chronic endometritis in patients with recurrent reproductive failure. Fertil Steril 2020;113(1):187–196.e1.

[50] Johnston-MacAnanny EB, Hartnett J, Engmann LL, Nulsen JC, Sanders MM, Benadiva CA. Chronic endometritis is a frequent finding in women with recurrent implantation failure after in vitro fertilization. Fertil Steril 2010;93(2):437–41.

[51] He Y, Wu H, He X, Xing Q, Zhou P, Cao Y, et al. Application of atosiban in frozen-thawed cycle patients with different times of embryo transfers. Gynecol Endocrinol J Int Soc Gynecol Endocrinol 2016;32(10):811–15.

[52] Mao X, Zhang J, Chen Q, Kuang Y, Zhang S. Short-term copper intrauterine device placement improves the implantation and pregnancy rates in women with repeated implantation failure. Fertil Steril 2017;108(1) 55–61.e1.

[53] Tehraninejad ES, Raisi E, Ghaleh FB, Rashidi BH, Aziminekoo E, Kalantari V, et al. The sequential embryo transfer compared to blastocyst embryo transfer in in vitro fertilization (IVF) cycle in patients with the three repeated consecutive IVF. A randomized controlled trial. Gynecol Endocrinol J Int Soc Gynecol Endocrinol 2019;35(11):955–9.

[54] Shahrokh Tehraninejad E, Azimi Nekoo E, Ghaffari F, Hafezi M, Karimian L, Arabipoor A. Zygote intrafallopian tube transfer vs intrauterine cleavage or blastocyst stage transfer after intracytoplasmic sperm injection cycles in patients with repeated implantation failure: a prospective follow-up study. J Obstet Gynaecol Res 2015;41(11):1779–84.

[55] Rubio C, Bellver J, Rodrigo L, Bosch E, Mercader A, Vidal C, et al. Pre-implantation genetic screening using fluorescence in situ hybridization in patients with repetitive implantation failure and advanced maternal age: two randomized trials. Fertil Steril 2013;99(5):1400–7.
[56] Blockeel C, Schutyser V, De Vos A, Verpoest W, De Vos M, Staessen C, et al. Prospectively randomized controlled trial of PGS in IVF/ICSI patients with poor implantation. Reprod Biomed Online 2008;17 (6):848–54.
[57] Simon A, Laufer N. Assessment and treatment of repeated implantation failure (RIF). J Assist Reprod Genet 2012;29(11):1227–39.
[58] Rufas-Sapir O, Stein A, Orvieto R, Avrech OM, Kotler N, Pinkas H, et al. Is assisted hatching beneficial in patients with recurrent implantation failures? Clin Exp Obstet Gynecol 2004;31(2):110–12.
[59] Tasaka A, Doshida M, Sato Y, Kyoya T, Nakajo Y, Kyono K. Outcome of IMSI (intracytoplasmic morphologically selected sperm injection) in patients with repeated ICSI failures. Fertil Steril 2009;92(3):S76.
[60] Delaroche L, Yazbeck C, Gout C, Kahn V, Oger P, Rougier N. Intracytoplasmic morphologically selected sperm injection (IMSI) after repeated IVF or ICSI failures: a prospective comparative study. Eur J Obstet Gynecol Reprod Biol 2013;167(1):76–80.
[61] Karabulut S, Aksunger O, Korkmaz O, Gozel HE, Keskin I. Intracytoplasmic morphologically selected sperm injection, but for whom? Zygote 2019;27(5):299–304.
[62] Sung N, Khan SA, Yiu ME, Jubiz G, Salazar MD, Skariah A, et al. Reproductive outcomes of women with recurrent pregnancy losses and repeated implantation failures are significantly improved with immunomodulatory treatment. J Reprod Immunol 2021;148:103369.
[63] Ho Y-K, Chen H-H, Huang C-C, Lee C-I, Lin P-Y, Lee M-S, et al. Peripheral CD56 + CD16 + NK cell populations in the early follicular phase are associated with successful clinical outcomes of intravenous immunoglobulin treatment in women with repeated implantation failure. Front Endocrinol 2019;10:937.
[64] Madkour A, Bouamoud N, Louanjli N, Kaarouch I, Copin H, Benkhalifa M, et al. Intrauterine insemination of cultured peripheral blood mononuclear cells prior to embryo transfer improves clinical outcome for patients with repeated implantation failures. Zygote Camb Engl 2016;24(1):58–69.
[65] Nobijari FF, Arefi SS, Moini A, Taheripanah R, Fazeli E, Kharazi H, et al. Endometrium immunomodulation by intrauterine insemination administration of treated peripheral blood mononuclear cell prior frozen/thawed embryos in patients with repeated implantation failure. Zygote Camb Engl 2019;27(4):214–18.
[66] Yu N, Zhang B, Xu M, Wang S, Liu R, Wu J, et al. Intrauterine administration of autologous peripheral blood mononuclear cells (PBMCs) activated by HCG improves the implantation and pregnancy rates in patients with repeated implantation failure: a prospective randomized study. Am J Reprod Immunol N Y N 1989 2016;76(3):212–16.
[67] Okitsu O, Kiyokawa M, Oda T, Miyake K, Sato Y, Fujiwara H. Intrauterine administration of autologous peripheral blood mononuclear cells increases clinical pregnancy rates in frozen/thawed embryo transfer cycles of patients with repeated implantation failure. J Reprod Immunol 2011;92(1–2):82–7.
[68] Li S, Wang J, Cheng Y, Zhou D, Yin T, Xu W, et al. Intrauterine administration of hCG-activated autologous human peripheral blood mononuclear cells (PBMC) promotes live birth rates in frozen/thawed embryo transfer cycles of patients with repeated implantation failure. J Reprod Immunol 2017;119:15–22.
[69] Yoshioka S, Fujiwara H, Nakayama T, Kosaka K, Mori T, Fujii S. Intrauterine administration of autologous peripheral blood mononuclear cells promotes implantation rates in patients with repeated failure of IVF-embryo transfer. Hum Reprod Oxf Engl 2006;21(12):3290–4.
[70] Noushin MA, Ashraf M, Thunga C, Singh S, Singh S, Basheer R, et al. A comparative evaluation of subendometrial and intrauterine platelet-rich plasma treatment for women with recurrent implantation failure. FS Sci 2021;2(3):295–302.
[71] Moldenhauer LM, Keenihan SN, Hayball JD, Robertson SA. GM-CSF is an essential regulator of T cell activation competence in uterine dendritic cells during early pregnancy in mice. J Immunol Baltim Md 1950 2010;185(11):7085–96.
[72] Kalem Z, Namli Kalem M, Bakirarar B, Kent E, Makrigiannakis A, Gurgan T. Intrauterine G-CSF administration in recurrent implantation failure (RIF): an RCT. Sci Rep 2020;10(1):5139.
[73] Urman B, Ata B, Yakin K, Alatas C, Aksoy S, Mercan R, et al. Luteal phase empirical low molecular weight heparin administration in patients with failed ICSI embryo transfer cycles: a randomized open-labeled pilot trial. Hum Reprod Oxf Engl 2009;24(7):1640–7.

[74] Noci I, Milanini MN, Ruggiero M, Papini F, Fuzzi B, Artini PG. Effect of dalteparin sodium administration on IVF outcome in non-thrombophilic young women: a pilot study. Reprod Biomed Online 2011;22 (6):615–20.

[75] Lodigiani C, Dentali F, Banfi E, Ferrazzi P, Librè L, Quaglia I, et al. The effect of parnaparin sodium on in vitro fertilization outcome: a prospective randomized controlled trial. Thromb Res 2017;159:116–21.

[76] Berker B, Taşkin S, Kahraman K, Taşkin EA, Atabekoğlu C, Sönmezer M. The role of low-molecular-weight heparin in recurrent implantation failure: a prospective, quasi-randomized, controlled study. Fertil Steril 2011;95(8):2499–502.

[77] Xiong. Low-molecularweight heparin in women with... Google Scholar [Internet]. [cited 2021 Sep 13]. <https://scholar.google.com/scholar_lookup?hl = en&volume = 31&publication_year = 2015&pages = 614-7 &journal = J + Pract + Obstet + Gynecol&author = ZF + Xiong&author = XH + Dang&author = B + Li&author = LY + Wang&title = Low%E2%80%90molecular%E2%80%90weight + heparin + in + women + with + repeated + implantation + failure + %28in + Chinese%29>.

[78] Yang X-L, Chen F, Yang X-Y, Du G-H, Xu Y. Efficacy of low-molecular-weight heparin on the outcomes of in vitro fertilization/intracytoplasmic sperm injection pregnancy in non-thrombophilic women: a *meta*-analysis. Acta Obstet Gynecol Scand 2018;97(9):1061–72.

[79] Huang P, Wei L, Li X, Qin A. Effects of intrauterine perfusion of human chorionic gonadotropin in women with different implantation failure numbers. Am J Reprod Immunol N Y N 1989 2018;79(2).

[80] Liu X, Ma D, Wang W, Qu Q, Zhang N, Wang X, et al. Intrauterine administration of human chorionic gonadotropin improves the live birth rates of patients with repeated implantation failure in frozen-thawed blastocyst transfer cycles by increasing the percentage of peripheral regulatory T cells. Arch Gynecol Obstet 2019;299(4):1165–72.

[81] Sher G, Fisch JD. Effect of vaginal sildenafil on the outcome of in vitro fertilization (IVF) after multiple IVF failures attributed to poor endometrial development. Fertil Steril 2002;78(5):1073–6.

[82] Kliman HJ, Frankfurter D. Clinical approach to recurrent implantation failure: evidence-based evaluation of the endometrium. Fertil Steril 2019;111(4):618–28.

[83] Tan J, Kan A, Hitkari J, Taylor B, Tallon N, Warraich G, et al. The role of the endometrial receptivity array (ERA) in patients who have failed euploid embryo transfers. J Assist Reprod Genet 2018;35(4):683–92.

[84] Young SL. Evaluation of endometrial function: a Heraclean or Sisyphean task? Fertil Steril 2017;108(4):604–5.

[85] Almquist LD, Likes CE, Stone B, Brown K, Savaris R, Forstein DA, et al. Endometrial BCL6 testing for the prediction of IVF outcomes: a cohort study. Fertil Steril 2017;108(6):1063–9.

[86] Kwak-Kim J, Skariah A, Wu L, Salazar D, Sung N, Ota K. Humoral and cellular autoimmunity in women with recurrent pregnancy losses and repeated implantation failures: a possible role of vitamin D. Autoimmun Rev 2016;15(10):943–7.

[87] Egerup P, Kolte AM, Larsen EC, Krog M, Nielsen HS, Christiansen OB. Recurrent pregnancy loss: what is the impact of consecutive vs non-consecutive losses? Hum Reprod Oxf Engl 2016;31(11):2428–34.

[88] Koot YEM, Hviid Saxtorph M, Goddijn M, de Bever S, Eijkemans MJC, Wely MV, et al. What is the prognosis for a live birth after unexplained recurrent implantation failure following IVF/ICSI? Hum Reprod Oxf Engl 2019;34(10):2044–52.

CHAPTER

16

The embryo in reproductive failure: immunological view

Tia Brodeur and Navid Esfandiari

Division of Reproductive Endocrinology and Infertility, Department of Obstetrics and Gynecology and Reproductive Sciences, Larner College of Medicine at the University of Vermont, Burlington, VT, United States

1 Introduction

In vitro fertilization (IVF) is one of the most significant medical advances of the 20th century. Approximately 2% of all infants born every year in the United States are conceived using assisted reproductive technology (ART) [1]. The age of a female partner or egg donor is the most important indicator of IVF success; however, since its inception about 40 years ago, IVF has become incrementally successful, largely attributable to advances in embryology laboratory techniques. IVF may be the first line of treatment for severe male factor infertility, including those requiring extraction of testicular sperm or epididymal aspiration. Moreover, IVF can provide additional diagnostic information important to the infertile couple, such as poor or failed fertilization after insemination of oocytes, oocyte morphological abnormalities, and poor embryo development.

Following oocyte retrieval, conventional insemination or intracytoplasmic sperm injection (ICSI) can be performed to fertilize the oocyte (s). The choice of fertilization technique depends on the diagnosis, source and quality of sperm, number and quality of oocytes, as well as fertilization results of prior cycles if available. The fertilization assessment typically takes place 16–18 hours after insemination of oocytes or ICSI. From this point forward, morphological assessments are used to grade the quality of developing embryos (Fig. 1). In contemporary practice, top-quality embryos are typically transferred into the uterus at either the cleavage stage (day 3 postfertilization) or blastocyst stage (day 5 or 6 postfertilization). It has been proposed that culture to the blastocyst stage may help to select embryos of higher reproductive potential, and blastocysts may have better synchrony with the endometrium, which results in higher implantation rates. However, a *meta*-analysis of clinical pregnancy rate and live birth rate shows marginal improvement when blastocysts are transferred [2].

Immunology of Recurrent Pregnancy Loss and Implantation Failure
DOI: https://doi.org/10.1016/B978-0-323-90805-4.00017-1

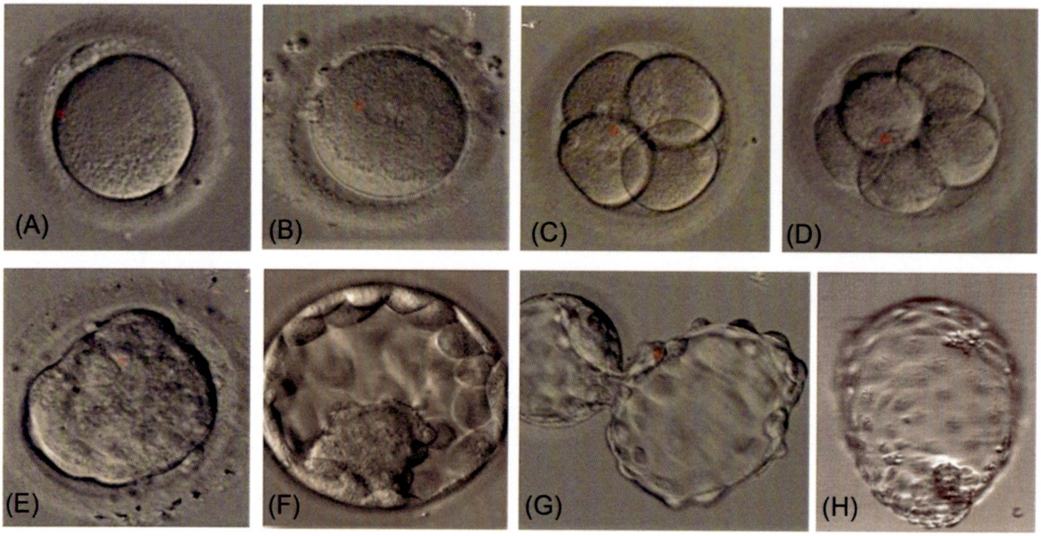

FIGURE 1 Human oocyte and embryos: (A) mature oocyte at metaphase II (MII), (B) fertilized oocyte with two pronuclei, (C) 4-cell stage embryo, (D) 8-cell stage embryo, (E) morula, (F) blastocyst, (G) hatching blastocyst, and (H) completely hatched blastocyst.

Grading systems exist for both cleavage stage embryos and blastocysts, and these grading systems help select embryos that lead to higher live birth rates [3]. Cleavage stage embryos are graded using criteria based on blastomere number, symmetry, and degree of fragmentation, and the Gardner system is a commonly used embryo grading system for blastocysts. The blastocyst grading is based on a three-part score which includes: (1) a numerical score from 1 to 6 describing the degree of expansion of the blastocyst; (2) a hybrid qualitative and quantitative assessment of the inner cell mass; (3) a qualitative assessment of the trophectoderm (TE). These composite scores are associated with implantation potential [4,5].

Biochemical and metabolomics approaches have also been suggested as methods to select high-quality embryos; however, these methods are currently still largely experimental and could contribute considerable cost and personnel hours. Time-lapse microscopy may help assess the reproductive potential of embryos by measuring the kinetics of embryonic development [6], and preimplantation genetic testing for aneuploidy (PGT-A) may also be useful in selecting embryos with a normal chromosomal complement. For patients who have had recurrent implantation failure (RIF) despite the transfer of good-quality or euploid embryos, investigating alternative assessments of reproductive potential of the embryos may inform future therapeutics.

2 Embryo-derived signals influencing the local immune environment

2.1 Human chorionic gonadotropin

Human chorionic gonadotropin (hCG) is the major serum protein marker of pregnancy produced by the embryo. hCG produced by the blastocyst can be detected in maternal serum as early as seven days after fertilization [7]. In addition to providing critical support to the corpus luteum for the maintenance of progesterone secretion, hCG induces proliferation and invasion of cytotrophoblasts cells. It also regulates endometrial microarchitecture through the control of matrix metalloproteinases, vascular endothelial growth factor (VEGF), and macrophage colony-stimulating factor (M-CSF). In addition, hCG influences the activity of innate and adaptive immune cells, including macrophages, natural killer (NK) cells, and T cells [7,8]. Although NK cells do not express a receptor for hCG, the hCG-mediated proliferation of these cells occurs via the mannose receptor [9].

2.2 Early pregnancy factor

Early pregnancy factor (EPF) is secreted by the early embryo within hours of fertilization [10] and has immunosuppressive properties [11,12]. An effect of EPF on embryo development and the promotion of implantation has also been suggested [13]. EPF was identified as a result of the ability of serum to augment the rosette inhibition test from recently inseminated women. Rosette inhibition test is an in vitro assay using antiserum to peripheral blood lymphocytes that inhibited the formation of rosettes when peripheral blood mononuclear cells were mixed with sheep red blood cells [14]. The mechanism by which EPF inhibits lymphocyte proliferation and/or activation has not been established. However, these studies suggest that the preimplantation embryo secretes EPF, thereby influencing the phenotype of the local immune cells and potentially limiting T cell expansion, which could lead to an overexuberant immune response.

2.3 Platelet-activating factor

Platelet-activating factor (PAF) belongs to a phospholipid family, which can mediate histamine release from platelets. PAF is produced by both the embryo and endometrium and has a wide range of biological functions. PAF is produced as early as fertilization in human IVF [15]. PAF has been suggested as a predictor of embryo viability, as PAF levels in culture media have been associated with clinical pregnancy rates [16,17]. PAF is implicated in implantation in part due to studies in the mouse in which PAF antagonists inhibit implantation. PAF also enhances Prostaglandin E2 (PGE_2) synthesis and $PGF_{2\alpha}$ release, suggesting one potential mechanism for PAF role in implantation [16–18]. PAF may also have activity in other areas of the reproductive tract and, interestingly, appears to have a promotility effect on sperm [16].

3 Embryo-endometrial dialogue

3.1 Inflammatory cytokines

Inflammatory cytokines, such as IL-1β and prostaglandins (PGs), are expressed at higher levels in early pregnancy endometrium compared to nonpregnant endometrium [19]. The fine balance of leukocytes and low proinflammatory cytokine levels secreted from both endometrium and local leukocytes are likely to promote implantation. Indeed, higher levels of IL-1β and TNF-α were found to correlate with the clinical pregnancy rate in IVF patients [20]. IL-1β is produced by both the blastocyst and the endometrium and can stimulate PG production by the endometrial cells in vitro [21].

3.2 Prostaglandins

PGs are mediators of acute inflammation that also amplify responses of immune cells at the transcriptional level [22]. Cyclooxygenases-1 (COX-1) and COX-2 are the enzymes that mediate the production of PG from arachidonic acid. PGs are produced by most tissues and can be expressed constitutively or in response to injurious stimuli [23]. PGs are required for implantation, and PGE2 is a mediator of the decidual reaction. PGE2 receptors in the endometrial stroma can be induced in a progesterone-dependent fashion [24]. COX-1 production is downregulated in response to progesterone and 17β-estradiol; thus its levels decrease prior to implantation. COX-2 expression is restricted to the site of implantation in mice [25], suggesting that PGs in the implantation microenvironment may be tightly controlled to prevent an overexuberant immune response.

3.3 Leukemia inhibitory factor

Leukemia inhibitory factor (LIF) is a member of the IL-6 family of cytokines and is highly pleiotropic [26]. LIF is expressed by endometrial glands and stroma and appears to have a role in implantation that is conserved across species. Reciprocal embryo transfer studies in mice have shown that maternal LIF is required for implantation. Embryos express LIF receptors, and the addition of exogenous LIF to the embryo culture medium can improve blastocyst formation in humans [27]. Knockdown of LIF in mouse embryos in vitro decreased blastulation rate and diameters of blastocysts [28]. However, LIF receptor-deficient embryos can develop to the blastocyst stage [29]. Decidual NK cells and T cells have been shown to express LIF messenger RNA [30]. Taken together, this suggests that LIF may mediate endometrial-leukocyte interactions as well as embryo-endometrium interactions during the process of implantation.

4 Recurrent implantation failure

Implantation is an intricate process requiring direction from both the embryo and the endometrium, and immune factors have mutual permissive and restrictive roles in this process. Decidualization is characterized by a marked increase in leukocytes, and up to 40% of cells comprising the decidua are leukocytes—predominantly NK cells [31]. In humans, decidualization is triggered by elevated progesterone and does not require embryo-derived signals [32]. During implantation, trophoblast cells of the blastocyst adhere to the epithelial layer of the endometrium and then proliferate and invade the endometrial stroma. Impaired trophoblast invasion also contributes to obstetric conditions such as fetal growth restriction and preeclampsia [7].

4.1 Definition

The definition of RIF varies and may continue to evolve as the field of ART advances. Variation in the definition of RIF leads to heterogeneity in this field of study. However, we propose that the failure of implantation following three embryo transfers (regardless of single or multiple embryo transfer) of good-quality embryos is one acceptable definition for RIF.

4.2 Prevalence

The prevalence of RIF is difficult to ascertain, and despite being presumably rare, it can be devastating to patients and perplexing to the IVF team. A recent study by Pirtea et al. demonstrated that 92.6% of patients would achieve pregnancy after up to three single euploid embryo transfers with live birth rates of 64.8%, 54.4%, and 54.1% after the first, second, and third single euploid embryo transfer, respectively [33], suggesting that true RIF is uncommon when euploid embryos are transferred.

4.3 Evaluation

Patients who have undergone multiple failed transfers of good quality embryos should be considered candidates for evaluation of RIF. Unfortunately, some patients who undergo numerous embryo transfers are unsuccessful due to poor embryo development, aneuploidy, or suboptimal endometrial lining. These patients may not fall under the diagnosis of RIF as poor embryo quality may ultimately explain implantation failure and potentially the basis of their infertility.

4.3.1 Preimplantation genetic testing

Aneuploidy is a deviation from the normal chromosome complement. Because many women undergoing IVF are of advanced reproductive age and thus are more prone to oocyte aneuploidy, a significant portion of implantation failure is related to the genetic competence of the embryo.

Preimplantation genetic testing (PGT) has been designed to screen for oocyte and embryo aneuploidies (PGT-A) in high-risk patients. PGT allows the selection of normal embryos to avoid failed implantation and pregnancy losses. There are two basic approaches to obtaining nuclear material for the genetic analysis; from oocytes and from in vitro developed embryos. For oocytes, it entails the aspiration of the first polar body from the mature metaphase-II oocyte, the second polar body from normally fertilized oocytes, or both. Polar body biopsies provide an indirect assessment of the chromosome complement of the associated oocyte. Therefore only the maternal component of the subsequent embryo could be tested (Fig. 2) [34]. For embryos, it involves removal of one or two blastomeres from a cleavage stage embryo (day-3 postfertilization), or a few TE cells from a blastocyst-stage embryo (day-5 or -6 postfertilization) (Fig. 3). The cleavage-stage embryos have higher rates of mosaicism, and a biopsy of one or two blastomeres for genetic testing could yield false results if the embryo was mosaic [35]. However, the ability

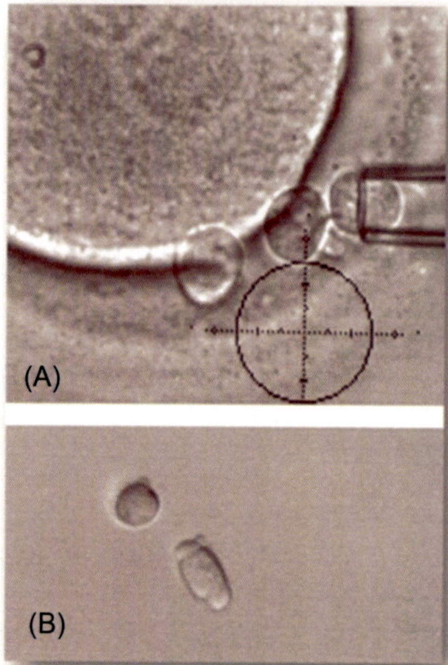

FIGURE 2 Polar body biopsy: (A) removing polar bodies from a fertilized oocyte and (B) polar bodies (×400 magnification).

to biopsy several TE cells provides more DNA for testing and may reduce the chance of misdiagnoses due to mosaicism.

There may be some benefit of PGT-A in patients with RIF or pregnancy losses; however, this has not been conclusively demonstrated [36]. Notably, there is no direct association between embryo morphology and chromosomal status as aneuploid embryos may have good quality embryo morphology, and euploid embryos may have less than optimal quality. A study by Sato et al. compared pregnancy rates in patients with RIF undergoing PGT-A or transfer of an untested embryo. A total of 92 RIF patients were enrolled, and 42 of these patients underwent PGT-A. In the PGT-A group, 24/42 patients received an embryo transfer, and 42/50 patients received an embryo transfer in the control group. The live birth rate *per patient* was 35.7% in the PGT-A group and 26% in patients receiving a transfer of an untested embryo, which was not statistically significant. However, the live birth rate per embryo transfer was 62.5% in the PGT-A group and 31.7% in the control group ($p = 0.016$) [37]. This study included patients up to age 42, who would be more prone to failed implantation related to aneuploidy, and it is unclear whether PGT-A is an optimal strategy for younger patients.

Genetic abnormalities such as balanced translocations or inversions may also lead to recurrent pregnancy loss. A study by Raziel et al. demonstrated that in 65 patients with high-order RIF (6–16 cycles without achieving clinical pregnancy with a mean of 7.8 cycles), 15.4% had chromosomal abnormalities to include translocation, mosaicism, inversion, and deletion. Interestingly, patients with chromosomal abnormalities did not have

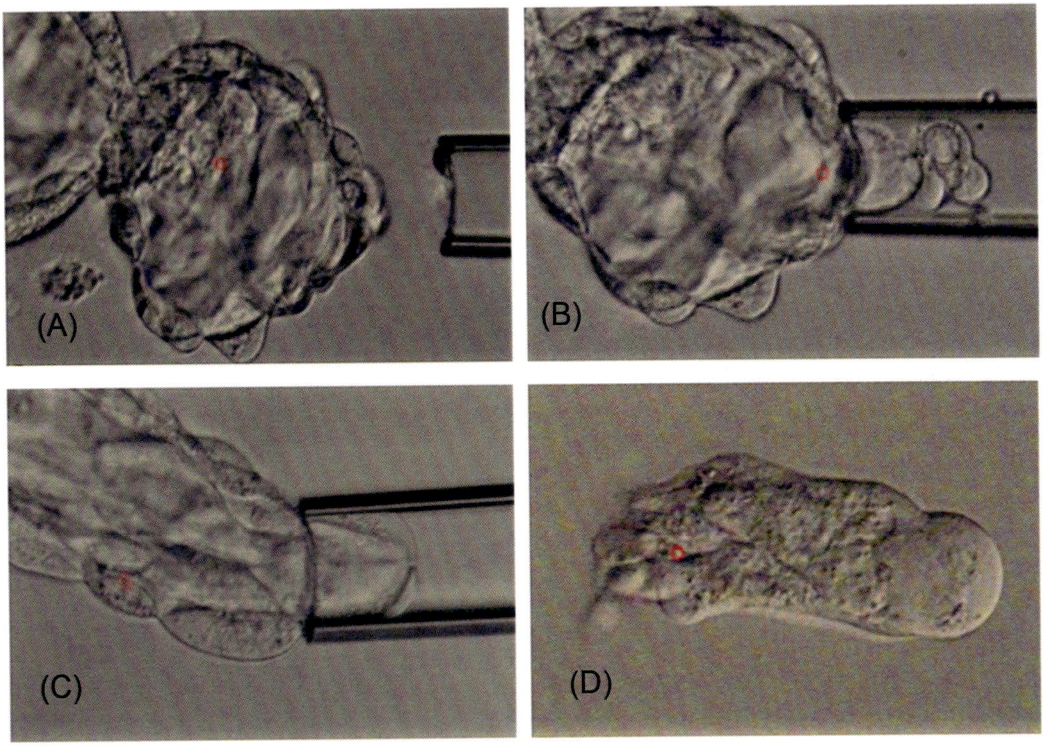

FIGURE 3 The process of trophectoderm (TE) biopsy: (A) Expanding blastocyst on day 5 with TE herniating through an opening made in the zona 2 days earlier using laser, (B) blastocyst is immobilized with the holding pipette, and few TE cells are drawn into the biopsy pipette, (C) 3–5 TE cells are biopsied from blastocyst using laser pulses, and (D) biopsied TE cells.

differences in embryo morphology [38]. The applicability of this small sample size with quite high-order RIF, however, may not be broad.

4.3.2 *Sperm DNA damage and aneuploidy*

The impact of DNA damage on sperm quality is an area of active investigation, though there is currently a dearth of information about sperm DNA damage in the setting of RIF. Interestingly, a study of 637 couples undergoing IUI (387 cycles), IVF (388 cycles), and ICSI (223 cycles) showed that patients with high levels of sperm DNA fragmentation had higher rates of biochemical pregnancy. In this study, the inclusion criteria were sperm concentration of 1 million/mL in unwashed semen, and for the female partner, age less than 40 years, BMI less than 30 kg/m^2, and baseline FSH less than 12 IU/L [39]. A case series of four couples with male factor infertility with RIF after ICSI using ejaculated sperm demonstrated that all four couples were able to achieve pregnancy when testicular sperm was used, with the authors proposing that the sperm may have suffered post-testicular DNA damage [40]. Sperm DNA integrity testing is not currently used broadly clinically, and the benefit may not outweigh the consumption of resources given a marginal improvement in outcomes [41]. The Practice Committee of the American Society for Reproductive Medicine does not recommend the routine testing of sperm DNA damage and suggests there is insufficient evidence to recommend the routine use of sperm DNA integrity tests in the evaluation and treatment of the infertile couple [42]. Treatment of varicocele or infection, if present, may improve sperm DNA integrity in infertile men while data on antioxidant supplementation and lifestyle changes remain less clear [43]. Compared to oocyte aneuploidy, sperm aneuploidy is uncommon but is more common among infertile men. Moreover, sperm aneuploidy is shown to be associated with embryo aneuploidy. However, aneuploid sperm exist in otherwise normal semen samples [44].

4.3.3 *Embryo-endometrial asynchrony*

The window of implantation begins 1–2 days earlier in patients undergoing ovarian stimulation with gonadotropins compared to natural cycles [45]. The choice of ovarian stimulation protocol [46,47] or GnRH agonist versus hCG trigger may influence endometrial receptivity [46]. The endometrial receptivity array (ERA) has been used in some clinical settings to assess embryo-endometrial asynchrony by measuring transcripts of genes expressed at different points in the endometrial cycle. This testing aims to identify patients with endometrial receptivity that is "displaced," meaning that an adjustment of the embryo transfer time may improve synchrony with the embryo that the endometrium will receive. When patients with RIF are compared to IVF control patients, the incidence of abnormal ERA is 25% versus 14%, respectively (four studies with 50, 62, 80, and 85 RIF patients, respectively) [48]. A retrospective study of 50 RIF patients with around 5 failed frozen blastocyst transfers showed that 24% of those patients were "nonreceptive (pre- or postreceptive)," and of those that were deemed "prereceptive" by the test, personalization of their embryo transfer timing led to a clinical pregnancy rate of 50% after first transfer compared to 35.3% in the receptive group [49]. A retrospective study by Tan et al. reviewed a total of 88 patients who had undergone ERA. Of these, 38 patients had prior failed euploid embryo transfer, which eliminated the contribution of aneuploidy to implantation failure. In this study, the percentage of patients with a displaced window of implantation by ERA was 22.5%. There was a trend, although not statistically significant, toward increasing live birth rate with personalization of embryo transfer timing [50].

4.3.4 *Endometriosis (B-cell CLL/lymphoma 6 assay)*

B-cell CLL/lymphoma 6 (BCL6) is a master transcriptional regulator expressed in germinal center B cells and T follicular helper cells [51]. BCL6 has been proposed as a marker of endometrial receptivity in patients with unexplained infertility and also a biomarker for endometriosis [52], with the suggestion that a proportion of unexplained infertility patients have subclinical endometriosis. Eutopic endometrium in endometriosis patients may be subjected to an inflammatory environment via a variety of pathways, including progesterone resistance since progesterone can decrease inflammation in the endometrium [53]. This could be a potential explanation for the association between endometrial receptivity and endometriosis. Women with low BCL6 expression had higher clinical pregnancy rates than women with high BCL6 expression (64.7% versus 17.3%), as well as higher live birth rates (58.8% versus 11.5%) [54].

4.3.5 Chronic endometritis and endometrial dysbiosis

Chronic endometritis is likely caused by endometrial dysbiosis. The hallmark pathologic feature is the presence of plasma cells in the immunohistochemistry study of an endometrial biopsy specimen [55,56]. A high rate of chronic endometritis has been seen in patients with RIF and recurrent pregnancy loss (14% and 27%, respectively) [57]. A study by Cicinelli et al. attempted to determine whether treatment of chronic endometritis improved pregnancy outcomes. In this study, 70 out of 106 RIF patients were diagnosed with chronic endometritis. All patients were treated with antibiotics; however, 24.6% had chronic endometritis on repeat assessment. The patients with resolved chronic endometritis had a live birth rate of 60.8%, and the refractory chronic endometritis patients had a live birth rate of 13.3% [58]. The refractory chronic endometritis as a comparison group is interesting; however, this group may have more severe pathology at baseline, which could contribute bias to the study.

5 Improving implantation rate

5.1 Embryo selection

While embryo morphology, including pronuclear, cleavage, and blastocyst stages, provides useful information about the reproductive potential of an individual embryo, more than half of embryo transfers will not result in pregnancy. Adjunct methods to assess the implantation potential of embryos are an active area of investigation.

5.2 Embryo morphokinetics

Time-lapse microscopy has been used to study embryo progression, and the morphokinetics of embryo development has been proposed as predictive of the reproductive potential of individual embryos; however, this has yet to be validated as a technique to improve implantation and live birth rate. Time-lapse microscopy has the benefit of being noninvasive as compared to PGT. However, the current state-of-the-art is unlikely to approach the sensitivity of PGT-A with regard to the ploidy status of the embryo [34,59].

5.3 Metabolomics and metabolite concentrations in embryos culture media

Metabolomic profiling of spent embryo culture media has been proposed as a noninvasive tool to assess the reproductive potential of embryos. A study by Seli et al. utilized spent embryo culture media from embryos that did or did not result in clinical pregnancy after single day-3 embryo transfer. They then subjected the media to near-infrared spectroscopy to generate a viability index that could predict the viability of day 2 and day 3 embryos independently from morphology [60]. Levels of glutamate, alanine, and leucine have also been used as markers of embryo quality [34]. Metabolomic studies at the serum level have also been performed, including an intriguing study by RoyChoudhury et al. comparing women with RIF with women who had successful implantation with each embryo transfer. This study found that patients with recurrent implantation success had significantly lower levels of valine, adipic acid, l-lysine, creatine, ornithine, glycerol, d-glucose, and urea than patients with RIF [61]. This study also demonstrated that serum endothelial nitric oxide synthase (eNOS) was lower in patients with RIF. eNOS is found in the endometrium during the luteal phase, and NOS inhibitors may impact endometrial receptivity and implantation [62].

5.4 Embryo culture conditions

Embryo co-culture systems were developed to provide early embryos with the source of growth factors and cytokines in vitro that would mimic the in vivo milieu to promote the ongoing development of the embryo [63]. A *meta*-analysis has reviewed 17 studies on embryo co-culture with multiple cell types and shows favorable results with regard to both implantation and clinical pregnancy [64]. There have been no reports of long-term health implications of offspring from co-cultured embryos in the 30 years since many of these studies have been done. However, this would be a difficult association to ascertain. The addition of cytokines and growth factors does have an advantage over co-culture systems where cytokines and growth factors released can be incompletely known. A randomized controlled trial comparing routine culture versus addition of the cytokines, such as GM-CSF, heparin-binding epidermal growth factor (HB-EGF), and LIF, demonstrated marginal improvement in blastocyst development and quality, increased implantation rate, and increased live birth rate (45% versus 33%) with no change in preterm birth or live birth weight [65]. While these results are encouraging, long-term health effects on offspring are difficult to assess.

5.5 Sequential embryo transfer

The use of sequential embryo transfer in patients with RIF has been described. Studies using sequential embryo transfer have used the sequential transfer of cleavage stage embryos followed by blastocysts, which is presumably less likely to improve endometrial-embryo asynchrony [66,67]. Using a different approach, a study by Esfandiari et al. describes a patient who had undergone five prior unsuccessful IVF cycles and underwent natural cycle frozen embryo transfer (FET) (with luteal progesterone). She underwent double embryo transfer using a total of four cleavage-stage day-3 embryos—two embryos were transferred each time for two times, 5 and 7 days after her LH surge. This led to the term singleton live birth. Notably, this patient had 21-day cycles; however, she had been unsuccessful in prior medicated FET cycles [68]. The use of sequential transfer in this fashion may overcome endometrial-embryo asynchrony as the embryos transferred were introduced at the same embryological stage into an endometrium with two different levels of advancement.

5.6 Assisted hatching

Assisted hatching includes a constellation of procedures that mechanically, chemically, or using laser, hatches, or thins the zona pellucida. Heterogeneity in methods, patient population, and diagnoses may confound the association between assisted hatching and live birth. However, a large *meta*-analysis has concluded insufficient evidence to perform assisted hatching routinely [69]. Data regarding the benefit of assisted hatching in RIF have been mixed [70,71]. A *meta*-analysis of RCTs has shown an increased clinical pregnancy rate in patients with RIF undergoing fresh embryo transfer with assisted hatching (clinical pregnancy rate RR 1.73, confidence interval 1.37–2.17), with no benefit in an unselected group undergoing assisted hatching [72]. This suggests that a subset of patients may have a zona pellucida-related impairment to implantation.

5.7 Therapies to improve sperm aneuploidy

Studies on therapies to improve sperm aneuploidy are limited. In a small study of 31 infertile male patients with a high frequency sperm aneuploidy, 22 patients received recombinant FSH (150 IU three times per week). Treatment with recombinant FSH decreased total sperm aneuploidy ($p < 0.001$); however, pregnancy rates were not reported [73]. A study of 97 healthy volunteers by Young et al. sought to determine the association between nutrition and sperm aneuploidy. The participants completed a dietary questionnaire which was then extrapolated to average daily nutrient intake. Their results showed a 3.6% decrease in aggregate sperm aneuploidy per 100 micrograms increase in daily folate intake (95% confidence interval: −6.3, −0.8). Notably, this was not an infertile population and pregnancy rates were not assessed [74]. A third study included couples undergoing ICSI cycles for severe oligoasthenoteratospermia. The 33 couples included in this study had a least one prior unsuccessful ICSI cycle. The treatment was a three-month course of *L*-carnitine 1 g given twice per day plus acetyl-*L*-carnitine 500 mg given twice per day plus one 30-mg cinnoxicam tablet every 4 days. Before and after treatment, multiple outcomes were assessed including sperm concentration, motility, and morphology, percentage of aneuploid sperm, number of oocytes fertilized, number of embryos transferred, and pregnancy and live birth rates. A total of 22 of 33 patients had reduced sperm aneuploidy (15% at baseline to 2% after the intervention, $p < 0.01$) and increased sperm morphology (3% at baseline to 7.5% after the intervention, $p < 0.01$). The clinical pregnancy rate in the group of 22 patients with improved sperm parameters was 50%, compared to 9.1% in the group of 11 patients who also received the intervention but with no significant change in their sperm parameters ($p < 0.01$). Live birth rates were 45.4% and 9.1%, respectively ($p < 0.01$) [75]. These pilot studies suggest potential therapeutics for sperm aneuploidy.

6 Immunomodulation

6.1 Granulocyte colony-stimulating factor

Manipulation of several different immune targets has been proposed as a therapeutic tactic over the last several decades with the purpose of improving implantation and ongoing pregnancy rates. Granulocyte colony-stimulating factor (G-CSF) is a cytokine primarily responsible for stem cell differentiation and migration and produced predominantly by hematopoietic cells and a small subset of nonhematopoietic cells. G-CSF has immunomodulatory functions during implantation, such as dampening leukocyte activity. A recent *meta*-analysis concluded that G-CSF via both local and systemic administration improves the clinical pregnancy rate [76]. On the contrary, a recent randomized clinical trial of patients with RIF (using the definition of less than 40 years of age who failed to achieve pregnancy after transfer of four good-quality embryos in a minimum of three fresh or frozen cycles) showed neither significant difference in clinical pregnancy nor live birth rate nor endometrial thickness [77].

6.2 Platelet-rich plasma

Platelet-rich plasma (PRP) has been used as a treatment for patients with RIF in several studies. The PRP contains multiple growth factors, including platelet-derived growth factor, insulin-like growth factor, and VEGF [78]. A small study of thirty patients with RIF undergoing FET did not show a difference between patients receiving PRP 48 hours prior to embryo transfer and those who did not [79]. A PRP single-arm study by Nazari et al. reported 18 pregnancies out of 20 patients with RIF, including 16 ongoing pregnancies, one molar pregnancy, and one miscarriage [80]. This study is quite limited without a control group and absence of pregnancy outcome data.

6.3 Granulocyte colony-stimulating factor and platelet-rich plasma

A recent pilot study using both PRP and G-CSF (protocol for endometrial receptivity improvement "PRIMER") on selected patients with RIF (defined as ≥2 embryo transfers and at least five morphologically good embryos transferred) and first-attempt IVF/ICSI controls receiving no adjunct medications was performed. There were 33 patients in each group. The ongoing pregnancy and live birth rate was 27.3% in both groups. The authors argue that their finding demonstrates that RIF patients using PRIMER achieve equivalent live birth rate to the control group; however, the rate of pregnancy loss in the PRIMER group was 25% compared to 9% in the controls. Although this was not statistically significant likely due to the small sample size [81], it raises questions about the efficacy of PRIMER among subgroups such as patients with high-order RIF.

6.4 Intralipid

Intralipid, a soybean oil emulsion initially developed and currently in use for parenteral nutrition, has had mixed reports as an immunomodulatory agent [82]. Intralipid has been used as a treatment for both RIF and pregnancy loss, with the thought that it decreases NK cell cytotoxicity [83]. NK cells are innate lymphocytes responsible for killing tumor cells and virus-infected cells and also produce cytokines to recruit and influence other lymphocytes [84,85]. NK cells have been implicated in both RIF and pregnancy loss [83]. A retrospective single-center study showed an increased cost of $681 per live birth with the use of Intralipid without benefit to the live birth rate [86]. In contrast, a single-blinded randomized clinical trial of IVF/ICSI patients with at least one prior implantation failure showed a live birth rate of 28.8% in the Intralipid group and 10% in a control group [87]. This population may not reflect a true RIF population due to criteria of at least one failed implantation. In a case series of RIF patients with a median of 9.5 (3–19) unsuccessful embryo transfers, 9 patients receiving Intralipid achieved a clinical pregnancy (56%). In this study, the live birth rate is reported as a percentage of those achieving a live birth after achieving clinical pregnancy rather than compared to a total number of patients receiving an embryo transfer. The live birth rate was 55% (five out of nine patients who became pregnant) in the Intralipid group as compared to 33% live birth rate (two out of six patients in the control group) but was not statistically significant. The literature review performed in this same publication using three studies (including their own data) with 107 recurrent miscarriage patients and 47 controls supported use of Intralipid therapy [88].

6.5 Human leukocyte antigen

Human leukocyte antigens (HLAs) are cell surface proteins with immunoregulatory functions. HLA expression in human syncytiotrophoblasts is highly regulated to preempt an immune response to the embryo, which is semiallogeneic by nature. HLA-G is expressed on the surface of trophoblasts and has a presumptive role in modulating interactions between macrophages, NK cells, and the embryo, as both cells express HLA-G receptors [25]. Measurement of soluble HLA-G has been proposed as a marker of reproductive potential. In separate studies by Sher et al. and Kotze et al., embryos with soluble HLA-G in their culture medium had higher implantation and clinical pregnancy rates [89] and ongoing pregnancy rates [90].

6.6 Intrauterine human chorionic gonadotropin

Because of the integral role of hCG in implantation, several studies have been conducted on patients receiving intrauterine hCG prior to embryo transfer. Results have been mixed in both primary analyses and *meta*-analyses. It has been proposed that intrauterine hCG may improve endometrial asynchrony that may occur in IVF cycles. In a study by Strug et al., oocyte donors underwent endometrial biopsy two days after intrauterine infusion of hCG. They had higher levels of α-smooth muscle actin than controls, with a perhaps more modest effect on Notch1 and C3 [91]. Because young oocyte donors were included in this study, it is possible that the endometrial effect may have been more pronounced due to a robust response and resultant higher estradiol levels. In prospective studies by Mansour et al. and Santibañez et al., there was an improvement in clinical pregnancy and live birth rates after transferring cleavage stage embryos following intrauterine hCG [92,93]. A randomized controlled trial by Zarei et al. also showed an increase in clinical pregnancy and live birth rates after intrauterine hCG administration prior to day 3 embryo transfer [94]. Studies of blastocyst transfer have not shown this beneficial effect of hCG on the clinical pregnancy rate [95,96]. Volovsky et al. demonstrated a negative effect on clinical pregnancy and live birth rates in patients without RIF in response to intrauterine hCG and no effect in patients suffering from RIF [97]. Given these conflicting results, further studies are needed to determine which, if any, patient population may benefit from hCG therapy.

6.7 Endometrial injury ("scratching")

Multiple studies evaluating pregnancy rates in both intrauterine insemination and IVF patients have shown a benefit to endometrial injury or "scratching," often achieved via endometrial biopsy with an endometrial biopsy curette [98–100]. A recent systematic review and *meta*-analysis of IVF patients undergoing fresh or FET found no significant difference in clinical pregnancy or live birth rates in patients undergoing endometrial scratch [101]. A large randomized controlled trial of women undergoing fresh or FET has shown that endometrial scratching is not associated with live birth rate [102]. There remains the possibility that this could benefit a specific subset of patients prone to perturbations of their endometrial microarchitecture; however, the identification of these patients may be elusive. Based on available evidence, endometrial scratching is not currently recommended as a routine treatment for RIF.

7 Summary and recommendations

1. Embryo and endometrial crosstalk may be altered by immunologic factors as well as intrinsic factors.
2. Aneuploidy may be a cause of RIF, and genetic testing of either embryos or the patient and the partner may be considered.
3. The use of immunomodulatory agents remains under active investigation, and rigorous study seems warranted in light of the prominent role of the immune system in implantation.

References

[1] ART Success Rates. CDC; 2021. <https://www.cdc.gov/art/artdata/index.html> [accessed 2.20.2021].
[2] Glujovsky D, Farquhar C, Quinteiro Retamar AM, Alvarez Sedo CR, Blake D. Cleavage stage vs blastocyst stage embryo transfer in assisted reproductive technology. Cochrane Database Syst Rev 2016;(6) CD002118.
[3] Balaban B, Urman B, Sertac A, Alatas C, Aksoy S, Mercan R. Blastocyst quality affects the success of blastocyst-stage embryo transfer. Fertil Steril 2000;74(2):282–7.
[4] Gardner DK, Lane M, Stevens J, Schlenker T, Schoolcraft WB. Blastocyst score affects implantation and pregnancy outcome: towards a single blastocyst transfer. Fertil Steril 2000;73(6):1155–8.
[5] Elder K, Cohen J. Human preimplantation embryo selection. Boca Raton, Fla: CRC Press; 2008.
[6] Armstrong S, Bhide P, Jordan V, Pacey A, Marjoribanks J, Farquhar C. Time-lapse systems for embryo incubation and assessment in assisted reproduction. Cochrane Database Syst Rev 2019;5 CD011320.
[7] Schumacher A, Zenclussen AC. Human chorionic gonadotropin-mediated immune responses that facilitate embryo implantation and placentation. Front Immunol 2019;10:2896.
[8] Gridelet V, Perrier d'Hauterive S, Polese B, Foidart JM, Nisolle M, Geenen V. Human chorionic gonadotrophin: new pleiotropic functions for an "old" hormone during pregnancy. Front Immunol 2020;11:343.
[9] Kane N, Kelly R, Saunders PT, Critchley HO. Proliferation of uterine natural killer cells is induced by human chorionic gonadotropin and mediated via the mannose receptor. Endocrinology 2009;150(6):2882–8.

[10] Fan XG, Zheng ZQ. A study of early pregnancy factor activity in preimplantation. Am J Reprod Immunol 1997;37(5):359–64.
[11] Bose R, Cheng H, Sabbadini E, McCoshen J, MaHadevan MM, Fleetham J. Purified human early pregnancy factor from preimplantation embryo possesses immunosuppresive properties. Am J Obstet Gynecol 1989;160 (4):954–60.
[12] Morton H. Early pregnancy factor: an extracellular chaperonin 10 homologue. Immunol Cell Biol 1998;76 (6):483–96.
[13] Athanasas-Platsis S, Corcoran CM, Kaye PL, Cavanagh AC, Morton H. Early pregnancy factor is required at two important stages of embryonic development in the mouse. Am J Reprod Immunol 2000;43(4):223–33.
[14] Bach JF, Dormont J, Dardenne M, Balner H. In vitro rosette inhibition by antihuman antilymphocyte serum. Correlation with skin graft prolongation in subhuman primates. Transplantation 1969;8(3):265–80.
[15] Nakatsuka M, Yoshida N, Kudo T. Platelet activating factor in culture media as an indicator of human embryonic development after in-vitro fertilization. Hum Reprod 1992;7(10):1435–9.
[16] Harper MJ. Platelet-activating factor: a paracrine factor in preimplantation stages of reproduction? Biol Reprod 1989;40(5):907–13.
[17] Harper MJ, Kudolo GB, Alecozay AA, Jones MA. Platelet-activating factor (PAF) and blastocyst-endometrial interactions. Prog Clin Biol Res 1989;294:305–15.
[18] Tiemann U. The role of platelet-activating factor in the mammalian female reproductive tract. Reprod Domest Anim 2008;43(6):647–55.
[19] Calleja-Agius J, Jauniaux E, Pizzey AR, Muttukrishna S. Investigation of systemic inflammatory response in first trimester pregnancy failure. Hum Reprod 2012;27(2):349–57.
[20] Boomsma CM, Kavelaars A, Eijkemans MJ, et al. Endometrial secretion analysis identifies a cytokine profile predictive of pregnancy in IVF. Hum Reprod 2009;24(6):1427–35.
[21] Huang JC, Liu DY, Yadollahi S, Wu KK, Dawood MY. Interleukin-1 beta induces cyclooxygenase-2 gene expression in cultured endometrial stromal cells. J Clin Endocrinol Metab 1998;83(2):538–41.
[22] Tilley SL, Coffman TM, Koller BH. Mixed messages: modulation of inflammation and immune responses by prostaglandins and thromboxanes. J Clin Invest 2001;108(1):15–23.
[23] Ricciotti E, FitzGerald GA. Prostaglandins and inflammation. Arterioscler Thromb Vasc Biol 2011;31 (5):986–1000.
[24] Psychoyos A, Nikas G, Gravanis A. The role of prostaglandins in blastocyst implantation. Hum Reprod 1995;10(Suppl 2):30–42.
[25] Norwitz ER, Schust DJ, Fisher SJ. Implantation and the survival of early pregnancy. N Engl J Med 2001;345 (19):1400–8.
[26] Nicola NA, Babon JJ. Leukemia inhibitory factor (LIF). Cytokine Growth Factor Rev 2015;26(5):533–44.
[27] Dunglison GF, Barlow DH, Sargent IL. Leukaemia inhibitory factor significantly enhances the blastocyst formation rates of human embryos cultured in serum-free medium. Hum Reprod 1996;11(1):191–6.
[28] Cheng TC, Huang CC, Chen CI, et al. Leukemia inhibitory factor antisense oligonucleotide inhibits the development of murine embryos at preimplantation stages. Biol Reprod 2004;70(5):1270–6.
[29] Paria BC, Song H, Dey SK. Implantation: molecular basis of embryo-uterine dialogue. Int J Dev Biol 2001;45 (3):597–605.
[30] Jokhi PP, King A, Sharkey AM, Smith SK, Loke YW. Screening for cytokine messenger ribonucleic acids in purified human decidual lymphocyte populations by the reverse-transcriptase polymerase chain reaction. J Immunol 1994;153(10):4427–35.
[31] Ander SE, Diamond MS, Coyne CB. Immune responses at the maternal-fetal interface. Sci Immunol 2019;4(31):1–22.
[32] Taylor HS, Pal L, Seli E, Fritz MA. Speroff's clinical gynecologic endocrinology and infertility. 9th ed. Philadelphia: Wolters Kluwer; 2020.
[33] Pirtea P, De Ziegler D, Tao X, et al. Rate of true recurrent implantation failure is low: results of three successive frozen euploid single embryo transfers. Fertil Steril 2021;115(1):45–53.
[34] Nagy ZP, Varghese, Alex C., Agarwal, A. In vitro fertilization. A Textb Curr Emerg Meth Dev. 2019.
[35] Esfandiari N, Bunnell ME, Casper RF. Human embryo mosaicism: did we drop the ball on chromosomal testing? J Assist Reprod Genet 2016;33(11):1439–44.
[36] Coughlan C. What to do when good-quality embryos repeatedly fail to implant. Best Pract Res Clin Obstet Gynaecol 2018;53:48–59.

[37] Sato T, Sugiura-Ogasawara M, Ozawa F, et al. Preimplantation genetic testing for aneuploidy: a comparison of live birth rates in patients with recurrent pregnancy loss due to embryonic aneuploidy or recurrent implantation failure. Hum Reprod 2019;34(12):2340–8.

[38] Raziel A, Friedler S, Schachter M, Kasterstein E, Strassburger D, Ron-El R. Increased frequency of female partner chromosomal abnormalities in patients with high-order implantation failure after in vitro fertilization. Fertil Steril 2002;78(3):515–19.

[39] Bungum M, Humaidan P, Axmon A, et al. Sperm DNA integrity assessment in prediction of assisted reproduction technology outcome. Hum Reprod 2007;22(1):174–9.

[40] Weissman A, Horowitz E, Ravhon A, Nahum H, Golan A, Levran D. Pregnancies and live births following ICSI with testicular spermatozoa after repeated implantation failure using ejaculated spermatozoa. Reprod Biomed Online 2008;17(5):605–9.

[41] Collins JA, Barnhart KT, Schlegel PN. Do sperm DNA integrity tests predict pregnancy with in vitro fertilization? Fertil Steril 2008;89(4):823–31.

[42] Schlegel PN, Sigman M, Collura B, et al. Diagnosis and treatment of infertility in men: AUA/ASRM guideline part I. Fertil Steril 2021;115(1):54–61.

[43] Esteves SC, Santi D, Simoni M. An update on clinical and surgical interventions to reduce sperm DNA fragmentation in infertile men. Andrology 2020;8(1):53–81.

[44] Garcia-Mengual E, Trivino JC, Saez-Cuevas A, Bataller J, Ruiz-Jorro M, Vendrell X. Male infertility: establishing sperm aneuploidy thresholds in the laboratory. J Assist Reprod Genet 2019;36(3):371–81.

[45] Nikas G, Develioglu OH, Toner JP, Jones Jr. HW. Endometrial pinopodes indicate a shift in the window of receptivity in IVF cycles. Hum Reprod 1999;14(3):787–92.

[46] Rackow BW, Kliman HJ, Taylor HS. GnRH antagonists may affect endometrial receptivity. Fertil Steril 2008;89(5):1234–9.

[47] Bashiri A, Halper KI, Orvieto R. Recurrent implantation failure-update overview on etiology, diagnosis, treatment and future directions. Reprod Biol Endocrinol 2018;16(1):121.

[48] Lessey BA, Young SL. What exactly is endometrial receptivity? Fertil Steril 2019;111(4):611–17.

[49] Hashimoto T, Koizumi M, Doshida M, et al. Efficacy of the endometrial receptivity array for repeated implantation failure in Japan: a retrospective, two-centers study. Reprod Med Biol 2017;16(3):290–6.

[50] Tan J, Kan A, Hitkari J, et al. The role of the endometrial receptivity array (ERA) in patients who have failed euploid embryo transfers. J Assist Reprod Genet 2018;35(4):683–92.

[51] Ise W, Fujii K, Shiroguchi K, et al. T follicular helper cell-germinal center b cell interaction strength regulates entry into plasma cell or recycling germinal center cell fate. Immunity 2018;48(4) 702–715 e704.

[52] Evans-Hoeker E, Lessey BA, Jeong JW, et al. Endometrial BCL6 overexpression in eutopic endometrium of women with endometriosis. Reprod Sci 2016;23(9):1234–41.

[53] Lessey BA, Kim JJ. Endometrial receptivity in the eutopic endometrium of women with endometriosis: it is affected, and let me show you why. Fertil Steril 2017;108(1):19–27.

[54] Almquist LD, Likes CE, Stone B, et al. Endometrial BCL6 testing for the prediction of in vitro fertilization outcomes: a cohort study. Fertil Steril 2017;108(6):1063–9.

[55] Liu Y, Ko EY, Wong KK, et al. Endometrial microbiota in infertile women with and without chronic endometritis as diagnosed using a quantitative and reference range-based method. Fertil Steril 2019;112(4) 707-717 e701.

[56] Liu L, Yang H, Guo Y, Yang G, Chen Y. The impact of chronic endometritis on endometrial fibrosis and reproductive prognosis in patients with moderate and severe intrauterine adhesions: a prospective cohort study. Fertil Steril 2019;111(5) 1002-1010 e1002.

[57] Bouet PE, El Hachem H, Monceau E, Gariepy G, Kadoch IJ, Sylvestre C. Chronic endometritis in women with recurrent pregnancy loss and recurrent implantation failure: prevalence and role of office hysteroscopy and immunohistochemistry in diagnosis. Fertil Steril 2016;105(1):106–10.

[58] Cicinelli E, Matteo M, Tinelli R, et al. Prevalence of chronic endometritis in repeated unexplained implantation failure and the IVF success rate after antibiotic therapy. Hum Reprod 2015;30(2):323–30.

[59] Zaninovic N, Irani M, Meseguer M. Assessment of embryo morphology and developmental dynamics by time-lapse microscopy: is there a relation to implantation and ploidy? Fertil Steril 2017;108(5):722–9.

[60] Seli E, Vergouw CG, Morita H, et al. Noninvasive metabolomic profiling as an adjunct to morphology for noninvasive embryo assessment in women undergoing single embryo transfer. Fertil Steril 2010;94(2):535–42.

[61] RoyChoudhury S, Singh A, Gupta NJ, et al. Repeated implantation failure vs repeated implantation success: discrimination at a metabolomic level. Hum Reprod 2016;31(6):1265–74.
[62] Chwalisz K, Garfield RE. Role of nitric oxide in implantation and menstruation. Hum Reprod 2000;15(Suppl 3):96–111.
[63] Vajta G, Rienzi L, Cobo A, Yovich J. Embryo culture: can we perform better than nature? Reprod Biomed Online 2010;20(4):453–69.
[64] Kattal N, Cohen J, Barmat LI. Role of coculture in human in vitro fertilization: a meta-analysis. Fertil Steril 2008;90(4):1069–76.
[65] Fawzy M, Emad M, Elsuity MA, et al. Cytokines hold promise for human embryo culture in vitro: results of a randomized clinical trial. Fertil Steril 2019;112(5) 849–857 e841.
[66] Ashkenazi J, Yoeli R, Orvieto R, Shalev J, Ben-Rafael Z, Bar-Hava I. Double (consecutive) transfer of early embryos and blastocysts: aims and results. Fertil Steril 2000;74(5):936–40.
[67] Loutradis D, Drakakis P, Dallianidis K, et al. A double embryo transfer on days 2 and 4 or 5 improves pregnancy outcome in patients with good embryos but repeated failures in IVF or ICSI. Clin Exp Obstet Gynecol 2004;31(1):63–6.
[68] Esfandiari N, Coogan-Prewer J, Gotlieb L, Claessens EA, Casper RF. Successful pregnancy following double-frozen embryo transfer in a patient with repeated implantation failure. Fertil Steril 2008;90(4) 1199 e1113–1195.
[69] Carney SK, Das S, Blake D, Farquhar C, Seif MM, Nelson L. Assisted hatching on assisted conception (in vitro fertilisation (IVF) and intracytoplasmic sperm injection (ICSI). Cochrane Database Syst Rev 2012;12 CD001894.
[70] Stein A, Rufas O, Amit S, et al. Assisted hatching by partial zona dissection of human pre-embryos in patients with recurrent implantation failure after in vitro fertilization. Fertil Steril 1995;63(4):838–41.
[71] Valojerdi MR, Eftekhari-Yazdi P, Karimian L, Ashtiani SK. Effect of laser zona pellucida opening on clinical outcome of assisted reproduction technology in patients with advanced female age, recurrent implantation failure, or frozen-thawed embryos. Fertil Steril 2008;90(1):84–91.
[72] Martins WP, Rocha IA, Ferriani RA, Nastri CO. Assisted hatching of human embryos: a systematic review and meta-analysis of randomized controlled trials. Hum Reprod Update 2011;17(4):438–53.
[73] Piomboni P, Serafini F, Gambera L, et al. Sperm aneuploidies after human recombinant follicle stimulating hormone therapy in infertile males. Reprod Biomed Online 2009;18(5):622–9.
[74] Young SS, Eskenazi B, Marchetti FM, Block G, Wyrobek AJ. The association of folate, zinc and antioxidant intake with sperm aneuploidy in healthy non-smoking men. Hum Reprod 2008;23(5):1014–22.
[75] Cavallini G, Magli MC, Crippa A, Ferraretti AP, Gianaroli L. Reduction in sperm aneuploidy levels in severe oligoasthenoteratospermic patients after medical therapy: a preliminary report. Asian J Androl 2012;14 (4):591–8.
[76] Zhang L, Xu WH, Fu XH, et al. Therapeutic role of granulocyte colony-stimulating factor (G-CSF) for infertile women under in vitro fertilization and embryo transfer (IVF-ET) treatment: a meta-analysis. Arch Gynecol Obstet 2018;298(5):861–71.
[77] Kalem Z, Namli Kalem M, Bakirarar B, Kent E, Makrigiannakis A, Gurgan T. Intrauterine G-CSF administration in recurrent implantation failure (RIF): an Rct. Sci Rep 2020;10(1):5139.
[78] Dawood AS, Salem HA. Current clinical applications of platelet-rich plasma in various gynecological disorders: an appraisal of theory and practice. Clin Exp Reprod Med 2018;45(2):67–74.
[79] Aghajanzadeh F, Esmaeilzadeh S, Basirat Z, Mahouti T, Heidari FN, Golsorkhtabaramiri M. Using autologous intrauterine platelet-rich plasma to improve the reproductive outcomes of women with recurrent implantation failure. JBRA Assist Reprod 2020;24(1):30–3.
[80] Nazari L, Salehpour S, Hoseini S, Zadehmodarres S, Ajori L. Effects of autologous platelet-rich plasma on implantation and pregnancy in repeated implantation failure: a pilot study. Int J Reprod Biomed 2016;14 (10):625–8.
[81] Dieamant F, Vagnini LD, Petersen CG, et al. New therapeutic protocol for improvement of endometrial receptivity (PRIMER) for patients with recurrent implantation failure (RIF)—a pilot study. JBRA Assist Reprod 2019;23(3):250–4.
[82] Wanten GJ, Calder PC. Immune modulation by parenteral lipid emulsions. Am J Clin Nutr 2007;85 (5):1171–84.

[83] Roussev RG, Acacio B, Ng SC, Coulam CB. Duration of intralipid's suppressive effect on NK cell's functional activity. Am J Reprod Immunol 2008;60(3):258–63.
[84] Poznanski SM, Ashkar AA. What defines NK cell functional fate: phenotype or metabolism? Front Immunol 2019;10:1414.
[85] Cooper MA, Fehniger TA, Caligiuri MA. The biology of human natural killer-cell subsets. Trends Immunol 2001;22(11):633–40.
[86] Martini AE, Jasulaitis S, Fogg LF, Uhler ML, Hirshfeld-Cytron JE. Evaluating the utility of intralipid infusion to improve live birth rates in patients with recurrent pregnancy loss or recurrent implantation failure. J Hum Reprod Sci 2018;11(3):261–8.
[87] Singh N, Davis AA, Kumar S, Kriplani A. The effect of administration of intravenous intralipid on pregnancy outcomes in women with implantation failure after IVF/ICSI with non-donor oocytes: a randomised controlled trial. Eur J Obstet Gynecol Reprod Biol 2019;240:45–51.
[88] Placais L, Kolanska K, Kraiem YB, et al. Intralipid therapy for unexplained recurrent miscarriage and implantation failure: case-series and literature review. Eur J Obstet Gynecol Reprod Biol 2020;252:100–4.
[89] Sher G, Keskintepe L, Fisch JD, et al. Soluble human leukocyte antigen G expression in phase I culture media at 46 hours after fertilization predicts pregnancy and implantation from day 3 embryo transfer. Fertil Steril 2005;83(5):1410–13.
[90] Kotze D, Kruger TF. HLA-G as a marker for embryo selection in assisted reproductive technology. Fertil Steril 2013;100(6):e44.
[91] Strug MR, Su R, Young JE, et al. Intrauterine human chorionic gonadotropin infusion in oocyte donors promotes endometrial synchrony and induction of early decidual markers for stromal survival: a randomized clinical trial. Hum Reprod 2016;31(7):1552–61.
[92] Santibanez A, Garcia J, Pashkova O, et al. Effect of intrauterine injection of human chorionic gonadotropin before embryo transfer on clinical pregnancy rates from in vitro fertilisation cycles: a prospective study. Reprod Biol Endocrinol 2014;12:9.
[93] Mansour R, Tawab N, Kamal O, et al. Intrauterine injection of human chorionic gonadotropin before embryo transfer significantly improves the implantation and pregnancy rates in in vitro fertilization/intracytoplasmic sperm injection: a prospective randomized study. Fertil Steril 2011;96(6) 1370–1374 e1371.
[94] Zarei A, Parsanezhad ME, Younesi M, et al. Intrauterine administration of recombinant human chorionic gonadotropin before embryo transfer on outcome of in vitro fertilization/ intracytoplasmic sperm injection: a randomized clinical trial. Iran J Reprod Med 2014;12(1):1–6.
[95] Wirleitner B, Schuff M, Vanderzwalmen P, et al. Intrauterine administration of human chorionic gonadotropin does not improve pregnancy and life birth rates independently of blastocyst quality: a randomised prospective study. Reprod Biol Endocrinol 2015;13:70.
[96] Hong KH, Forman EJ, Werner MD, et al. Endometrial infusion of human chorionic gonadotropin at the time of blastocyst embryo transfer does not impact clinical outcomes: a randomized, double-blind, placebo-controlled trial. Fertil Steril 2014;102(6) 1591–1595 e1592.
[97] Volovsky M, Healey M, MacLachlan V, Vollenhoven BJ. Should intrauterine human chorionic gonadotropin infusions ever be used prior to embryo transfer? J Assist Reprod Genet 2018;35(2):273–8.
[98] Nastri CO, Lensen SF, Gibreel A, et al. Endometrial injury in women undergoing assisted reproductive techniques. Cochrane Database Syst Rev 2015;(3) CD009517.
[99] Goel T, Mahey R, Bhatla N, Kalaivani M, Pant S, Kriplani A. Pregnancy after endometrial scratching in infertile couples undergoing ovulation induction and intrauterine insemination cycles-a randomized controlled trial. J Assist Reprod Genet 2017;34(8):1051–8.
[100] Barash A, Dekel N, Fieldust S, Segal I, Schechtman E, Granot I. Local injury to the endometrium doubles the incidence of successful pregnancies in patients undergoing in vitro fertilization. Fertil Steril 2003;79(6):1317–22.
[101] Vitagliano A, Andrisani A, Alviggi C, et al. Endometrial scratching for infertile women undergoing a first embryo transfer: a systematic review and meta-analysis of published and unpublished data from randomized controlled trials. Fertil Steril 2019;111(4) 734–746 e732.
[102] Lensen S, Venetis C, Ng EHY, et al. Should we stop offering endometrial scratching prior to in vitro fertilization? Fertil Steril 2019;111(6):1094–101.

CHAPTER

17

Natural killer cell pathology and repeated implantation failures

Atsushi Fukui[1,2], *Ayano Yamaya*[1], *Shinichiro Saeki*[1], *Ryu Takeyama*[1], *Toru Kato*[1], *Yu Wakimoto*[1] *and Hiroaki Shibahara*[1]

[1]Department of Obstetrics and Gynecology, School of Medicine, Hyogo Medical University, Nishinomiya, Hyogo, Japan [2]Fukushima Medical Center for Child and Woman, Fukushima Medical University, Fukushima, Japan

1 Introduction

1.1 Repeated implantation failure

Repeated implantation failure (RIF) is often defined as a lack of implantation after repeated good-quality embryo transfers. However, there is no universally accepted definition for RIF regarding the number of failed embryo transfer cycles, the number of embryos transferred, the developmental stage of transferred embryos (early cleavage vs. blastocyst stage), morphological quality of the transferred embryos, or whether or not including embryos with preimplantation genetic diagnosis, such as preimplantation genetic testing for aneuploidies (PGT-A).

There have been multiple attempts to include these variables in the diagnosis of RIF. For example, Polanski et al. defined RIF as the absence of implantation after two consecutive cycles of fresh or frozen embryo transfers with a cumulative number of transferred embryos of four or more cleavage-stage good quality embryos or two or more good quality blastocysts [1]. Coughlan et al. also defined RIF as the failure to achieve a clinical pregnancy after transferring at least four good-quality embryos in a minimum of three fresh or frozen embryo transfers in women under age 40 [2]. Recently, the Canadian guideline for RIF [3] and the ESHRE special interest group [4] defined RIF as implantation failures after 2–3 times or more embryo transfers or 3–4 or more good quality embryos transferred.

Immunology of Recurrent Pregnancy Loss and Implantation Failure
DOI: https://doi.org/10.1016/B978-0-323-90805-4.00007-9

Women with RIF share some causes of recurrent pregnancy loss (RPL). Hence, RIF patients have been investigated in the same manner as RPL patients generally. However, recent studies indicate that RPL and RIF are somewhat similar but distinct pathological conditions. Therefore it is recommended not to apply RPL screening methods or treatment modalities for women with RIF without proper investigation.

1.2 Prevalence of repeated implantation failure

Since there is no worldwide consensus for a definition of the RIF so far, it is difficult to estimate the true prevalence of RIF. However, the prevalence of RIF was reported to be 10% of patients undergoing IVF-ET worldwide. Recently, Busnelli et al. showed that the estimated prevalence of RIF is 15% using a definition of RIF based on three failed attempts for good quality embryo transfer [5].

When the PGT-A performed good quality euploid embryos are transferred, the potential causes of RIF can be restricted to uterine or maternal etiologies by eliminating genetic etiologies of embryos, so-called "poor oocyte or embryo quality." A large-scale study of women who had three embryo transfers using euploid embryos after PGT-A was reported [6]. According to the study, the accumulated pregnancy and live birth rates were 69.9% and 64.8%, respectively, by the first transfer, 87.9% and 83.9% by the second transfer, and 95.2% and 92.6% by the third transfer. Thus, per this study, the frequency of true RIF, excluding cases with implantation failures due to embryonic aneuploidy, can be less than 5%.

2 Natural killer cell and pregnancy

Natural killer (NK) cells play an important role in human pregnancy, and systemic regulation of NK cells contributes to reproductive success. The majority of peripheral blood NK (pNK) cells are conventional NK cells, and tissue-resident NK cells are present in the uterus, called uterine NK (uNK), endometrial NK (eNK), or decidual NK (dNK) cells. NK cells compose 5%–10% of peripheral blood lymphocytes (PBLs), and 70%–90% of decidual immune cells during early pregnancy are distinguished from other cell types by expressing NK-specific surface markers, that is, CD56. Human NK cell subsets include $CD56^{bright}$ and $CD56^{dim}$ NK cells. The main function of $CD56^{dim}$ cells is cytotoxicity to the target cells such as tumor cells or foreign pathogens, while $CD56^{bright}$ NK cells mainly produce cytokines such as tumor necrosis factor (TNF)-α and IFN-γ. $CD56^{bright}$ cells are mostly in the uterine endometrium and decidua, and $CD56^{dim}$ cells are mainly in peripheral blood.

NK cells are innate immune cells that can attack target cells without memory using their activating and inhibitory receptors. Cells expressing major histocompatibility complex (MHC) molecules escape from the NK cell attack since MHC binding to inhibitory receptors on NK cells suppresses NK cell cytotoxicities. In humans, extravillous cytotrophoblasts express classical MHC class I antigen, human leukocyte antigen (HLA)-C, and nonclassical class I antigen, HLA-G, F, and E. Hence, these cells can escape from NK cell cytotoxicity. However, syncytiotrophoblasts do not express MHC molecules, making them vulnerable to NK cell cytotoxicity.

MHC class I chain-related protein A and B (MIC-A and -B) and soluble HLA-G have been reported to participate in controlling pNK cell cytotoxicity. Hence, failed regulation of NK cells, not only uNK cells but pNK cells, or other NK cell controlling mechanisms may cause reproductive failures and obstetrical complications, such as RPL, implantation failures, or pre-eclampsia [7].

Increased cytotoxic NK cells in the decidua can be detrimental to fetal survival during implantation and early pregnancy. Abnormalities in the numbers and proportions of NK cells, aberrant expression of NK cell surface receptors, and dysregulated cytokine production have been reported in various reproductive failures [8–14]. Therefore it is convincing that the appropriate regulation of immune cells, especially NK cells and their functions, is critical to have reproductive success [15].

3 Underlying immunopathologies related to reproduction

The most widespread method to evaluate the NK cell function seems to be NK cell cytotoxicity using pNK cells. It is usually measured by the $_{51}Cr$ releasing assay or flow cytometry using the K562 cell line as target cells. The other method to evaluate the NK cell function is to analyze the NK cell subpopulation using several NK cell surface markers (CD16, CD56, and others) and to measure cytokines, chemokines, and other factors secreted from pNK cells and uNK cells.

NK cells are abundant in the uterus, where pregnancy is established and maintained. The main population of uNK cells is $CD56^{bright}$ NK cells. Unlike pNK ($CD56^{dim}$) cells, $CD56^{bright}$ uNK cells are not cytotoxic and involved in placentation and vascular remodeling through the production of cytokines, growth-promoting proteins, and angiogenic factors. It has been reported that increased cytotoxicity of uNK cells and abnormal cytokine production lead to reproductive failures, such as RIF [9,10]. Therefore NK cell regulation by the immunomodulatory treatment seems to be critical for successful reproductive outcomes in women with RIF.

3.1 Peripheral blood natural killer cell cytotoxicities and proportions

It has been shown that if the prepregnancy pNK cell cytotoxicity exceeds 46% using $_{51}Cr$ releasing assay and the proportion of pNK cells ($CD56^+$ cells) measured by flow cytometry exceeds 16.4%, pregnancy may end in a biochemical pregnancy or a miscarriage in the study of RPL women [16]. Additionally, pNK cell cytotoxicity or abnormal proportion of pNK cells at the time of embryo transfer predicts subsequent pregnancy outcomes in women undergoing IVF cycles [17]. The percentages of $CD56^+$ and $CD56^+/CD16^+$ NK cells in patients with failed implantation were significantly higher than those with successful implantation [17]. Besides, NK cell cytotoxicity by $_{51}Cr$ release assay showed lower NK cell cytotoxicities at the time of embryo transfer in patients with successful implantation as compared to those who had a failed previous cycle or with failed implantation [17].

On the other hand, Moffett et al. reported that examination of pNK cells would not tell what is happening in the uterus [18]. On the contrary, Park et al. reported that peripheral blood pNK cells reflect changes in dNK cells in women with RPL [19]. NKp46 is one of the natural cytotoxic receptors, and its expression is involved with the cytokine production of both pNK and uNK cells. The percentage of peripheral blood $CD56^{+}/NKp46^{+}$ cells is significantly correlated with that of endometrial $CD56^{bright}/NKp46^{+}$ cells during the mid-luteal phase [20,21]. This indicates that the expression of specific NK cell markers correlates between pNK and uNK cells. Hence, evaluation of pNK cell function may reflect that of uNK cells, and selecting a treatment strategy based on the pNK cell evaluation may improve the reproductive outcome of women with reproductive failures.

3.2 Endometrial and decidual natural killer cell cytotoxicities and proportions

Uterine NK cell or dNK cell proportions can be measured using flow cytometry. The endometrial tissues are collected in the mid-luteal phase, which is the window of implantation, using an endometrial sampler. uNK or dNK cells are mechanically dispersed, stained with monoclonal antibodies, and measured by flow cytometric analysis [17]. In cases of RIF or RPL, $CD16^{+}/CD56^{dim}$ cells, which are highly cytotoxic, are increased in the uterine endometrium or decidua [17,22]. Tuckerman et al. reported that the proportion of endometrial $CD56^{+}$ cells from women with RIF [median (range) $CD56^{+}$ cell density = 14.5% (1.5%–71.4%)] was significantly higher than controls [5% (2.1%–19.2%)] using immunohistochemistry assay [23]. The study concluded that the increased $CD56^{+}$ NK cells in the endometrium of women with RIF are directly involved in the implantation failure process. A higher proportion of mid-luteal $CD56^{+}$ uNK cells was reported in women with idiopathic RIF than controls or known cause RIF such as chromosomal abnormality, thyroid disorder, thrombophilia [24], and uNK cells were correlated with $CD56^{+}$ pNK cells ($r = 0.707$, $p < 0.001$) [24]. Based on these findings, the analysis of NK cells was recommended in women with idiopathic RIF. Since the association between NK cells and RIF was stronger in the endometrium, an endometrial biopsy for uNK cells assessment should be considered during the mid-luteal phase as well.

On the other hand, the uNK cell is vital for the decidualization and formation and remodeling of spiral arteries by secreting cytokines, chemokines, and angiogenic factors [25,26]. Kofod et al. reported that uNK cell abundance at the preimplantation endometrium seemed to be important for normal fertility and successful pregnancy [27]. Thus, functional analysis of uNK cells, in addition to the simple enumeration of uNK cells, seems to be critical for assessing NK cell-related pathologies.

3.3 Natural killer cell activating and inhibitory receptors

NK cells express various activating and inhibitory receptors. The balance between these receptors determines NK cell function, namely NK cell cytotoxicity. As for the activating receptors, studies have been performed on the expression levels of killer immunoglobulin-like receptors (KIR: KIR2DS1, KIR2DS2, KIR2DL4, and others), c-type lectin receptors (CD94/NKG2C and CD94/NKG2D), natural cytotoxicity receptors (NCRs: NKp46, NKp44, and NKp30), and CD16 in healthy and aberrant pregnancies, including RPL and RIF. The expression of inhibitory receptors on NK cells, for example, KIRs [KIR2DL1 (CD158a), KIR2DL2/3 (CD158b), and others] and c-type lectin receptors (CD94/NKG2A), was also evaluated during healthy and aberrant pregnancies, such as RPL and RIF [28–36]. Moreover, NKp46, one of the NK cell-activating receptors, is the primary triggering receptor involved in cytotoxicity and cytokine production [8,10–12,37,38]. $NKp46^+$ NK cells can be subdivided into $NKp46^{bright}$ and $NKp46^{dim}$ NK cells. $NKp46^{bright}$ NK cells, especially $NKp46^{bright}/CD56^{bright}$ NK cells, produce a significantly higher level of IFN-γ compared to $NKp46^{dim}$ NK cells [37]. The proportion of peripheral blood $NKp46^{bright}$ NK cells was significantly lower, and that of $NKp46^{dim}$ NK cells was significantly higher in women who failed to conceive than successfully conceived women within one year trial by natural conception or IVF-ET, according to the causes of infertility. $NKp46^{bright}$ NK cells were positively correlated with TGF-β producing NK cells, and $NKp46^{dim}$ NK cells were positively correlated with type 1 (IFN-γ and TNF-α) cytokine-producing NK cells, suggesting that lower production of TGF-β and higher production of type 1 cytokines in women who failed to conceive [11]. These findings suggest that NKp46 might be a potentially useful marker for predicting reproductive outcomes, such as RIF and RPL.

3.4 Natural killer cell-producing cytokines

NK cells can produce various cytokines but mainly type 1 cytokines, such as IFN-γ and TNF-α. Nevertheless, NK cells produce other cytokines such as IL-4, IL-10, TGF-β, GM-CSF, M-CSF, LIF, and others. Pregnancy is associated with a shift away from T helper 1 (Th1) immune responses and a bias toward Th2 responses [39]. Similar to the Th1/Th2 concept, NK1/NK2 model has been suggested for NK cells, showing comparable polarity in cytokine secretion profiles [9,40–42]. NK1 cytokines are essential for maintaining pregnancy [43–46] via angiogenesis and arterial remodeling. Incomplete remodeling of the spiral artery causes RPL and preeclampsia [47]. We reported significantly greater numbers of TNF-α and IFN-γ producing NK cells in women with reproductive failures, such as RPL and RIF. Hence, dysregulation of inflammatory cytokine production by NK cells, especially TNF-α and IFN-γ, seems to be deeply related to the immunopathology of reproductive failures [9].

4 Candidates for evaluation

A unique endometrial immune cell-based score was recently reported to predict pregnancy outcomes for embryo transfer [48,49]. The expression of endometrial immune cell markers, including $CD56^+$ NK cell, $CD68^+$ pan-macrophage, $CD163^+$ M2 macrophage, $Foxp3^+$ regulatory T cell, $CD1a^+$ immature dendritic cell, $CD83^+$ mature dendritic cell, $CD8^+$ cytotoxic T cell, and $CD57^+$ mature NK and T cells, was examined using immunohistochemistry technique. Endometrial samples were collected at the mid-luteal phase of the menstrual cycle before the controlled ovarian stimulation cycle. There were no differences in the expression of $CD56^+$ NK cell and regulatory T cells between pregnant women and implantation failure women at their first IVF-ET attempt. However, the nomogram using pan macrophages and M2 macrophages was developed for the embryo transfer strategy [48].

Mid-luteal endometrial biopsy samples were analyzed with targeted RNA sequencing via next-generation sequencing between patients with RPL and unexplained infertility and control women. The expression of NK cell-related genes (FOXO1, GZMB, IL15, SCNN1A, SGK1, and SLC2A1) necessary for endometrial transformation during decidualization was measured and scored. A combined evaluation of these 6 genes was designated as a decidualization score in which the maximal score was "6," and the minimal was "0." The decidualization score <4 provided evidence of abnormality in the decidualization process [49].

The NK cell pathologies can be evaluated by the following strategies; pNK cell cytotoxicity using $_{51}$Cr release assay or flow cytometry, and pNK and uNK cell proportion analysis using various NK cell markers. These tests are fairly well standardized and straightforward. Moreover, the results are consistent and helpful in determining therapeutic indications. Furthermore, the expression of NKp46 on pNK and uNK cells and NK1/NK2 ratio, like the Th1/Th2 ratio, are potential biomarkers waiting for further investigation and clinical confirmation.

5 Treatment

5.1 Anticoagulants (aspirin and heparin)

Although it is not clear whether thrombophilia or antiphospholipid syndrome is a cause of RIF, it has been reported that preimplantation aspirin and heparin therapy may effectively prevent RIF [50]. In addition, the uterine radial artery resistance index (URa-RI) in women with RPL was higher, and its levels were increased with the increased number of pNK cells [51]. Moreover, low molecular weight heparin treatment effectively decreased URa-RI with improved pregnancy outcomes in women with RPL and elevated URa-RI [51]. Besides, positive effects of pregnancy outcomes for women with RPL and higher URa-RI were reported [52]. Patients were on preconception low molecular weight heparin (Enoxaparin, 20–40 mg (Lovenox) = 2,000–4,000 IU (Clexane)/day, determined by patient's body weight), low-dose aspirin (81 mg/day), and prednisone (10 mg/day) treatment [53]. Hence, aspirin-heparin therapy may be effective for RIF.

The package inserts state that aspirin should not be administered after 28 weeks of gestation due to the possibility of prolonged gestation, early closure of ductus arteriosus, inhibition of uterine contractions, and increased hemorrhage during delivery. However, aspirin at 81 mg/day dose for certain pregnancy-related conditions at any point of pregnancy is an exception to the FDA recommendations to avoid the use of NSAIDs in pregnancy at 20 weeks or later. At the same time, heparin can be used until the onset of labor pain.

5.2 Corticosteroids

Prednisolone has been shown to decrease NK cell cytotoxicity and inhibit TNF-α production and is used in RIF women with a high Th1/Th2 ratio and high NK cell cytotoxicity. However, a Cochrane review reported that there was no clear evidence that glucocorticoids significantly improved the clinical outcome of IVF-ET [54]. The American

Society for Reproductive Medicine guidelines state that glucocorticoid administration from controlled ovarian stimulation to the peri-embryo transfer period increases the live birth rate [55]. Prednisolone has low placental transfer from mother to fetus and no reports about its teratogenicity. Therefore it can be used safely at 10 mg/day or less, but caution should be exercised at 20 mg/day or more because of concerns for increased gestational diabetes, hypertension, and infection [56].

5.3 Immunoglobulins

Immunoglobulins are known as immunomodulators and have been used in RIF for their useful function in establishing and maintaining pregnancy through immunomodulatory effects, such as normalizing NK cell cytotoxicity and Th1/Th2 ratio. The Korean Society for Reproductive Immunology guidelines about intravenous immunoglobulin G (IVIG) treatment for reproductive abnormalities such as RIF and RPL state that immunological work up such as NK cell levels and cytotoxicity, and Th1/Th2 ratio, should be performed to decide immunoglobulin administration and follow up (the interval and duration) for women with RIF [57].

Immunoglobulin-G should be administered cautiously in patients with IgA deficiency because it may cause hypersensitivity reactions in patients with anti-IgA antibodies. Therefore serum IgA levels should be measured prior to administration. Since the half-life of immunoglobulin is 18–25 days, it should be administered at 400 mg/kg every 3–4 weeks starting before embryo transfer until at least 10 weeks of pregnancy. Depending on the patient's state, the endpoint of IVIG treatment and the need for further laboratory tests can be determined. The changes of NK cell cytotoxicities after administration of IVIG for women with reproductive failure such as RPL or RIF are shown in Fig. 1. NK cell cytotoxicity generally decreases with the course of pregnancy. In patients with reproductive failures who have high NK cell cytotoxicity, NK cell cytotoxicity decreases after IVIg but increases again after the half-life of IVIG, indicating the need for repeated administration of IVIG. The effectiveness of IVIG therapy for RIF women was reviewed [58]. In this review, IVIG treatment was effective for RIF women for the pregnancy rate and live birth rate. Thus, IVIG may be a beneficial therapeutic strategy in RIF patients selected according to relevant immunological disturbances such as NK, NKT, and T cells. Besides, the IVF-ET success rates of women with RPL and RIF with cellular immune abnormalities and thrombophilia were significantly improved by immunomodulatory treatment, including prednisone, IVIG, and anticoagulation treatment with low molecular weight heparin and low dose aspirin [15]. The pregnancy rate of IVF cycles and the live birth rates per ET cycle with immunomodulatory treatment were significantly increased in all patients as compared to historical controls [15]. Recently, Saab et al. reviewed *meta*-analyses for the IVIG treatment in women with RIF [59]. They reviewed four *meta*-analyses and concluded that IVIG is a valuable treatment for increasing the clinical pregnancy rate and LBR in women with RIF. However, the American Society for Reproductive Medicine guidelines states that there is not sufficient evidence to recommend immunoglobulin to be administered to patients undergoing IVF and embryo transfer, and further studies are needed to clarify the cases in which immunoglobulin is indicated [55].

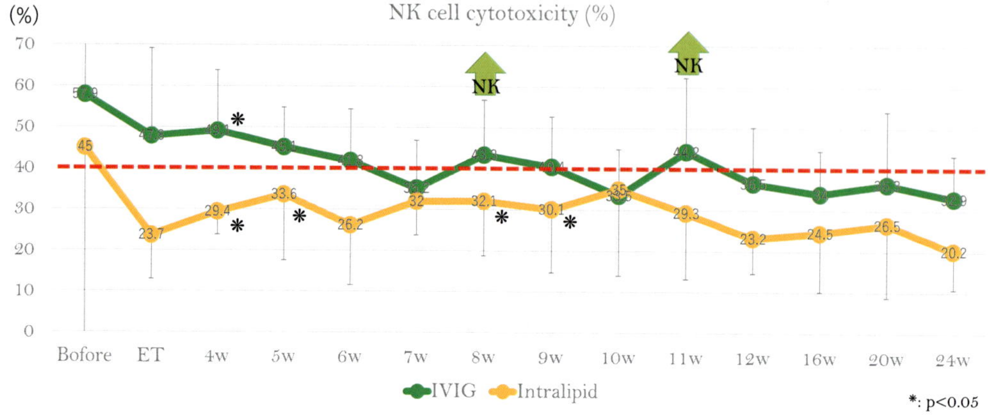

FIGURE 1 **Natural killer (NK) cell cytotoxicity in repeated implantation failure (RIF) women with intravenous immunoglobulin G (IVIG) and fat emulsion (Intralipid) treatment.** IVIG suppressed peripheral blood NK cell cytotoxicities (*green line*) in women with RIFs. The half-life of IVIG is 3 weeks, and NK cell cytotoxicities were increased again 4 weeks after the IVIG infusion. Therefore periodic administration of IVIG is imperative. Administration of fat emulsion (Intralipid) during pregnancy also suppressed peripheral blood NK cell cytotoxicities (*yellow line*), and the repeated administration of fat emulsion could maintain NK cell cytotoxicity.

5.4 Fat emulsion infusion therapy

Fat emulsion infusion therapy (lipid emulsion, Intralipid), consisting of soybean oil and egg yolk lecithin, has been used for pre and postoperative nutritional support and RIF due to the immunosuppressive effect on NK cell cytotoxicity and inflammatory cytokines [60,61]. The *meta*-analysis [62] and the review [63] on fat emulsion infusion therapy have shown that fat emulsion infusion therapy in RIF improves clinical pregnancy rates and live birth rates. On the other hand, contradictory result [64] was also reported. The American Society for Reproductive Medicine guideline states that there is insufficient evidence for the use of fat emulsion infusion treatment in patients undergoing IVF-ET. Further studies are needed to clarify the indications for the use of fat emulsions [55]. Fat emulsion infusion therapy is indicated to patients with immune abnormalities such as NK cell abnormalities. NK cell cytotoxicities in women with reproductive failures undergoing fat emulsion treatment are shown in Fig. 1.

Undiluted fat emulsion (10%–20% Intralipid) is commercially available. 2–4 mL of 20% fat emulsion solution is mixed with 250 mL of sterile IV saline and infused to the patients at the speed of 250 mL per hour [65]. Dilution should be made carefully to avoid changes in formulation. So, not-diluted fat emulsion (20% intralipid, 250 ml) is also available [66–68]. The fat emulsion is administered every two to three weeks before embryo transfer until 10 weeks gestation or longer [66–68].

5.5 Hydroxychloroquine

Hydroxychloroquine has been used as an antimalarial drug for many years and is also used for autoimmune diseases, including systemic lupus erythematosus, cutaneous lupus erythematosus, and rheumatoid arthritis. Hydroxychloroquine has attracted attention for its effects on infertility and RIF. A study of RIF patients with high Th1/Th2 levels [69] showed a decrease in the blood and endometrial Th1/Th2 [69] and Th17/regulatory T cell ratios [70] after hydroxychloroquine administration. Hydroxychloroquine does not have a concern for teratogenicity when used during pregnancy. While chloroquine, which has a similar chemical structure and pharmacological action to hydroxychloroquine, has genotoxicity and reproductive toxicity. Hence, it is associated with the increased risk of teratogenicity and fetal toxicity during pregnancy.

5.6 Tumor necrosis factor inhibitors

TNF inhibitors are used to treat rheumatoid arthritis, Behcet's disease, Kawasaki disease, Crohn's disease, and ulcerative colitis, and an increasing number of these patients are becoming pregnant while receiving TNF inhibitor therapy. Although an increased rate of miscarriage was noticed in pregnant women with rheumatic diseases treated with infliximab and adalimumab compared to those without, there was no difference in the rate of preterm birth and low birth weight, birth weight, or the prevalence of fetal malformations between patients with and without these drugs [71]. In addition, the prevalence of fetal malformations was reported not to be different in a study of pregnancies that occurred during accidental use of TNF inhibitors [72,73]. In RIF patients with high Th1/Th2 ratios, pregnancies with TNF inhibitors have been associated with reduced Th1/Th2 cell ratios and significantly increased implantation, clinical pregnancy and live birth rates [74]. Fu et al. reported decreased NK cytotoxicities and TNF-α levels after etanercept administration in women with RPL compared to controls with a placebo [75]. Based on these studies, TNF-α inhibitor treatment might be helpful in RIF women with a high Th1/Th2 ratio or NK cell cytotoxicity, and further investigation of TNF-α inhibitor treatment is warranted in women with RIF.

5.7 Vitamin D

Vitamin D deficiency (25-hydroxyvitamin D <30 ng/mL) may induce NK cell abnormalities, such as increased cytotoxic NK cells and increased NK cell cytotoxicity, and suppressed NK cell cytotoxicity, decreased expression of NK cell-activating receptor, and NK cell type 1 cytokine production has been shown with vitamin D administration (1000–2000 IU/day) [76,77]. In addition, active vitamin D (1, 25-dihydroxy vitamin D) has also been reported to decrease NK cell cytotoxicity and type 1 cytokine production by NK cells [77]. At the very least, vitamin D is vital for pregnancy and the fetus as it controls aberrant pro-inflammatory immune responses. At present, vitamin D is readily available as a supplement and may be administered when vitamin D levels are low.

6 Summary and recommendations

1. There is no universally accepted definition for RIF.
2. RIF is generally defined as 2–3 times of failed embryo transfers or failures after 3–4 good-quality embryos are transferred.
3. The frequency of true RIF seems to be less than 5%.

4. NK cells play an important role in human pregnancy, and the regulation of NK cells is important for reproductive success.
5. The general function of NK cells can be measured by NK cell cytotoxicity and cytokines production.
6. Evaluating NK cell levels (proportions) and cytotoxicity using uterine endometrial tissue and peripheral blood are good options for the evaluation and diagnosis of RIF of immune etiologies.
7. For RIF women with higher NK cell cytotoxicity or abnormal cytokines production, various immunomodulation treatments are available.
8. Aspirin and heparin treatment is a good option for RIF women with abnormal blood flow in the uterine radial artery measured by a Doppler blood flow study.
9. The mainstay of treatment for RIF women with NK cell abnormalities is IVIG and/or steroid therapy.

Fat emulsion, hydroxychloroquine, TNF inhibitors, and vitamin D have been utilized in RIF women with NK cell abnormalities (Fig. 2).

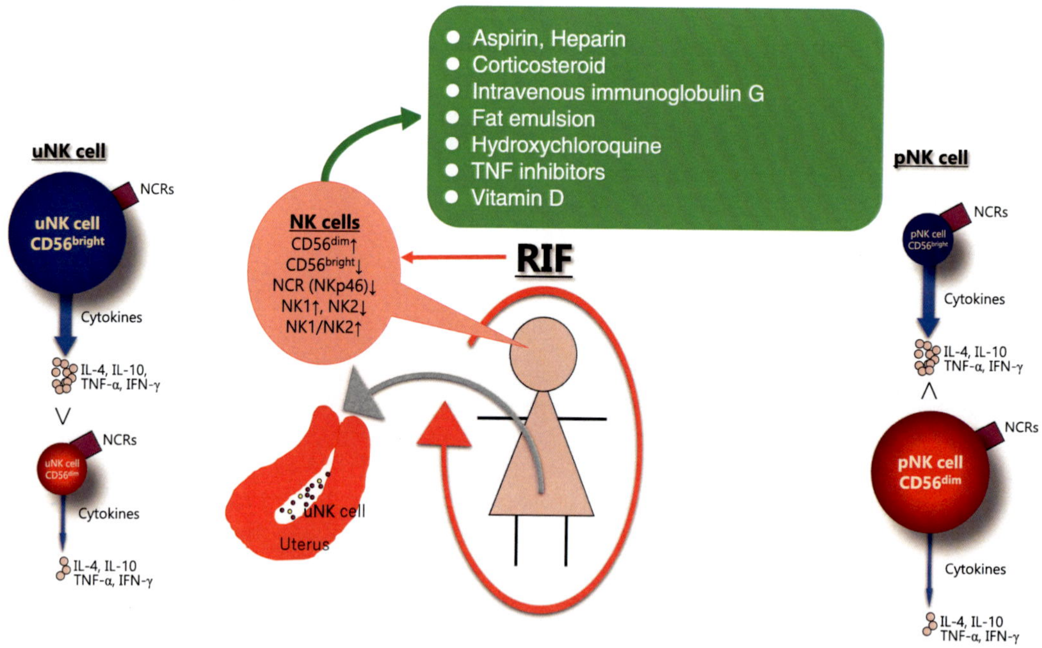

FIGURE 2 **Natural killer (NK) cell abnormalities in repeated implantation failure women and their treatments.** There are two types of NK cells, $CD56^{bright}$ NK cells, and $CD56^{dim}$ NK cells. The primary function of $CD56^{bright}$ NK cell is cytokine production, and $CD56^{dim}$ NK cell is cellular cytotoxicity. In women with RIF, cytotoxic $CD56^{dim}$ NK cells are increased, while cytokine-producing $CD56^{bright}$ NK cells are decreased. The expression of NKp46 on NK cells is decreased. Type 1 cytokine-producing NK cells (NK1) are increased, while type 2 cytokine-producing NK cells (NK2) are decreased. As a result, the NK1/NK2 ratio is increased in women with RIF. Immunological treatments such as heparin, aspirin, corticosteroid, intravenous immunoglobulin G, fat emulsion, hydroxychloroquine, tumor necrosis factor inhibitor, and vitamin D may be useful to normalize NK cell abnormalities in women with RIF.

References

[1] Polanski LT, Baumgarten MN, Quenby S, Brosens J, Campbell BK, Raine-Fenning NJ. What exactly do we mean by 'recurrent implantation failure'? A systematic review and opinion. Reprod Biomed Online 2014;28:409–23.

[2] Coughlan C, Ledger W, Wang Q, Liu F, Demirol A, Gurgan T, et al. Recurrent implantation failure: definition and management. Reprod Biomed Online 2014;28:14–38.

[3] Shaulov T, Sierra S, Sylvestre C. Recurrent implantation failure in IVF:a Canadian fertility and andrology society clinical practice guideline. Reprod Biomed Online 2020;41:819–33.

[4] Cimadomo D, Craciunas L, Vermeulen N, Vomstein K, Toth B. Definition, diagnostic and therapeutic options in recurrent implantation failure: an international survey of clinicians and embryologists. Hum Reprod 2021;36:305–17.

[5] Busnelli A, Reschini M, Cardellicchio L, Vegetti W, Somigliana E, Vercellini P. How common is real repeated implantation failure? An indirect estimate of the prevalence. Reprod Biomed Online 2020;40:91–7.

[6] Pirtea P, De Ziegler D, Tao X, Sun L, Zhan Y, Ayoubi JM, et al. Rate of true recurrent implantation failure is low: results of three successive frozen euploid single embryo transfers. Fertil Steril 2021;115:45–53.

[7] Fukui A, Yokota M, Funamizu A, Nakamua R, Fukuhara R, Yamada K, et al. Changes of NK cells in preeclampsia. Am J Reprod Immunol 2012;67:278–86.

[8] Fukui A, Ntrivalas E, Gilman-Sachs A, Kwak-Kim J, Lee SK, Levine R, et al. Expression of natural cytotoxicity receptors and a2V-ATPase on peripheral blood NK cell subsets in women with recurrent spontaneous abortions and implantation failures. Am J Reprod Immunol 2006;56:312–20.

[9] Fukui A, Kwak-Kim J, Ntrivalas E, Gilman-Sachs A, Lee SK, Beaman K. Intracellular cytokine expression of peripheral blood natural killer cell subsets in women with recurrent spontaneous abortions and implantation failures. Fertil Steril 2008;89:157–65.

[10] Fukui A, Funamizu A, Fukuhara R, Shibahara H. Expression of natural cytotoxicity receptors and cytokine production on endometrial natural killer cells in women with recurrent pregnancy loss or implantation failure, and the expression of natural cytotoxicity receptors on peripheral blood natural killer cells in pregnant women with a history of recurrent pregnancy loss. J Obstet Gynaecol Res 2017;43:1678–86.

[11] Mai C, Fukui A, Takeyama R, Yamamoto M, Saeki S, Yamaya A, et al. NK cells that differ in expression of NKp46 might play different roles in endometrium. J Reprod Immunol 2021;147:103367.

[12] Takeyama R, Fukui A, Mai C, Yamamoto M, Saeki S, Yamaya A, et al. Co-expression of NKp46 with activating or inhibitory receptors on, and cytokine production by, uterine endometrial NK cells in recurrent pregnancy loss. J Reprod Immunol 2021;145:103324.

[13] Lachapelle MH, Miron P, Hemmings R, Roy DC. Endometrial T, B, and NK cells in patients with recurrent spontaneous abortion. Altered profile and pregnancy outcome. J Immunol 1996;156:4027–34.

[14] Gilman-Sachs A, DuChateau BK, Aslakson CJ, Wohlgemuth GP, Kwak JY, Beer AE, et al. Natural killer (NK) cell subsets and NK cell cytotoxicity in women with histories of recurrent spontaneous abortions. Am J Reprod Immunol 1999;41:99–105.

[15] Sung N, Khan SA, Yiu ME, Jubiz G, Salazar MD, Skariah A, et al. Reproductive outcomes of women with recurrent pregnancy losses and repeated implantation failures are significantly improved with immunomodulatory treatment. J Reprod Immunol 2021;148:103369.

[16] Yamada H, Morikawa M, Kato EH, Shimada S, Kobashi G, Minakami H. Pre-conceptional natural killer cell activity and percentage as predictors of biochemical pregnancy and spontaneous abortion with normal chromosome karyotype. Am J Reprod Immunol 2003;50:351–4.

[17] Fukui A, Fujii S, Yamaguchi E, Kimura H, Sato S, Saito Y. Natural killer cell subpopulations and cytotoxicity for infertile patients undergoing in vitro fertilization. Am J Reprod Immunol 1999;41:413–22.

[18] Moffett A, Regan L, Braude P. Natural killer cells, miscarriage, and infertility. BMJ 2004;329:1283–5.

[19] Park DW, Lee HJ, Park CW, Hong SR, Kwak-Kim J, Yang KM. Peripheral blood NK cells reflect changes in decidual NK cells in women with recurrent miscarriages. Am J Reprod Immunol 2010;63:173–80.

[20] Fukui A. Uterine and circulating natural killer cells and their roles in women with recurrent pregnancy losses, implantation failures or preeclampsia. J Reprod Immunol 2010;86:14.

[21] Fukui A. NK cells and its role in reproduction. Am J Reprod Immunol 2010;64:1.

[22] Fukui A. Role of NK cells in women with recurrent pregnancy loss and implantation failure. Acta obstet gynaec Jpn 2011;63:2167–84.

[23] Tuckerman E, Mariee N, Prakash A, Li TC, Laird S. Uterine natural killer cells in peri-implantation endometrium from women with repeated implantation failure after IVF. J Reprod Immunol 2010;87:60–6.
[24] Santillan I, Lozano I, Illan J, Verdu V, Coca S, Bajo-Arenas JM, et al. Where and when should natural killer cells be tested in women with repeated implantation failure? J Reprod Immunol 2015;108:142–8.
[25] Russell P, Anderson L, Lieberman D, Tremellen K, Yilmaz H, Cheerala B, et al. The distribution of immune cells and macrophages in the endometrium of women with recurrent reproductive failure I: techniques. J Reprod Immunol 2011;91:90–102.
[26] Kofod L, Lindhard A, Hviid TVF. Implications of uterine NK cells and regulatory T cells in the endometrium of infertile women. Hum Immunol 2018;79:693–701.
[27] Kofod L, Lindhard A, Bzorek M, Eriksen JO, Larsen LG, Hviid TVF. Endometrial immune markers are potential predictors of normal fertility and pregnancy after in vitro fertilization. Am J Reprod Immunol 2017;78.
[28] Yamada H, Shimada S, Morikawa M, Iwabuchi K, Kishi R, Onoe K, et al. Divergence of natural killer cell receptor and related molecule in the decidua from sporadic miscarriage with normal chromosome karyotype. Mol Hum Reprod 2005;11:451–7.
[29] Verma S, King A, Loke YW. Expression of killer cell inhibitory receptors on human uterine natural killer cells. Eur J Immunol 1997;27:979–83.
[30] Fukui A, Fujii S, Kasai G, Kimura H, Yamaguchi E, Mizunuma H. Correlation with endometrial NK cell, KIRs and NKT cells. IFFS 2001 Sel Commun 2001;65–71.
[31] Acar N, Ustunel I, Demir R. Uterine natural killer (uNK) cells and their missions during pregnancy: a review. Acta Histochem 2011;113:82–91.
[32] Varla-Leftherioti M, Spyropoulou-Vlachou M, Niokou D, Keramitsoglou T, Darlamitsou A, Tsekoura C, et al. Natural killer (NK) cell receptors' repertoire in couples with recurrent spontaneous abortions. Am J Reprod Immunol 2003;49:183–91.
[33] Ntrivalas EI, Bowser CR, Kwak-Kim J, Beaman KD, Gilman-Sachs A. Expression of killer immunoglobulin-like receptors on peripheral blood NK cell subsets of women with recurrent spontaneous abortions or implantation failures. Am J Reprod Immunol 2005;53:215–21.
[34] Hong Y, Wang X, Lu P, Song Y, Lin Q. Killer immunoglobulin-like receptor repertoire on uterine natural killer cell subsets in women with recurrent spontaneous abortions. Eur J Obstet Gynecol Reprod Biol 2008;140:218–23.
[35] Yan WH, Lin A, Chen BG, Zhou MY, Dai MZ, Chen XJ, et al. Possible roles of KIR2DL4 expression on uNK cells in human pregnancy. Am J Reprod Immunol 2007;57:233–42.
[36] Wang XL, Wang Q, Sun CJ, Zhang WY. Genetic polymorphisms of killer cell immunoglobulin-like receptor 3DL2 in preeclampsia. J Perinat Med 2011;39:273–8.
[37] Yokota M, Fukui A, Funamizu A, Nakamura R, Kamoi M, Fuchinoue K, et al. Role of NKp46 expression in cytokine production by CD56-positive NK cells in the peripheral blood and the uterine endometrium. Am J Reprod Immunol 2013;69:202–11.
[38] Fukui A, Kamoi M, Funamizu A, Fuchinoue K, Chiba H, Yokota M, et al. NK cell abnormality and its treatment in women with reproductive failures such as recurrent pregnancy loss, implantation failures, preeclampsia, and pelvic endometriosis. Reprod Med Biol 2015;14:151–7.
[39] Kwak-Kim JY, Gilman-Sachs A, Kim CE. T helper 1 and 2 immune responses in relationship to pregnancy, nonpregnancy, recurrent spontaneous abortions and infertility of repeated implantation failures. Chem Immunol Allergy 2005;88:64–79.
[40] Carter LL, Dutton RW. Relative perforin- and Fas-mediated lysis in T1 and T2 CD8 effector populations. J Immunol 1995;155:1028–31.
[41] Peritt D, Robertson S, Gri G, Showe L, Aste-Amezaga M, Trinchieri G. Differentiation of human NK cells into NK1 and NK2 subsets. J Immunol 1998;161:5821–4.
[42] Higuma-Myojo S, Sasaki Y, Miyazaki S, Sakai M, Siozaki A, Miwa N, et al. Cytokine profile of natural killer cells in early human pregnancy. Am J Reprod Immunol 2005;54:21–9.
[43] Bulmer JN, Lash GE. Human uterine natural killer cells: a reappraisal. Mol Immunol 2005;42:511–21.
[44] Ashkar AA, Di Santo JP, Croy BA. Interferon gamma contributes to initiation of uterine vascular modification, decidual integrity, and uterine natural killer cell maturation during normal murine pregnancy. J Exp Med 2000;192:259–70.
[45] Terranova PF, Hunter VJ, Roby KF, Hunt JS. Tumor necrosis factor-alpha in the female reproductive tract. Proc Soc Exp Biol Med 1995;209:325–42.

[46] Toder V, Fein A, Carp H, Torchinsky A. TNF-alpha in pregnancy loss and embryo maldevelopment: a mediator of detrimental stimuli or a protector of the fetoplacental unit? J Assist Reprod Genet 2003;20:73–81.

[47] Brosens IA, Robertson WB, Dixon HG. The role of the spiral arteries in the pathogenesis of preeclampsia. Obstet Gynecol Annu 1972;1:177–91.

[48] Diao L, Cai S, Huang C, Li L, Yu S, Wang L, et al. New endometrial immune cell-based score (EI-score) for the prediction of implantation success for patients undergoing IVF/ICSI. Placenta 2020;99:180–8.

[49] Dambaeva S, Bilal M, Schneiderman S, Germain A, Fernandez E, Kwak-Kim J, et al. Decidualization score identifies an endometrial dysregulation in samples from women with recurrent pregnancy losses and unexplained infertility. F S Rep 2021;2:95–103.

[50] Bohlmann MK. Effects and effectiveness of heparin in assisted reproduction. J Reprod Immunol 2011;90:82–90.

[51] Koo HS, Kwak-Kim J, Yi HJ, Ahn HK, Park CW, Cha SH, et al. Resistance of uterine radial artery blood flow was correlated with peripheral blood NK cell fraction and improved with low molecular weight heparin therapy in women with unexplained recurrent pregnancy loss. Am J Reprod Immunol 2015;73:175–84.

[52] Bao SH, Chigirin N, Hoch V, Ahmed H, Frempong ST, Zhang M, et al. Uterine radial artery resistance index predicts reproductive outcome in women with recurrent pregnancy losses and thrombophilia. BioMed Res Int 2019;2019:8787010.

[53] Han AR, Ahn H, Vu P, Park JC, Gilman-Sachs A, Beaman K, et al. Obstetrical outcome of anti-inflammatory and anticoagulation therapy in women with recurrent pregnancy loss or unexplained infertility. Am J Reprod Immunol 2012;68:418–27.

[54] Boomsma CM, Keay SD, Macklon NS. Peri-implantation glucocorticoid administration for assisted reproductive technology cycles. Cochrane database Syst Rev 2012;CD005996.

[55] Practice Committee of the American Society for Reproductive Medicine. Electronic address Aao, Practice Committee of the American Society for Reproductive M: the role of immunotherapy in in vitro fertilization: a guideline. Fertil Steril 2018;110:387–400.

[56] Mekinian A, Cohen J, Alijotas-Reig J, Carbillon L, Nicaise-Roland P, Kayem G, et al. Unexplained recurrent miscarriage and recurrent implantation failure: is there a place for immunomodulation? Am J Reprod Immunol 2016;76:8–28.

[57] Sung N, Han AR, Park CW, Park DW, Park JC, Kim NY, et al. Ivig Task Force KSfRI: Intravenous immunoglobulin G in women with reproductive failure: the Korean Society for reproductive immunology practice guidelines. Clin Exp Reprod Med 2017;44:1–7.

[58] Abdolmohammadi-Vahid S, Pashazadeh F, Pourmoghaddam Z, Aghebati-Maleki L, Abdollahi-Fard S, Yousefi M. The effectiveness of IVIG therapy in pregnancy and live birth rate of women with recurrent implantation failure (RIF): a systematic review and meta-analysis. J Reprod Immunol 2019;134–135:28–33.

[59] Saab W, Seshadri S, Huang C, Alsubki L, Sung N, Kwak-Kim J. A systemic review of intravenous immunoglobulin G treatment in women with recurrent implantation failures and recurrent pregnancy losses. Am J Reprod Immunol 2021;85:e13395.

[60] Achilli C, Duran-Retamal M, Saab W, Serhal P, Seshadri S. The role of immunotherapy in in vitro fertilization and recurrent pregnancy loss: a systematic review and meta-analysis. Fertil Steril 2018;110:1089–100.

[61] Fukui A, Funamizu A, Yokota M, Yamada K, Nakamua R, Fukuhara R, et al. Uterine and circulating natural killer cells and their roles in women with recurrent pregnancy loss, implantation failure and preeclampsia. J Reprod Immunol 2011;90:105–10.

[62] Zhou P, Wu H, Lin X, Wang S, Zhang S. The effect of intralipid on pregnancy outcomes in women with previous implantation failure in in vitro fertilization/intracytoplasmic sperm injection cycles: a systematic review and meta-analysis. Eur J Obstet Gynecol Reprod Biol 2020;252:187–92.

[63] Coulam CB. Intralipid treatment for women with reproductive failures. Am J Reprod Immunol 2020;e13290.

[64] Canella P, Barini R, Carvalho PO, Razolli DS. Lipid emulsion therapy in women with recurrent pregnancy loss and repeated implantation failure: the role of abnormal natural killer cell activity. J Cell Mol Med 2021.

[65] Roussev RG, Acacio B, Ng SC, Coulam CB. Duration of intralipid's suppressive effect on NK cell's functional activity. Am J Reprod Immunol 2008;60:258–63.

[66] Meng L, Lin J, Chen L, Wang Z, Liu M, Liu Y, et al. Effectiveness and potential mechanisms of intralipid in treating unexplained recurrent spontaneous abortion. Arch Gynecol Obstet 2016;294:29–39.

[67] W E-K, M ES. Intralipid for repeated implantation failure (RIF): a randomized controlled trial. Fertil Steril 2015;104:e26.

[68] I G, MF G, A S, V A. Intralipid infusion does not improve live birth rates in women with unexplained recurrent implantation failure and may increase the risk of congenital malformations, a double-blinded randomised controlled trial. BJOG-Int J Obstet Gynaecol 2018;125.

[69] Ghasemnejad-Berenji H, Ghaffari Novin M, Hajshafiha M, Nazarian H, Hashemi SM, Ilkhanizadeh B, et al. Immunomodulatory effects of hydroxychloroquine on Th1/Th2 balance in women with repeated implantation failure. Biomed Pharmacother 2018;107:1277–85.

[70] Sadeghpour S, Ghasemnejad Berenji M, Nazarian H, Ghasemnejad T, Nematollahi MH, Abroon S, et al. Effects of treatment with hydroxychloroquine on the modulation of Th17/Treg ratio and pregnancy outcomes in women with recurrent implantation failure: clinical trial. Immunopharmacol Immunotoxicol 2020;42:632–42.

[71] Komoto S, Motoya S, Nishiwaki Y, Matsui T, Kunisaki R, Matsuoka K, et al. Japanese study group for pregnant women with IBD: pregnancy outcome in women with inflammatory bowel disease treated with anti-tumor necrosis factor and/or thiopurine therapy: a multicenter study from Japan. Intest Res 2016;14:139–45.

[72] Roux CH, Brocq O, Breuil V, Albert C, Euller-Ziegler L. Pregnancy in rheumatology patients exposed to anti-tumour necrosis factor (TNF)-alpha therapy. Rheumatology (Oxf) 2007;46:695–8.

[73] Hyrich KL, Symmons DP, Watson KD, Silman AJ. British Society fRBR: Pregnancy outcome in women who were exposed to anti-tumor necrosis factor agents: results from a national population register. Arthritis Rheum 2006;54:2701–2.

[74] Winger EE, Reed JL, Ashoush S, Ahuja S, El-Toukhy T, Taranissi M. Treatment with adalimumab (Humira) and intravenous immunoglobulin improves pregnancy rates in women undergoing IVF. Am J Reprod Immunol 2009;61:113–20.

[75] Fu J, Li L, Qi L, Zhao L. A randomized controlled trial of etanercept in the treatment of refractory recurrent spontaneous abortion with innate immune disorders. Taiwan J Obstet Gynecol 2019;58:621–5.

[76] Ota K, Dambaeva S, Han AR, Beaman K, Gilman-Sachs A, Kwak-Kim J. Vitamin D deficiency may be a risk factor for recurrent pregnancy losses by increasing cellular immunity and autoimmunity. Hum Reprod 2014;29:208–19.

[77] Ota K, Dambaeva S, Kim MW, Han AR, Fukui A, Gilman-Sachs A, et al. 1,25-dihydroxy-vitamin D3 regulates NK-cell cytotoxicity, cytokine secretion and degranulation in women with recurrent pregnancy losses. Eur J Immunol 2015.

CHAPTER

18

Helper T cell pathology and repeated implantation failures

Koji Nakagawa[1], *Keiji Kuroda*[2] *and Rikikazu Sugiyama*[3]

[1]Director, Center for Reproductive Medicine and Implantation Research, Sugiyama Clinic Shinjuku, Tokyo, Japan [2]Division Manager of Implantation Research and Endoscopy, Center for Reproductive Medicine and Implantation Research, Sugiyama Clinic Shinjuku, Tokyo, Japan [3]CEO, Center for Reproductive Medicine and Implantation Research, Sugiyama Clinic Shinjuku, Tokyo, Japan

1 The role of helper T cells in reproductive medicine

T helper (Th) cells play an important role in immune responses by producing several factors that activate other cells working within immune systems. These cells include B cells, cytotoxic T cells, macrophages, etc. Th cells express CD4 on their surface, and this protein plays a critical role in Th cell activation by binding to human leukocyte antigen (HLA) class II molecules, which aid the immune system in recognizing foreign substances. Th cells can be divided into two general subgroups with different characteristics and functions: Th1 and Th2 cells. These subpopulations can be differentiated by the pattern of cytokine production (chemical messengers). Th1 cells primarily synthesize several cytokines such as interferon gamma (INF-ϒ), tumor necrosis factor-beta (TNF-), and interleukin-2 (IL-2). In contrast, Th2 cells mainly produce interleukin-4 (IL-4), IL-5, IL-6, IL-9, IL-10, and IL-13. The primary role of Th1 cells is to stimulate cell-mediated responses involving cytotoxic T cells and macrophages, while Th2 cells primarily assist in stimulating B cells to produce antibodies.

Several steps are needed for Th cell activation. The first step begins with antigen-presenting cells such as macrophages. These cells ingest infectious agents or foreign particles, and export fragments of those, so-called antigens, to the cell surface. A surface receptor of the Th cells binds to the HLA complex. In the next step, Th cell activation occurs either through cytokines stimulation or the costimulatory reactions of a signaling protein, B7. B7 is found on the surface of the antigen-presenting cells [1], which binds to CD28, on the surface of the Th cell [2]. As a result of Th cell activation, the number of Th cells that recognize antigens increases, and

Immunology of Recurrent Pregnancy Loss and Implantation Failure
DOI: https://doi.org/10.1016/B978-0-323-90805-4.00010-9

Th cell cytokines are produced. Th cells do not directly kill infected cells or attack other cells (foreign cells or hatched blastocysts), as cytotoxic T cells do. Th cells activate cytotoxic T cells, help macrophages to attack infected or foreign cells, or they stimulate B cells to secrete antibodies.

The first step in achieving implantation is to establish a hatched blastocyst that will be accepted by the decidualized uterine endometrium. When maternal Th cells recognize a hatched blastocyst, rather than directly attacking it, Th cells produce various cytokines, such as INF-ϒ, TNF-, or IL-2. At the implantation site, maternal–fetal immune interactions involve both inhibitions of rejection immunity and immune tolerance induction. This phenomenon has also been documented in the fields of organ transplantation and cancer cell biology [3–6].

Immunological reactions during pregnancy resemble those during organ transplantation when an embryo is considered a transplanted organ. However, in pregnancy, T-cell activation by dendritic cells (DCs) is weaker than organ transplantation. The immunological mechanisms during a normal pregnancy are not yet well understood, but, immune tolerance must be addressed both before and after embryo embedding. Therefore, the inhibition of immunological rejection is important for both the establishment and the maintenance of a successful pregnancy. The rejection reaction from the mother's natural immunity must be suppressed during pregnancy. Extravillous trophoblasts are derived from an implanted embryo that does not express the histocompatibility antigens HLA-A, HLA-B, and HLA-D [7]. Therefore, extravillous trophoblasts can be targeted by natural killer (NK) cells but not targeted by cytotoxic T cells. However, extravillous trophoblasts express HLA-C, E and HLA-G [8–10]. Although fetal–maternal HLA-C mismatch might evoke maternal NK cell activation, the maternal immune response can be inhibited via the HLA-G expression on the exrtavillous trophoblast, which induces evasion from NK cell attack and inhibition of macrophage activation [8–10].

The induction of immune tolerance must be well organized and convenient. Although maternal NK and T-cell activation is evoked by a fetal–maternal HLA mismatch, maternal immune response switches to immune tolerance under conditions such as seminal plasma priming, adequate progesterone levels, or the expression of HLA-G [10–13]. This reaction differentiates tolerogenic DCs from immature DCs, and this differentiation subsequently induces immune tolerance via regulatory T cells [14]. Based on this chain of flow, immunological tolerance could be established to maintain pregnancy. Additionally, suppressed NK and NKT cell functions, downregulated type 1 cytokine production, and repressed Th1 cell differentiation, all of which are working against immunological rejection.

Th cells are believed to command the physical defense system. When Th cells receive signals for foreign subjects (coming from outside of the body) such as bacteria, viruses, or hatched blastocysts via DCs, Th1 cells produce cytokines, including INF-γ and TNF-α in response to intrusion by the foreign subjects. Cytokines then enhance inflammation in the local area and stimulate macrophage activation that involves phagocytosis. It is well known that proinflammatory cytokines suppress trophoblast growth and promote inflammatory responses at the area of maternal uterine endometrium where embryos are attached and implanted into and adversely affect embryo implantation. By contrast, IL-4, IL-6, and IL-10, which are produced by Th2 cells, can inhibit Th1 cell-induced tissue factor production via monocytes. Indeed, women who have experienced recurrent implantation failure (RIF) show high TNF-α/IL-4 ratios in their peripheral blood samples [15].

2 Th cell immunopathology and repeated implantation failure

The generic definition of RIF has been used to refer to women who have experienced three failures of fresh or cryopreserved-thawed embryo transfer (ET) with good-quality embryos [16]. It is well known that there are three requirements for embryo implantation: **a good embryo, adequate decidualization, and appropriate maternal response**. The etiologies of RIF have been thoroughly evaluated in the literature [17]. Table 1 shows the etiologies of RIF in the literature according to the three factors for embryo implantation mentioned above. The treatment strategies for each etiology in Table 1 are summarized in Fig. 1.

Th lymphocytes are usually classified as either Th1 or Th2 cells according to their ability to produce cytokines [18,19]. Th1 and Th2 cells are the leading characters in the immunological processes that determine either rejection or tolerance [20]. Common sense dictates that the establishment of pregnancy is an immunological condition wherein Th2 has shown superiority over Th1. By contrast, when Th1 shows superiority over Th2 the result is immunological rejection. As a consequence, the Th1 superiority can reject transferred embryos [20,21]. Based on this theory, implantation failures are thought to be a case of allograft rejection [22].

Th cell-mediated RIF has been defined as increased Th1/Th2 cell ratios, because either elevated Th1 or decreased Th2, or both elevated Th1 and decreased Th2 can result in increased Th1/Th2 cell ratios. Increased Th1 cell can evoke cell-mediated immunity and phagocyte-dependent inflammation. In contrast, decreased Th2 cell weakens inhibition of several functions of phagocytes [23]. Th1 and Th2 cells are classified by their cytokine profiles. Therefore, several combinations of the Th1/Th2 cell ratios have been reported with their cut-off values; TNF-α/IL-10 (>30.6), INF-Υ/IL-10 (>20.5) [24], and INF-Υ/IL-4 (≥ 10.3) [25] producing Th cell ratios.

3 Prevalence

Depending on the definition of RIF mentioned above, 10% of couples undergoing IVF will experience RIFs. Unfortunately, some of those who achieve implantation may face the disappointment of early pregnancy loss [26]. The patients who fail to achieve a pregnancy even when having received serial transfers of high-quality embryos need to be evaluated for underlying etiologies for their failure.

TABLE 1 The etiologies of repeated implantation failures are grouped by three underlying etiological categories.

Uterine environment	Embryo	Maternal condition
Uterine cavity abnormalities	Genetic abnormalities	Endometriosis
Endometrial polyps	Zona hardening	Immunological factors
Uterine myoma	Suboptimal culture effectors	Suboptimal ovarian stimulation
Thin endometrium	Altered expression of adhesive molecules	Thrombophilia
Inadequate endometrial change		Lifestyle
Hydrosalpinges		
Cesarean scar syndrome		
Chronic endometritis		

Source: *Modified from Margalioth (2006, p. 17).*

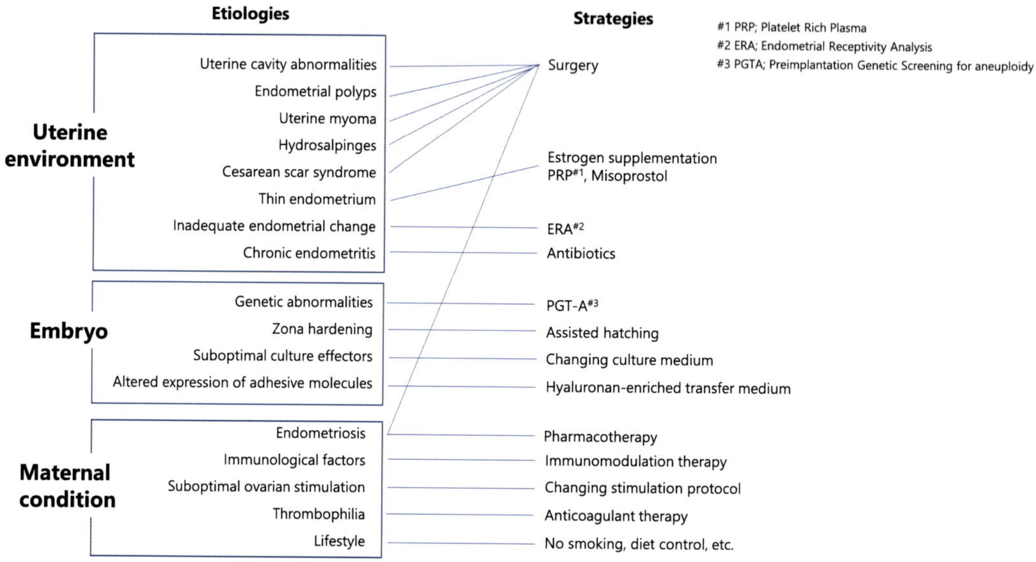

FIGURE 1 Various strategies for dealing with underlying etiologies of repeated implantation failures are listed.

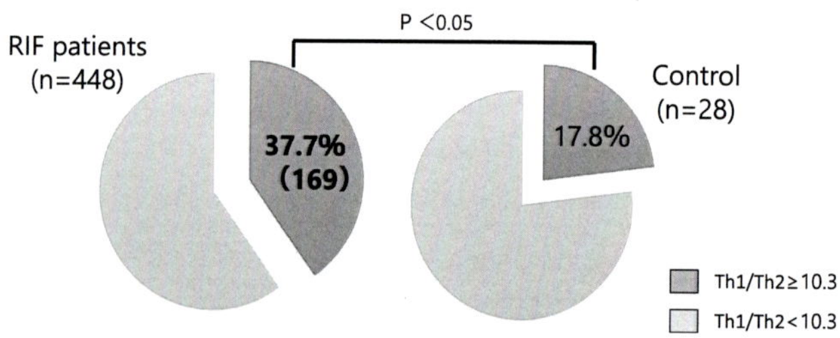

FIGURE 2 The prevalence of elevated Th1/Th2 cell ratios in women with repeated implantation failures and the control group is plotted. Around 37.7% of women with repeated implantation failure showed elevated Th1/Th2 cell ratios, which was significantly higher than that in the control group.

According to the cut-off value of INF-ϒ/IL-4 producing Th cell ratios [20], the prevalence of the women showing a Th1/Th2 cell ratio 10.3 or above among the RIF group was 37.7%, and this was significantly higher than that in the control group, which includes 28 women who had a history of normal pregnancy (Fig. 2) [25]. By contrast, cellular immune abnormalities (increased level of TNF-α/IL-10 producing Th1/Th2 cell ratio) were present in 111 of 189 recurrent pregnancy loss (RPL) women (58.7%) [26]. Therefore, about 38%–59% of the women with RIF or RPL showed Th cell immune abnormalities.

4 The theoretical understanding for the application of immunomodulation therapy in recurrent implantation failure

The successful implantation has been interpreted as a process whereby hatched blastocysts attach to the surface of the uterine endometrium and then invade into the decidualized endometrium. While the blastocyst is invading, vascularization and remodeling of the spiral arteries occur around the hatched blastocyst, establishing maternal–fetal circulation. Maternal immune effectors play a vital role in response to blastocyst invasion. Many studies have focused on the importance of a balanced cytokine environment, particularly the balance between type 1 and 2 cytokines produced by Th1 and Th2 cells. Th1 and Th2 cell ratios have helped evaluate the presence of either robust rejection of reduced tolerance. A shift in the ratio towards Th1 cells leads to an increase in proinflammatory cytokines such as TNF-α, INF-γ, and IL-2, all of which play essential roles in cytotoxic immune responses, including phagocytosis and inflammation. On the other hand, Th2 cells produce a variety of interleukins that activate the humoral immune responses and inhibit phagocyte functions, one of which is the antiinflammatory response [15]. Many observational studies have demonstrated theoretical mechanisms that explain why women with a history of RPL show an increased expression of proinflammatory cytokines [27].

NK cells are one of the immune cells responsible for innate immunity, and proinflammatory cytokines such as INF-γ are elevated when NK cell levels are increased in the peripheral blood. Inflammatory immune response has a significant with associated with several systemic inflammatory diseases, such as cardiovascular, degenerative, and neuroendocrine diseases and reproductive failures [28]. Increased NK cell levels and cytotoxicity, and increased Th 1 immune responses have been reported in women with RIF and RPL [29,30]. Therefore, the systemic and local inflammatory immune responses at the maternal–fetal interface are thought to play important roles in women with RIF. The inflammatory immune response can induce thrombosis in the microenvironment, leading to implantation failure of the hatched blastocyst.

Opinions differ on the local balance between pro- and antiinflammation, induced by several cytokines in the receptive uterine endometrium and decreasing antiinflammatory cytokines at the maternal–fetal interface has been dismissed as a simple-minded idea [31–33]. The cytokine concentrations of uterine-cavity fluid taken before ET in 210 women showed a positive relationship between IL-10 and TNF-α around the periods of both implantation and early pregnancy. By contrast, a negative relationship was observed between the secretions of IL-1β and monocyte chemoattractant protein 1 (MCP-1) levels during the same period [34].

5 Evaluation

The women who experienced three or more failures of implantation after either fresh or thawed ETs using cleavage-stage embryos or blastocysts are recommended to receive immunological testing, such as Th1/Th2 cell ratios. Peripheral blood Th1/Th2 cell ratios can be measured by flowcytometric analysis. The specific staining of lymphocytes was performed by incubating whole blood with anti-CD4–PC5- or anti-CD8–PC5-conjugated monoclonal antibodies (mAbs). After surface staining, specific intracellular staining using FastImmune™ IFN-γ-FITC/IL-4-PE was subsequently performed. Th1 cells were defined as CD4 + lymphocytes with intracellular IFN-γ but without IL-4. Th2 cells were detected as CD4 + lymphocytes with intracellular IL-4 but without IFN-γ. The ratio of IFN-γ to IL-4 positive Th cells was expressed as the Th1/Th2 ratio [20].

Several reports have indicated a relationship between an increase in peripheral blood NK cells and the Th1/Th2 cell ratio [15,35–37] among women showing RIF or with RPL. These studies motivated us to measure peripheral NK cells, but the relationship between the peripheral blood NK cells and uterine NK (uNK) cells seemed insignificant [33,38]. Contrarily, Tuckerman and coworkers found that RIF patients show a higher count of uNK cells in their endometrium [39]. The evaluation of NK cells from the peripheral blood or uterine endometrium has become helpful in searching for the underlying etiologies for RIF or RPL in the maternal site. These laboratory investigations aid clinicians in supporting RIF or RPL patients to select proper candidates for specific immunomodulation treatments or drugs. In women with RIF, immunomodulation treatments have been given to enhance or suppress immune responses, resulting in suppression of rejection or strengthening tolerance and consequently enhancing the uterine endometrium to accept a blastocyst without rejection.

6 Immunomodulation treatments targeting T helper cell pathology

Immunomodulation treatments with glucocorticoids, tacrolimus, intravenous immunoglobulin G, anti-TNF-α, and intralipid infusion have been utilized to improve the reproductive outcomes, including pregnancy rates for RIF patients, particularly when immunological disorders are present (Table 2) [40–43]. In this review, glucocorticoid and tacrolimus treatment, are discussed.

6.1 Glucocorticoids

Corticosteroids are potent antiinflammatory drugs that suppress both immune cell numbers and activity when used during implantation and early pregnancy. Corticosteroids are usually in the form of either glucocorticoids or mineralocorticoids. Corticosteroids are frequently prescribed for RIF treatment due to their rapid action and safety during reproduction and early pregnancy. Prednisolone or prednisone (glucocorticoid) is commonly used for RIF treatment rather than dexamethasone (mineralocorticoid) because dexamethasone shows approximately 10 times stronger vasoconstrictive effect than prednisolone or prednisone, and this strong effect might interfere with implantation.

6.1.1 Indication

Glucocorticoids have been proposed as a therapeutic option for women undergoing assisted reproduction technology (ART) treatment who show positive autoantibodies associated with uteroplacental thrombosis and vasoconstriction [44,45]. The RIF women showing overactive NK cells are also good candidates for glucocorticoids treatment. The administration of glucocorticoids has been proposed to normalize NK cell activity and cytokine production and suppress the overproduction of inflammatory mediators to improve endometrial receptivity [46]. Glucocorticoids have been used as a treatment for rheumatoid arthritis, which is often associated with elevated Th1/Th2 cell ratios. The literature indicated that glucocorticoids could inhibit the production of IFN-γ, TNF-α, and

TABLE 2 Characteristics of various immunomodulation treatments for abnormal cellular immune responses in women with repeated implantation failures and recurrent pregnancy losses.

	IVIG[a]	Anti-TNF-α	Intralipid	Tacrolimus
Route of administration	Intravenous	Intravenous	Intravenous	Oral
Dose	400–500 mg/kg/day	40 mg every 2 weeks	10% 250 mL/day	1–4 mg/day
Start	2–4 weeks prior to FET cycle or the day of ET	90–120 days before ET	After oocyte retrieval	2 days before ET day
After a positive PT[b]	Continued, every 3–4 weeks	-	Continued, every 2 weeks	Continued
Monitor	Peripheral NK cell count, NK cell cytotoxicity, Th1/Th2 ratio	Th1/Th2 ratio	Peripheral NK cell count	Th1/Th2 ratio, Th1 level
Concern	The majority of side effects are transient and mild [40] Rare side effects include – Renal impairment – Thrombosis – Arrhythmia – Aseptic meningitis – Hemolytic anemia – Transfusion-related acute lung injury	The most common side effects are injection site reaction [41] Severe side effects are rare, including – Lymphomas – Serious infections – Reactivation of latent tuberculosis – Congestive heart failure	More frequently encountered side effects [42] are due to – Contamination of the IVF catheter, which results in sepsis – Vein irritation and thrombophlebitis – Less frequent side effects are more directly related to Intralipid. – Dyspnea – Cyanosis – Allergic reaction (Soybean allergy) – Hyperlipemia – Hypercoagulability – delayed adverse reactions; hepatomegaly, jaundice, splenomegaly – Thrombocytopenia – Leukopenia – Increase in liver enzyme – Overloading syndrome	Most side effects are not severe [43] Severe side effects are – Infection – Decreased serum Mg – Increased serum potassium – Anemia – Neutropenia – A toxic effect on the CNS – Pleurisy – Decreased renal function – Seizure – Insomnia – Diabetic changes

[a]IVIG: *intravenous immunoglobulin.*
[b]PT; *pregnancy test.*

other Th1 cells but upregulate the production of IL-4, IL-10, and IL-13 by Th2 cells [47]. Therefore, glucocorticoids can act antiinflammatory effect and moderate Th1/Th2 cell ratios. The RIF women showing elevated Th1/Th2 cell ratios are also good candidates for glucocorticoids treatment.

6.1.2 Dosing

The administration is relatively simple since it is an oral medication. The cost is reasonable compared to other immunomodulation drugs. However, the timing and duration of this drug are very important due to its potent antiinflammatory effect. For the treatment of RIF women, prednisolone should be taken regularly and persistently. According to previous reports [48], between 15 and 60 mg of prednisolone is used as the treatment for women with autoantibodies. In a study by Han et al., women with unexplained infertility and autoantibodies elevated NK cells assays or Th1/Th2 ratios were treated with prednisone 10 mg/day starting 48 hours after ovulation of natural cycle or 2 weeks before the ART cycles, in conjunction with IVIg treatment if necessary. Prednisone treatment was adjusted based on Th1/Th2 ratio, NK cell levels, or cytotoxicity and gradually tapered off started 12–14 weeks gestation [49].

6.1.3 Side effect

Glucocorticoids can cause immunosuppression, decreasing function and the number of lymphocytes, including both T and B cells and other immunocompetent cells. Therefore, susceptibility to infection can be increased with prolonged use. Moreover, glucocorticoids can induce several systemic changes such as hyperglycemia, osteoporotic changes, and peptic ulcers.

Since glucocorticoids are potent antiinflammatory drugs, there may be implantation disturbance for women without elevated Th cell immunity when glucocorticoids are used in the peri-implantation periods. Hence, the selection of patients with immune abnormalities, such as the presence of autoimmunity, increased NK cell number and cytotoxicity, or increased Th1 or Th1/Th2 cell ratios for the prednisone treatment is of utmost importance for successful reproductive outcomes. Additionally, the regular monitoring of immune and other pertinent parameters while on corticosteroid treatment is critical to determine adequate dosing and avoid the possible side effects of the medication.

6.2 Tacrolimus

6.2.1 Action mechanism

Recent advances in new immunosuppressive drugs have resulted in higher graft survival rates for organ transplantation [50]. Tacrolimus, an immunosuppressant, was developed in Japan and has become a standard regimen for allogenic organ transplantation. It works by suppressing the recipient's immune reaction and effectively reduces the rejection of transferred organs [51].

Tacrolimus is a calcineurin inhibitor, and the working mechanism of action is inhibiting the differentiation of the lymphocytes that react to alloantigens, diminishing cytotoxic T cell proliferation, and promoting IL-2 receptor development and the production of IL-2 and IFN-γ from T cell-derived soluble mediators [52]. The effectiveness of tacrolimus is well known for patients with organ transplantation and other diseases of immune abnormalities, such as rheumatoid arthritis, which is induced by T cell activation [53,54].

6.2.2 Indication

Patients who have experienced RIF with elevated Th1/Th2 levels may benefit from tacrolimus treatment [20]. We recently published two prospective cohort studies using tacrolimus for patients with RPL and RIF [25,55,56]. For these studies, a total of 124 RIF women showing increased Th1/Th2 ratios (≥10.3) received immunomodulation treatment by tacrolimus. The patients began tacrolimus 1 or 2 days before the ET day and kept receiving it until the pregnancy test. Out of 124 patients, 58 got pregnant (pregnancy rate, 46.8%) [55]. Initially, between 1 and 3 mg of tacrolimus was used daily depending on the Th1/Th2 cell ratio, but the patients who received the 1 mg dose showed significantly lower clinical pregnancy rates compared with those receiving 3 mg [57]. Our recent study demonstrated that Th1 cell percentage alone predicted the ART results, particularly the ongoing pregnancy rate. The patients with a high Th1 cell percentage (>28.8%) received an additional 1 mg dosage of tacrolimus (2–4 mg/day), even though they showed a moderate Th1/Th2 ratio [56]. According to the adjustment of the tacrolimus dosage, the clinical pregnancy rate in a high Th1 cell parentage was improved (from 21.4% to 44.4%) [57]. Therefore, the daily dosage of tacrolimus may be determined by the Th1 cell levels only for the treatment of RIF patients.

The concentration of tacrolimus was used to determine the optimal dose in organ transplantation cases. Contrary to the expectations that the effectiveness of tacrolimus was concentration-dependent, the initial results for the group receiving 2 mg daily were similar to those of the group receiving a 3 mg daily. It is suspected the low dose tacrolimus treatment (1–3 mg/day) may not be as effective as higher doses (≥4 mg/day) to prevent RIF. In transplantation cases, the optimal concentration if tacrolimus was reported to be 5–15 ng/mL/day. In our study, ≥4 mg/day dosing reached over 5 ng/mL concentration when considering +2 SD ranges (Fig. 3). Treatment with tacrolimus has been quite effective for patients with RIF

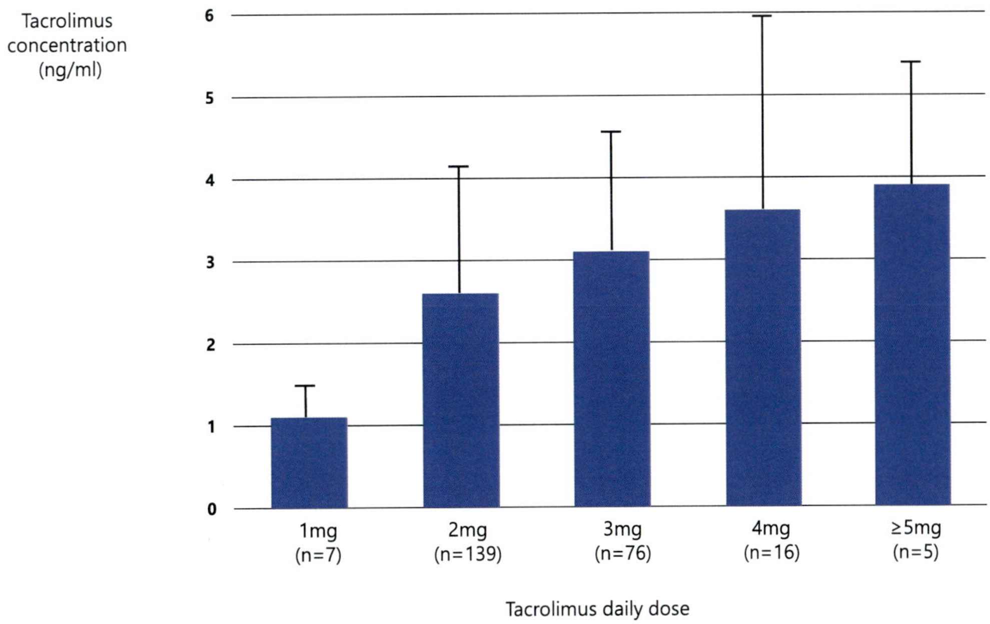

FIGURE 3 The serum concentrations of each tacrolimus daily dose (ng/mL) are plotted. The serum concentrations of tacrolimus in women with daily dosage of 1, 2, 3, 4, and ≥5 mg of tacrolimus are daily dosage of 1, 2, 3, 4, and ≥5 mg are 1.1 ± 0.4, 2.6 ± 1.6, 3.1 ± 1.5, 3.6 ± 2.3, and 3.9 ± 1.4 ng/mL, respectively (mean ± SD).

showing elevated Th1/Th2 cell ratios. Moreover, this drug is also effective for RPL patients [58]. RCT trials are expected to further establish the effectiveness of tacrolimus for RIF patients, which should increase the confidence in this treatment.

7 Reproductive outcome

The effect of tacrolimus treatment on the RIF women with elevated Th1/Th2 cell ratios has been demonstrated. Our report indicated that between 40% and 50% of the RIF women with elevated Th1/Th2 cell ratios could get pregnant with tacrolimus administration, and those RIF women without tacrolimus treatment have little chance to achieve pregnancy [25,57]. Few reports evaluate the reproductive outcome of the glucocorticoids treatment between women with RIF showing abnormal Th cell immunity and the control. The randomized-controlled trial investigating the clinical effects of prednisone treatment for RIF patients is ongoing now [59], expecting more evidence for the effect of prednisone on pregnancy outcomes in patients with RIF.

8 Summary and recommendations

Th cells are indispensable for the protection of self from the invasion of pathogens or foreign cells, but the inhibition of immunological rejection is also important for both the establishment and the maintenance of a successful pregnancy. Women showing higher Th1 cell immunity have an increased risk of RIF. Dysregulated Th cell immune responses, such as increased Th1 cell and Th1/Th2 cell ratios, are associated with RIF. Immunomodulation treatments targeting these inflammatory T cells have been reported to improve the reproductive outcomes of RIF patients.

Glucocorticoids are frequently prescribed for RIF treatment due to their rapid action and safety during reproduction and early pregnancy. Tacrolimus is another effective immunomodulation drug, which can inhibit the production of IL-2 and IFN-γ produced from T cell-derived soluble mediators. Since an adequate inflammatory reaction is needed during the implantation period, these medications should be indicated for RIF women with unbalanced of Th1/Th2 cell ratios or increased Th1 immunity and should not be given empirically to RIF patients.

- A significant proportion (37.7%) of women with RIF have unbalanced Th1/Th2 cell ratios.
- Glucocorticoid treatment can be considered in women with RIF and the evidence of autoimmune and cellular immune abnormalities.
- Tacrolimus, a calcineurin inhibitor, has been reported to be effective in RIF
- Of cellular immune abnormalities, particularly in Th cell pathologies.
- Immunosuppression treatment for RIF should not be given empirically.
- Serial monitoring of immune and other pertinent parameters is needed for adequate dosing and prevention of possible side effects while the RIF patients are on immunosuppressive treatment.

References

[1] McAdam AJ, Schweitzer AN, Sharpe AH. The role of B7 co-stimulation in activation and differentiation of CD4 + and CD8 + T cells. Immunol Rev 1988;165:231–47.

[2] Linterman MA, Denton AE, Divekar DP, et al. CD28 expression is required after T cell priming for helper T cell responses and protective immunity to infection. ELife. 2014. Available from: https://doi.org/10.7554/eLife.03180.

[3] Colf LA, Bankovich AJ, Hanick NA, et al. How a single T cell receptor recognizes both self and foreign. Cell. 2007;129:135–46.

[4] Borbulevych OY, Piepenbrick KH, Gloor BE, et al. T cell receptor cross-reactivity directed by antigen-dependent tuning of peptide-MHC molecular flexibility. Immunity 2009;31:885–96.

[5] Borst J, Ahrends T, Bąbała N. CD4 + T cell help in cancer immunology and immunotherapy. Nat Rev Immunol 2018;18:635–47.

[6] Ahrends T, Brost J. The opposing roles of CD4 + T cells in anti-tumor immunity. Immunology 2018;154:582–92.

[7] Loke YW, King A. Human implantation: Cell Biology and Immunology. Cambridge, UK: Cambridge University Press; 1995.

[8] Kovates S, Main EK, Librach C. A class I antigen, HLA-G, expressed human trophoblasts. Science 1990;248:220–3.

[9] Hiby SE, King A, Sharkey A. Molecular studies of trophoblast HLA-G: polymorphism, isoforms, imprinting and expression in preimplantation embryo. Tissue Antigens 1999;53:1–13.

[10] King A, Burrows TD, Hiby SE, et al. Surface expression of HLA-C antigen by human extravillous trophoblast. Placenta 2000;21:376–87.

[11] Kahn DA, Baltimore D. Pregnancy induces a fetal antigen-specific maternal T regulatory cell response that contributes to tolerance. Proc Natl Acad Sci USA 2000;18:9299–304.

[12] Triburgs T, Scherjon SA, van der Mast BJ, et al. Fetal-maternal HLA-C mismatch is associated with decidual T cell activation and induction of functional T regulatory cells. J Reprod Immunol 2009;82:148–57.

[13] Shakwat A, Shaikly V, Elzatma E. Interaction between HLA-G and monocyte/macrophage in human pregnancy. J Reprod Immunol 2010;85:40–6.

[14] Maldonado RA, von Andrian UH. How tolegenic dendtric cells induce regulatory T cell. Adv Immunol 2010;108:111–65.

[15] Kwak-Kim JYH, Chung-Bang HS, Ng SC, et al. Increased T helper 1 cytokine responses by circulating T cells are present in women with recurrent pregnancy losses and in infertile women with multiple implantation failures after IVF. Hum Reprod 2003;18:767–73.

[16] Coughlan C, Ledger W, Wang Q, et al. Recurrent implantation failure: definition and management. Reprod Biomed Online 2014;28:14–38.

[17] Margalioth EJ, Ben-Chetrit A, Gal M. Investigation and treatment of repeated implantation failure following IVF-ET. Hum Reprod 2006;21:3036–46.

[18] Hviid MM, Macklon N. Immune modulation treatments—where is the evidence? Fertil Steril 2017;107:1284–93.

[19] Chaouat G, Ledee-Bataille N. Th1/Th2 paradigm in pregnancy: paradigm last? Int Arch Immunol. 2004;134:93–119.

[20] Saito S, Nakashima A. Th1/Th2/Th17 and regulatory T-cell paradigm in pregnancy. Am J Reprod Immunol 2010;63:601–10.

[21] Ng SC, Gilman-Sachs A, Thakar P. Expression of intracellular Th1 and Th2 cytokines in women with recurrent spontaneous abortion, implantation failures after IVF-ET or normal pregnancy. Am J Reprod Immunol 2002;48:77–86.

[22] Riley JK. Trophoblast immune receptors in maternal-fetal tolerance. Immunol Invest 2008;37:395–426.

[23] Mosmann J, Moore K. The role of IL-10 in crossregulation ofTH1 and TH2 response. Immunol Today 1991;12:A49.

[24] Winger EE, Reed JL, Ashoush S. Elevated preconception CD56 + 16 + and/or Th1:Th2 levels predict benefit from IVIG therapy in subfertile women undergoing IVF. Am J Reprod Immunol 2011;66:394–403.

[25] Nakagawa K, Kwak-Kim J, Ota K. Immunosuppression with Tacrolimus improved reproductive outcome of women with repeated implantation failure and elevated peripheral blood Th1/Th2 cell ratios. Am J Reprod Immunol 2015;73:353–61.

[26] Lee SK, Kim JY, Han AR, et al. Intravenous immunoglobulin improves pregnancy outcome in women with recurrent pregnancy losses with cellular immune abnormalities. Am J Reprod Immunol 2016;75:59–68.
[27] Raghupathy R, Makhseed M, Azizieh F. Maternal Th1- and Th2-type reactivity to placental antigens in normal human pregnancy and unexplained recurrent spontaneous abortions. Cell Immunol 1999;196:122–30.
[28] Kawk-Kim J, Bao S, Lee SK. Immunological modes of pregnancy loss: inflammation, immune effectors, and stress. Am J Reprod Immunol 2014;72:129–40.
[29] Kwak-Kim JY, Beaman KD, Gilman-Sachs A. Up-regulated expression of CD56 +, CD56 + /CD16 +, and CD19 + cells in peripheral blood lymphocytes in pregnant women with recurrent pregnancy losses. Am J Reprod Immunol 1995;34:93–9.
[30] Lee SK, Na BJ, Kim JY, et al. Determination of clinical cellular immune markers in women with recurrent pregnancy loss. Am J Reprod Immunol 2013;70:398–411.
[31] Chaouat G, Zourbas S, et al. New insights into maternal-fetal interactions at implantation. Reprod Biomed Online 2001;2:198–203.
[32] Mekinian A, Cohen J, Alijotas-Reig J, et al. Unexplained recurrent miscarriage and recurrent implantation failure: is there a place for immunomodulation? Am J Reprod Immunol 2016;76:8–28.
[33] Robertson SA, Jin M, Yu D, et al. Corticosteroid therapy in assisted reproduction -immune suppression is a faulty premise. Hum Reprod 2016;31:2164–73.
[34] Boomsma CM, Kavelaars A, Eijkemans MJC, et al. Endometrial secretion analysis identifies a cytokine profile predictive of pregnancy in IVF. Hum Reprod 2009;24:1427–35.
[35] Matsubayashi H, Hosaka T, Sugiyama Y, et al. Increased natural killer–cell activity is associated with infertile women. Am J Reprod Immunol 2001;46:318–22.
[36] Emmer PM, Nelen WL, Steegers EA. Peripheral natural killer cytotoxicity and CD56(pos)CD16(pos) cells increase during early pregnancy in women with a history of recurrent spontaneous abortion. Hum Reprod 2000;15:1163–9.
[37] King K, Smith S, Chapman M. Detailed analysis of peripheral blood natural killer (NK) cells in women with recurrent miscarriage. Hum Reprod 2010;25:52–8.
[38] Moffett A, Shreeve N. First do no harm: uterine natural killer (NK) cells in assisted reproduction. Hum Reprod 2015;30:1519–25.
[39] Tuckerman E, Mariee N, Prakash A. Uterine natural killer cells in peri-implantation endometrium from women with repeated implantation failure after IVF. J Reprod Immunol 2010;87:60–6.
[40] Guo Y, Tian X, Wang X. Adverse effects of immunoglobulin therapy. Front Immunol 2018;9:1299. Available from: https://doi.org/10.3389/fimmu.2018.01299.
[41] Scheinfeld N. Adalimumab: a review of side effects. Expert Opin Drug Saf 2005;4:637–41. Available from: https://doi.org/10.1517/14740338.4.4.637. PMID: 16011443.
[42] https://www.drugs.com/pro/intralipid.html.
[43] Kim YH, Shin HY, Kim SM. Long-term safety and efficacy of tacrolimus in *Myasthenia gravis*. Yonsei Med J 2019;60:633–9. Available from: https://doi.org/10.3349/ymj.2019.60.7.633.
[44] Geva E, Amit A, Lerner-Geva L, et al. Prednisone and aspirin improve pregnancy rate in patients with reproductive failure and autoimmune antibodies: a prospective study. Am J Reprod Imunol. 2000;43:36–40.
[45] Hasegawa I, Yamamoto Y, Suzuki M, et al. Prednisolone plus low-dose aspirin improves the implantation rate in women with autoimmune conditions who are undergoing in vitro fertilization. Fertil Steril 1998;70:1044–8.
[46] Boomsma CM, Keay SD, Macklon NS. Peri-implantation glucocorticoid administration for assisted reproductive technology cycles. Cochrane Database Syst Rev 2012;Cd005996.
[47] Elenkov IJ. Glucocorticoids and the Th1/Th2 balance. Ann N Y Acad Sci 2004;1024:138–46.
[48] Taniguchi F. Results of prednisolone given to improve the outcome of in vitro fertilization-embryo transfer in women with antinuclear antibodies. J Reprod Med 2005;50:383–8.
[49] Han AR, Ahn H, Vu P, et al. Obstetrical outcome of anti-inflammatory and anticoagulation therapy in women with recurrent pregnancy loss or unexplained infertility. Am J Reprod Immunol 2012;68:418–27.
[50] Goring SM, Lew AR, Ghement I. A network meta-analysis of the efficacy of belatacept, cyclosporine and tacrolimus for immunosuppression therapy in adult renal transplant recipients. Curr Ned Res Opin. 2014;30:1473–87.

[51] Uchida K. Long-term Prograf multicenter retrospective study in kidney transplantation: seven-year follow-up. Transplant Now. 2006;19:380–9.

[52] Kino T, Hatanaka H, Miyata S. FK-506, a novel immunosuppressant isolated from a *Streptomyces*. II. Immunosuppressive effect of FK-506 in vitro. J Antibiot. 1987;40:1256–65.

[53] Ram R, Gafter-Gvili A, Yeshuru M. Prophylaxis regimens for GVHD: systematic review and meta-analysis. Bone Marrow Transplant 2009;43:643–53.

[54] Ramiro S, Gaujoux–Viala C, Nam JL, et al. Safety of synthetic and biological DMARDs: a systematic literature review informing the 2013 update of the EULAR recommendations for management of rheumatoid arthritis. Ann Rheum Dis 2014;73:529–35.

[55] Nakagawa K, Kwak-Kim J, Kuroda K. Immunosuppressive treatment using tacrolimus promotespregnancy outcome in infertile women with repeated implantation failures. Am J Reprod Immunol 2017. Available from: https://doi.org/10.1111/aji.12682.

[56] Nakagawa K, Kwak-Kim J, Hisano M, et al. Obstetrical and perinatal outcome of the women with repeated implantation failures or recurrent pregnancy losses who received pre- and post- conception tacrolimus treatment. Am J Reprod Immunol 2019;82:e13142.

[57] Nakagawa K, Sugiyama R. Immunomodulating treatment for the patients with repeated implantation failures caused by immunological rejection. In: Kuroda K, Brosens J, Quenby S, Takeda S, editors. Treatment Strategy for Unexplained Infertility and Recurrent Miscarriage. Singapore: Springer; 2018. p. 460.

[58] Nakagawa K, Kuroda K, Sugiyama R. After 12 consecutive miscarriages, a patient received immunosuppressive treatment and delivered an infant. Reprod Med Biol. 2017. Available from: https://doi.org/10.1002/rmb2.12040.

[59] Wei D, Chen ZJ, Sun Y. Prednisone for patients with recurrent implantation failure: study protocol for a double-blind, multicenter, randomized, placebo-controlled trial. Trials 2020;21:719.

CHAPTER

19

B-cell pathology and repeated implantation failures

Shihua Bao, Mengyang Du and Xiao Wang

Department of Reproductive Immunology, Shanghai First Maternity and Infant Hospital, Tongji University School of Medicine, Shanghai, P.R. China

1 Introduction

Repeated implantation failure (RIF) is a highly frustrating condition for both the affected couples and clinicians. Its treatment constitutes one of the most difficult challenges in the field of in vitro fertilization (IVF) and embryo transfer (ET). Advanced maternal age, smoking status of both parents, elevated body mass index, stress, poor-quality gametes (particularly in older women), thrombophilia, uterine factors (i.e., congenital uterine anomalies, endometrial polyps, submucosal fibroids, intrauterine adhesions), and adnexal pathologies (i.e., hydrosalpinx) have been reported to be possible causes of RIF [1–3]. However, in the great majority of cases, the etiology remains unknown.

Pioneering studies show that the immune system plays a crucial role in successful implantation, and disruption of this system leads to implantation failures. For a successful pregnancy, the maternal immune system is required to perform paradoxical functions. It must be immunologically tolerant to a growing semiallogeneic fetus while maintaining the capacity to recognize and respond to foreign antigens. Immune profile changes necessary to support a healthy gestation have been well documented in innate and adaptive immune cells, including suppression of T effector cells specific for fetal antigens, generation of regulatory T cells (Tregs), and proportional and functional changes of natural killer (NK) cells, macrophage, monocyte, and dendritic cells (DCs) [4,5].

However, historically, B cells have not received equal attention like other primary immune cells in adaptive immunity concerning their role in pregnancy. Early studies primarily explored the traditional function of B cells, antibody production and only hinted that B cells might contribute to pregnancy in multifaceted ways. However, studies of other diseases have expanded the understanding of B-cell function. It is now widely accepted that B cells exhibit functional promiscuity, from operating as effectors of adaptive

Immunology of Recurrent Pregnancy Loss and Implantation Failure
DOI: https://doi.org/10.1016/B978-0-323-90805-4.00022-5

immunity to performing innate-like suppressive functions [6,7]. The knowledge of B cell-mediated immune mechanisms is critical to designing innovative therapies for immune-mediated pregnancy pathologies since many studies have repeatedly shown that immune dysfunction contributes to pregnancy disorders, such as recurrent pregnancy loss (RPL), preeclampsia, and immune-based infertility [8–11].

While knowledge of the nature and mechanism of action remain limited for regulatory B cells (Breg) in the context of pregnancy, their importance has been revealed by seminal studies using the early abortion mouse model, demonstrating that the Breg cells could restore maternal immune tolerance and support healthy pregnancies [12–15]. Furthermore, clinical studies have established a correlation between Breg impairment and poor pregnancy outcomes such as RIF and RPL [10,16–18], Thus far, there is mounting evidence suggesting a pivotal role of Breg within the pregnancy immune network, which can potentially be harnessed into a form of immunotherapy to treat pregnancy pathologies. Current immunotherapies addressing unexplained RPL, such as lymphocyte immunotherapy, trophoblast membrane immunization, and intravenous immunoglobulin G (IVIgG) infusion treatment, have improved obstetric outcomes [19]. However, the development of effective targeted immunotherapy that may further improve pregnancy outcomes is an unexplored avenue of research. In this chapter, we review the latest research exploring B-cell pathology to underline RIF. Furthermore, we highlight the potential function of B-cell subsets, which can be utilized as an immunotherapeutic agent for averting or treating pregnancy pathologies and complications.

2 B-cell immunopathology and reproduction

2.1 B-cell subsets

B cells have been considered immune effectors, which are solely responsible for adaptive immunity. Recently, the role of B cell subpopulations has been acknowledged as well for making a bridge between innate and adaptive immunity. Proper orchestration of B1 and B2 cells plays a vital role in maintaining protective immunity and preventing autoimmune disorders [20]. Hence, studying the role of B-cell subpopulations during pregnancy deserves proper consideration.

B-cell subpopulations in murine have been extensively studied, which helped to differentiate B1 and B2 subpopulations in terms of functionality and ontogeny [6]. Unlike mice, human B-cell subpopulations based on their phenotypic expression is still inconclusive and a matter of continuous debate [21–25]. Despite having controversies, $CD20^+CD27^+CD43^+CD38^{lo/int}$ cells have been defined as human B-1 cell subpopulations recently [26,27]. In mice fetuses, B1 cells develop from hematopoietic stem cells (HSCs) in the fetal liver and divide into B-1a or B-1b according to CD5 expression. However, the expression of CD5 in human B-1 cells is not distinct. Human B-1 cells also express CD20, CD27, and CD43 [28]. Unlike mice, B1 and B2 in humans develop from $Lin^-CD34^+CD38^{lo}$ stem cell population. In mice, B1 cells are predominant in the peritoneal cavity and found in the pleural cavities, spleen, and the intestine [29], whereas in humans, they are primarily found in the peripheral circulation [30].

B2 cells are differentiated from their bone marrow (BM) precursors since postnatal life, producing cells expressing surface immunoglobulin M (sIgM) [31]. These cells leave the BM and migrate to the spleen. In the spleen, they mature into the marginal zone or follicular B cells [32]. Somatic hypermutation in B cells and its Ig-class switching are achieved through exposure to antigens and signaling from T helper (Th) cells [33]. In mice, T-cell-independent somatic hypermutation outside the germinal center in the absence of any immune response has also been reported [34–36]. Furthermore, Toll-like receptor-mediated stimulation of human and mouse B cells can induce T-cell-independent activation and Ig-class switching [37]. The activated B cells then differentiate into plasma and memory cells [33].

2.2 Antibody-producing B cells in pregnancy

Traditionally, B-cell function is primarily associated with humoral immunity, and as such, B cells were thought to play a role in pregnancy by virtue of antibody production. Experiments conducted during the 1970s elegantly showed that the IgG fraction of maternal serum prevented maternal effector T cells from delivering a functional cytotoxic response against allogeneic trophoblast tissue [38]. Moreover, activation of spontaneous humoral responsiveness during pregnancy as demonstrated by an increase in the levels of both IgG- and IgM-secreting cells in the uterine draining lymph nodes was reported early on as one of the possible suppressive mechanisms in place during pregnancy [39]. More recently, murine and clinical studies provided supporting evidence, featuring significantly enhanced IgM and IgA serum levels in pregnant individuals compared to nonpregnant controls [14,18,40]. An elegant study expounding the human memory B-cell phenotype during pregnancy has likewise shown that the expression level of memory marker CD27 was markedly increased during gestation, which directly impacted the profile of immunoglobulin isotypes being produced. In that study, upregulated CD27 expression was associated with increased IgG and IgA isotype production [41]. Lastly, splenic B cells from a murine pregnant model with poor gestational outcomes have been reported to produce a significantly higher titer of IgG isotypes as well as IgA and IgM relative to titers obtained from mice with good gestational outcomes [42]. Thus far, activating the B-cell humoral response leading to antibody production is necessary to achieve a healthy pregnancy.

2.3 Nonprecipitating asymmetric antibodies

The 1970s was the period during which a subset of IgG called "asymmetric antibodies" was discovered. Asymmetric antibodies are IgG possessing a mannose-rich oligosaccharide residue bound to one Fab region, thus rendering them incapable of activating immune effector functions such as complement fixation, phagocytosis, and cytotoxicity [43]. In the context of pregnancy, serum from pregnant women contained higher titers of asymmetric antibodies compared to nonpregnant women, mostly with paternal antigen specificity [44,45]. Conversely, women experiencing recurrent spontaneous abortions had significantly lower proportions of asymmetric antibodies in serum than healthy pregnant women [45,46]. Therefore asymmetric antibodies were suggested to have a protective role during pregnancy and referred to as protective maternal antibodies; however, the mechanisms underpinning this response have not been explored further in either subsequent preclinical or clinical studies.

2.4 Autoantibodies

Contrary to the protective antibodies during pregnancy, the presence of autoantibodies has been implicated in pregnancy failure. The majority of these studies have investigated the impact of antiphospholipid (aPL) antibodies and agonistic autoantibodies against the angiotensin II type I receptor [47–49]. High levels of aPL have been implicated in the etiology of aPL syndrome, an autoimmune disease characterized by a hypercoagulable state that induces systemic blood clotting (thrombosis). In pregnancy, aPL syndrome leads to compromised reproductive outcomes such as miscarriages, fetal loss, and severe preeclampsia [47,50,51]. β2 glycoprotein I is the cofactor for anticardiolipin antibodies. Studies have shown that 30% of patients with RIF tested positive for criteria or noncriteria aPL antibodies, such as antiphosphatidylinositol (aPI), antiphosphatidylethanolamine (aPE), and antiphosphatidylserine (aPS) antibodies, in comparison with fertile controls who have a 16% prevalence of these antibodies ($p = 0\,0.019$) [52,53]. Due to the presence of specific antibodies implicated in both APS and RIF, it is important to consider if RIF should be added to the clinical criteria for aPL syndrome [54].

In addition, angiotensin II type I receptor autoantibodies have a significant role in the possible initiation and immunopathology of preeclampsia. These agonistic autoantibodies mimic the natural ligand of the angiotensin type I receptor and induce downstream production of antiangiogenic factors, such as s-Flt1 and Endoglin, which lead to the onset of preeclamptic symptoms, including hypertension, proteinuria, glomerular endotheliosis, resulting in placental and embryonic abnormalities [55,56].

2.5 Breg cells in pregnancy

In recent years, a subgroup of B cells, the Breg, have emerged as a potential key player in immunological tolerance. Breg, named due to their capacity to regulate other immune cell types negatively, have been extensively studied in the fields of autoimmunity, transplantation, immune tolerance, and cancer, with the consensus that despite their low numbers, Bregs can potently modulate immune responses [7]. Breg suppress the immune response by inhibiting the differentiation of DCs, suppressing the proliferation of Th1 and Th17 cells, and inducing the differentiation of forkhead transcription factor P3 (FoxP3) positive regulatory T cells (FoxP3$^+$ Treg) (Fig. 1) [57]. Clear phenotypes and characteristic markers of Breg are not yet defined. The molecular architecture and the differential function of B-cell phenotypes are heavily influenced by the external milieu and the presence of foreign antigens [6]. Therefore, in the dynamic scenario of pregnancy, where the expression of paternal antigens increases over time, it is reasonable to expect corresponding changes in the B-cell architecture to support adaptation of the maternal immune response to pregnancy [58].

B-cell subsets that have been recognized as Breg in murine pregnancy include IL-10$^+$, CD5$^+$CD1d$^+$, CD80$^+$CD86$^+$, CD80$^+$CD86$^+$CD27$^+$IL-10$^+$, and IL-35$^+$ B cells and PIBF1$^+$ choriodecidual B cells. Breg subsets identified in human pregnancy are IL-10$^+$, and CD24^{hi}CD27$^+$ B cells, CD24^{hi}CD38^{hi} transitional B cells, CD27$^+$IgM$^+$ memory B cells, CD38^{hi}CD27^{hi} plasmablasts, marginal zone B cells, and IL-35$^+$ B cells. In both mice and humans, the capacity for IL-10 and a lesser extent, TGF-β production remains a hallmark for Breg identification. However, additional markers such as IL-35 and PIBF1 have been reported to instill a regulatory capacity in B cells during pregnancy [15,59,60]. The inherent difficulty of precise identification

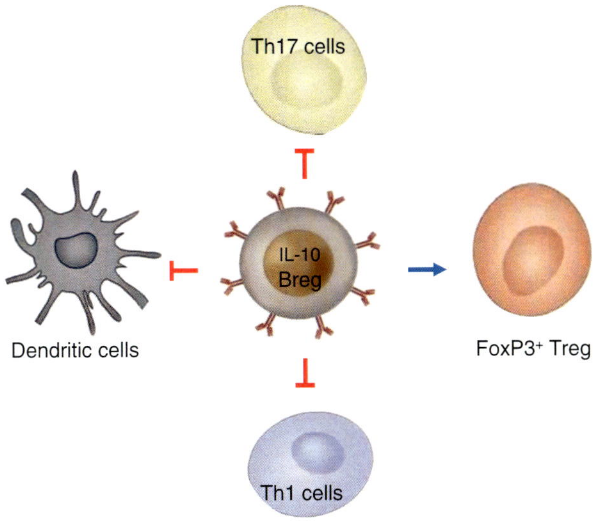

FIGURE 1 B regulatory (Breg) cells suppress the proliferation of T helper (Th)1 and Th17 cells, inhibit the differentiation of dendritic cells, and induce the differentiation of forkhead transcription factor P3 (FoxP3) positive regulatory T (Treg) cells.

of the Breg by cytokine production has led to the surrogate markers including $CD5^{+}CD1d^{+}$ for IL-10-producing B cells (B10 cells) in mice, and $CD24^{hi}CD27^{+}$ for the human equivalent [61,62].

The regulatory role of Breg during pregnancy has yet to be fully explored. However, in studies performed to date, several recurring themes are apparent; (1) reduced levels and/or dysfunction of circulating the Breg may indicate poor obstetric outcomes, (2) Breg may be involved in fostering the appropriate conditions necessary for implantation, and (3) the Breg can restore fetal tolerance in conditions afflicted with immune-mediated pregnancy complications.

Pioneering studies in mice have definitively shown that reduced frequencies of Breg cells during early pregnancy are associated with spontaneous abortion [13,17]. Repopulation of Breg cells by passive transfer of splenic $CD5^{+}CD1d^{hi}$ B10 cells rescued pregnancies from failure by inhibiting DC maturation and expanding Treg cell populations [13]. Similarly, transfer of B10 cells, but not IL-10-negative B effector cells or B cells from IL-10-deficient mice, prevented fetal loss induced by LPS-driven inflammation by reducing IL-17A and IL-6 production from T cells and expanding the Treg population [12]. In humans, IL-10-producing $CD19^{+}CD24^{hi}CD27^{+}$ B cells in peripheral blood are significantly higher in women undergoing normal pregnancies as opposed to nonpregnant women or women who suffer spontaneous abortions [17,60]. Furthermore, $CD19^{+}CD24^{hi}CD27^{+}$ Breg isolated from peripheral blood taken from women in their first trimester of pregnancy could successfully inhibit TNF secretion by activated T effector cells ex vivo [17] (Fig. 2).

Recent studies in mice have demonstrated the participation of Breg during the peri-implantation period of pregnancy. Breg-associated markers TLR9 and CD86 are upregulated on splenic B cells 3.5 days postcopulation, while significant expansion of uterine B cells and alterations in the B-cell phenotype were observed from 2.5 to 8.5 days postcopulation [63,64]. Moreover, these uterine B cells, enriched with $CD80^{+}$, $CD86^{+}$, and $IL\text{-}10^{+}$ B-cell subsets, effectively suppress the proliferation and activation of syngeneic $CD4^{+}$ effector T cells [64].

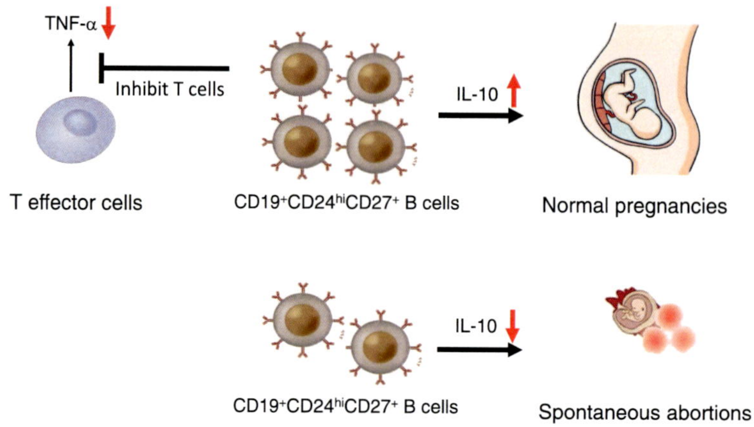

FIGURE 2 The proportion of IL-10-producing $CD19^+CD24^{hi}CD27^+$ B cells in peripheral blood is significantly higher in normal pregnancy than in spontaneous abortion, and $CD19^+CD24^{hi}CD27^+$ B cells isolated from normal women's pregnancy could inhibit tumor necrosis factor secretion by activated T effector cells ex vivo.

The modification of B-cell phenotypes presumably happening in the earliest phase of pregnancy may well be a result of trophoblast education [61]. Lastly, impairment of IL-10 production by B cells has been reported as one of the key pathological mechanisms in RIF, since peripheral blood B cells from women who have experienced RIF expressed lower IL-10 mRNA levels and secreted significantly lower levels of IL-10 upon ex vivo restimulation, relative to healthy controls [16]. These studies suggest a beneficial role of Breg for appropriate embryo implantation during early pregnancy. Although limited in number, these studies highlight the critical regulatory role of Breg in pregnancy, which parallels with that observed in autoimmune disease, graft tolerance, and cancer. In all conditions, Breg aid in the control of the tissue environment via induction and maintenance of Treg, modification of the Th cell response, and inhibition of effector cell responses by cytotoxic T, NK, and DCs.

3 Dysregulation of B cells in women with repeated implantation failure

Investigation of B cells and their subsets have been reported in women with reproductive failures. Peripheral blood $CD19^+$ B cells in women who experienced miscarriage and RIF were significantly lower than those who subsequently had a live birth [65,66]. In the subset of women with multiple IVF failures, women with poor ovarian responses (POR) and low vitamin D (VD) levels, had significantly higher $CD19^+$ B and $CD19^+/5^+$ B1 cell levels than those of normal ovarian response and low VD and POR and normal VD ($p < 0.05$, respectively) [67]. Additionally, in a case report of a woman with a history of RPL and infertility, increased peripheral blood $CD19^+$ B cells were associated with persistently decreased levels of memory B-cell subsets [10]. The measurement of peripheral $CD19^+$ B-cell subsets may help predict the pregnancy outcome in women with RIF.

In women undergoing the IVF cycle, the number of mature oocytes and the blastocyst formation rate were significantly higher in women with high VD levels. In women with low VD levels undertaking progestin-primed ovarian stimulation protocol, peripheral blood B and NK cell proportions and Th/Tc cell ratios were significantly higher than women with high VD levels ($p < 0.05$, respectively). In women undertaking agonist protocol, again, peripheral blood B-cell proportions and Th/Tc cell ratios were higher in women with low VD compared to those with high VD ($p < 0.05$, respectively) [68]. Low VD levels were reported to induce excessive B-cell responses in multiple sclerosis patients by enriching antigen-experienced class-switched memory B cells and antibody-secreting plasma cells [69]. Hence, VD deficiency on B cells may exacerbate the maternal proinflammatory milieu in the growing follicles and uterine lining.

Increased numbers of circulating $CD5^+$ B cells correlate with high levels of aPL in patients with aPL syndrome and participate in altered immunity of women with RPL. This distinct subset of innate-like B cells is characterized by the production of multispecific antibodies that could interact with oxidized phospholipid epitopes [70]. Interestingly, in both human and murine studies, B cell-directed therapies appear to have clinical and serological benefits, including decreased aPL titers [71,72]. Therefore B-cell immunopathology involved in aPL may have much broader clinical and immunological manifestations than what we have recognized so far [73]. Persistent IgM aPL without isotype switch may be a rare variant form of aPL manifestation associated with dysregulated B-cell subsets [66].

4 Clinical translational perspectives of B-cell immunopathology

4.1 Prednisone

Prednisolone is frequently used in inflammatory and autoimmune diseases due to its antiinflammatory and immunosuppressive properties. Several pharmacological mechanisms have been explained the efficacy of prednisolone in patients with RIF and RPL. The excessive number of uNK cells increased NK cell cytotoxicity, and dominant Th1 immunity are associated with RIF and pregnancy losses. Prednisolone administration was reported to significantly reduce uNK cells (from 14% to 9%). In addition, inhibiting Th1 cytokine secretion, increasing the number of regulatory T cells, and inhibiting NK cells cytotoxicity are other mechanisms through which prednisolone modulates the immune system [74–76]. In a nonrandomized clinical trial, increased pregnancy and live birth rates have been reported in women with recurrent IVF failures after administration of combined low molecular weight heparin and prednisolone [77]. A prospective, multicenter, randomized, double-blind, placebo-controlled, clinical trial (1:1 ratio of prednisone vs placebo) is ongoing in China (ChiCTR1800018783) to determine if prednisone can enhance live births as the primary outcome in infertile patients with RIF [78].

4.2 Hydroxychloroquine

Hydroxychloroquine, an antimalarial drug, is widely used in patients with various autoimmune diseases, particularly systemic lupus erythematosus (SLE). Besides antithrombotic effects, hydroxychloroquine has immunoregulatory properties and can block the production of proinflammatory cytokines [79]. Few in vitro studies showed a potential benefit of hydroxychloroquine in early trophoblastic migration and implantation in aPL syndrome. Hydroxychloroquine restored trophoblastic fusion and differentiation decreased by aPL antibodies and decreased TLR4 expression [80]. Hydroxychloroquine reduced the aPL antibody binding on syncytiotrophoblasts and restored the annexin A5 expression [81]. Physiopathological studies are lacking to demonstrate the effect of hydroxychloroquine in patients with unexplained RPL and RIF.

Currently, limited clinical data are demonstrating the efficacy of hydroxychloroquine in patients with unexplained RPL and RIF. A small retrospective study (30 patients with 35 pregnancies) of refractory obstetrical aPL syndrome showed that the outcome of pregnancies treated by the addition of hydroxychloroquine compared to previous pregnancies under the conventional treatment, pregnancy losses decreased from 81% to 19% ($p < 0.05$) [82]. Several patients with chronic intervillositis had successful pregnancy outcomes after the addition of hydroxychloroquine [83]. The immunomodulatory effects of hydroxychloroquine could be interesting to further investigate in patients with unexplained RPL and RIF presumed to be related to the immunological disturbances.

4.3 Intravenous immunoglobulin

Studies have explained the mechanisms of IVIG to improve the pregnancy, including (1) inhibition of B-cell autoantibody production, (2) decreasing balance of Th1/Th2 cytokines, (3) decreasing percentages of peripheral blood NK cells, (4) suppression of NK cell cytotoxicity, and (5) increasing T suppressor activity in women with RPL and RIF [84–86]. Heilmann et al. (2010) [87] in a retrospective study on 188 patients with RIF (three or more RIF) who were treated with 0.2 g/kg immunoglobulin G, seven days before ET, showed that implantation and live birth rates were 21.0% and 42.0%, respectively in the study group. Whereas, in the control group, implantation rate and live birth rate were 9.3% and 16.1%. A study in 2017 also reported an increased Foxp3 mRNA, which is expressed by regulatory T cells, and decreased Th17 cell frequencies in peripheral blood following IVIG therapy in women with RIF [88]. A systematic review also reported that IVIG in IVF/ICSI cycles is beneficial in these patients [89]. The use of IVIG was significantly associated with a higher implantation rate, and RR was 2.708 (95%CI: 1.302–5.629) compared with the placebo. The clinical pregnancy rate and the live birth rate were significantly increased in patients randomized to IVIG; RR was 1.475 (95%CI: 1.191–1.825) for the clinical pregnancy rate, and RR was 1.616 (95%CI: 1.243–2.101) for the live birth rate. In general, the recent *meta*-analysis supports the IVIG treatment; however, due to the significant heterogeneity of the study population, the paucity of recommending IVIG treatment is present [86].

4.4 Rituximab

Rituximab is a chimeric murine-human monoclonal antibody against the CD20 antigen expressed on pre-B, immature, and mature B cells. It has been used to treat diseases characterized by excessive numbers of B cells, overactive B cells, or dysfunctional B cells, including several autoimmune diseases, B-cell malignancies, and transplant rejection [90]. When rituximab was introduced to normal pregnant rats, B-cell depletion and reduced uterine perfusion pressure (RUPP) were noticed [91]. RUPP rats treated with rituximab exhibited less blood pressure increases in response to induced placental ischemia [92]. This data provides potential insight into human adverse pregnancy outcomes, leading to a new therapeutic approach for pregnancy failures. As such, a recent case report has described a successful pregnancy after rituximab treatment in a patient with a history of IVF failures (two times) and positive anticardiolipin antibody (ACA). Following a course of rituximab, the patient's ACA became negative, and she successfully conceived with IVF treatment [93]. Rituximab destroys B cells that have CD20 on their surfaces, both normal and malignant B cells. However, CD20 is not expressed on stem cells and most plasma cells. Thus B-cell regeneration is possible, and immunoglobulin synthesis by plasma cells is not affected after treatment with anti-CD20 antibodies [94,95].

4.5 Breg-based therapy

From a clinical perspective, the role of Breg as a vital contributor to clinical immune homeostasis has been acknowledged and appreciated, especially in the field of autoimmunity, where it was first demonstrated that pan-B-cell depletion by rituximab had limited effectiveness due to the concomitant depletion of beneficial Bregs [96]. However, as the Breg field remains relatively unexplored, innovative clinical translational studies have been few and far between. Clinical studies involving the Breg over the past 10 years have focused primarily on determining Breg's potential as a prognostic or diagnostic marker of disease [58]. Moreover, these studies were limited to autoimmune diseases such as chronic immune thrombocytopenia, inflammatory rheumatic diseases, and SLE (clinicaltrials.gov). Intellectual property-wise, a wide-ranging patent covering IL-10-producing B cells as tools for manipulating autoimmune or inflammatory diseases and conditions and treatment for infectious disease and/or cancer was granted in 2018 [97]. Also patented is a method of generating IL-10-producing B cells [97], but other than that, no significant clinical applications for a Breg-based therapy used as an intervention for any kind of condition have commenced.

5 Summary and recommendations

In conclusion, B cells play an immunoregulatory role in the pregnancy process in addition to producing antibodies. As it stands, research on the Breg and its role in healthy pregnancies remains in its infancy. Although progress has been achieved, many questions remain unanswered, particularly surrounding the interactions that this population of immune cells has with other more well-studied populations and how they orchestrate the induction and maintain maternal immune tolerance together. Furthermore, as many pregnancy pathologies have their basis in suboptimal implantation, which is a critical phase in gestation and highly immune-regulated, insufficient knowledge of the implantation events becomes an impediment to translational clinical developments. We predict, however, that further discoveries of the fundamental immune mechanisms behind maternal tolerance, combined with the rapid and exciting advances in cell-based immunotherapies in the cancer domain, can be harnessed, and innovative immunotherapeutic interventions can be designed and applied to many pregnancy pathologies. Certainly, the Bregs have the desirable attributes to be the basis of such a therapeutic intervention and one that it is currently pursuing.

- Abnormalities in B cells and their subsets, such as B1 and B2 cells, are associated with reproductive failures.
- Activation of the B-cell humoral response leading to antibody production is a necessary step toward achieving a healthy pregnancy.
- The presence of autoantibodies is implicated in pregnancy failure.
- aPL syndrome leads to compromised reproductive outcomes such as miscarriages, fetal loss, and severe preeclampsia.
- The measurement of peripheral CD19 + B-cell subsets may help to predict the pregnancy outcome in women with RIF.
- Reduced levels and/or dysfunction of circulating the Breg may predict reproductive failures and poor obstetric outcomes.
- Breg cells may be involved in fostering the appropriate conditions necessary for implantation.
- Bregs can restore fetal tolerance in conditions afflicted with immune-mediated pregnancy complications.
- The effect of prednisolone and hydroxychloroquine in patients with RPL and RIF is promising and needs to be further evaluated.
- The use of IVIG in IVF/ICSI cycles is a beneficial treatment in patients with RIF.

References

[1] Somigliana E, Vigano P, Busnelli A, Paffoni A, Vegetti W, Vercellini P. Repeated implantation failure at the crossroad between statistics, clinics and over-diagnosis. Reprod Biomed Online 2018;36(1):32–8. Available from: https://doi.org/10.1016/j.rbmo.2017.09.012 (In eng).

[2] Cakiroglu Y, Tiras B. Determining diagnostic criteria and cause of recurrent implantation failure. Curr Opin Obstet Gynecol 2020;32(3):198–204. Available from: https://doi.org/10.1097/gco.0000000000000620 (In eng).

[3] Sheikhansari G, Pourmoghadam Z, Danaii S, Mehdizadeh A, Yousefi M. Etiology and management of recurrent implantation failure: a focus on intra-uterine PBMC-therapy for RIF. J Reprod Immunol 2020;139:103121. Available from: https://doi.org/10.1016/j.jri.2020.103121 (In eng).

[4] Arck PC, Hecher K. Fetomaternal immune cross-talk and its consequences for maternal and offspring's health. Nat Med 2013;19(5):548–56. Available from: https://doi.org/10.1038/nm.3160 (In eng).

[5] Mor G, Cardenas I. The immune system in pregnancy: a unique complexity. Am J Reprod Immunol 2010;63 (6):425–33. Available from: https://doi.org/10.1111/j.1600-0897.2010.00836.x (In eng).

[6] LeBien TW, Tedder TF. B lymphocytes: how they develop and function. Blood 2008;112(5):1570–80. Available from: https://doi.org/10.1182/blood-2008-02-078071 (In eng).

[7] Boothby MR, Hodges E, Thomas JW. Molecular regulation of peripheral B cells and their progeny in immunity. Genes Dev 2019;33(1-2):26–48. Available from: https://doi.org/10.1101/gad.320192.118 (In eng).

[8] Ghaebi M, Nouri M, Ghasemzadeh A, et al. Immune regulatory network in successful pregnancy and reproductive failures. Biomed Pharmacother 2017;88:61–73. Available from: https://doi.org/10.1016/j.biopha.2017.01.016 (In eng).

[9] Marron K, Walsh D, Harrity C. Detailed endometrial immune assessment of both normal and adverse reproductive outcome populations. J Assist Reprod Genet 2019;36(2):199–210. Available from: https://doi.org/10.1007/s10815-018-1300-8 (In eng).

[10] Sung N, Byeon HJ, Garcia MDS, et al. Deficiency in memory B cell compartment in a patient with infertility and recurrent pregnancy losses. J Reprod Immunol 2016;118:70–5. Available from: https://doi.org/10.1016/j.jri.2016.09.003 (In eng).
[11] Yang F, Zheng Q, Jin L. Dynamic function and composition changes of immune cells during normal and pathological pregnancy at the maternal-fetal interface. Front Immunol 2019;10:2317. Available from: https://doi.org/10.3389/fimmu.2019.02317 (In eng).
[12] Busse M, Campe KJ, Nowak D, et al. IL-10 producing B cells rescue mouse fetuses from inflammation-driven fetal death and are able to modulate T cell immune responses. Sci Rep 2019;9(1):9335. Available from: https://doi.org/10.1038/s41598-019-45860-2 (In eng).
[13] Jensen F, Muzzio D, Soldati R, Fest S, Zenclussen AC. Regulatory B10 cells restore pregnancy tolerance in a mouse model. Biol Reprod 2013;89(4):90. Available from: https://doi.org/10.1095/biolreprod.113.110791 (In eng).
[14] Muzzio DO, Ziegler KB, Ehrhardt J, Zygmunt M, Jensen F. Marginal zone B cells emerge as a critical component of pregnancy well-being. Reproduction 2016;151(1):29–37. Available from: https://doi.org/10.1530/rep-15-0274 (In eng).
[15] Slawek A, Lorek D, Kedzierska AE, Chelmonska-Soyta A. Regulatory B cells with IL-35 and IL-10 expression in a normal and abortion-prone murine pregnancy model. Am J Reprod Immunol (New York, NY: 1989) 2020;83(3):e13217. Available from: https://doi.org/10.1111/aji.13217 (In eng).
[16] Koushaeian L, Ghorbani F, Ahmadi M, et al. The role of IL-10-producing B cells in repeated implantation failure patients with cellular immune abnormalities. Immunol Lett 2019;214:16–22. Available from: https://doi.org/10.1016/j.imlet.2019.08.002 (In eng).
[17] Rolle L, Memarzadeh Tehran M, Morell-García A, et al. Cutting edge: IL-10-producing regulatory B cells in early human pregnancy. Am J Reprod Immunol 2013;70(6):448–53. Available from: https://doi.org/10.1111/aji.12157 (In eng).
[18] Ziegler KB, Muzzio DO, Matzner F, et al. Human pregnancy is accompanied by modifications in B cell development and immunoglobulin profile. J Reprod Immunol 2018;129:40–7. Available from: https://doi.org/10.1016/j.jri.2018.07.003 (In eng).
[19] Wong LF, Porter TF, Scott JR. Immunotherapy for recurrent miscarriage. Cochrane Database Syst Rev 2014;2014(10):Cd000112. Available from: https://doi.org/10.1002/14651858.CD000112.pub3 (In eng).
[20] Quách TD, Hopkins TJ, Holodick NE, et al. Human B-1 and B-2 B cells develop from Lin-CD34 + CD38lo stem cells. J Immunol 2016;197(10):3950–8. Available from: https://doi.org/10.4049/jimmunol.1600630 (In eng).
[21] Perez-Andres M, Grosserichter-Wagener C, Teodosio C, van Dongen JJ, Orfao A, van Zelm MC. The nature of circulating CD27 + CD43 + B cells. J Exp Med 2011;208(13):2565–6. Available from: https://doi.org/10.1084/jem.20112203 (In eng).
[22] Descatoire M, Weill JC, Reynaud CA, Weller S. A human equivalent of mouse B-1 cells? J Exp Med 2011;208(13):2563–4. Available from: https://doi.org/10.1084/jem.20112232 (In eng).
[23] Griffin DO, Holodick NE, Rothstein TL. Human B1 cells are CD3-: a reply to "a human equivalent of mouse B-1 cells?" and "the nature of circulating CD27 + CD43 + B cells.". J Exp Med 2011;208(13):2566–9. Available from: https://doi.org/10.1084/jem.20111761 (In eng).
[24] Reynaud CA, Weill JC. Gene profiling of $CD11b^+$ and $CD11b^-$ B1 cell subsets reveals potential cell sorting artifacts. J Exp Med 2012;209(3):433–4. Available from: https://doi.org/10.1084/jem.20120402 (In eng).
[25] Rothstein TL, Griffin DO, Holodick NE, Quach TD, Kaku H. Human B-1 cells take the stage. Ann N Y Acad Sci 2013;1285:97–114. Available from: https://doi.org/10.1111/nyas.12137 (In eng).
[26] Quách TD, Rodríguez-Zhurbenko N, Hopkins TJ, et al. Distinctions among circulating antibody-secreting cell populations, including B-1 cells, in human adult peripheral blood. J Immunol 2016;196(3):1060–9. Available from: https://doi.org/10.4049/jimmunol.1501843 (In eng).
[27] Engelbertsen D, Vallejo J, Quách TD, et al. Low levels of IgM antibodies against an advanced glycation endproduct-modified apolipoprotein B100 peptide predict cardiovascular events in nondiabetic subjects. J Immunol 2015;195(7):3020–5. Available from: https://doi.org/10.4049/jimmunol.1402869 (In eng).
[28] Griffin DO, Holodick NE, Rothstein TL. Human B1 cells in umbilical cord and adult peripheral blood express the novel phenotype CD20 + CD27 + CD43 + CD70. J Exp Med 2011;208(1):67–80. Available from: https://doi.org/10.1084/jem.20101499 (In eng).

[29] Montecino-Rodriguez E, Dorshkind K. New perspectives in B-1 B cell development and function. Trends Immunol 2006;27(9):428–33. Available from: https://doi.org/10.1016/j.it.2006.07.005 (In eng).
[30] Griffin DO, Rothstein TL. A small CD11b(+) human B1 cell subpopulation stimulates T cells and is expanded in lupus. J Exp Med 2011;208(13):2591–8. Available from: https://doi.org/10.1084/jem.20110978 (In eng).
[31] Fettke F, Schumacher A, Costa SD, Zenclussen AC. B cells: the old new players in reproductive immunology. Front Immunol 2014;5:285. Available from: https://doi.org/10.3389/fimmu.2014.00285 (In eng).
[32] Monroe JG, Dorshkind K. Fate decisions regulating bone marrow and peripheral B lymphocyte development. Adv Immunol 2007;95:1–50. Available from: https://doi.org/10.1016/s0065-2776(07)95001-4 (In eng).
[33] Pillai S, Cariappa A, Moran ST. Marginal zone B cells. Annu Rev Immunol 2005;23:161–96. Available from: https://doi.org/10.1146/annurev.immunol.23.021704.115728 (In eng).
[34] William J, Euler C, Christensen S, Shlomchik MJ. Evolution of autoantibody responses via somatic hypermutation outside of germinal centers. Science 2002;297(5589):2066–70. Available from: https://doi.org/10.1126/science.1073924 (In eng).
[35] Weller S, Braun MC, Tan BK, et al. Human blood IgM "memory" B cells are circulating splenic marginal zone B cells harboring a prediversified immunoglobulin repertoire. Blood 2004;104(12):3647–54. Available from: https://doi.org/10.1182/blood-2004-01-0346 (In eng).
[36] Toellner KM, Jenkinson WE, Taylor DR, et al. Low-level hypermutation in T cell-independent germinal centers compared with high mutation rates associated with T cell-dependent germinal centers. J Exp Med 2002;195(3):383–9. Available from: https://doi.org/10.1084/jem.20011112 (In eng).
[37] He B, Qiao X, Cerutti A. CpG DNA induces IgG class switch DNA recombination by activating human B cells through an innate pathway that requires TLR9 and cooperates with IL-10. J Immunol 2004;173(7):4479–91. Available from: https://doi.org/10.4049/jimmunol.173.7.4479 (In eng).
[38] Taylor PV, Hancock KW. Antigenicity of trophoblast and possible antigen-masking effects during pregnancy. Immunology 1975;28(5):973–82 (In eng).
[39] Carter J, Dresser DW. Pregnancy induces an increase in the number of immunoglobulin-secreting cells. Immunology 1983;49(3):481–90 (In eng).
[40] Muzzio DO, Soldati R, Ehrhardt J, et al. B cell development undergoes profound modifications and adaptations during pregnancy in mice. Biol Reprod 2014;91(5):115. Available from: https://doi.org/10.1095/biolreprod.114.122366 (In eng).
[41] Grimsholm O, Piano Mortari E, Davydov AN, et al. The interplay between CD27(dull) and CD27(bright) B cells ensures the flexibility, stability, and resilience of human B cell memory. Cell Rep 2020;30(9):2963–77. Available from: https://doi.org/10.1016/j.celrep.2020.02.022 e6. (In eng).
[42] Fröhlich C, Ehrhardt J, Krüger D, Trojnarska D, Zygmunt M, Muzzio DO. Pregnancy status alters IL-21-mediated effects on murine B lymphocytes. Reproduction 2020;159(3):351–9. Available from: https://doi.org/10.1530/rep-19-0407 (In eng).
[43] Margni RA, Paz CB, Cordal ME. Immunochemical behavior of sheep non-precipitating antibodies isolated by immunoadsorption. Immunochemistry 1976;13(3):209–14. Available from: https://doi.org/10.1016/0019-2791(76)90217-2 (In eng).
[44] Malan Borel I, Gentile T, Angelucci J, et al. IgG asymmetric molecules with antipaternal activity isolated from sera and placenta of pregnant human. J Reprod Immunol 1991;20(2):129–40. Available from: https://doi.org/10.1016/0165-0378(91)90029-p (In eng).
[45] Zenclussen AC, Gentile T, Kortebani G, Mazzolli A, Margni R. Asymmetric antibodies and pregnancy. Am J Reprod Immunol 2001;45(5):289–94. Available from: https://doi.org/10.1111/j.8755-8920.2001.450504.x (In eng).
[46] Barrientos G, Fuchs D, Schröcksnadel K, et al. Low levels of serum asymmetric antibodies as a marker of threatened pregnancy. J Reprod Immunol 2009;79(2):201–10. Available from: https://doi.org/10.1016/j.jri.2008.11.002 (In eng).
[47] Di Prima FA, Valenti O, Hyseni E, et al. Antiphospholipid syndrome during pregnancy: the state of the art. J Prenatal Med 2011;5(2):41–53 (In eng).
[48] Herse F, LaMarca B. Angiotensin II type 1 receptor autoantibody (AT1-AA)-mediated pregnancy hypertension. Am J Reprod Immunol 2013;69(4):413–18. Available from: https://doi.org/10.1111/aji.12072 (In eng).
[49] Salmon JE, Girardi G. Antiphospholipid antibodies and pregnancy loss: a disorder of inflammation. J Reprod Immunol 2008;77(1):51–6. Available from: https://doi.org/10.1016/j.jri.2007.02.007 (In eng).

[50] Heilmann L, Schorsch M, Hahn T, Fareed J. Antiphospholipid syndrome and preeclampsia. Semin Thromb Hemost 2011;37(2):141–5. Available from: https://doi.org/10.1055/s-0030-1270341 (In eng).
[51] Vinatier D, Dufour P, Cosson M, Houpeau JL. Antiphospholipid syndrome and recurrent miscarriages. Eur J Obstet Gynecol Reprod Biol 2001;96(1):37–50. Available from: https://doi.org/10.1016/s0301-2115(00)00404-8 (In eng).
[52] Stern C, Chamley L, Hale L, Kloss M, Speirs A, Baker HW. Antibodies to beta2 glycoprotein I are associated with in vitro fertilization implantation failure as well as recurrent miscarriage: results of a prevalence study. Fertil Steril 1998;70(5):938–44. Available from: https://doi.org/10.1016/s0015-0282(98)00312-4 (In eng).
[53] Kwak-Kim J, Skariah A, Wu L, Salazar D, Sung N, Ota K. Humoral and cellular autoimmunity in women with recurrent pregnancy losses and repeated implantation failures: a possible role of vitamin D. Autoimmun Rev 2016;15(10):943–7. Available from: https://doi.org/10.1016/j.autrev.2016.07.015 (In eng).
[54] Bashiri A, Halper KI, Orvieto R. Recurrent implantation failure-update overview on etiology, diagnosis, treatment and future directions. Reprod Biol Endocrinol 2018;16(1):121. Available from: https://doi.org/10.1186/s12958-018-0414-2 (In eng).
[55] Robinson CJ, Johnson DD, Chang EY, Armstrong DM, Wang W. Evaluation of placenta growth factor and soluble Fms-like tyrosine kinase 1 receptor levels in mild and severe preeclampsia. Am J Obstet Gynecol 2006;195(1):255–9. Available from: https://doi.org/10.1016/j.ajog.2005.12.049.
[56] Zhou CC, Zhang Y, Irani RA, et al. Angiotensin receptor agonistic autoantibodies induce preeclampsia in pregnant mice. Nat Med 2008;14(8):855–62. Available from: https://doi.org/10.1038/nm.1856 (In eng).
[57] Cai X, Zhang L, Wei W. Regulatory B cells in inflammatory diseases and tumor. Int Immunopharmacol 2019;67:281–6. Available from: https://doi.org/10.1016/j.intimp.2018.12.007 (In eng).
[58] Guzman-Genuino RM, Hayball JD, Diener KR. Regulatory B cells: dark horse in pregnancy immunotherapy? J Mol Biol 2021;433(1):166596. Available from: https://doi.org/10.1016/j.jmb.2020.07.008 (In eng).
[59] Huang B, Faucette AN, Pawlitz MD, et al. Interleukin-33-induced expression of PIBF1 by decidual B cells protects against preterm labor. Nat Med 2017;23(1):128–35. Available from: https://doi.org/10.1038/nm.4244 (In eng).
[60] Liu J, Chen X, Hao S, et al. Human chorionic gonadotropin and IL-35 contribute to the maintenance of peripheral immune tolerance during pregnancy through mediating the generation of IL-10(+) or IL-35(+) Breg cells. Exp Cell Res 2019;383(2):111513. Available from: https://doi.org/10.1016/j.yexcr.2019.111513 (In eng).
[61] Iwata Y, Matsushita T, Horikawa M, et al. Characterization of a rare IL-10-competent B-cell subset in humans that parallels mouse regulatory B10 cells. Blood 2011;117(2):530–41. Available from: https://doi.org/10.1182/blood-2010-07-294249 (In eng).
[62] Yanaba K, Bouaziz JD, Haas KM, Poe JC, Fujimoto M, Tedder TF. A regulatory B cell subset with a unique CD1dhiCD5+ phenotype controls T cell-dependent inflammatory responses. Immunity 2008;28(5):639–50. Available from: https://doi.org/10.1016/j.immuni.2008.03.017 (In eng).
[63] Lorek D, Kedzierska AE, Slawek A, Chelmonska-Soyta A. Expression of Toll-like receptors and costimulatory molecules in splenic B cells in a normal and abortion-prone murine pregnancy model. Am J Reprod Immunol (New York, NY: 1989) 2019;82(2):e13148. Available from: https://doi.org/10.1111/aji.13148 (In eng).
[64] Guzman-Genuino RM, Eldi P, Garcia-Valtanen P, Hayball JD, Diener KR. Uterine B cells exhibit regulatory properties during the peri-implantation stage of murine pregnancy. Front Immunol 2019;10:2899. Available from: https://doi.org/10.3389/fimmu.2019.02899 (In eng).
[65] Tu W, Li Y, Ding Q, et al. Association between peripheral CD19+ B cells and reproductive outcome in women with recurrent implantation failure. Clin Lab 2020;66(1). Available from: https://doi.org/10.7754/Clin.Lab.2019.190510 (In eng).
[66] Ota K, Dambaeva S, Lee J, Gilman-Sachs A, Beaman K, Kwak-Kim J. Persistent high levels of IgM antiphospholipid antibodies in a patient with recurrent pregnancy losses and rheumatoid arthritis. Am J Reprod Immunol 2014;71(3):286–92. Available from: https://doi.org/10.1111/aji.12196 (In eng).
[67] Wu L, Vendiola JA, Salazar Garcia MD, et al. Poor ovarian response is associated with serum vitamin D levels and pro-inflammatory immune responses in women undergoing in-vitro fertilization. J Reprod Immunol 2019;136:102617. Available from: https://doi.org/10.1016/j.jri.2019.102617 (In eng)(In eng).
[68] Wu L, Kwak-Kim J, Zhang R, et al. Vitamin D level affects IVF outcome partially mediated via Th/Tc cell ratio. Am J Reprod Immunol 2018;80(6):e13050. Available from: https://doi.org/10.1111/aji.13050 (In eng).

[69] Haas J, Schwarz A, Korporal-Kuhnke M, Faller S, Jarius S, Wildemann B. Hypovitaminosis D upscales B-cell immunoreactivity in multiple sclerosis. J Neuroimmunol 2016;294:18–26. Available from: https://doi.org/10.1016/j.jneuroim.2016.03.011 (In eng).

[70] Youinou P, Renaudineau Y. The antiphospholipid syndrome as a model for B cell-induced autoimmune diseases. Thromb Res 2004;114(5-6):363–9. Available from: https://doi.org/10.1016/j.thromres.2004.06.019 (In eng).

[71] Binder CJ, Silverman GJ. Natural antibodies and the autoimmunity of atherosclerosis. Springer Semin Immunopathol 2005;26(4):385–404. Available from: https://doi.org/10.1007/s00281-004-0185-z (In eng).

[72] Hörkkö S, Binder CJ, Shaw PX, et al. Immunological responses to oxidized LDL. Free Radic Biol Med 2000;28 (12):1771–9. Available from: https://doi.org/10.1016/s0891-5849(00)00333-6 (In eng).

[73] Khattri S, Zandman-Goddard G, Peeva E. B-cell directed therapies in antiphospholipid antibody syndrome—new directions based on murine and human data. Autoimmun Rev 2012;11(10):717–22. Available from: https://doi.org/10.1016/j.autrev.2011.12.011 (In eng).

[74] Quenby S, Kalumbi C, Bates M, Farquharson R, Vince G. Prednisolone reduces preconceptual endometrial natural killer cells in women with recurrent miscarriage. Fertil Steril 2005;84(4):980–4. Available from: https://doi.org/10.1016/j.fertnstert.2005.05.012 (In eng).

[75] Schiessl B, Innes BA, Bulmer JN, et al. Localization of angiogenic growth factors and their receptors in the human placental bed throughout normal human pregnancy. Placenta 2009;30(1):79–87. Available from: https://doi.org/10.1016/j.placenta.2008.10.004 (In eng).

[76] Huang Q, Wu H, Li M, Yang Y, Fu X. Prednisone improves pregnancy outcome in repeated implantation failure by enhance regulatory T cells bias. J Reprod Immunol 2021;143:103245. Available from: https://doi.org/10.1016/j.jri.2020.103245 (In eng).

[77] Siristatidis C, Chrelias C, Creatsa M, et al. Addition of prednisolone and heparin in patients with failed IVF/ICSI cycles: a preliminary report of a clinical trial. Hum Fertil (Camb) 2013;16(3):207–10. Available from: https://doi.org/10.3109/14647273.2013.803608 (In eng).

[78] Lu Y, Yan J, Liu J, et al. Prednisone for patients with recurrent implantation failure: study protocol for a double-blind, multicenter, randomized, placebo-controlled trial. Trials 2020;21(1):719. Available from: https://doi.org/10.1186/s13063-020-04630-6 (In eng).

[79] Mekinian A, Costedoat-Chalumeau N, Masseau A, et al. Obstetrical APS: is there a place for hydroxychloroquine to improve the pregnancy outcome? Autoimmun Rev 2015;14(1):23–9. Available from: https://doi.org/10.1016/j.autrev.2014.08.040 (In eng).

[80] Marchetti T, Ruffatti A, Wuillemin C, de Moerloose P, Cohen M. Hydroxychloroquine restores trophoblast fusion affected by antiphospholipid antibodies. J Thromb Haemost 2014;12(6):910–20. Available from: https://doi.org/10.1111/jth.12570 (In eng).

[81] Wu XX, Guller S, Rand JH. Hydroxychloroquine reduces binding of antiphospholipid antibodies to syncytiotrophoblasts and restores annexin A5 expression. Am J Obstet Gynecol 2011;205(6):576. Available from: https://doi.org/10.1016/j.ajog.2011.06.064 e7-14. (In eng).

[82] Mekinian A, Lazzaroni MG, Kuzenko A, et al. The efficacy of hydroxychloroquine for obstetrical outcome in anti-phospholipid syndrome: Data from a European multicenter retrospective study. Autoimmun Rev 2015;14(6):498–502. Available from: https://doi.org/10.1016/j.autrev.2015.01.012 (In eng).

[83] Mekinian A, Costedoat-Chalumeau N, Masseau A, et al. Chronic histiocytic intervillositis: outcome, associated diseases and treatment in a multicenter prospective study. Autoimmunity 2015;48(1):40–5. Available from: https://doi.org/10.3109/08916934.2014.939267 (In eng).

[84] Coulam CB, Acacio B. Does immunotherapy for treatment of reproductive failure enhance live births? Am J Reprod Immunol (New York, NY: 1989) 2012;67(4):296–304. Available from: https://doi.org/10.1111/j.1600-0897.2012.01111.x (In eng).

[85] Yamada H, Morikawa M, Furuta I, et al. Intravenous immunoglobulin treatment in women with recurrent abortions: increased cytokine levels and reduced Th1/Th2 lymphocyte ratio in peripheral blood. Am J Reprod Immunol 2003;49(2):84–9. Available from: https://doi.org/10.1034/j.1600-0897.2003.01184.x (In eng).

[86] Saab W, Seshadri S, Huang C, Alsubki L, Sung N, Kwak-Kim J. A systemic review of intravenous immunoglobulin G treatment in women with recurrent implantation failures and recurrent pregnancy losses. Am J Reprod Immunol 2021;85(4):e13395. Available from: https://doi.org/10.1111/aji.13395 (In eng).

[87] Heilmann L, Schorsch M, Hahn T. CD3-CD56 + CD16 + natural killer cells and improvement of pregnancy outcome in IVF/ICSI failure after additional IVIG-treatment. Am J Reprod Immunol 2010;63(3):263–5. Available from: https://doi.org/10.1111/j.1600-0897.2009.00790.x (In eng).

[88] Ahmadi M, Abdolmohammadi-Vahid S, Ghaebi M, et al. Regulatory T cells improve pregnancy rate in RIF patients after additional IVIG treatment. Syst Biol Reprod Med 2017;63(6):350 9. Available from: https://doi.org/10.1080/19396368.2017.1390007 (In eng).

[89] Li J, Chen Y, Liu C, Hu Y, Li L. Intravenous immunoglobulin treatment for repeated IVF/ICSI failure and unexplained infertility: a systematic review and a *meta*-analysis. Am J Reprod Immunol 2013;70(6):434–47. Available from: https://doi.org/10.1111/aji.12170 (In eng).

[90] Chambers SA, Isenberg D. Anti-B cell therapy (rituximab) in the treatment of autoimmune diseases. Lupus 2005;14(3):210–14. Available from: https://doi.org/10.1191/0961203305lu2138oa (In eng).

[91] Parrish MR, Murphy SR, Rutland S, et al. The effect of immune factors, tumor necrosis factor-alpha, and agonistic autoantibodies to the angiotensin II type I receptor on soluble fms-like tyrosine-1 and soluble endoglin production in response to hypertension during pregnancy. Am J Hypertens 2010;23(8):911–16. Available from: https://doi.org/10.1038/ajh.2010.70 (In eng).

[92] LaMarca B, Wallace K, Herse F, et al. Hypertension in response to placental ischemia during pregnancy: role of B lymphocytes. Hypertension 2011;57(4):865–71. Available from: https://doi.org/10.1161/hypertensionaha.110.167569 (In eng).

[93] Ng CT, O'Neil M, Walsh D, Walsh T, Veale DJ. Successful pregnancy after rituximab in a women with recurrent in vitro fertilisation failures and anti-phospholipid antibody positive. Ir J Med Sci 2009;178(4):531–3. Available from: https://doi.org/10.1007/s11845-008-0265-5 (In eng).

[94] Cianchini G, Corona R, Frezzolini A, Ruffelli M, Didona B, Puddu P. Treatment of severe pemphigus with rituximab: report of 12 cases and a review of the literature. Arch Dermatol 2007;143(8):1033–8. Available from: https://doi.org/10.1001/archderm.143.8.1033 (In eng).

[95] Fatourechi MM, el-Azhary RA, Gibson LE. Rituximab: applications in dermatology. Int J Dermatol 2006;45 (10):1143–55. Available from: https://doi.org/10.1111/j.1365-4632.2006.03007.x quiz 1155. (In eng).

[96] Jamin C, Morva A, Lemoine S, Daridon C, de Mendoza AR, Youinou P. Regulatory B lymphocytes in humans: a potential role in autoimmunity. Arthritis Rheum 2008;58(7):1900–6. Available from: https://doi.org/10.1002/art.23487 (In eng).

[97] Thomas F. Tedder, Ayumi Yoshizaki, Tomomitsu Miyagaki, Evgueni Kountikov, Jonathan C. Poe. Methods of expanding and assessing B cells and using expanded B cells to treat disease. U.S. Patent 10611999B2. 2018-6-5

CHAPTER

20

Endometrial pathology and repeated implantation failures

Maud Lansiaux[1], *Virginie Vaucoret*[1] *and Nathalie Lédée*[1,2]

[1]Hôpital des Bluets, Centre de PMA, Paris, France [2]MatriceLab Innove, Pépinière Paris-Santé- Cochin, Paris, France

1 Introduction

The human endometrium is a unique tissue in our species. It is a dynamic tissue that undergoes a continuous process of proliferation, differentiation, destruction, and renewal, menstrual cycle after menstrual cycle for up to 45 years. This complex transformation is the result of ovarian-derived estrogen and progesterone. Estrogen induces endometrial proliferation, whereas the postovulatory rise in progesterone levels triggers a coordinated program of differentiation, characterized by proliferative arrest and secretory transformation of the epithelial cells, an influx of uterine natural killer (uNK) cells, vascular remodeling, and differentiation of stromal fibroblasts into specialized decidual cells. A functional consequence of this coordinated endometrial remodeling is that it transiently becomes receptive to embryo implantation. This phenomenon is referred to as the "window of implantation (WOI)."

The WOI is probably one of the most crucial events in assisted reproductive technologies (ART). During the last three decades, many improvements have been introduced to make ART cycles more successful, including better embryo culture conditions and selection criteria, extending embryo culture to the blastocyst stage, and allowing the best quality embryos to transfer. However, despite expanding global experience with ART, the majority of in vitro fertilization (IVF) attempts do not result in a successful pregnancy, foremost as a result of implantation failure. As a matter of fact, the success rates of IVF remain modest as the probability of live birth is approximately 25%–30% per initiated cycle [1].

The process of implantation involves two main components: (1) a competent embryo that should have the potential to implant and (2) a receptive endometrium that should enable implantation [2]. The "cross-talk" between the embryo and the endometrium that finally leads to apposition, attachment, and invasion of embryos is mandatory for successful

Immunology of Recurrent Pregnancy Loss and Implantation Failure
DOI: https://doi.org/10.1016/B978-0-323-90805-4.00001-8

implantation and subsequent normal placentation. These processes are under thorough investigation and seem to involve many mediators originating from the embryo, endometrium, and maternal immune system [3–5]. Thus any abnormalities attributed to the embryo, endometrium, or maternal immune system would result in implantation failure. It explains the sheer complexity of assessing the causes of repeated implantation failure (RIF).

In this review, we exclusively focus on the endometrium and the endometrial immune environment. In addition, we discuss endometrial pathologies that might contribute to the failure of implantation and review treatments that could potentially repair an endometrium that is not receptive to embryo implantation.

2 Endometrial pathologies

2.1 Anatomical disturbances

A variety of endometrial structural anomalies associated with RIF can be identified. Indeed, anatomical malformations of the uterus, either congenital or acquired, may interfere with normal implantation [6–8]. When possible, intrauterine abnormalities should be corrected [9], and hydrosalpinx should be removed to enhance implantation [6].

Septate uterus is the most common congenital uterine anomaly, which may contribute to RIF. Lavergne et al. [10] reported poor outcomes following IVF in women with untreated septate uterus compared to those who underwent metroplasty before IVF treatment. Another study by Ban-Frangež et al. [11] also revealed a higher miscarriage rate (80%) following IVF/ ICSI in women with a septate uterus, which was decreased to 30% after surgical treatments.

Submucosal and intramural fibroids that distort the endometrial cavity are associated with decreased pregnancy and implantation rates in women who attempt to conceive naturally or proceed with IVF treatment [12,13]. Several studies suggest that pregnancy rates improve following the resection of these fibroids [14,15]. Pritts et al. [15] conducted a systematic review of the literature and *meta*-analysis of existing controlled studies and concluded that removing submucous myomas improved clinical pregnancy and implantation rates. Intramural fibroids appear to decrease fertility, but the effects of myomectomy were unclear. On the contrary, subserosal fibroids did not affect fertility outcomes, and removal did not confer benefit. There are several mechanisms by which fibroids can adversely affect implantation, including increased uterine contractility, deranged cytokine profile, abnormal vascularization, and chronic endometrial inflammation [16,17]. Overall, myomectomy should be performed before ART in case of submucosal myoma (FIGO type 0, 1, 2, and 3 abutting the endometrium) and intramural myomas (FIGO 4) >5 cm, mostly distorting the cavity [18]. Subserosal (FIGO type 5, 6, 7) should not be treated.

Endometrial polyps may also interfere with embryo implantation. The removal of endometrial polyps has been found to result in improved spontaneous pregnancy rates in three nonrandomized studies [13,19,20]. A more recent systematic review found that the hysteroscopic removal of endometrial polyps doubled the clinical pregnancy rate in women undergoing intrauterine insemination treatment [21]. Thus it seems highly likely that endometrial polyps contribute to RIF.

The presence of adhesions within the uterine cavity may interfere with successful implantation. Intrauterine adhesions often occur following curettage of the gravid uterus to terminate an unwanted pregnancy or in cases of retained products of conception after a pregnancy or miscarriage. Intrauterine surgery or intrauterine infection of the nongravid uterus may also lead to the formation of intrauterine adhesions. Demirol and Gurgan reported that intrauterine adhesions were found in 8.5% of women with RIF [9]. The evidence available so far suggests that hysteroscopic removal of intrauterine adhesions improves fertility outcomes [22,23].

2.2 Displaced window of implantation

Regardless of its quality, an embryo is only able to implant during a very short period of time, starting 4–5 days after ovulation and lasting for 4 days [24]. For many years, the WOI had been assumed to be constant for all women. Accordingly, the transfer of blastocysts after IVF is routinely performed on day 5 after oocyte retrieval, or day 5–6 of progesterone therapy, based on classic studies of the timing of implantation [25]. However, investigators have recently demonstrated the existence of a "displaced WOI" [2,26,27] that has been considered an important factor contributing to RIF [28]. To determine the moment of the endometrial cycle that corresponds to the start of WOI, diagnostic tools have been developed (WIN test and ERA test) and aim to personalize the embryo transfer (ET) according to the results.

2.2.1 *The window implantation test*

The window implantation (WIN) Test was developed by the University of Montpellier, France. Using a transcriptomic approach, Haouzi et al. [29] identified 11 genes of which mRNAs were strongly overexpressed in receptive endometrium (LH + 7) in comparison to secretory prereceptive (LH + 2) endometriums. An endometrium is defined as "receptive" when the mean expression of the 11 genes is ≥ 70%, "partially receptive" when the mean expression is between 50% and 70% and "nonreceptive" when the mean expression is <50% of the expression level of the positive control which is meant to be 100%.

Endometrial biopsies have to be collected during the implantation window: 6–9 days after the spontaneous luteinizing hormone (LH + 6 to LH + 9) surge during natural cycles or 5–9 days after progesterone administration during hormone replacement therapy cycles (Pg + 5 to Pg + 9). A blastocyst is transferred when the endometrium is receptive. A 2- or 3-day embryo is transferred 48–72 h before the receptive time. In 2021, Haouzi et al. [30] reported that patients with RIF ($n = 217$) who underwent a WIN Test showed that both implantation (22.7% vs 7.2%) and live birth rates per patient (31.8% vs 8.3%) were significantly higher in the customized ET than in the classic ET procedure group.

2.2.2 *The endometrial receptivity analysis test*

The endometrial receptivity analysis (ERA) Test is a diagnostic tool developed in Spain and patented by the Igenomix R&D department since 2009. It analyzes 248 genes related to endometrial receptivity. The ERA test classifies the results of the endometrial biopsies into receptive and nonreceptive and further into prereceptive or postreceptive. The biopsy is performed in a substitute cycle, 5 days after progesterone uptake. In a study published in 2013, Ruiz-Alonso et al. [31] reported that 3 in every 10 implantation failure patients have a displaced WOI. Around 22 RIF patients ($n = 22/85$; 25.9%) were nonreceptive, and in 15 of them, a second ERA validated a displacement of the WOI. In eight of them, personalized ET was performed on the day designated by the ERA, resulting in a 50.0% pregnancy rate and 38.5% implantation rate. More cohort and randomized controlled studies with a larger population are necessary to demonstrate the interest in the ERA test.

3 Endometrial immunopathology

3.1 How to detect endometrial immune pathology?

The concept of uterine immune profiling (UtimPro) test is an innovative concept that relies on the analysis of the local immune reaction occurring within the endometrium during the implantation window. A crucial immune endometrial switch should occur during the implantation window not only to avoid the rejection of the semiallogenic embryo but to promote its growth and nutrition [32]. Since embryo implantation is the crux of assisted reproduction, it requires understanding the local immune environment and the interplay between the endometrium and the embryo. During this time, important immune cells leave the uterus (such as B lymphocytes and some CD8 + lymphocytes) while others enter the endometrium. During the implantation window, almost all the immune cells belonging to our adaptive immunity escape from the endometrium while innate immune cells (macrophage, uNK cells, and dendritic cells) invade into the endometrium [33]. The newly created immune environment plays a key role in embryo implantation [33]. uNK cells differ from circulating NK cells—by their phenotype, their repertoire of activating and inhibiting receptors, the cytokines they secrete, and their low cytotoxic potential. Regulatory T cells make the link between adaptive and local immune expression. Underactive immune cells fail to create the necessary implantation reaction; conversely, overactive immune cells can lead to the destruction of the endometrium and eventually the rejection of the embryo. This unique immune reaction is essential for promoting embryo adhesion, and its disruption is likely to impede implantation.

Early on, the ideal environment during the implantation window was thought to contain mainly Th2 (compared with Th1) cytokines, which would selectively allow the development of local mechanisms that promote immunotropism and angiogenesis at the same time that they downregulate inflammation and cytotoxic pathways [34]. Over time, the concept of pregnancy as a Th2 phenomenon has evolved into a concept of Th1/Th2 balance since both the absence and a large excess of Th1 cytokines are thought to be deleterious for implantation and placentation, as is the absence of Th2 cytokines [35]. The transient immune switch to Th2, together with adequate uNK cell activation, appears fundamental in establishing local maternal tolerance and survival of the fetus. Interleukin (IL)-15 is directly involved in the postovulatory recruitment and maturation of uNK cells in the uterus [36], and under their control, it is essential for adequate Th2 cytokine production. Its effects on blood NK cells are different than on uNK cells. It does not convert them into potent cytolytic cells but instead participates directly in their maturation [37]. IL-18 is a Th2-promoting cytokine that, through the action of angiopoietin-2, affects the crucial destabilization of spiral arteries [38,39]. In human endometrium, IL-18 expression increases during the implantation window. Its main role is to remodel the maternal-side vasculature. However, IL-15 and IL-18, produced by either epithelial or stromal endometrial cells, are both bivalents: at high levels, and when not immuno-regulated, they behave as proinflammatory Th1 cytokines. They reflect the local production of IFN-γ and TNF-α, which can activate uNK cells to become cytotoxic [40]. We also focus on the role of TWEAK (TNF weak inducer of apoptosis) and its receptor, Fn-14 (fibroblast growth factor-inducible molecule 14), as immunoregulators of the local immune equilibrium. Using an animal model, we observed that TWEAK offers protection against the deleterious effects of a Th1 dominant (TNF-rich) environment during implantation and thus increases embryo survival [41]. TWEAK and its ligand, Fn-14, act as immune regulators of the Th1/Th2 cytokine balance in the human endometrium [42].

3.2 The UtimPro test

The UtimPro test is indicated for patients with implantation failures or unexplained recurrent miscarriages (RMs). From a biopsy of endometrium collected by aspiration in the mid-luteal phase before assisted reproductive treatment, the UtimPro quantifies targeted biomarkers (IL-18, IL-15, TWEAK, Fn-14, CD56) by quantitative real-time PCR to document the immune endometrial environment in which the embryo is transferred. This immune endometrial profile method (Patent No. PCT/EP2013/065355) is a technique for increasing implantation success in assisted fertilization. The norms of expression for these biomarkers in a fertile cohort were documented, and an immune profile was reproducible from one cycle to another over a period of 6 months if no surgery or pregnancy occurred in the meantime. Based on the immune check-up, several suggestions for personalized care can be applied at the subsequent ET. The objective is to reequilibrate the immune environment.

3.3 UtimPro test outcomes

Four types of endometrial immune profiles can be diagnosed:

1. A BALANCED endometrial immune activation characterized by IL-18/TWEAK and IL-15/Fn-14 mRNA ratios and CD56 + cell count in the same range as previously defined in the fertile cohort.
2. A LOW endometrial immune activation profile is characterized by low mRNA ratios for IL-15/Fn-14 (reflecting immature uNK cells) and/or IL-18/TWEAK or an absence of uNK recruitment.
3. An OVER endometrial immune activation is characterized by high mRNA ratios of IL-18/TWEAK and/or IL-15/Fn-14 and/or a high CD56 + cell count.
4. A MIXED endometrial immune profile is characterized by a high mRNA ratio of IL-18/TWEAK (excess of Th1 cytokines) simultaneously with a low IL-15/Fn-14 ratio (reflecting immature NK).

3.4 Incidence of abnormal UtimPro test

The endometrial immune profile is prone to change each time an intrauterine event occurs, including any intrauterine surgery or pregnancy. In a cohort study including 1738 infertile patients published in 2020, Lédée et al. [43] documented endometrial immune deregulation in 83.5% (overimmune local activation in 45%, low-immune local activation in 28%, and mixed profile in 10.5%). Based on the immune profile, a personalized IVF treatment was proposed either to control uNK cells activation or to promote local mechanisms of embryo adhesion. Pregnancy rates were significantly higher for patients with a history of RIF or RMs after well-diagnosed deregulation and applied personalized care, compared to nonderegulated patients (respectively 37.7% and 56% vs 26.9 and 24%, $p < 0.001$, respectively).

4 Immune treatments for women with abnormal immune profile

4.1 The personalization and its rationale

Adjuvant therapies aim to achieve an "ideal" immune environment according to the endometrial immune profile, and the details are summarized in Table 1. No one size fits all. Hence, precision medicine is clearly the future of reproductive medicine. Nevertheless, despite the use of classical molecules (compatible with pregnancy), much remains to be discovered about their indication and precise mechanism of action.

4.2 Glucocorticoids

Using the Utimpro, Glucocorticoids are recommended as a first-line treatment to women with an over endometrial immune activation or mixed immune profile. In 2018, Lédée and al. [44] documented the impact of prednisone on the endometrial expression of

TABLE 1 Summary of suggested therapies according to the endometrial immune profile.

Immune endometrial profile	Glucocorticoids	Intralipids	Luteal hCG	LMWH	Scratching	Supplementation in progesterone	Exposure to seminal plasma after ET
No dysregulation	–	–	–	–	–	–	–
Low immune activation	–	–	X	–	X	–	X
Over immune activation	X	X	–	X	–	X	–
Mixed profile	X	X	X	X	X	X	–

LMWH, Low molecular weight heparin; *hCG*, human chorionic gonadotropin; *ET*, embryo transfer.

immune biomarkers at the time of endometrial receptivity in a RIF population. Interestingly, both immune biomarkers (mRNA ratios of IL-18/TWEAK and IL-15/Fn-14) were normalized in 54.5% of the cases, and only one was normalized in 16.5% with prednisone treatment. The IL-18/TWEAK mRNA ratio reflecting the Th1/Th2 local equilibrium was significantly reduced, through a significant increase of TWEAK expression, in patients who were subsequently pregnant under prednisone. The significant increase of TWEAK appears as a potential immune mechanism able to explain the control of local cytotoxicity. Notably, it was previously documented that local high IL-18/TWEAK endometrial expression reflected a deleterious Th1 dominant environment and was associated with cytotoxic activation of innate immune cells [40]. Furthermore, Mas et al. [41] showed that TWEAK offers protection against the deleterious effects of a Th1 dominant (TNF-rich) environment during implantation and thus increases embryo survival.

The treatment consists of daily oral corticoids (i.e., prednisone 20 mg/day if over immune profile and 10 mg/day if mixed immune profile). The treatment begins on day 3 of the cycle prior to ET and is then continued until the first trimester if pregnant. A cycle test under corticoids is required to evaluate if corticoids are able to normalize the immune environment. Indeed, in our experience, 30% of RIF patients with overimmune activation were resistant to corticoids. The treatment must be gradually decreased to avoid adrenal insufficiency. Published data in pregnant women treated with corticoids are reassuring. However, patients should be warned of the risk of bone density loss (prevention with vitamin D supplementation during the treatment).

4.3 Intralipids

Intralipid therapy (ILT) is proposed as a second-line treatment for women with immune overactivation profiles who either fail to become pregnant with glucocorticoids or whose immune profile remains dysregulated despite glucocorticoids. Whilst the exact mechanism in which immune modulation is achieved remains unclear, some authors have suggested that ILT could inhibit proinflammatory mediators, specifically Th1 cells [45], and has immunosuppressive properties on NK cells [46,47].

ILT is a 20% fat emulsion typically comprising 20% soybean oil, 1.2% egg yolk phospholipids, 2.25% glycerin, and water. It is given as a slow intravenous (IV) infusion made up of 100 mL of intralipids diluted in 400 mL of saline water over 90 minutes in an inpatient hospital environment. The ILT is given on day 8 of the cycle of ET. If pregnancy occurs, the procedure is reapplied at 5 and 9 weeks of amenorrhea. All infusions are performed in a hospital under medical supervision. The tolerance to the infusion is reported to be excellent. In a study published by Lédée et al. [47], none of the 94 women receiving a slow ILT reported side effects during or after the infusion. Notably, they did not observe any allergic reaction or casual symptoms like headache, dizziness, flushing, drowsiness, nausea, or sweating, as it is described with nondiluted intralipids infusion. On another note, the safety of intralipids during the initial period of implantation and during pregnancy has been indirectly evaluated, insofar as intralipid is diluted with propofol, the principal anesthetic drug used during general anesthesia for oocyte retrieval and during surgery for pregnancy [48].

4.4 Luteal human chorionic gonadotropin supplementation

Luteal human chorionic gonadotropin (hCG) supplementation is recommended for women with low local endometrial immune activation during the implantation window. The embryo physiologically produces hCG during pregnancy. It is directly involved in the local reaction inducing an immunological tolerance with adequate angiogenesis and activation of uNK cells at the maternal–fetal interface [49–52]. Kane et al. reported in 2008 that hCG is a hormone triggering the proliferation and maturation of uNK cells by activating the mannose receptor [53]. Thus, HCG supplementation would correct the observed uNK dysregulation, such as deficient recruitment and/or immaturity. Low doses of HCG (1500 IU subcutaneously) are given during the mid-luteal phase following the ET: three injections at days 4, 6, and 8 after oocyte retrieval for a fresh ET, or three injections at days progesterone +4, +6, and +8 for a frozen ET.

4.5 Low molecular weight heparin

In case of resistance to corticosteroids and ILT, low molecular weight heparin (LMWH) may be proposed as a third-line treatment for women with an over endometrial immune activation or mixed immune profile. LMWH is used for the prevention of pregnancy complications associated with thrombophilic disorders, particularly antiphospholipid syndrome. Nevertheless, recent studies have suggested that heparin may directly affect the placental trophoblasts, independently of its anticoagulant activity. In addition, heparin has a well-documented anticomplement effect [54,55].

A prophylactic dosage of LMWH (i.e., enoxaparin 40 mg daily) on the day of oocyte retrieval or starting from the beginning of the implantation window of the ET cycle is recommended. Women self-administer the LMWH subcutaneously, and LMWH is continued up to the 12th week of pregnancy if the pregnancy test 14 days after ET is positive or discontinued if it is negative. Monitoring of the platelet count should be performed on the day of oocyte retrieval and two times weekly.

4.6 Scratching

Endometrial scratching is recommended in case of a local low activation profile or mixed activation profile with a low IL-15/Fn-14 mRNA ratio, interpreted as an immaturity of uNK cells.

Endometrial scratching could enhance uNK cell maturation, which strongly depends on the adequate expression of IL-15. Any type of local injuries (as endometrial biopsy) performed during the mid-luteal phase of the cycle activates and stimulates, at the next luteal phase, the subsequent expression of adhesion molecules, chemokines, and interestingly, IL-15 via toll-like receptor pathways [56]. In 2015, Gnainsky et al. [57] demonstrated that endometrial scratching upregulated the expression of proinflammatory cytokines involved in the recruitment of monocytes and their differentiation into macrophages and dendritic cells. While their role in human endometrium at the time of implantation is still unclear, they seem to play a major role in tissue remodeling and angiogenesis, required for the endometrial decidualization and regulation of trophoblast invasion. Likewise, Yu Liang et al. [58]

were able to confirm these findings and reported that the scratching induced increased vascular endothelial growth factor, a central growth factor involved in local angiogenesis and placentation.

Endometrial scratching, therefore, must be performed in the mid-luteal phase of the cycle prior to fresh or frozen ET. It involves obtaining an endometrial biopsy sample with a small 3 mm in width catheter known as a pipelle. Usually, without hooking on the cervix, the catheter is pushed forward through the cervix to the fundus and then retracted in circular movements in order to stimulate the endometrium. It is considered to be a low-risk procedure with a low rate of complications. Scratching can be performed on an outpatient basis without anesthesia and is associated with minor pain for most patients.

4.7 Supplementation in progesterone

An adaptation of luteal support may be proposed in local high activation profiles and mixed profiles. Progesterone, besides its endocrine role, is a crucial mediator of the local immune tolerance required for a successful pregnancy. Progesterone influences the maternal immune system through distinct pathways [56]: (1) Via the production of progesterone-induced blocking factor (PIBF), which inhibits NK cell activity and leads to Th2- dominant cytokine production by maternal lymphocytes and (2) Via the induction of galectin-1, a progesterone-induced molecule, essential for inducing tolerogenic dendritic cells, which in return promote in vivo expansion of IL-10 secreting Treg cells.

When a local immune overactivation is observed, high daily vaginal doses of progesterone (1200 mg) may be recommended for its immunosuppressive properties or association of vaginal and IM or subcutaneous progesterone. Daily vaginal micronized progesterone or dual administration (vaginal and subcutaneous) is given starting from the day of oocyte retrieval or starting from the beginning of the implantation window of the ET cycle. It is then continued through 8 weeks after ET for pregnant women. Monitoring of plasma progesterone levels is crucial to ensure correct absorption by the vaginal tissues. A blood sample should be collected at least 48 hours after the initiation of progesterone intake.

Compared to oral administration, vaginal progesterone bypasses hepatic first-pass effects and, therefore, has better bioavailability. It has no undesirable side effects, such as sleepiness, fatigue, and headaches [59]. Compared with intramuscular (IM) administration, endometrial bioavailability after vaginal progesterone use is also higher due to the direct transport of progesterone from the vagina to the uterus—the so-called "uterine first-pass effect." The IM route has mild side effects, generally restricted to the injection site [60,61]. One major disadvantage of vaginal progesterone formulation is that it can be messy and associated with vaginal discharge and irritation.

5 Key points

1. RIFs in infertile patients submitted to ART cycles are complex problems due to the different factors involved in embryo development and implantation.

2. Despite expanding global experience with ART, the majority of IVF attempts do not result in a successful pregnancy, foremost as a result of implantation failure.
3. Immune endometrial receptivity is described as the "last barrier" for embryo implantation.
4. All explorations should begin after a precise check-up regarding anatomical disturbances and adequate endometrial proliferation.
5. Endometrial immune profiling is a key factor in planning personalized IVF/ICSI treatments.
6. The objective of personalization aims to achieve an "ideal" immune environment to enhance the potentiality of embryo implantation.
7. In a low immune endometrial environment, it is suggested to perform a local injury (scratching) in the mid-luteal phase of the cycle preceding the ET cycle, supplement the luteal phase with human Chorionic Gonadotropin and have sexual intercourse after the ET.
8. In an over activation immune endometrial environment, it is suggested to administer corticoids from day -1 of the IVF cycle as the first line of treatment or slow infusion of intralipids as the second line of treatment (previous IVF failure under corticoids or absence of normalization of the uterine check-up under corticoids), high dose of progesterone (1200 mg vaginally) in the luteal phase and no sexual intercourse post-transfer.
9. Only a normalization of the immune profile under adjuvant therapies may attest to their efficacy.

References

[1] Toftager M, Bogstad J, Løssl K, Prætorius L, Zedeler A, Bryndorf T, et al. Cumulative live birth rates after one ART cycle including all subsequent frozen-thaw cycles in 1050 women: secondary outcome of an RCT comparing GnRH-antagonist and GnRH-agonist protocols. Hum Reprod Oxf Engl 2017;32(3):556–67 Mar 1.

[2] Lessey BA. Assessment of endometrial receptivity. Fertil Steril 2011;96(3):522–9 Sep 1.

[3] Achache H, Revel A. Endometrial receptivity markers, the journey to successful embryo implantation. Hum Reprod Update 2006;12(6):731–46 Dec.

[4] Fukui A, Funamizu A, Yokota M, Yamada K, Nakamua R, Fukuhara R, et al. Uterine and circulating natural killer cells and their roles in women with recurrent pregnancy loss, implantation failure and preeclampsia. J Reprod Immunol 2011;90(1):105–10 Jun.

[5] Gonen-Gross T, Goldman-Wohl D, Huppertz B, Lankry D, Greenfield C, Natanson-Yaron S, et al. Inhibitory NK receptor recognition of HLA-G: regulation by contact residues and by cell specific expression at the fetal-maternal interface. PLoS One 2010;5(1):e8941 Jan 28.

[6] Practice Committee of American Society for Reproductive Medicine in Collaboration with Society of Reproductive Surgeons. Myomas and reproductive function. Fertil Steril 2008;90(5 Suppl):S125–30 Nov.

[7] Sanders B. Uterine factors and infertility. J Reprod Med 2006;51(3):169–76 Mar.

[8] Strandell A. The influence of hydrosalpinx on IVF and embryo transfer: a review. Hum Reprod Update 2000;6(4):387–95 Aug.

[9] Demirol A, Gurgan T. Effect of treatment of intrauterine pathologies with office hysteroscopy in patients with recurrent IVF failure. Reprod Biomed Online 2004;8(5):590–4 Jan.

[10] Lavergne N, Aristizabal J, Zarka V, Erny R, Hedon B. Uterine anomalies and in vitro fertilization: what are the results? Eur J Obstet Gynecol Reprod Biol 1996;68(1–2):29–34 Sep.

[11] Ban-Frangez H, Tomazevic T, Virant-Klun I, Verdenik I, Ribic-Pucelj M, Bokal EV. The outcome of singleton pregnancies after IVF/ICSI in women before and after hysteroscopic resection of a uterine septum compared to normal controls. Eur J Obstet Gynecol Reprod Biol 2009;146(2):184–7 Oct.

[12] Bernard G, Darai E, Poncelet C, Benifla J-L, Madelenat P. Fertility after hysteroscopic myomectomy: effect of intramural myomas associated. Eur J Obstet Gynecol Reprod Biol 2000;88(1):85–90 Jan 1.

[13] Varasteh NN, Neuwirth RS, Levin B, Keltz MD. Pregnancy rates after hysteroscopic polypectomy and myomectomy in infertile women. Obstet Gynecol 1999;94(2):168–71 Aug.

[14] Fernandez H. Hysteroscopic resection of submucosal myomas in patients with infertility. Hum Reprod 2001;16(7):1489–92 Jul 1.
[15] Pritts EA, Parker WH, Olive DL. Fibroids and infertility: an updated systematic review of the evidence. Fertil Steril 2009;91(4):1215–23 Apr.
[16] Donnez J, Jadoul P. What are the implications of myomas on fertility? A need for a debate? Hum Reprod Oxf Engl 2002;17(6):1424–30 Jun.
[17] Taylor E, Gomel V. The uterus and fertility. Fertil Steril 2008;89(1):1–16 Jan.
[18] Hart R, Khalaf Y, Yeong CT, Seed P, Taylor A, Braude P. A prospective controlled study of the effect of intramural uterine fibroids on the outcome of assisted conception. Hum Reprod Oxf Engl 2001;16 (11):2411–17 Nov.
[19] Shokeir TA, Shalan HM, El-Shafei MM. Significance of endometrial polyps detected hysteroscopically in eumenorrheic infertile women. J Obstet Gynaecol Res 2004;30(2):84–9.
[20] Spiewankiewicz B, Stelmachów J, Sawicki W, Cendrowski K, Wypych P, Swiderska K. The effectiveness of hysteroscopic polypectomy in cases of female infertility. Clin Exp Obstet Gynecol 2003;30(1):23–5.
[21] Bosteels J, Weyers S, Puttemans P, Panayotidis C, Van Herendael B, Gomel V, et al. The effectiveness of hysteroscopy in improving pregnancy rates in subfertile women without other gynaecological symptoms: a systematic review. Hum Reprod Update 2010;16(1):1–11 Feb.
[22] Yasmin H, Nasir A, Noorani K. Hystroscopic management of Ashermans Syndrome. JPMA J Pak Med Assoc 2007;57:553–5 Dec 1.
[23] Zikopoulos KA, Kolibianakis EM, Platteau P, de Munck L, Tournaye H, Devroey P, et al. Live delivery rates in subfertile women with Asherman's syndrome after hysteroscopic adhesiolysis using the resectoscope or the Versapoint system. Reprod Biomed Online 2004;8(6):720–5 Jan 1.
[24] Psychoyos A. From Lataste to the "window of implantation": 100 years of fascinating discoveries. Contracept Fertil Sex 1992;21(4):333–8.
[25] Wilcox AJ, Baird DD, Weinberg CR. Time of implantation of the conceptus and loss of pregnancy. N Engl J Med 1999;340(23):1796–9 Jun 10.
[26] Galliano D, Bellver J, Díaz-García C, Simón C, Pellicer A. ART and uterine pathology: how relevant is the maternal side for implantation? Hum Reprod Update 2015;21(1):13–38 Feb.
[27] Kliman HJ, Frankfurter D. Clinical approach to recurrent implantation failure: evidence-based evaluation of the endometrium. Fertil Steril 2019;111(4):618–28 Apr.
[28] Valdes CT, Schutt A, Simon C. Implantation failure of endometrial origin: it is not pathology, but our failure to synchronize the developing embryo with a receptive endometrium. Fertil Steril 2017;108(1):15–18 Jul.
[29] Haouzi D, Mahmoud K, Fourar M, Bendhaou K, Dechaud H, De Vos J, et al. Identification of new biomarkers of human endometrial receptivity in the natural cycle. Hum Reprod 2009;24(1):198–205 Jan 1.
[30] Haouzi D, Entezami F, Torre A, Innocenti C, Antoine Y, Mauries C, et al. Customized frozen embryo transfer after identification of the receptivity window with a transcriptomic approach improves the implantation and live birth rates in patients with repeated implantation failure. Reprod Sci Thousand Oaks Calif 2021;28 (1):69–78 Jan.
[31] Ruiz-Alonso M, Blesa D, Díaz-Gimeno P, Gómez E, Fernández-Sánchez M, Carranza F, et al. The endometrial receptivity array for diagnosis and personalized embryo transfer as a treatment for patients with repeated implantation failure. Fertil Steril 2013;100(3):818–24 Sep.
[32] Liu S, Diao L, Huang C, Li Y, Zeng Y, Kwak-Kim JYH. The role of decidual immune cells on human pregnancy. J Reprod Immunol 2017;124:44–53 Nov.
[33] Lee JY, Lee M, Lee SK. Role of endometrial immune cells in implantation. Clin Exp Reprod Med 2011;38 (3):119–25 Sep.
[34] Wegmann TG, Lin H, Guilbert L, Mosmann TR. Bidirectional cytokine interactions in the maternal-fetal relationship: is successful pregnancy a TH2 phenomenon? Immunol Today 1993;14(7):353–6 Jul.
[35] Chaouat G. The Th1/Th2 paradigm: still important in pregnancy? Semin Immunopathol 2007;29(2):95–113 Jun.
[36] Kitaya K, Yamaguchi T, Honjo H. Central role of interleukin-15 in post-ovulatory recruitment of peripheral blood CD16(-) natural killer cells into human endometrium. J Clin Endocrinol Metab 2005;90 (5):2932–40 May.
[37] Verma S, Hiby SE, Loke YW, King A. Human decidual natural killer cells express the receptor for and respond to the cytokine interleukin 15. Biol Reprod 2000;62(4):959–68 Apr.

[38] Croy BA, Esadeg S, Chantakru S, van den Heuvel M, Paffaro VA, He H, et al. Update on pathways regulating the activation of uterine natural killer cells, their interactions with decidual spiral arteries and homing of their precursors to the uterus. J Reprod Immunol 2003;59(2):175–91 Aug.
[39] Goldman-Wohl DS, Ariel I, Greenfield C, Lavy Y, Yagel S. Tie-2 and angiopoietin-2 expression at the fetal-maternal interface: a receptor ligand model for vascular remodelling. Mol Hum Reprod 2000;6(1):81–7 Jan.
[40] Petitbarat M, Rahmati M, Sérazin V, Dubanchet S, Morvan C, Wainer R, et al. TWEAK appears as a modulator of endometrial IL-18 related cytotoxic activity of uterine natural killers. PLoS One 2011;6(1):e14497 Jan 7.
[41] Mas AE, Petitbarat M, Dubanchet S, Fay S, Ledée N, Chaouat G. Immune regulation at the interface during early steps of murine implantation: involvement of two new cytokines of the IL-12 family (IL-23 and IL-27) and of TWEAK. Am J Reprod Immunol N Y N 1989 2008;59(4):323–38 Apr.
[42] Petitbarat M, Serazin V, Dubanchet S, Wayner R, de Mazancourt P, Chaouat G, et al. Tumor necrosis factor-like weak inducer of apoptosis (TWEAK)/fibroblast growth factor inducible-14 might regulate the effects of interleukin 18 and 15 in the human endometrium. Fertil Steril 2010;94(3):1141–3 Aug.
[43] Lédée N, Petitbarat M, Chevrier L, Vitoux D, Vezmar K, Rahmati M, et al. The uterine immune profile may help women with repeated unexplained embryo implantation failure after in vitro fertilization. Am J Reprod Immunol N Y N 1989 2016;75(3):388–401.
[44] Lédée N, Prat-Ellenberg L, Petitbarat M, Chevrier L, Simon C, Irani EE, et al. Impact of prednisone in patients with repeated embryo implantation failures: beneficial or deleterious? J Reprod Immunol 2018;127:11–15 Jun.
[45] Granato D, Blum S, Rössle C, Boucher JL, Malnoë A, Dutot G. Effects of parenteral lipid emulsions with different fatty acid composition on immune cell functions in vitro. J Parenter Enter Nutr 2000;24(2):113–18.
[46] Roussev RG, Acacio B, Ng SC, Coulam CB. Duration of intralipid's suppressive effect on NK cell's functional activity. Am J Reprod Immunol N Y N 1989 2008;60(3):258–63.
[47] Lédée N, Vasseur C, Petitbarat M, Chevrier L, Vezmar K, Dray G, et al. Intralipid® may represent a new hope for patients with reproductive failures and simultaneously an over-immune endometrial activation. J Reprod Immunol 2018;130:18–22 Nov.
[48] Jarahzadeh MH, Jouya R, Mousavi FS, Dehghan-Tezerjani M, Behdad S, Soltani HR. Propofol or Thiopental sodium in patients undergoing reproductive assisted technologies: Differences in hemodynamic recovery and outcome of oocyte retrieval: a randomized clinical trial. Iran J Reprod Med 2014;12(1):77–82 Jan.
[49] Perrier d'Hauterive S, Berndt S, Tsampalas M, Charlet-Renard C, Dubois M, Bourgain C, et al. Dialogue between blastocyst hCG and endometrial LH/hCG receptor: which role in implantation? Gynecol Obstet Invest 2007;64(3):156–60.
[50] Bansal AS, Bora SA, Saso S, Smith JR, Johnson MR, Thum M-Y. Mechanism of human chorionic gonadotrophin-mediated immunomodulation in pregnancy. Expert Rev Clin Immunol 2012;8(8):747–53 Nov.
[51] Berndt S, Blacher S, d'Hauterive S, Thiry M, Tsampalas M, Cruz A, et al. Chorionic gonadotropin stimulation of angiogenesis and pericyte recruitment. J Clin Endocrinol Metab 2009;94:4567–74 Nov 1.
[52] Tsampalas M, Gridelet V, Berndt S, Foidart J-M, Geenen V, Perrier d'Hauterive S. Human chorionic gonadotropin: a hormone with immunological and angiogenic properties. J Reprod Immunol 2010;85(1):93–8 May.
[53] Kane N, Kelly R, Saunders PTK, Critchley HOD. Proliferation of uterine natural killer cells is induced by human chorionic gonadotropin and mediated via the mannose receptor. Endocrinology 2009;150(6):2882–8 Jun.
[54] Oberkersch R, Attorresi AI, Calabrese GC. Low-molecular-weight heparin inhibition in classical complement activation pathway during pregnancy. Thromb Res 2010;125(5):e240–5 May.
[55] Girardi G, Redecha P, Salmon JE. Heparin prevents antiphospholipid antibody–induced fetal loss by inhibiting complement activation. Nat Med 2004;10(11):1222–6 Nov.
[56] Lédée N, Petitbarat M, Prat-Ellenberg L, Dray G, Cassuto GN, Chevrier L, et al. Endometrial immune profiling: a method to design personalized care in assisted reproductive medicine Front Immunol [Internet] 2020;[cited 2021 Apr 27];11. Available from: https://www.frontiersin.org/articles/10.3389/fimmu.2020.01032/full.
[57] Gnainsky Y, Granot I, Aldo P, Barash A, Or Y, Mor G, et al. Biopsy-induced inflammatory conditions improve endometrial receptivity: the mechanism of action. Reprod Camb Engl 2015;149(1):75–85 Jan.
[58] Liang Y., Han J., Jia C., Ma Y., Lan Y., Li Y., et al. Effect of endometrial injury on secretion of endometrial cytokines and ivf outcomes in women with unexplained subfertility. Mediators Inflamm [Internet]. 2015 [cited 2021 Apr 27];2015. Available from: <https://www.ncbi.nlm.nih.gov/pmc/articles/PMC4637501/>.

[59] Friedler S, Raziel A, Schachter M, Strassburger D, Bukovsky I, Ron-El R. Luteal support with micronized progesterone following in-vitro fertilization using a down-regulation protocol with gonadotrophin-releasing hormone agonist: a comparative study between vaginal and oral administration. Hum Reprod Oxf Engl 1999;14(8):1944–8 Aug.
[60] Miles RA, Paulson RJ, Lobo RA, Press MF, Dahmoush L, Sauer MV. Pharmacokinetics and endometrial tissue levels of progesterone after administration by intramuscular and vaginal routes: a comparative study. Fertil Steril 1994;62(3):485–90 Sep.
[61] Cicinelli E, de Ziegler D, Bulletti C, Matteo MG, Schonauer LM, Galantino P. Direct transport of progesterone from vagina to uterus. Obstet Gynecol 2000;95(3):403–6 Mar.

CHAPTER

21

Thrombophilia, antiphospholipid antibodies, and anticoagulation in recurrent implantation failure

Marcelo Borges Cavalcante[1,2] *and Ricardo Barini*[3]

[1]Postgraduate Program in Medical Sciences, Universidade de Fortaleza (UNIFOR), Fortaleza, Ceará, Brazil [2]CONCEPTUS – Reproductive Medicine, Fortaleza, Ceará, Brazil [3]Department of Obstetrics and Gynecology, Campinas University (UNICAMP), Campinas, São Paulo, Brazil

1 Introduction

During the limited implantation window, intimate interaction between the blastocyst and the receptive endometrium takes place with numerous cellular and molecular events, which are yet to be fully unraveled. In a spontaneous conception, approximately 30% of embryos fail to implant into the endometrium, and 10% of them end as chemical pregnancies [human chorionic gonadotropin (hCG) levels are detectable without clinical pregnancy diagnosis] [1,2]. Implantation failure occurs during the early implantation stage, such as embryo apposition or adhesion, resulting in a negative pregnancy test. It may also occur later after successful embryo invasion through the luminal surface of the endometrium. In the latter case, β-hCG can be detected in blood or urine, but the pregnancy fails before an intrauterine gestational sac can be seen, which is defined as a biochemical pregnancy. Clinical pregnancy is diagnosed when an intrauterine gestational sac (approximately 5 weeks of gestation) is detected by ultrasound.

The term "embryo implantation failure" can be used in patients who are trying to conceive naturally or undergoing assisted reproductive technology (ART) cycles. However, the term "recurrent implantation failure" (RIF) is only applicable in patients undergoing ART cycles. Several definitions are proposed for RIF. Currently, the most popular definition for RIF is the failure to achieve a clinical pregnancy after the transfer of at least four good-quality embryos (based on a morphological assessment) in a minimum of three fresh or frozen cycles in a woman under 40 years of age [2,3]. RIF is a challenge for specialists

Immunology of Recurrent Pregnancy Loss and Implantation Failure
DOI: https://doi.org/10.1016/B978-0-323-90805-4.00020-1

in the realm of reproductive medicine. Even with recent advances in ART, a significant number of couples are unable to conceive despite multiple attempts of in vitro fertilization (IVF) and embryo transfer (ET), and frozen ET (FET). In addition, a lack of logical explanations for RIF made them more frustrated since the etiology of RIF is poorly understood. The underlying causes of RIF can be divided into embryonic and maternal factors. The implantation potential of a euploid embryo depends on the quality of the sperm and oocyte and the laboratory factors involved in the micromanipulation of gametes and embryo culture. Age is another main factor responsible for women's reproductive capacity, reflecting the embryo quality, endometrial receptivity, and gestational success. Other factors that are involved in implantation failures include uterine factors (congenital anomalies, fibroids, adenomyosis, endometritis, polyps, synechiae, and hydrosalpinx), endometriosis, systemic infections, overweight/obesity, stress, sleep disorders, lifestyle, environmental factors, hormonal conditions, immunological factors (auto and alloimmune), and thrombophilic conditions (acquired and inherited) [4–6].

2 Definition of thrombophilia

Optimal blood flow to the uterus, particularly endometrium/decidua, and placenta seems essential for embryo implantation and successful obstetrical outcomes. Historically, the Virchow triad, including hypercoagulability, hemodynamic changes (stasis and turbulence), and endothelial injury/dysfunction, described factors that can induce changes in blood flow [7]. Thrombophilia can be defined as a group of conditions that increase the risk of thrombosis (abnormal "blood-clotting") in veins or arteries. Thrombophilia can be inherited (hereditary) or acquired (not hereditary). Inherited thrombophilias occur due to mutations in genes responsible for producing clotting factors (known as genetic mutations). Acquired thrombophilias leading to blood clots are due to the imbalance of procoagulants, anticoagulants, and fibrinolytic substances in the blood, adverse vascular conditions, or the presence of specific antibodies, such as antiphospholipid antibodies [6].

The association between thrombophilias (inherited and/or acquired) and reproductive failure has been described for decades, especially regarding the relationship between antiphospholipid syndrome (APS) and obstetric complications, including recurrent miscarriages, intrauterine fetal death, preeclampsia, and placental insufficiency. However, the relationship between thrombophilias and RIF has not been well elucidated [8–10]. Possible mechanisms of developing unwanted reproductive outcomes in women with thrombophilias include microvascular occlusion at the decidua and thrombosis of maternal vessels. This would subsequently lead to reduced perfusion of the intervillous space, leading to embryo implantation failure.

2.1 Acquired thrombophilia

APS is an acquired thrombophilia, first described as patients with systemic lupus erythematosus and positive lupus anticoagulant complicated with thrombosis, recurrent miscarriage, and/or thrombocytopenia by Hughes in 1983 [11]. It is an autoimmune disease associated with the presence of autoantibodies. APA criteria include antibodies of anticardiolipin (aCL), anti-β2-glycoprotein I (anti-β2GPI), and lupus anticoagulant (LAC). Noncriteria APAs (phosphatidic acid, phosphatidylethanolamine, phosphatidylinositol, phosphatidylserine, phosphatidylglycerol, and phosphatidylcholine) are also described, but they are not considered for APS diagnosis [8,12–15].

The association between APS and recurrent miscarriage is well established in the literature. APS and APAs have been widely studied in association with RIF. Studies observed a higher prevalence of APAs in women with a history of implantation failures after IVF cycles. However, other studies did not find the association [13,14]. Further studies are needed to explore an association between APAs and implantation failure, with further attention to the standardization of the laboratory methodology and diagnostic criteria of APAs, preferably following the Sapporo criteria, as well as evaluating different groups of patients undergoing ART such as implantation failure and RIF [8,12–15].

2.2 Inherited thrombophilias

Inherited thrombophilia is a heterogeneous hereditary condition that increases the risk of thrombosis in its carriers. Genetic polymorphisms can manifest in forms of heterozygous (mutation in only one allele) or homozygous mutation (mutation in both alleles). The risk of a thromboembolic event may vary according to inherited thrombophilia and the number of variant alleles. Homozygous or a combination of two or more heterozygous mutations can lead to clinically apparent thrombotic disorders at an early age. Inherited thrombophilia increases the risk of maternal venous thromboembolism (VTE) and the risk of thrombosis at the maternal–placental interface, resulting in obstetrical complications, such as pregnancy loss, preeclampsia, intrauterine growth restriction, and placental abruption [16–18].

Factor V Leiden (FVL), prothrombin (PT) G20210A mutation, methylenetetrahydrofolate reductase (MTHFR) mutations, deficiency of protein C (PC) and protein S (PS), and antithrombin (AT), and plasminogen activator inhibitor-1 (PAI-1) 4G/5G polymorphism are inherited thrombophilias commonly related to adverse obstetrical outcomes. Inherited thrombophilia can be divided into high or low risk for VTE. High-risk thrombophilias for VTE include AT deficiency, FVL homozygous mutation, PT homozygous mutation, and a combination of heterozygous mutations in FVL and PT genes. Low risk inherited thrombophilias for VTE are FVL and PT heterozygous mutations and deficiencies in PC and PS [16–18].

2.2.1 FV Leiden

FV, a cofactor for activated factor X (FXa), is an important regulating factor in blood coagulation. Activated factor V (FVa) and FXa form the prothrombinase complex, which converts prothrombin to active thrombin. In 1994, Bertina et al. first described a defect in the FV gene, located on chromosome 1 (1q24.2), making it less susceptible to inactivation by activated PC (APC). FVL is a gene variant that is not susceptible to cleavage at position 506 by APC. The transition from guanine (G) to adenine (A) at nucleotide 1691 in exon 10 results in the synthesis of an FV molecule variant (FVL, FVR506Q, or FV: Q506) with the replacement of arginine with guanine at position 506 of the amino acid. This mutation confers partial resistance of FVa to activated PC and reduces the degradation of factor VIIIa. Therefore the defective FVa retains its procoagulant activity, as it does not undergo a natural blocking of PC, which predisposes to form abnormal and potentially harmful blood clots [16–18].

2.2.2 *Prothrombin gene mutation*

Prothrombin, coagulation factor II, when activated (thrombin), promotes the conversion of fibrinogen into fibrin and actively participates in clot formation with platelets. First described by Poort et al. in 1996, the G20210A mutation consists of the exchange of G to A in nucleotide 20210 in the prothrombin gene, located on chromosome 11 (11p11.2). The individual gene mutation may be homozygous involving two alleles or heterozygous involving one allele. The G allele is nonmutant, whereas the A allele corresponds to the mutant allele. Individuals carrying this mutation have high prothrombin levels and consequently manifest an exacerbated coagulation status with an increased risk for thromboembolic events [16–18].

2.2.3 *Methylenetetrahydrofolate reductase mutations*

MTHFR is encoded by a gene located on chromosome 1 (1p36.3). Among the dozens of MTHFR gene polymorphisms described in the literature, the two most commonly studied variants are C677T and A1298C: Cytosine is substituted to thymine in nucleotide 677 (C677T) in the first variant, and adenosine is exchanged to cytosine in nucleotide 1298 (A1298C) in the second. In both variants, a decrease in the activity of the MTHFR enzyme is noted [18].

MTHFR acts to regulate cellular methylation reactions, catalyzing the conversion of 5–10 methylenetetrahydrofolate to 5-methyltetrahydrofolate, where the methyl radical is donated for the remethylation of homocysteine to methionine (Fig. 1). The reduced MTHFR activity results in an increased need for folic acid, which keeps normal homocysteine to methionine remethylations. In the absence of sufficient folic acid, intracellular homocysteine is accumulated, and methionine resynthesis is reduced, compromising the main methylation reactions [16–18]. The decreased enzyme activity results in a high level of homocysteine. Hyperhomocysteinemia is a risk factor for thromboembolic events due to several mechanisms, such as increased vascular oxidative stress, local inflammatory reaction, and decreased relaxation of endothelial cells [16–18].

2.2.4 *Deficiency of protein S, protein C, and antithrombin*

Natural anticoagulants are needed to help stop the clotting process. Thus a deficiency of these substances alters the balance, leading to a thrombophilic state. PC, PS, and AT are the main natural anticoagulants involved in this process. Deficiencies of PC, PS, and AT are often caused by mutations in genes located on chromosome 2 [*PROC* gene (2q13–q14)], chromosome 3 [*PROS* gene (3q11–q11.2)], and chromosome 1 [*SERPINC1* gene (1q23 gene–25)], respectively. The transmission of these three conditions is autosomal recessive [16–18].

2.2.5 *Plasminogen activator inhibitor-1 4G/5G polymorphism*

Fibrinolysis is the process by which a fibrin clot (blood-clotting product) is resolved. The fibrinolytic system or plasminogen/plasmin system is composed of several proteins

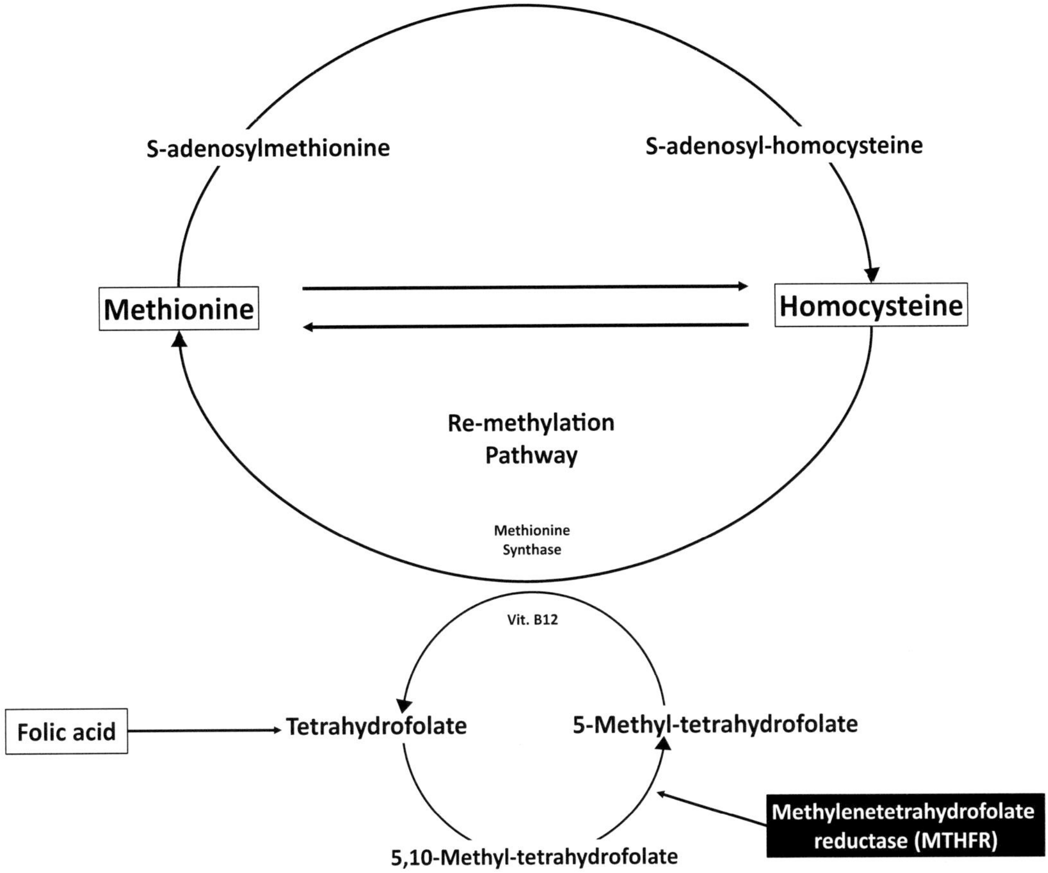

FIGURE 1 Methylenetetrahydrofolate reductase (MTHFR) action in homocysteine metabolism (Remethylation pathway). MTHFR mutation increases the risk of hyperhomocysteinemia due to failure of methionine resynthesis.

(serum proteases and inhibitors) that regulate the generation of plasmin from plasminogen. Plasmin functions to degrade fibrin and activate extracellular matrix metalloproteinases. Two physiological plasminogen activators are known: the tissue-type plasminogen activator (tPA) and the urokinase-type plasminogen activator (u-PA). Fibrinolysis is controlled by inhibitors of the fibrinolytic system, such as PAI-1. PAI-1 is the primary inhibitor of the physiological activator of plasminogen (tPA) (Fig. 2). Elevated plasma levels of PAI-1 can impair the fibrinolytic system and promote fibrin clot permanence [6,16,17].

PAI-1 is secreted in different tissues such as vascular endothelium, liver, and adipose tissue. The PAI-1 4G/5G polymorphism, also known as the −675 polymorphism of the SERPINE1 gene, located on chromosome 7 (7q21.3−22), is consisted of an insertion or deletion of guanosine 675 bp from the translation start site, affecting the transcription of this gene and resulting in altered plasma concentration of PAI-1. The 4G allele has a binding site for a transcription activator, which reflects higher concentrations of PAI-1, whereas the 5G allele has an additional binding site intended for a transcriptional repressor,

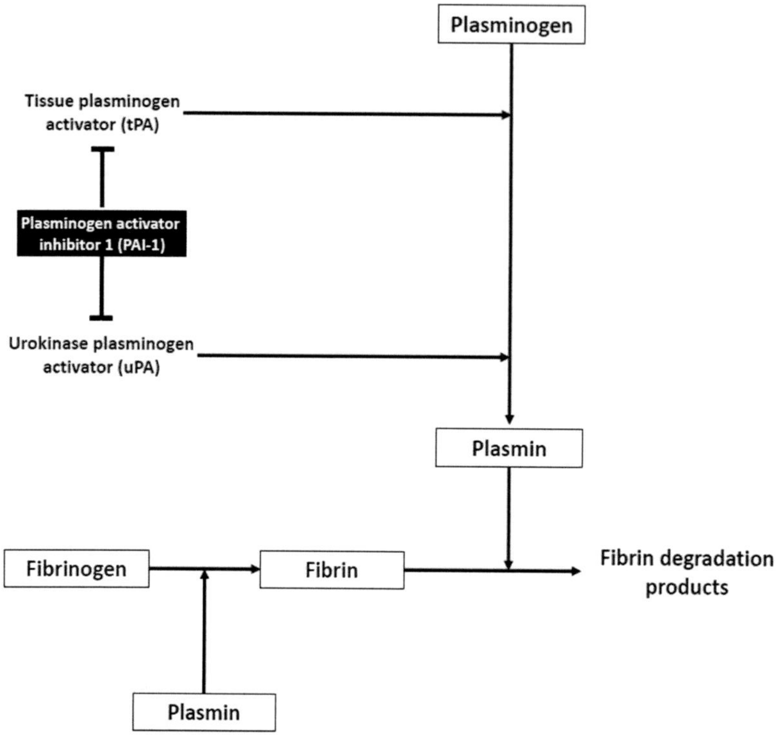

FIGURE 2 Urokinase plasminogen activator (u-PA) and tissue plasminogen activator (tPA) are plasminogen activators. Elevated plasma levels of plasminogen activator inhibitor-1 (PAI-1) can impair the fibrinolytic system and promote fibrin clot permanence.

resulting in lower circulating PAI-1 levels. Homozygotes for the 4G allele have 25% higher concentrations of PAI-1 than homozygotes for 5G. The presence of the 4G allele is associated with an increased risk of thromboembolic events and cardiovascular diseases. In addition, a high concentration of PAI-1 is found in women with an early abortion of unknown etiologies, as impaired fibrinolysis promotes fibrin deposition in the placental circulation [6,16–18].

3 Prevalence

3.1 Antiphospholipid antibodies

The prevalence of APS and inherited thrombophilia in the general population is unknown. In addition to ethnic and environmental factors, unstandardized diagnostic criteria, especially to define cases of APS (based on the Sydney 2006 criteria), have been the biggest limitation to determine the precise prevalence. Duarte-García et al. observed that the annual incidence of APS in the general population is 2.1 [95% confidence interval (CI): 1.4–2.8] per 100,000 population aged ≥18 years and the incidences are similar in both sexes. Moreover, an estimated prevalence of APS was reported to be 50 per 100,000 (95% CI: 42–58) [19]. In this study, the most frequent autoantibodies in APS were LAC (75% of cases), followed by aCL IgM (42%), aCL IgG (39%), anti-β2GPI IgG (22%), and anti-β2GPI IgM (11%) [19].

Often, women with a history of infertility and implantation failures do not present with clinical criteria for APS. Therefore most studies are limited to assessing the frequency of APAs in this population. Buckingham and Chamley compiled data from 27 studies that assessed the association of APAs and female infertility and reported a prevalence of APAs of 22.2% (range of 0%–66%) in this population. In addition, a higher prevalence of APAs was observed among women with RIF (14 studies, 29.7% range of 4%–66%) [12]. The association of noncriteria APAs with reproductive failures has also been described. Varla-Leftherioti et al. observed an increase in the frequency of APAs among women with recurrent miscarriage (>2 spontaneous losses) and RIF. A total of 26.9% had at least one positive APA. The aCL antibodies were more frequently found (56.1%) followed by anti-β2GPI (52.6%), anti-PS (28.9%), anti-PI (13.1%), anti-PA (10.8%), and LAC (8.6%) [12].

3.2 Inherited thrombophilia

The prevalence of inherited thrombophilia varies widely among healthy populations and between specific patient groups, especially those with a history of VTE. The prevalence of thrombophilia in Caucasians is approximately 10–7000 per 100,000 individuals worldwide. In a healthy Caucasian population, FVL is the most prevalent inherited thrombophilia (FVL heterozygous: 2%–7%; FVL homozygous: 0.06%–0.25%), followed by G20210A mutation (heterozygous: 1%–2%; homozygous: rare), hyperhomocysteinemia (<5%), PC deficiency (0.2%–0.5%), PS deficiency (0.1%–0.7%), combined FVL and G20210A mutation heterozygosity (0.1%), and AT deficiency (0.02%) [20,21]. In individuals with a history of the incident and recurrent VTE, thrombophilia is considerably more prevalent. In this group, FVL is also more prevalent (12%–50%), followed by G20210A mutation (3%–20%), hyperhomocysteinemia (<10%), PC deficiency (2%–10%), PS deficiency (1%–10%), and AT deficiency (1%–5%) [20–22].

The association between inherited thrombophilia and RIF has been described. Safdarian et al. evaluated the frequency of FVL, MTHFR C677T mutation, prothrombin 20210A mutation, homocysteine level, PAI-1 mutation, PS, PC, and AT deficiency among women with a history of RIF. The prevalence of at least one thrombophilia was 61.5% (59/96) of patients, with FVL being the most prevalent (16.7%), followed by MTHFR C677T mutation (homozygote: 6.3%; heterozygote: 11.5%), prothrombin 20210A mutation (9.4%), hyperhomocysteinemia (7.3%), PC deficiency (4.2%), PS deficiency (4.2%), AT deficiency (2.1%), and PAI-1 mutation (homozygote: 2.1%; heterozygote: 10.4%). The presence of at least one thrombophilia (95% CI = 1.74–5.70, OR = 3.15, $p = 0.00$), FVL (95% CI = 1.26–10.27, OR = 3.06, $p = 0.01$), and MTHFR C677T mutation (95% CI = 1.55–97.86, OR = 12.33, $p = 0.05$) showed a statistically significant association with embryo implantation failure [23]. Multiple studies corroborate these results, however, others have failed to confirm [6,24–26].

4 Underlying immunopathologies related to reproduction

Coagulation and consequential thrombotic events at the maternal–fetal interface can lead to pregnancy losses in patients with APS. However, the development of the intervillous space occurs after 10 weeks of gestation, making it difficult to explain the implantation failure only with microthrombosis in decidual vessels. It is speculated that other mechanisms are involved in the pathophysiology of implantation failure in patients with inherited or acquired thrombophilia or both [8,13,27]. The proinflammatory effect of APAs could affect fertility by disrupting oocyte development or uterine decidualization [13,28,29]. APAs were shown to reduce the proliferation, invasion, and syncytialization of trophoblasts, as well as hCG production. In addition, reproductive failure may occur due to activation of vascular cells, including endothelial cells, platelets, and monocytes, inducing procoagulant effects of extracellular vesicles and disruption of the anticoagulant Annexin A5 shield on cellular surfaces [13,29,30]. APAs are associated with complement activation since complement and coagulation pathways are strongly connected [31,32]. Activation of the complement system also has a strong association with reproductive failures [33].

The association between prothrombin 20210A mutation and implantation failure has not been well defined. Soluble fms-like tyrosine kinase-1 (sFlt-1 or sVEGFR-1) is a protein with antiangiogenic properties, which binds to angiogenic factors such as vascular endothelial growth factor (VEGF) and placental growth factor, reducing neoangiogenesis. High levels of thrombin in decidual cells induce a greater production of sFlt-1, which inhibits enzymes related to extravillous cytotrophoblast invasion [34,35].

Recently, studies have established a connection between folic acid metabolism and the embryonic implantation process. Folic acid and homocysteine are present in the follicular fluid. Elevated levels of homocysteine in the follicular fluid are associated with poor oocyte quality [36]. Moreover, women with MTHFR 677TT (homozygous mutation, TT) have significantly higher homocysteine and NK cell cytotoxicities than women with MTHFR 677CC and 677 CT [37] and increased NK cell cytotoxicities have been reported to be associated with RIFs [38]. Histopathological studies found that pregnant women with hyperhomocysteinemia have reduced total chorionic villus area in addition to poor vascularization [39]. These mechanisms strengthen a possible association between MTHFR gene mutation with or without hyperhomocysteinemia and RIF.

Fibrinolysis is crucial for the implantation process. The activation of plasminogen facilitates cell migration through proteolysis and dissolution of the basement membrane. Inhibition of fibrinolysis by increased plasminogen activator inhibitors such as PAI-1 may impair the depth of adequate trophoblast invasion [40,41]. Therefore hypofibrinolysis due to 4G/5G polymorphism appears to be a possible risk factor for implantation failure.

5 Diagnosis

5.1 Antiphospholipid syndrome

The first diagnostic criterion for APS, Sapporo Criteria, was proposed in 1999. Subsequently, the Sapporo criteria were revised in 2006 at an international congress held in Sydney. Currently, the diagnosis of APS requires the presence of at least one clinical and one laboratory criterion (Table 1) [42].

TABLE 1 Diagnosis criteria for antiphospholipid syndrome (Sydney, 2006).

Clinical criteria
1. Vascular thrombosis One or more clinical episodes of arterial, venous, or small vessel thrombosis, in any tissue or organ. Thrombosis must be confirmed by objective validated criteria (i.e. unequivocal findings of appropriate imaging studies or histopathology). For histopathologic confirmation, thrombosis should be present without significant evidence of inflammation in the vessel wall
2. Pregnancy morbidity a. One or more unexplained deaths of a morphologically normal fetus at or beyond the 10th week of gestation, with normal fetal morphology documented by ultrasound or by direct examination of the fetus, or b. One or more premature births of a morphologically normal neonate before the 34th week of gestation because of (i) eclampsia or severe pre-eclampsia defined according to standard definitions or (ii) recognized features of placental insufficiency, or c. Three or more unexplained consecutive spontaneous abortions before the 10th week of gestation, with maternal anatomic or hormonal abnormalities and paternal and maternal chromosomal causes excluded
Laboratory criteria
1. Lupus anticoagulant (LA) present in plasma, on two or more occasions at least 12 weeks apart, detected according to the guidelines of the International Society on Thrombosis and Hemostasis (Scientific Subcommittee on LAs/phospholipid-dependent antibodies) 2. aCL antibody of IgG and/or IgM isotype in serum or plasma, present in medium or high titer (i.e., >40 GPL or MPL, or >the 99th percentile), on two or more occasions, at least 12 weeks apart, measured by a standardized ELISA 3. Anti-$\beta_2$glycoprotein-I antibody of IgG and/or IgM isotype in serum or plasma (in titer >the 99th percentile), present on two or more occasions, at least 12 weeks apart, measured by a standardized ELISA, according to recommended procedures

Ref. Miyakis et al. (2006).

5.2 Inherited thrombophilia

The diagnosis of inherited thrombophilia (FVL, MTHFR C677T mutation, prothrombin 20210A mutation, and PAI-1 4G/5G polymorphism) is through molecular genetic technology in search of polymorphisms, determining the presence and quantity of mutated alleles [9,18].

The diagnosis of PC deficiency is based on the quantification of PC levels measured by ELISAs, radioimmunoassays, or electroimmunoassays. PC activity levels vary from 0% to 30% in case of severe deficiencies and from 30% to 70% in case of partial deficiencies. PC deficiency has two biological forms. Type I deficiency is characterized by a consistent reduction in PC and antigen activity. In type II deficiency, PC activity is reduced, but PC antigen is normal. Molecular testing is available but not necessary for the diagnosis [43].

The diagnosis of PS deficiency is based on the quantification of antigen levels (total protein S or free protein S) and anticoagulant activity. Free and total PS is measured by ELISA, with three biological forms. Types I and III are the most common types with quantitative deficiencies and low levels of free antigens (with normal total PS levels in type III and decreased total PS levels in type I). Type II (rare type) is a qualitative deficiency with normal levels of total and free PS. Molecular testing is available but not necessary for diagnosis [43].

Two types of AT deficiency are as follows: type I (quantitative, characterized by reduced plasma levels of the antigen and functional activity of AT) and type II (qualitative, characterized by the presence of variant AT in plasma, with normal antigen levels and decreased activity). Type II can be further subdivided into the reactive site, heparin-binding site, and pleiotropic defects (mutations clustered in a region called s1C–s4B). The diagnosis and classification of AT deficiency are performed by plasma investigation of antigen activity and concentrations, using functional and immunological methods [44].

Abnormal results of PC, PS, and AT deficiency should be repeated in one or more separate samples to confirm. A single abnormal result is not sufficient to diagnose these deficiencies. Laboratory investigation is not reliable during pregnancy with vitamin K intake, heparin use, acute phase of a thrombotic event, or presence of liver disease [43,44].

Approximately 70% of plasma homocysteine is bound to albumin and other proteins, and a small portion is free in the form of homocysteine and homocysteine-cysteine disulfide. Methods that are used to measure free amino acids cannot detect small amounts of free homocysteine. Therefore reducing agents are added, such as dithiothreitol or beta-mercaptoethanol, in the sample to release homocysteine from plasma proteins and determine the total plasma homocysteine (tHcy). All methods that determine tHcy find similar physiological levels (5–15 μmol/L or μM). Currently, most laboratories use methods based on high-performance liquid chromatography or stable isotope-dilution gas chromatography-mass spectrometry. Other methods to measure tHcy include fluorescence polarization immunoassay, immunoenzymatic assay by chemiluminescence, and chromatographic method with fluorescence detection [45].

6 Treatment

Given the controversial data in the literature, screening and treatment of inherited thrombophilias to improve pregnancy outcomes in women with a history of obstetric complications such as fetal loss, preeclampsia, fetal growth restriction, placental abruption, infertility, or failed embryo implantation, has no international consensus for the evaluation of thrombophilias yet. Currently, only women with a personal history of VTE or a first-degree relative (e.g., parent or sibling) with a history of high-risk inherited thrombophilia are candidates for inherited thrombophilia screening despite the accumulated evidence of thrombophilia in these populations [46].

Traditionally, APS treatment is based on antiplatelet and anticoagulant therapies and, in advanced cases, immune-suppressant or modulatory drugs such as prednisone, hydroxychloroquine, intravenous immunoglobulins, rituximab, and eculizumab. Currently, a therapeutic proposal for patients diagnosed with pure obstetric APS (defined by the presence of at least one clinical obstetric and one laboratory criterion) is a combination of low-dose aspirin (LDA, 75–100 mg/day) and heparin [unfractionated heparin (UFH) or low molecular weight heparin (LMWH)] in prophylactic dose [47]. This recommendation is based on two randomized controlled trials in which women with recurrent first-trimester pregnancy losses and APAs were randomized to LDA or a combination of aspirin and UFH [48,49]. A significantly higher rate of live births was reported in the group treated with LDA plus UFH compared to isolated use of LDA (71% vs 42%, OR: 3.37 95% CI 1.40–8.10) [48]. The association of LDA plus low dose (UFH 10,000 units every 12 hours) of heparin was as effective as the LDA plus high dose (UFH 20,000 units every 12 hours) of heparin for the treatment of recurrent pregnancy loss associated with antiphospholipid antibodies [49].

Despite contradictory evidence of the association between thrombophilia (acquired and inherited) and embryo implantation failure, heparin (UFH or LMWH) alone or in combination with LDA has been proposed as adjuvant therapy in women undergoing ART. The Cochrane Library reported a beneficial effect of using heparin on clinical outcomes of ART cycles regardless of thrombophilia diagnosis. In the treated group, a higher live birth rate (LBR) was observed compared to a placebo (OR 1.77, 95% CI 1.07–2.90, three studies, 386 women, $I^2 = 51\%$, low-quality evidence) or without any intervention (OR 1.61, 95% CI 1.03–2.53, three studies, 386 women, $I^2 = 29\%$, low-quality evidence) and an improvement in the clinical pregnancy rate [50]. Another metaanalysis by Potdar et al. corroborated these results. Women with a history of RIF and heparin treatment (with or without thrombophilia) had improved LBR (RR 1.79, 95% CI 1.10–2.90, $p = 0.02$) and reduced miscarriage rate (RR 0.22, 95% CI 0.06–0.78, $p = 0.02$) [51]. Contrarily, other studies, including one metaanalysis reported prior to the Cochrane report, did not observe improvements in clinical outcomes of patients undergoing IVF/ICSI (with or without a history of implantation failures) and heparin treatment with and without LDA [52–54].

The beneficial effect of heparin in improving the embryo implantation process was explained not only due to anticoagulant activity but an antiinflammatory effect to improve the invasion of extravillous trophoblast cells and decrease trophoblast apoptosis [55,56]. In addition, heparin suppressed complement activation (at various points in the classical, alternative, and terminal pathways) in vivo and protected the mother from pregnancy complications induced by APAs antibodies in the mouse model [57,58]. Heparin is a widely used drug during pregnancy, especially for the prevention and treatment of thromboembolic events [21] and its use for APS treatment has been established [8]. The Food and Drug Administration has classified heparin in category B. Due to its large molecular weight, it does not cross the placenta and is not excreted in breast milk. The use of heparin, preferably LMWH, during pregnancy and breastfeeding is considered safe. The risk of bleeding in patients using LMWH even at the prophylactic dose during pregnancy and postpartum hemorrhage (0.5% and 1%, respectively) or as therapeutic doses (1.5% and 2%, respectively) should be recognized as well as other complications, such as thrombocytopenia and osteoporosis when using heparin [56].

7 Summary and recommendations

Studies suggest that changes in endometrial/decidual blood flow due to microthrombosis induced by thrombophilic conditions can impair embryo implantation. Other mechanisms, including exacerbated inflammatory uterine environment, reduced trophoblastic invasion, and activation of the complement system, also seem to reduce implantation rates in patients with thrombophilia. Currently, the association between APS, APAs, and inherited thrombophilia with RIF is still unclear. Consequently, screening these conditions is not routinely recommended for women with a history of embryo implantation failures. However, despite the lack of consensus in evaluating these women, recent metaanalyses demonstrated improved reproductive outcomes in IVF/ET and FET cycles in women with RIF when heparin alone or heparin and LDA adjuvant therapies were given. Well-designed randomized and double-blind clinical trials with strict criteria for clinical and laboratory diagnosis are needed.

- Thrombophilia can be classified as acquired (such as APS) or inherited.
- Reduction in endometrial/decidual blood flow caused by microthrombosis induced by thrombophilia seems to impair embryo implantation.
- Inflammatory uterine environment, complement activation, and impaired trophoblastic invasion are underlying mechanisms of implantation failure induced by thrombophilia.
- No general consensus was made whether women with RIF have a higher prevalence of APS, APAs, and inherited thrombophilia.
- The use of antiplatelet (low dose aspirin) and anticoagulant agents (heparin) was proposed as an adjuvant therapy in ART to improve clinical results of women with a history of implantation failures.

References

[1] Ashary N, Tiwari A, Modi D. Embryo implantation: war in times of love. Endocrinology 2018;159:1188–98.
[2] Coughlan C, Ledger W, Wang Q, et al. Recurrent implantation failure: definition and management. Reprod Biomed Online 2014;28:14–38.
[3] Kolte AM, Bernardi LA, Christiansen OB, et al. Terminology for pregnancy loss prior to viability: a consensus statement from the ESHRE early pregnancy special interest group. Hum Reprod 2015;30:495–8.
[4] Shaulov T, Sierra S, Sylvestre C. Recurrent implantation failure in IVF: a Canadian fertility and andrology society clinical practice guideline. Reprod Biomed Online 2020;41:819–33.
[5] Bashiri A, Halper KI, Orvieto R. Recurrent implantation failure-update overview on etiology, diagnosis, treatment and future directions. Reprod Biol Endocrinol 2018;16:121.
[6] Ivanov P, Tsvyatkovska T, Konova E, Komsa-Penkova R. Inherited thrombophilia and IVF failure: the impact of coagulation disorders on implantation process. Am J Reprod Immunol 2012;68:189–98.
[7] Devis P, Knuttinen MG. Deep venous thrombosis in pregnancy: incidence, pathogenesis and endovascular management. Cardiovasc Diagn Ther. 2017;7:S309–19.
[8] Garcia D, Erkan D. Diagnosis and management of the antiphospholipid syndrome. N Engl J Med 2018;378:2010–21.
[9] Arachchillage DRJ, Makris M. Inherited thrombophilia and pregnancy complications: should we test? Semin Thromb Hemost 2019;45:50–60.
[10] Simcox LE, Ormesher L, Tower C, Greer IA. Thrombophilia and pregnancy complications. Int J Mol Sci. 2015;16:28418–28.
[11] Hughes GR. Thrombosis, abortion, cerebral disease, and the lupus anticoagulant. Br Med J 1983;287:1088–9.
[12] Varla-Leftherioti M, Keramitsoglou T, Gourniezaki G, et al. Retrospective analysis on the incidence of the types and isotypes of antiphospholipid antibodies in subfertile women. J Reprod Immunol 2014;101–102:59.
[13] Ata B, Urman B. Thrombophilia and assisted reproduction technology-any detrimental impact or unnecessary overuse? J Assist Reprod Genet 2016;33:1305–10.
[14] Stern C, Chamley L. Antiphospholipid antibodies and coagulation defects in women with implantation failure after IVF and recurrent miscarriage. Reprod Biomed Online 2006;13:29–37.
[15] Kwak-Kim J, Park JC, Ahn HK, Kim JW, Gilman-Sachs A. Immunological modes of pregnancy loss. Am J Reprod Immunol 2010;63:611–23.
[16] Middeldorp S. Inherited thrombophilia: a double-edged sword. Hematology Am Soc Hematol Educ Program 2016;2016:1–9.
[17] Phillippe HM, Hornsby LB, Treadway S, Armstrong EM, Bellone JM. Inherited thrombophilia. J Pharm Pract. 2014;27:227–323.

[18] Dautaj A, Krasi G, Bushati V, et al. Hereditary thrombophilia. Acta Biomed 2019;90:44–6.
[19] Duarte-García A, Pham MM, Crowson CS, et al. The epidemiology of antiphospholipid syndrome: a population-based study. Arthritis Rheumatol. 2019;71:1545–52.
[20] Seligsohn U, Lubetsky A. Genetic susceptibility to venous thrombosis. N Engl J Med 2001;344:1222–31.
[21] Stevens SM, Woller SC, Bauer KA, et al. Guidance for the evaluation and treatment of hereditary and acquired thrombophilia. J Thromb Thrombolysis 2016;41:154–64.
[22] Cohoon KP, Heit JA. Inherited and secondary thrombophilia. Circulation 2014;129:254–7.
[23] Safdarian L, Najmi Z, Aleyasin A, Aghahosseini M, Rashidi M, Asadollah S. Recurrent IVF failure and hereditary thrombophilia. Iran J Reprod Med. 2014;12:467–70.
[24] Qublan HS, Eid SS, Ababneh HA, et al. Acquired and inherited thrombophilia: implication in recurrent IVF and embryo transfer failure. Hum Reprod 2006;21:2694–8.
[25] Steinvil A, Raz R, Berliner S, et al. Association of common thrombophilias and antiphospholipid antibodies with success rate of in vitro fertilisation. Thromb Haemost 2012;108:1192–7.
[26] Han AR, Han JW, Lee SK. Inherited thrombophilia and anticoagulant therapy for women with reproductive failure. Am J Reprod Immunol 2021;85:e13378.
[27] Rey E, Kahn SR, David M, Shrier I. Thrombophilic disorders and fetal loss: a meta-analysis. Lancet 2003;361:901–8.
[28] Tomassetti C, Meuleman C, Pexsters A, et al. Endometriosis, recurrent miscarriage and implantation failure: is there an immunological link? Reprod Biomed Online 2006;13:58–64.
[29] Kwak-Kim J, Agcaoili MS, Aleta L, et al. Management of women with recurrent pregnancy losses and antiphospholipid antibody syndrome. Am J Reprod Immunol 2013;69:596–607.
[30] de Laat B, Mertens K, de Groot PG. Mechanisms of disease: antiphospholipid antibodies-from clinical association to pathologic mechanism. Nat Clin Pract Rheumatol 2008;4:192–9.
[31] Chaturvedi S, Brodsky RA, McCrae KR. Complement in the pathophysiology of the antiphospholipid syndrome. Front Immunol 2019;10:449.
[32] Foley JH, Conway EM. Cross talk pathways between coagulation and inflammation. Circ Res 2016; 118:1392–408.
[33] Tincani A, Cavazzana I, Ziglioli T, Lojacono A, De Angelis V, Meroni P. Complement activation and pregnancy failure. Clin Rev Allergy Immunol 2010;39:153–9.
[34] Lockwood CJ, Krikun G, Rahman M, Caze R, Buchwalder L, Schatz F. The role of decidualization in regulating endometrial hemostasis during the menstrual cycle, gestation, and in pathological states. Semin Thromb Hemost 2007;33:111–17.
[35] Lockwood CJ, Toti P, Arcuri F, et al. Thrombin regulates soluble fms-like tyrosine kinase-1 (sFlt-1) expression in first trimester decidua: implications for preeclampsia. Am J Pathol 2007;170:1398–405.
[36] Szymański W, Kazdepka-Ziemińska A. Wpływ stezenia homocysteiny w płynie pecherzykowym na stopień dojrzałości komórki jajowej [effect of homocysteine concentration in follicular fluid on a degree of oocyte maturity]. Ginekol Pol 2003;74:1392–6.
[37] Ota K, Takahashi T, Han A, Damvaeba S, Mizunuma H, Kwak-Kim J. Effects of MTHFR C677T polymorphism on vitamin D, homocysteine and natural killer cell cytotoxicity in women with recurrent pregnancy losses. Hum Reprod 2020;35:1276–87.
[38] Sung N, Khan SA, Yiu ME, Jubiz G, Salazar MD, Skariah A, et al. Reproductive outcomes of women with recurrent pregnancy losses and repeated implantation failures are significantly improved with immunomodulatory treatment. J Reprod Immunol 2021;148:103369.
[39] Nelen WL, Bulten J, Steegers EA, Blom HJ, Hanselaar AG, Eskes TK. Maternal homocysteine and chorionic vascularization in recurrent early pregnancy loss. Hum Reprod 2000;15:954–60.
[40] Lang IM, Moser KM, Schleef RR. Elevated expression of urokinase-like plasminogen activator and plasminogen activator inhibitor type 1 during the vascular remodeling associated with pulmonary thromboembolism. Arterioscler Thromb Vasc Biol 1998;18:808–15.
[41] Gris JC, Ripart-Neveu S, Maugard C, et al. Respective evaluation of the prevalence of haemostasis abnormalities in unexplained primary early recurrent miscarriages. The Nimes Obstetricians and Haematologists (NOHA) study. Thromb Haemost 1997;77:1096–103.
[42] Miyakis S, Lockshin MD, Atsumi T, et al. International consensus statement on an update of the classification criteria for definite antiphospholipid syndrome (APS). J Thromb Haemost 2006;4:295–306.

[43] Wypasek E, Undas A. Protein C and protein S deficiency—practical diagnostic issues. Adv Clin Exp Med. 2013;22:459–67.
[44] Khor B, Van Cott EM. Laboratory tests for antithrombin deficiency. Am J Hematol 2010;85:947–50.
[45] American Society of Human Genetics/American College of Medical Genetics Test and Transfer Committee Working Group. Measurement and use of total plasma homocysteine. Am J Hum Genet 1998;63:1541–3.
[46] American College of Obstetricians and Gynecologists' Committee on Practice Bulletins–Obstetrics. ACOG practice bulletin No. 197: inherited thrombophilias in pregnancy. Obstet Gynecol 2018;132:e18–34.
[47] Ghembaza A, Saadoun D. Management of antiphospholipid syndrome. Biomedicines. 2020;8:508.
[48] Rai R, Cohen H, Dave M, Regan L. Randomised controlled trial of aspirin and aspirin plus heparin in pregnant women with recurrent miscarriage associated with phospholipid antibodies (or antiphospholipid antibodies). BMJ 1997;314:253–7.
[49] Kutteh WH, Ermel LD. A clinical trial for the treatment of antiphospholipid antibody-associated recurrent pregnancy loss with lower dose heparin and aspirin. Am J Reprod Immunol 1996;35:402–7.
[50] Akhtar MA, Sur S, Raine-Fenning N, Jayaprakasan K, Thornton JG, Quenby S. Heparin for assisted reproduction. Cochrane Database Syst Rev 2013;8:CD009452.
[51] Potdar N, Gelbaya TA, Konje JC, Nardo LG. Adjunct low-molecular-weight heparin to improve live birth rate after recurrent implantation failure: a systematic review and meta-analysis. Hum Reprod Update 2013;19:674–84.
[52] Akhtar MA, Eljabu H, Hopkisson J, Raine-Fenning N, Quenby S, Jayaprakasan K. Aspirin and heparin as adjuvants during IVF do not improve live birth rates in unexplained implantation failure. Reprod Biomed Online 2013;26:586–94.
[53] Siristatidis C, Dafopoulos K, Salamalekis G, et al. Administration of low-molecular-weight heparin in patients with two or more unsuccessful IVF/ICSI cycles: a multicenter cohort study. Gynecol Endocrinol 2018;34:747–51.
[54] Seshadri S, Sunkara SK, Khalaf Y, El-Toukhy T, Hamoda H. Effect of heparin on the outcome of IVF treatment: a systematic review and meta-analysis. Reprod Biomed Online 2012;25:572–84.
[55] Tersigni C, Marana R, Santamarìa A, Castellani R, Scambia G, Simone ND. In vitro evidences of heparin's effects on embryo implantation and trophoblast development. Reprod Sci 2012;19:454–62.
[56] Mulloy B, Hogwood J, Gray E, Lever R, Page CP. Pharmacology of heparin and related drugs. Pharmacol Rev 2016;68:76–141.
[57] Girardi G, Redecha P, Salmon JE. Heparin prevents antiphospholipid antibody-induced fetal loss by inhibiting complement activation. Nat Med 2004;10:1222–6.
[58] Ninomiya H, Kawashima Y, Nagasawa T. Inhibition of complement-mediated haemolysis in paroxysmal nocturnal haemoglobinuria by heparin or low-molecular weight heparin. Br J Haematol 2000;109:875–81.

SECTION IV

Dysregulated neuroimmune-endocrine network in reproductive failures

CHAPTER

22

The ovarian immune pathology and reproductive failures

Li Wu[1], *Xuhui Fang*[1], *Yanshi Wang*[1] *and Joanne Kwak-Kim*[2]

[1]Reproductive Medicine Center, Department of Obstetrics and Gynecology, The First Affiliated Hospital of USTC, Division of Life Sciences and Medicine, University of Science and Technology of China, Hefei, Anhui, P.R. China [2]Reproductive Medicine and Immunology, Obstetrics and Gynecology, Clinical Sciences Department, Chicago Medical School, Rosalind Franklin University of Medicine and Science, Vernon Hills, IL, United States

1 Introduction

The human ovary is a common target for autoimmune attacks, resulting in immune inflammation and consequential ovarian dysfunction, such as abnormal follicular development, ovulatory failure, insufficient corpus luteal function, infertility, and even pregnancy failures. Both humoral and cellular autoimmune responses can be involved in the etiopathogenesis of ovarian autoimmunity. Antiovarian antibodies (AOAs) are primarily associated with premature ovarian insufficiency (POI), premature ovarian failure (POF), and unexplained infertility (UI) [1]. Indeed, in POF, an autoimmune mechanism has been suggested in approximately 3.0%–66.6% of cases [2–6]. Ovarian lymphocytic infiltration, "lymphocytic oophoritis," is a typical POF pathology that is often associated with infiltration of mononuclear cells in the ovary, presence of autoantibodies to ovarian antigens, and other autoimmune disorders [7]. Notably, in POF women with autoimmune etiology, restoration of ovarian function is often seen after regression of autoimmune status [8].

Women with ovarian autoimmunity may receive various clinical diagnoses based on clinical manifestation or laboratory findings, such as POI, POF, poor ovarian reserve (POR), diminished ovarian reserve (DOR), autoimmune oophoritis, and UI [9]. Frequently, ovarian autoimmunity is associated with other underlying conditions, including PCOS and endometriosis (EMS), in which autoimmunity is a common underlying etiology [9,10] Interestingly, women with autoimmune oophoritis may present multifollicular ovaries with increased serum LH levels, which is erroneously recognized as a polycystic ovarian syndrome

Immunology of Recurrent Pregnancy Loss and Implantation Failure
DOI: https://doi.org/10.1016/B978-0-323-90805-4.00018-3

(PCOS), but they do not manifest hyperandrogenemia [11]. Notably, the autoimmune mechanism has been considered one of the underlying etiologies for PCOS. Contrariwise, low progesterone levels due to oligo/anovulation and relative hyperestrogenism in PCOS overstimulate the immune system, leading to the production of autoantibodies [12].

Timely shifting and balancing the pro and antiinflammatory immune responses are critical for the successful implantation and maintenance of pregnancy by establishing maternal immune tolerance to the fetus, while unbalanced immune response results in inflammatory process and pathology, leading to infertility, especially UI, and repeated implantation failures (RIFs) [13]. For example, in women with ovarian endometrioma, increased levels of inflammatory cytokines and autoantibodies in sera and peritoneal fluid are associated with infertility and even pregnancy losses [14]. Furthermore, autoimmunity, including autoimmune oophoritis, antiphospholipid syndrome (APS), and autoimmune thyroid disease, preferentially affects women throughout their reproductive age and impacts various stages of human reproduction, leading to ovarian failures, implantation failures, and pregnancy losses. Hence, this chapter comprehensively reviews immunopathologies related to ovarian autoimmunity, clinical manifestations, and management.

2 Ovarian immunology

An ovarian cycle includes folliculogenesis, ovulation, and either menstruation or implantation. Neuroendocrine hormones, immune effectors, and related factors control the physiological changes during an ovarian cycle. Hence, it is critical to comprehend the immune responses within an ovary and endometrium during an ovarian cycle in human reproduction. Peripheral blood immune effectors are recruited to the ovarian stroma, follicles, and endometrium during an ovarian cycle and actively involved in folliculogenesis, ovulation, corpus luteum (CL) formation, luteolysis, and upcoming implantation [14]. Hence, abnormal immune responses in the ovary and endometrium often result in reproductive failures, such as poor folliculogenesis, poor ovarian response, poor egg quality, RIF, and recurrent pregnancy losses (RPLs) [14].

Two nonconceptual events during an ovarian cycle, such as ovulation and menstruation, have similarities to inflammatory processes [15]. It is reported that women with more follicular waves had higher serum C-reactive protein (CRP) levels. Indeed, higher CRP levels were correlated with increased progesterone levels, suggesting that the normal ovarian cycle may be related to inflammation [12,15]. However, these "physiological" inflammations are different from pathological inflammation, which often ends in autoimmunity. Additionally, menstruation and implantation have similar characteristic inflammatory features, such as tissue edema and the infiltration of many inflammatory cells. After ovulation, the endometrium prepares for successful implantation with immunotropism and angiogenesis [16]. Various immune effectors appear in the endometrium even before the embryonic implantation and contribute to implantation by participating in tissue remodeling and angiogenesis, while without implantation, menstruation is initiated. It is suggested that decreased progesterone level induces breakdown, revival, and restoration of endometrial tissue, suggesting progesterone could inhibit inflammatory responses. Additionally, estrogen has both pro- and antiinflammatory roles depending on the immune stimulus and subsequent antigen-specific immune responses, cell types, target organs with the specific microenvironment, reproductive status, estrogen concentration, estrogen receptor expression, and intracellular metabolism of estrogen [17]. Thus failed or dysregulated ovulation and ovarian hyperstimulation may induce ovarian autoimmunity and systemic proinflammatory responses, resulting in potential obstetrical complications, including RIF, RPL, and preeclampsia [18].

3 Ovarian autoimmunity

3.1 Autoimmune lymphocytic oophoritis

Autoimmune lymphocytic oophoritis, a known specific cause of primary ovarian insufficiency (POI) and POF, was first described in women with Addison's disease and adrenal autoimmunity. It is characterized by mononuclear inflammatory cell infiltration in the ovarian theca cell layers of growing follicles and CL while sparing primordial and primary follicles [19]. Histological inflammatory features in ovarian biopsy samples and circulating ovarian and adrenal autoantibodies are diagnostic features of autoimmune POI [20]. Cellular and humoral immunity may play a vital role in autoimmune lymphocytic oophoritis. Infiltrated inflammatory cells are predominantly T (CD4 + , CD8 +), natural killer (NK), B, and plasma cells, mainly producing IgG antibodies [21]. Inflammation is often present in perivascular and perineural regions and affects steroid-producing cells in the preovulatory follicles, notably the internal and external layers of theca, corpora luteum, and occasionally granulosa cells [22]. These distinctive immunopathological findings present a theoretical possibility that an immunosuppressive treatment might restore fertility in autoimmune POI cases. In addition, regulatory T cells (Treg) have a critical role in systemic tolerance of ovarian antigens, and the amelioration of autoimmune ovarian disease was reported with the induction of mouse zona pellucida (ZP) protein 3 specific Treg cells in the mouse model [23].

Besides, in women with autoimmune oophoritis, the presence of large multifollicular ovaries combined with elevated serum LH levels has been mistaken for PCOS [24]. However, in contrast to women with PCOS, women with autoimmune oophoritis would not be expected to demonstrate signs of androgen excess. Instead, women with POI based on autoimmune oophoritis are likely to have clinical and laboratory evidence of androgen deficiency, presumably related to impaired production of theca cell and adrenal androgens. In most cases, any question regarding a differential diagnosis between autoimmune oophoritis and PCOS could be resolved by testing circulating adrenal cortex autoantibodies and measuring FSH and serum androgen levels.

3.2 Molecular and cellular targets for ovarian autoimmunity

The involvement of serum AOA in ovarian pathology remains controversial. It has been reported that AOA are found in 24%–73% of women with POI [24]. The presence of positive AOA is associated with one-third of women with UI. Various autoantibodies participate in ovarian insufficiency, and a wide range of target antigens in the ovary have been reported (Fig. 1). These autoantibodies bind to steroid hormone-producing cells, including granulosa

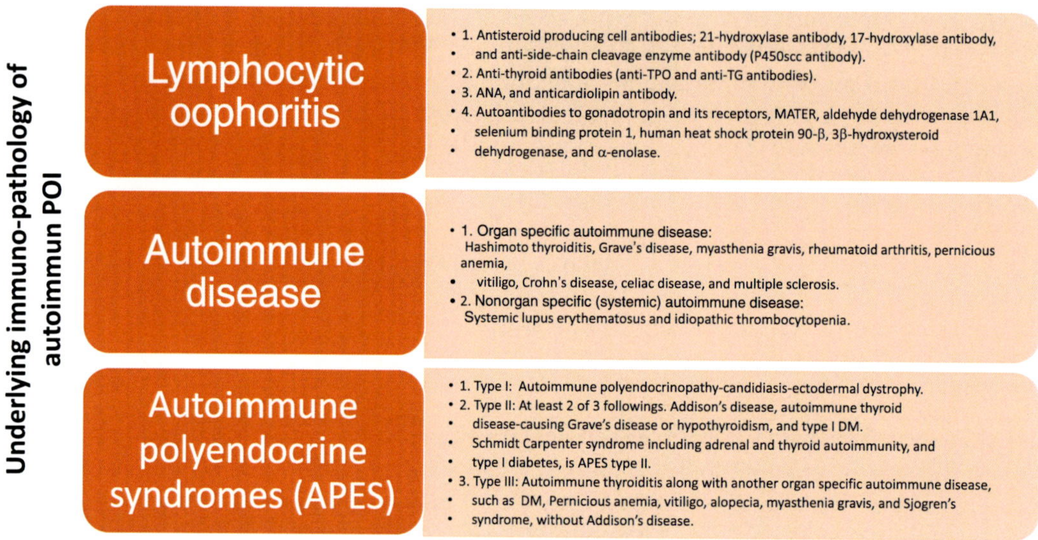

FIGURE 1 Underlying immuno-pathology of autoimmune premature ovarian insufficiency.

cells, oocytes, gonadotropins, gonadotropin-receptors, ZP of oocytes, and CL [3]. The most common targets are gonadotropins and their receptors, especially β-subunit of FSH [25].

The most autoimmune POI is commonly associated with clinical or preclinical autoimmune Addison's disease. Autoantibodies to steroidogenic enzymes, targeting steroidogenic cytochrome P450 enzymes, such as 21-hydroxylase (21OH), 17-hydroxylase (17OH), side-chain cleavage enzyme (P450scc), are invariantly present in affected women [11,25] and autoantibodies to these molecules identify POI due to steroid-producing cell autoimmunity (SCA-POI). The other molecular targets for AOA include maternal antigen that embryos require (MATER), aldehyde dehydrogenase-1A1 (ALDH1A1), selenium binding protein 1 (SBP1), human heat shock protein 90-β (HSP90 β), 3beta-hydroxysteroid dehydrogenase (3b-HSD), and α-enolase [22,26]. In addition, several autoantibodies, such as anticardiolipin and antinuclear antibodies have been reported as markers for ovarian autoimmunity. The diversity of suspected autoantigens and related antibodies indicates the multiplicity of pathological processes leading to ovarian damage [3].

A high prevalence of antioocyte cytoplasm antibodies has been reported in fertilization failure and oocyte retrieval failure cases [27,28]. These antibodies recognize MATER, which is expressed predominantly in parathyroid tissue and lesser in the oocyte and persists until the blastocyst stage [29]. The knock-out mouse model shows that inactivation of the gene results in the sterility of female mice [30]. In humans, MATER (the human homolog as NALP5)-autoantibodies were detected in 58% of autoimmune polyendocrine syndrome (APES) type 1 patients, 7% of Addison's disease patients, 12% of steroidogenic cell autoimmune POI (SCA-POI) cases, 0% of idiopathic POI, and <1% of healthy control [31]. Hence, NALP5 autoantibodies are an indirect sign of ovarian autoimmunity in a subgroup of women undergoing the autoimmune attack against theca cells. For this reason, anti21OH antibodies remain the most accurate diagnostic marker for autoimmune POI. Contrarily, the absence of anti-NALP5 antibodies may indicate idiopathic POI [31].

ZP is an extracellular substance around oocytes composed of glycoproteins with potential antigenicity. ZP plays an indispensable role in sperm attachment, fertilization, and supporting the communication between oocytes and follicles. Based on these functions, it is speculated that anti-ZP antibodies may impair the communication between oocyte and granulosa cells [32]. Prevalence of anti-ZP antibodies has been consistently reported in women with poor IVF response and low fertilization rates [33]. However, women with anti-ZP were reported not to have consistent patterns suggesting the presence of antibodies directed against the oocyte cytoplasm or the steroid-producing cells of the ovary [32]. To clarify these controversies, highly sensitive assays have been proposed, but no significant association between anti-ZP antibodies and UI has been established yet.

The anticorpus luteal antibody was also reported to be associated with POI. A 67 kDa antigen was found in bovine and human CL, which interacted with anticorpus luteal antibodies in female patients with systemic lupus erythematosus (SLE) under 40 years old. SLE patients with anticorpus luteal antibodies had high serum FSH levels, representing the early stage of ovarian dysfunction [34]. Hence, SLE patients, particularly reproductive-age women, should be carefully evaluated for the presence of autoimmune POI.

AOA might be the cause or the result of autoimmune POI. Conversely, its titer is poorly correlated with the clinical severity of the disease [32]. Additionally, AOA was reported to have a high false positivity rate with a significant positivity in normal controls [32], which could be partially due to naturally occurring antialbumin antibodies [35]. In patients with a normal ovarian function who have various organ-specific or systemic autoimmune diseases, or unexplained repeated reproductive failures, it is even more challenging to determine the significance of AOA for several reasons: (1) AOA recognize multiple antigenic targets in the ovary; (2) AOA response may be transient or variable with time, and (3) the presence of AOA does not imply their etiopathogenic role in the disease manifestation [34]. Further studies are needed to determine specific molecular targets for AOA.

4 Primary ovarian insufficiency

4.1 Definition

POI, often expressed as POF or premature menopause is a common cause of female infertility. POI/POF is defined as loss of ovarian function before the age of 40, influencing around 1% of reproductive-age women [36]. The main symptom is an absence of regular menstrual cycles, infertility, and decreased health-related quality of life [36]. The diagnosis of POI/POF [based on the European Society of Human Reproduction and Embryology (ESHRE) guidelines] can be made if women experience: (1) amenorrhea or oligomenorrhea for at least four months and (2) an elevated follicle-stimulating hormone (FSH) level >25 IU/l on two occasions at least 4 weeks apart [36,37]. The term "insufficiency" is preferable to "failure" since it includes subfertility with the residual follicular activity until complete follicular exhaustion. In contrast, DOR is characterized as regular menses with initial alterations of ovarian reserve tests, evaluated by anti-Müllerian hormone (AMH) and antral follicle count (AFC) [38].

4.2 Clinical manifestation

The clinical symptoms of POI are primarily the result of estrogen deficiency and may include amenorrhea, hypomenorrhea, vascular motor instability (hot flashes, night sweats), sleep disorders, vulvovaginal atrophy, changes in urinary frequency, and repeated genitourinary tract infection. Emotional disorders include irritability and emotional instability. Other hormonal imbalances, such as androgen deficiency, also significantly impact the POI symptoms. Androgen deficiency is characterized by low sexual desire, difficult sexual intercourse, and lack of vitality [39]. Women with POI also have an increased risk of cardiovascular disease, dementia, cognitive decline, Parkinson's disease, and osteoporosis [37]. Notably, these patients experience reproductive failures such as implantation failures and RPLs [37].

4.3 The histopathological types of primary ovarian insufficiency

POI can be divided into nonfollicular and follicular forms by the histopathological types. The nonfollicular form is characterized by the depletion of ovarian follicles and the loss of ovarian function. Underlying conditions, such as gonadal dysgenesis, mixed gonadoblastoma, and hermaphroditism, are the primary causes of complete follicular depletion. Contrarily, the histological follicular patterns demonstrate spontaneous or induced recovery of ovarian function, and it is associated with the following three ovarian conditions: (1) oophoritis (follicular inflammation), (2) decreased follicle counts in the ovary, or (3) numerous primordial follicles (so-called resistant ovary syndrome) [40]. POI is neither irreversible nor permanent because residual oocytes in the ovary can be supplemented, ovulated, and fertilized. The prevalence of spontaneous pregnancies in women with POI/POF was reported to be 2.17% [41]. With the progression of the disease, the follicular pattern may change into a nonfollicular pattern.

4.4 The etiologies of primary ovarian insufficiency

There are many etiologies for POI. The potential etiologies can be divided into different groups: genetics, autoimmune, environmental, iatrogenic, and idiopathic [42]. In most cases, however, no precise cause can be identified and remained idiopathic.

4.4.1 Autoimmune etiologies for primary ovarian insufficiency

Approximately 4% of spontaneous primary POI women have autoimmune oophoritis [11]. Autoimmune etiologies of POI can be determined by the presence of lymphocytic oophoritis, autoantibodies to ovarian antigens, and other autoimmune diseases characterized by serum ovarian, adrenocortical or steroidogenic cell autoantibodies (Fig. 1). POI of autoimmune origin can be divided into adrenal autoimmunity, nonadrenal autoimmunity, and isolated. Among all, specific antiadrenal antibodies are most frequent and represent 60%–80% of cases. Additional 10%–30% of women with POI of autoimmune origin have concomitant autoimmune diseases, especially autoimmune thyroiditis and type 1 diabetes mellitus [43]. While 10%–20% of Addison's disease patients have POI, 2.5%–20% of women with POI demonstrate some evidence of adrenal autoimmunity, including high serum levels of antibodies targeted to steroid-producing cells [44].

Autoimmune disorders unusually occur in women of reproductive age, which would increase the POI risk. Autoimmune diseases can be characterized by the presence of organ and nonorgan-specific autoantibodies. Organ-specific autoimmune diseases, such as Hashimoto thyroiditis, Grave's disease, myasthenia, rheumatoid arthritis, pernicious anemia, vitiligo, Crohn's disease, celiac disease, and multiple sclerosis, as well as some nonorgan-specific disorders, such as SLE, and idiopathic thrombocytopenia, are associated with POI [45].

POI is often associated with more than one autoimmune disease in the same patient, known as APESs [46–48]. Three different types of APES have been described: (1) APES-I, also called autoimmune polyendocrinopathy-candidiasis-ectodermal dystrophy (APECED), includes adrenal and parathyroid autoimmunity [49]. It is caused by the mutation of the autoimmune regulator gene and has characteristics of T cell-mediated autoimmunity. 45%–60% of APES type 1 patients develop POF. (2) APES-II, also called Schmidt Carpenter syndrome, includes adrenal and thyroid autoimmunity as well as type I diabetes. These patients develop autoimmunity due to specific HLA alleles that determine the target tissues by autoreactive T cells [47]. 5%–50% of them have early ovarian failures. (3) APES-III includes autoimmune diseases, such as pernicious anemia or vitiligo, without Addison's disease. The prevalence of APES type III in POI patients is 33.7% [48].

4.4.2 *Other etiologies*

POI can be familial or sporadic. It is believed that about 7% of cases are purely genetic. Genetic mechanisms include specific genetic mutations in oocytes, enzymes, or hormone receptors, nonspecific effects of chromosomes by the inhibition of meiosis, reduced pool of primordial follicles, and increased follicular atresia due to apoptotic granulosa cells or incomplete maturation of the follicle [50].

A considerable part of POI is caused by environmental factors. The most common relevant environmental factor is smoking, which is a significant risk factor for future POI [51]. Besides, the most-reported substances are phthalates, bisphenol A, and pesticides, which decrease ovarian function by increasing follicular depletion, resulting in an earlier onset of menopause. These effects were found when exposure occurred at any time throughout life. The primary mechanism seemed to be an increase in atresia of preantral follicles [52].

As mentioned above, the POI can be associated with localized or systemic nonadrenal disorders, such as thyroid diseases, hypoparathyroidism, hypophysitis, type 1 diabetes mellitus, and nonendocrine autoimmune diseases.

Iatrogenic POI is often observed in patients treated for malignant tumors. The most common causes of POI in this group were ovarian surgery, chemotherapy, and/or pelvic radiotherapy [53].

5 Diseases associated with ovarian autoimmunity

5.1 Polycystic ovary syndrome

PCOS is a metabolic syndrome characterized by chronic oligo/anovulation, clinical and/or biochemical hyperandrogenism, and polycystic ovaries on vaginal ultrasonography. PCOS is common among women of reproductive age with a 6%–10% prevalence [54]. The clinical symptoms of PCOS include oligomenorrhea, infertility, acne, hirsutism, obesity, and acanthosis nigricans. In addition, women with PCOS may suffer from other endocrine and metabolic disorders and have increased risks of breast or endometrial cancer, impaired glucose tolerance, insulin resistance (IR), vitamin D deficiency, diabetes, and cardiovascular diseases [55,56]. It is reported that long-term estrogen stimulation without progesterone may increase the risk of breast or endometrial cancer in women with PCOS [57].

The study on the pathogenesis of PCOS mainly focuses on two interrelated metabolic factors, IR and hyperandrogenemia [58]. Inflammatory and autoimmune factors have been enounced for the etiopathogenesis of PCOS for its close association with IR and thyroid disorders. It is reported that the prevalence of thyroid autoimmunity in PCOS population is higher than that in the general population of women. Anti-thyroperoxidase antibodies (anti-TPO) and antithyroglobulin antibodies (anti-TG) are primary biomarkers of thyroid autoimmunity, Hashimoto's thyroiditis (HT) [59]. Many studies have shown a significant correlation between thyroid antibodies and specific causes of infertility, such as PCOS, EMS, and reduced ovarian reserve [60]. For example, PCOS and HT are related to reproductive problems in women of reproductive age, such as infertility, spontaneous abortion, and obstetrical complications, such as gestational hypertension, preeclampsia, preterm birth, low birth weight, and postpartum hemorrhage [61]. It is suggested that functional autoantibodies leading to hypo and hyperthyroidism may cause similar damages to the ovary. Autoantibodies that lead to hyperthyroidism may induce PCOS, while those of hypothyroidism may lead to POI [9,62]. The copresence of PCOS and HT was associated with lower AMH levels, suggesting that the reduction in the ovarian reserve may be faster and more aggressive in these women than women with only PCOS [9,62]. However, the duration of autoimmune thyroiditis and autoantibody exposure to reduce ovarian reserve has not been elucidated well.

5.2 Ovarian endometriosis

EMS is an estrogen-dependent chronic inflammatory disease that affects around 5%–10% of reproductive-age women [10]. EMS is defined by the presence of endometrial glands and stroma at extra-uterine sites [63]. Retrograde menstruation theory has been widely accepted as an underlying etiology of EMS. More than one-third of EMS patients with pelvic pain, dyspareunia, and dysmenorrhea suffer from pregnancy failure or infertility [10,64]. The potential mechanisms for reproductive failures include distortions with pelvic adhesions, occlusions of the tubal ostia, and decreased folliculogenesis. Another mechanism to explain its relation to infertility is chronic intraperitoneal inflammation with increased proinflammatory cytokines. The inflammatory condition could influence the oocyte and sperm interactions, embryo development, and implantation. Recent studies reported that the endocrine factors via autocrine manners and immune-related factors particularly, innate immunity, participate in the pathophysiology of EMS-related infertility. Moreover, EMS is associated with the presence of autoantibodies. In peritoneal fluid and sera of infertile EMS women, various autoantibodies, such as antinuclear (ANA), anti-SSA/Ro anti-SSB/la, and antiphospholipid antibodies (aPL), were present more frequently in women with EMS than control women [10].

5.3 Recurrent pregnancy losses

It has been speculated that oocyte-specific genetic factors might play a role in the pathogenesis of RPL. In fact, the risk of pregnancy losses increases with maternal age [65]. Similarly, an exponential relationship between maternal age and chromosomal abnormalities in oocytes has also been reported. Since the physiological age of the ovary determines the result of pregnancy, the decline in ovarian function in women with POI leads to a decrease in the quality and quantity of oocytes and an increase in meiosis, embryonic aneuploidy, and subsequent miscarriages [66,67]. Studies found that about 40%–50% of miscarriages in the first trimester resulted from chromosomal abnormalities in the conceptuses [68]. In aged women undergoing assisted reproductive technology (ART), donor eggs are often preferred to increase reproductive outcomes than their own eggs since the decreased fertility in these women is mainly related to the quality and quantity of oocytes rather than the uterine environment [69]. In women with physiological (normal aging) and pathological ovarian failures (POI), the correlations exist between oocyte quality and RPL, suggesting they may have a common etiological pathway. Exploring molecular pathways associated with physiological aging and pathological disorders affecting oocyte quality may enable researchers and clinicians to improve fertility and pregnancy outcomes.

The whole process of ovarian reserve, including establishment, activation during adolescence, and age-dependent or pathological depletion, is largely affected by genetic predispositions. Genetic defects involving these processes may lead to a series of ovarian malfunction, including POI. So far, several pathogenic mutations in various oocyte-specific genes have been associated with POI, such as BMP15 and GDF9 genes, and these genes are positively correlated with oocyte maturation, fertilization, embryo qualities and pregnancy outcomes in humans [70,71]. Thus the intraovarian BMP/GDF system is utmostly important in regulating ovarian function for developing a competent oocyte for embryo generation, and dysregulation of this system may result in infertility, subfertility, and miscarriages [72]. Since POI is associated with infertility, factors leading to POI may be related to RPL. Indeed, women with infertility have an increased risk of miscarriage and vice versa [73]. Additionally, a high frequency of miscarriages has been reported after ART treatment [74].

Besides, FOXO3 and FOXL2, which belong to the FOXO (Forkhead box O) family of transcription factors, negatively regulate follicular maturation and maintain proper ovarian function. Therefore activation of these proteins could block the primordial follicle growth and result in infertility [75]. It is also proved that the FOX factors, including FOXO1 and FOXL2, play a central role in endometrial decidualization, which is indispensable for implantation and placentation by maintaining an appropriate microenvironment for embryo implantation and growth [75]. Due to the high expression levels and multiple functions of these FOX proteins in uterine tissues, it is speculated that the mutations of these genes may be of great significance in RPL and may play a role in POI [76,77].

5.4 Unexplained infertility

UI accounts for 10%–17% of infertile women [78]. Unexplained or idiopathic infertility is a condition in which couples cannot conceive without definite causes. 68% of women with UI had positive AOA, and ALDH1A1 and SBP1 were reported to be unique antigens to ovarian autoimmunity in women with unknown infertility [8]. ALDH1A1 is the cytosolic form of aldehyde dehydrogenase, which participates in retinoid metabolism [79]. Since retinoids are involved in steroid production [80], oocyte maturation [81], early embryo development [82], reduced FSH-induced FSH and LH receptor expression [83], and granulosa cell maturation [84], altered ALDH1A1 expressions could affect reproductive processes. SBP1, a cytoplasmic protein, is expressed on the surface epithelium of the ovary [85]. Decreased selenium in follicular fluid was reported in women with UI [86], and SP56, a conjugated peptide of SBP1, reduced fertility in mice model [87]. Based on these findings, autoantibodies to SBP1 were speculated to have a detrimental effect on infertility. Autoantibodies such as AOA, antiphospholipid antibody, antithyroid antibody, and dysregulated cytokines and chemokines may result in UI. Therefore the altered adaptive immune responses may lead to infertility, especially UI [88,89].

6 Diagnosis of primary ovarian insufficiency

The diagnosis of POI can be made by careful history taking, hormonal checkups, and ultrasound evaluation (Fig. 2). There is currently no consensus regarding the best diagnostic criteria of POI. ESHRE guideline recommends the following diagnostic criteria: (1) oligomenorrhea or amenorrhea for at least 4 months, and (2) increased FSH levels >25 IU/L

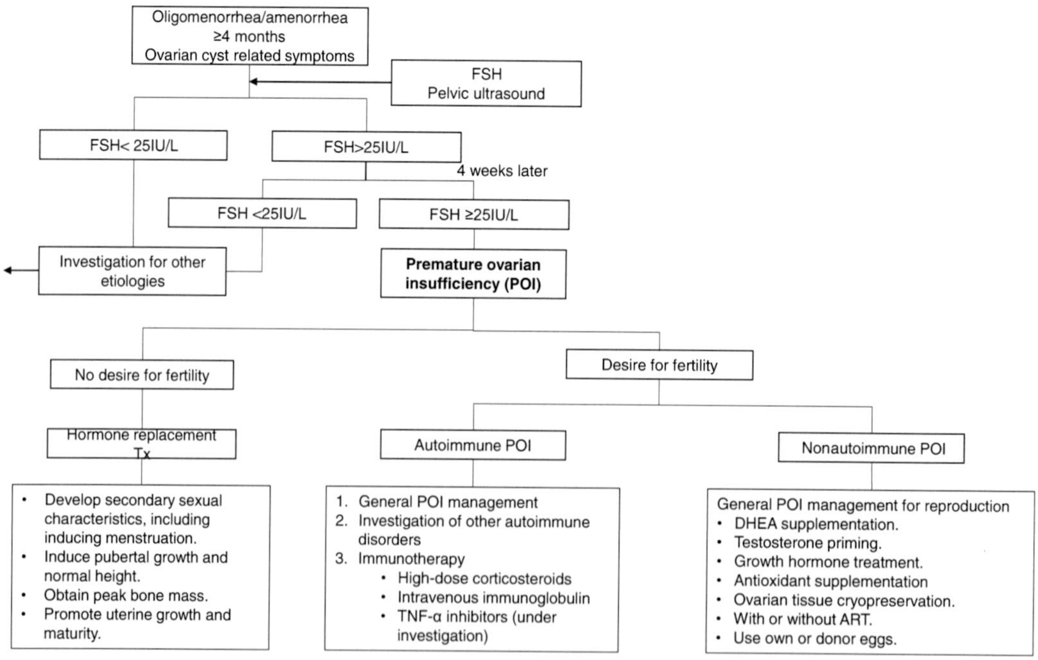

FIGURE 2 Diagnostic strategies and management of POI.

repeated at least 4 weeks intervals [36]. The American College of Obstetricians and Gynecologists recommends diagnostic criteria as (1) menstrual irregularity for at least three consecutive months, and (2) FSH and estradiol levels (generally confirmed with increased FSH levels $>30-40$ IU/L and decreased estrogen levels ≤ 50 pg/mL, or depending on the laboratory standard) repeated at least 1-month intervals, and (3) prolactin and thyroid function tests [90,91]. Although anti-Müllerian hormone (AMH) has also been proposed as a diagnostic criterion, it is of great significance only when interpreted together with FSH and estrogen levels [92]. Inhibin B is not suitable as a diagnostic indicator because it changes with menstrual cycle days [93]. In addition, pelvic ultrasound can be helpful in diagnosis. Typical findings are small ovaries, few follicles in the cavity, and no growth of follicles [94].

6.1 Diagnostic approaches

Although there have been many explorations for the underlying etiologies for POI, there is still a lack of solid evidence to prove a causal relationship between POI and related factors. Therefore the following screening tests are recommended to diagnose POI.

1. Chromosomal abnormalities:
   a. Fragile X testing is indicated for women with POI.
   b. Screening for Turner syndrome, trisomy X, FMR1, and BMP15 premutation located on the X chromosome should be considered since these conditions are common chromosomal and gene abnormalities associated with POI.
   c. Once the Y chromosome is detected in women, gonadectomy is recommended because these women have a high risk of developing gonadal neoplasia.
2. Autoimmune abnormalities:
   a. Screening for adrenal autoantibody, 21-hydroxylase antibody (21OH-Ab), alternatively adrenocortical antibodies (ACA), is recommended to identify patients with possible Addison's disease.
   b. Screening for thyroid antibodies, such as TPO-Ab, seems reasonable since POI is often associated with hypothyroidism.
   c. Screening for other autoimmune diseases, such as autoimmune polyglandular syndromes (APS), Sjögren's syndrome, and type I diabetes, is recommended in women with POI.
3. Screening for medical or surgical interventions history should be made because of the possible causality with POI [36]. For example, patients with ovarian mass may receive chemotherapy, radiotherapy, and surgical manipulation of ovaries, and have decreased gonadal blood supply.
4. Screening for infections, including epidemic parotitis and human immunodeficiency viruses, can be considered. The current evidence, however, is not sufficient for routine screening [36,95].
5. Screening for pollutants and toxins can be considered since they may affect the ovaries.

6.2 Diagnosis of autoimmune oophoritis

Autoimmune oophoritis is often challenging to diagnose since no single test can diagnose autoimmune oophoritis, and clinical symptoms may vary based on its prognosis and other underlying conditions. Antiadrenal antibodies (most commonly against the 21-hydroxylase enzyme, CYP21) and antithyroid antibodies (anti-TPO and anti-TG antibodies) should be tested in women with POI. The existence of these antibodies may suggest the autoimmune cause of POI. In addition, further evaluation and treatment of adrenal and thyroid abnormalities should take into account to improve women's fertility, if necessary [96].

7 Treatment for autoimmune primary ovarian insufficiency

7.1 Women with a desire for reproduction

Most women with autoimmune oophoritis may still have a certain degree of ovarian function but received the diagnosis, such as POI, POF, poor ovarian response (POR), and DOR, before undergoing ART cycles. Immunosuppressive therapy is a possible approach to restore ovarian function in women with autoimmune oophoritis with spared primary and primordial follicles. It is reported that using high-dose corticosteroids and intravenous immunoglobulin treatment may normalize ovarian function in women with POI [97,98]. Other not commonly used experimental immune treatment might be monoclonal antibodies [such as tumor necrosis factor (TNF)-α inhibitors] when treating POI with autoimmune ovarian damage to restore the immune response, especially Th2 dominance [99]. Yet, no systematic studies were reported for a possible role of immune therapy in women with autoimmune oophoritis undergoing ART cycles. However, considering the underlying etiology, there is a potential to improve reproductive outcomes by applying immunotherapy for this population.

Autoimmune attacks on the adrenal reticular zone can affect ovarian function by decreasing adrenal androgen production, resulting in low androgen levels. Low androgen levels have an adverse impact on ovarian function. Therefore DHEA supplementation should benefit these women with adrenal and thyroid autoimmunities and the combination of low testosterone and low-DHEAS [100]. In the past decades, various studies have evaluated the role of testosterone supplement before or during controlled ovarian hyperstimulation in POR women, with inconsistent results [101]. In a recent *meta*-analysis of randomized controlled trials, testosterone supplementation was reported to be beneficial for POR by significantly increasing the number of oocytes, embryos, and MII oocytes [102]. A great quantity of evidence has confirmed the crucial role of testosterone in early follicular growth and development and granulosa cell proliferation, such as promoting early follicular development, increasing the number and susceptibility of preantral and antral follicles, and stimulating the transition of primary follicles to secondary follicles [102,103].

GH treatment was reported to promote ovarian tissue repair, estrogen release, and oocyte maturation via activation of the Notch-1 signaling pathway in ovarian tissue in the mouse model [104]. Additionally, GH treatment combined with gonadotropins may upregulate intraovarian IGF-I and augment the stimulatory effect of FSH on folliculogenesis [105,106]. However, reproductive outcomes of GH treatment have been equivocal in POR cases, and currently, it is only recommended to the GH-deficient (the clonidine test negative) POR patients [106,107].

Lastly, in women with POI who need fertility preservation, ovarian tissue cryopreservation using vitrification followed by in vitro activation of the dormant/suppressed follicles has been reported [108].

7.2 Women without a desire for reproduction

Since ovarian autoimmunity has a wide range of clinical manifestations and reproductive concerns, it is important to understand patients' needs based on their age, clinical symptoms, and desire for reproduction. For patients with delayed puberty caused by chemotherapy or radiotherapy, the treatment goals of inducing puberty are (1) to develop secondary sexual characteristics, including inducing menstruation, (2) induce pubertal growth and normal height, (3) obtain peak bone mass, and (4) promote uterine growth and maturity [91]. Hormone replacement therapy (HRT) is recommended to alleviate low estrogen symptoms in patients with POI. However, periodic follow-up is needed since HRT can cause adverse risks, including breast cancer, and women with POI have various underlying etiologies. For example, (1) HRT is recommended in the normal reproductive lifespan of women with Turner syndrome; (2) women with a breast cancer history, HRT is contraindicated; (3) HRT may be effective in women with EMS; (4) in women with hypertension, obesity or a history of venous thromboembolism, transdermal estradiol is the preferred method of administration [109–112]; (5) women with migraine or fibroids, HRT is not contraindicated. Regarding androgen therapy, it is only supported by limited data, and the long-term health effects are unclear. It is also encouraged to maintain an active lifestyle, avoid or stop smoking, and maintain adequate body weight to optimize bone health [36,37]. No systematic study has been reported for a possible role of immune therapy in this population.

8 Reproductive outcomes

Women with POI may have different reproductive outcomes based on the remnant ovarian follicles at the time of diagnosis. If women with POI have spared primary and primordial follicles and immediately initiate treatment, reproductive outcomes are more favorable than those without primary and primordial follicles or delayed treatment. A few POI patients could have their own babies with hormonal and immune therapy with and without ART. Women with advanced POI often have no chance to obtain their own eggs and are directed to donor egg cycles.

9 Summary and recommendations

1. Autoimmunity, including antiphospholipid syndrome (APS), thyroid and ovarian autoimmunities, preferentially affects women throughout their reproductive age.

2. The diagnosis of POI can be made based on menstrual disturbance, hormone levels, and biochemical confirmations, including serum AMH and AFC with ultrasound evaluation.
3. The potential etiologies for POI are genetics, environmental, iatrogenic, autoimmune, and idiopathic factors.
4. Autoimmune etiologies of POI can be determined by the presence of lymphocytic oophoritis, autoantibodies to ovarian antigens, and other autoimmune diseases characterized by serum ovarian, adrenocortical or steroidogenic cell autoantibodies.
5. In women with autoimmune oophoritis, the presence of large multifollicular ovaries combined with elevated serum LH levels has been mistaken for the PCOS.
6. Before treating women with POI, antiadrenal and antithyroid antibodies should be tested.
7. The diversity of suspected autoantigens and related antibodies in autoimmune oophoritis shows the multiplicity of autoimmune pathologies leading to ovarian damage.
8. High-dose corticosteroids and intravenous immunoglobulin treatment may normalize ovarian function in women with autoimmune POI.
9. No systematic studies were reported for a possible role of immune therapy in women with autoimmune oophoritis undergoing ART cycles.
10. Considering the autoimmune etiology, there is a potential to improve reproductive outcomes by applying immunotherapy for women with autoimmune POI.

References

[1] Luborsky J. Ovarian autoimmune disease and ovarian autoantibodies. J Womens Health Gend Based Med 2002;11(7):585–99.

[2] Falorni A, Laureti S, Candeloro P, et al. Steroid-cell autoantibodies are preferentially expressed in women with premature ovarian failure who have adrenal autoimmunity. Fertil Steril 2002;78(2):270–9.

[3] Forges T, Monnier-Barbarino P, Faure GC, Bene MC. Autoimmunity and antigenic targets in ovarian pathology. Hum Reprod Update 2004;10(2):163–75.

[4] Kelkar RL, Meherji PK, Kadam SS, Gupta SK, Nandedkar TD. Circulating auto-antibodies against the zona pellucida and thyroid microsomal antigen in women with premature ovarian failure. J Reprod Immunol 2005;66(1):53–67.

[5] Luborsky J, Llanes B, Davies S, Binor Z, Radwanska E, Pong R. Ovarian autoimmunity: greater frequency of autoantibodies in premature menopause and unexplained infertility than in the general population. Clin Immunol 1999;90(3):368–74.

[6] Yan G, Schoenfeld D, Penney C, Hurxthal K, Taylor AE, Faustman D. Identification of premature ovarian failure patients with underlying autoimmunity. J Womens Health Gend Based Med 2000;9(3):275–87.

[7] Mande PV, Parikh FR, Hinduja I, et al. Identification and validation of candidate biomarkers involved in human ovarian autoimmunity. Reprod Biomed Online 2011;23(4):471–83.

[8] Edassery SL, Shatavi SV, Kunkel JP, et al. Autoantigens in ovarian autoimmunity associated with unexplained infertility and premature ovarian failure. Fertil Steril 2010;94(7):2636–41.

[9] Serin AN, Birge O, Uysal A, Gorar S, Tekeli F. Hashimoto's thyroiditis worsens ovaries in polycystic ovary syndrome patients compared to anti-Mullerian hormone levels. BMC Endocr Disord 2021;21(1):44.

[10] de Ziegler D, Borghese B, Chapron C. Endometriosis and infertility: pathophysiology and management. Lancet. 2010;376(9742):730–8.

[11] Bakalov VK, Anasti JN, Calis KA, et al. Autoimmune oophoritis as a mechanism of follicular dysfunction in women with 46,XX spontaneous premature ovarian failure. Fertil Steril 2005;84(4):958–65.

[12] Jilma B, Dirnberger E, Loscher I, et al. Menstrual cycle-associated changes in blood levels of interleukin-6, alpha1 acid glycoprotein, and C-reactive protein. J Lab Clin Med 1997;130(1):69–75.
[13] Wang W, Sung N, Gilman-Sachs A, Kwak-Kim JT. Helper (Th) cell profiles in pregnancy and recurrent pregnancy losses: Th1/Th2/Th9/Th17/Th22/Tfh Cells. Front Immunol 2020;11:2025.
[14] Yang X, Gilman-Sachs A, Kwak-Kim J. Ovarian and endometrial immunity during the ovarian cycle. J Reprod Immunol 2019;133:7–14.
[15] Clancy KB, Baerwald AR, Pierson RA. Systemic inflammation is associated with ovarian follicular dynamics during the human menstrual cycle. PLoS One 2013;8(5):e64807.
[16] Ledee N, Petitbarat M, Chevrier L, et al. The uterine immune profile may help women with repeated unexplained embryo implantation failure after in vitro fertilization. Am J Reprod Immunol 2016;75(3):388–401.
[17] Straub RH. The complex role of estrogens in inflammation. Endocr Rev 2007;28(5):521–74.
[18] Kwak-Kim JY, Chung-Bang HS, Ng SC, et al. Increased T helper 1 cytokine responses by circulating T cells are present in women with recurrent pregnancy losses and in infertile women with multiple implantation failures after IVF. Hum Reprod 2003;18(4):767–73.
[19] Welt CK. Autoimmune oophoritis in the adolescent. Ann N Y Acad Sci 2008;1135:118–22.
[20] La Marca A, Brozzetti A, Sighinolfi G, Marzotti S, Volpe A, Falorni A. Primary ovarian insufficiency: autoimmune causes. Curr Opin Obstet Gynecol 2010;22(4):277–82.
[21] Kalantaridou SN, Braddock DT, Patronas NJ, Nelson LM. Treatment of autoimmune premature ovarian failure. Hum Reprod 1999;14(7):1777–82.
[22] Ebrahimi M, Akbari Asbagh F. The role of autoimmunity in premature ovarian failure. Iran J Reprod Med 2015;13(8):461–72.
[23] Li J, Jin H, Zhang F, et al. Treatment of autoimmune ovarian disease by co-administration with mouse zona pellucida protein 3 and DNA vaccine through induction of adaptive regulatory T cells. J Gene Med 2008;10 (7):810–20.
[24] Xiong J, Tan R, Wang W, Wang H, Pu D, Wu J. Evaluation of CD4(+)CD25(+)FOXP3(+) regulatory T cells and FOXP3 mRNA in premature ovarian insufficiency. Climacteric. 2020;23(3):267–72.
[25] Bakalov VK, Vanderhoof VH, Bondy CA, Nelson LM. Adrenal antibodies detect asymptomatic auto-immune adrenal insufficiency in young women with spontaneous premature ovarian failure. Hum Reprod 2002;17 (8):2096–100.
[26] Kobayashi M, Nakashima A, Yoshino O, et al. Decreased effector regulatory T cells and increased activated CD4(+) T cells in premature ovarian insufficiency. Am J Reprod Immunol 2019;81(6):e13125.
[27] Luborsky JL, Visintin I, Boyers S, Asari T, Caldwell B, DeCherney A. Ovarian antibodies detected by immobilized antigen immunoassay in patients with premature ovarian failure. J Clin Endocrinol Metab 1990;70(1):69–75.
[28] Pires ES. Multiplicity of molecular and cellular targets in human ovarian autoimmunity: an update. J Assist Reprod Genet 2010;27(9–10):519–24.
[29] Tong ZB, Nelson LM. A mouse gene encoding an oocyte antigen associated with autoimmune premature ovarian failure. Endocrinology. 1999;140(8):3720–6.
[30] Nelson LM. Autoimmune ovarian failure: comparing the mouse model and the human disease. J Soc Gynecol Investig 2001;8(1 Suppl Proceedings):S55–7.
[31] Brozzetti A, Alimohammadi M, Morelli S, et al. Autoantibody response against NALP5/MATER in primary ovarian insufficiency and in autoimmune 'Addison's disease. J Clin Endocrinol Metab 2015;100(5):1941–8.
[32] Novosad JA, Kalantaridou SN, Tong ZB, Nelson LM. Ovarian antibodies as detected by indirect immunofluorescence are unreliable in the diagnosis of autoimmune premature ovarian failure: a controlled evaluation. BMC Womens Health 2003;3(1):2.
[33] Mardesic T, Ulcova-Gallova Z, Huttelova R, et al. The influence of different types of antibodies on in vitro fertilization results. Am J Reprod Immunol 2000;43(1):1–5.
[34] Pasoto SG, Viana VS, Mendonca BB, Yoshinari NH, Bonfa E. Anti-corpus luteum antibody: a novel serological marker for ovarian dysfunction in systemic lupus erythematosus? J Rheumatol 1999;26(5):1087–93.
[35] Pires ES, Parte PP, Meherji PK, Khan SA, Khole VV. Naturally occurring anti-albumin antibodies are responsible for false positivity in diagnosis of autoimmune premature ovarian failure. J Histochem Cytochem 2006;54(4):397–405.
[36] European Society for Human R, Embryology Guideline Group on POI, Webber L, et al. ESHRE guideline: management of women with premature ovarian insufficiency. Hum Reprod 2016;31(5):926–37.

[37] Faubion SS, Kuhle CL, Shuster LT, Rocca WA. Long-term health consequences of premature or early menopause and considerations for management. Climacteric. 2015;18(4):483–91.
[38] Practice Committee of the American Society for Reproductive M. Testing and interpreting measures of ovarian reserve: a committee opinion. Fertil Steril 2012;98(6):1407–15.
[39] Domniz N, Meirow D. Premature ovarian insufficiency and autoimmune diseases. Best Pract Res Clin Obstet Gynaecol 2019;60:42–55.
[40] Hoek A, Schoemaker J, Drexhage HA. Premature ovarian failure and ovarian autoimmunity. Endocr Rev 1997;18(1):107–34.
[41] Gargus E, Deans R, Anazodo A, Woodruff TK. Management of primary ovarian insufficiency symptoms in survivors of childhood and adolescent cancer. J Natl Compr Canc Netw 2018;16(9):1137–49.
[42] Laven JS. Primary ovarian insufficiency. Semin Reprod Med 2016;34(4):230–4.
[43] Ellis JS, Wan X, Braley-Mullen H. Transient depletion of CD4 + CD25 + regulatory T cells results in multiple autoimmune diseases in wild-type and B-cell-deficient NOD mice. Immunology. 2013;139(2):179–86.
[44] Irvine WJ, Chan MM, Scarth L, et al. Immunological aspects of premature ovarian failure associated with idiopathic 'Addison's disease. Lancet. 1968;2(7574):883–7.
[45] Aikawa NE, Sallum AM, Pereira RM, et al. Subclinical impairment of ovarian reserve in juvenile systemic lupus erythematosus after cyclophosphamide therapy. Clin Exp Rheumatol 2012;30(3):445–9.
[46] Ahonen P, Myllarniemi S, Sipila I, Perheentupa J. Clinical variation of autoimmune polyendocrinopathy-candidiasis-ectodermal dystrophy (APECED) in a series of 68 patients. N Engl J Med 1990;322(26):1829–36.
[47] Michels AW, Gottlieb PA. Autoimmune polyglandular syndromes. Nat Rev Endocrinol 2010;6(5):270–7.
[48] Szlendak-Sauer K, Jakubik D, Kunicki M, Skorska J, Smolarczyk R. Autoimmune polyglandular syndrome type 3 (APS-3) among patients with premature ovarian insufficiency (POI). Eur J Obstet Gynecol Reprod Biol 2016;203:61–5.
[49] Perheentupa J. APS-I/APECED: the clinical disease and therapy. Endocrinol Metab Clin North Am 2002;31 (2):295–320 vi.
[50] Vujovic S. Aetiology of premature ovarian failure. Menopause Int 2009;15(2):72–5.
[51] Jin M, Yu Y, Huang H. An update on primary ovarian insufficiency. Sci China Life Sci 2012;55(8):677–86.
[52] Hassold T, Hall H, Hunt P. The origin of human aneuploidy: where we have been, where we are going. Hum Mol Genet 2007;R203–8 16 Spec No. 2.
[53] Bricaire L, Laroche E, Bourcigaux N, Donadille B, Christin-Maitre S. Premature ovarian failures. Presse Med 2013;42(11):1500–7.
[54] Azziz R, Woods KS, Reyna R, Key TJ, Knochenhauer ES, Yildiz BO. The prevalence and features of the polycystic ovary syndrome in an unselected population. J Clin Endocrinol Metab 2004;89(6):2745–9.
[55] Wild RA, Carmina E, Diamanti-Kandarakis E, et al. Assessment of cardiovascular risk and prevention of cardiovascular disease in women with the polycystic ovary syndrome: a consensus statement by the androgen excess and polycystic ovary syndrome (AE-PCOS) society. J Clin Endocrinol Metab 2010;95(5):2038–49.
[56] Fux Otta C, Fiol de Cuneo M, Szafryk de Mereshian P. Polycystic ovary syndrome: physiopathology review. Rev Fac Cien Med Univ Nac Cordoba 2013;70(1):27–30.
[57] Hardiman P, Pillay OC, Atiomo W. Polycystic ovary syndrome and endometrial carcinoma. Lancet. 2003;361 (9371):1810–12.
[58] Buccola JM, Reynolds EE. Polycystic ovary syndrome: a review for primary providers. Prim Care 2003;30 (4):697–710.
[59] Hepsen S, Karakose M, Cakal E, et al. The assessment of thyroid autoantibody levels in euthyroid patients with polycystic ovary syndrome. J Turk Ger Gynecol Assoc 2018;19(4):215–19.
[60] Ho CW, Chen HH, Hsieh MC, et al. Increased risk of polycystic ovary syndrome and it's comorbidities in women with autoimmune thyroid disease. Int J Env Res Public Health 2020;17(7).
[61] Medenica S, Nedeljkovic O, Radojevic N, Stojkovic M, Trbojevic B, Pajovic B. Thyroid dysfunction and thyroid autoimmunity in euthyroid women in achieving fertility. Eur Rev Med Pharmacol Sci 2015;19(6):977–87.
[62] Gleicher N, Barad D, Weghofer A. Functional autoantibodies, a new paradigm in autoimmunity? Autoimmun Rev 2007;7(1):42–5.
[63] Bulun SE. Endometriosis. N Engl J Med 2009;360(3):268–79.
[64] Tanbo T, Fedorcsak P. Endometriosis-associated infertility: aspects of pathophysiological mechanisms and treatment options. Acta Obstet Gynecol Scand 2017;96(6):659–67.

[65] Nybo Andersen AM, Wohlfahrt J, Christens P, Olsen J, Melbye M. Maternal age and fetal loss: population based register linkage study. BMJ. 2000;320(7251):1708–12.
[66] El-Toukhy T, Khalaf Y, Hart R, Taylor A, Braude P. Young age does not protect against the adverse effects of reduced ovarian reserve—an eight year study. Hum Reprod 2002;17(6):1519–24.
[67] Dean DD, Agarwal S, Tripathi P. Connecting links between genetic factors defining ovarian reserve and recurrent miscarriages. J Assist Reprod Genet 2018;35(12):2121–8.
[68] Choi TY, Lee HM, Park WK, Jeong SY, Moon HS. Spontaneous abortion and recurrent miscarriage: a comparison of cytogenetic diagnosis in 250 cases. Obstet Gynecol Sci 2014;57(6):518–25.
[69] Wang YA, Farquhar C, Sullivan EA. Donor age is a major determinant of success of oocyte donation/recipient programme. Hum Reprod 2012;27(1):118–25.
[70] Gode F, Gulekli B, Dogan E, et al. Influence of follicular fluid GDF9 and BMP15 on embryo quality. Fertil Steril 2011;95(7):2274–8.
[71] Wu YT, Tang L, Cai J, et al. High bone morphogenetic protein-15 level in follicular fluid is associated with high quality oocyte and subsequent embryonic development. Hum Reprod 2007;22(6):1526–31.
[72] Li Y, Li RQ, Ou SB, et al. Increased GDF9 and BMP15 mRNA levels in cumulus granulosa cells correlate with oocyte maturation, fertilization, and embryo quality in humans. Reprod Biol Endocrinol 2014;12:81.
[73] Coulam CB. Association between infertility and spontaneous abortion. Am J Reprod Immunol 1992;27 (3–4):128–9.
[74] Torrealday S, Kodaman P, Pal L. Premature ovarian insufficiency—an update on recent advances in understanding and management. F1000Res. 2017;6:2069.
[75] Liu L, Rajareddy S, Reddy P, et al. Infertility caused by retardation of follicular development in mice with oocyte-specific expression of Foxo3a. Development. 2007;134(1):199–209.
[76] Elbaz M, Hadas R, Bilezikjian LM, Gershon E. Uterine Foxl2 regulates the adherence of the trophectoderm cells to the endometrial epithelium. Reprod Biol Endocrinol 2018;16(1):12.
[77] Popovici RM, Betzler NK, Krause MS, et al. Gene expression profiling of human endometrial-trophoblast interaction in a coculture model. Endocrinology. 2006;147(12):5662–75.
[78] Gelbaya TA, Potdar N, Jeve YB, Nardo LG. Definition and epidemiology of unexplained infertility. Obstet Gynecol Surv 2014;69(2):109–15.
[79] Marchitti SA, Brocker C, Stagos D, Vasiliou V. Non-P450 aldehyde oxidizing enzymes: the aldehyde dehydrogenase superfamily. Expert Opin Drug Metab Toxicol 2008;4(6):697–720.
[80] Clagett-Dame M, DeLuca HF. The role of vitamin A in mammalian reproduction and embryonic development. Annu Rev Nutr 2002;22:347–81.
[81] Mohan M, Thirumalapura NR, Malayer J. Bovine cumulus-granulosa cells contain biologically active retinoid receptors that can respond to retinoic acid. Reprod Biol Endocrinol 2003;1:104.
[82] Mark M, Ghyselinck NB, Chambon P. Function of retinoic acid receptors during embryonic development. Nucl Recept Signal 2009;7:e002.
[83] Minegishi T, Hirakawa T, Kishi H, et al. The mechanisms of retinoic acid-induced regulation on the follicle-stimulating hormone receptor in rat granulosa cells. Biochim Biophys Acta 2000;1495(3):203–11.
[84] Hattori M, Takesue K, Nishida N, Kato Y, Fujihara N. Inhibitory effect of retinoic acid on the development of immature porcine granulosa cells to mature cells. J Mol Endocrinol 2000;25(1):53–61.
[85] Huang KC, Park DC, Ng SK, et al. Selenium binding protein 1 in ovarian cancer. Int J Cancer 2006;118 (10):2433–40.
[86] Paszkowski T, Traub AI, Robinson SY, McMaster D. Selenium dependent glutathione peroxidase activity in human follicular fluid. Clin Chim Acta 1995;236(2):173–80.
[87] Hardy CM, Clydesdale G, Mobbs KJ. Development of mouse-specific contraceptive vaccines: infertility in mice immunized with peptide and polyepitope antigens. Reproduction. 2004;128(4):395–407.
[88] An LF, Zhang XH, Sun XT, Zhao LH, Li S, Wang WH. Unexplained infertility patients have increased serum IL-2, IL-4, IL-6, IL-8, IL-21, TNFalpha, IFNgamma and increased Tfh/CD4 T cell ratio: increased Tfh and IL-21 strongly correlate with presence of autoantibodies. Immunol Invest 2015;44(2):164–73.
[89] Sauer R, Roussev R, Jeyendran RS, Coulam CB. Prevalence of antiphospholipid antibodies among women experiencing unexplained infertility and recurrent implantation failure. Fertil Steril 2010;93 (7):2441–3.
[90] Nelson LM. Clinical practice. Primary ovarian insufficiency. N Engl J Med 2009;360(6):606–14.

[91] Collins G, Patel B, Thakore S, Liu J. Primary ovarian insufficiency: current concepts. South Med J 2017;110(3):147–53.
[92] Rudnicka E, Kruszewska J, Klicka K, et al. Premature ovarian insufficiency - aetiopathology, epidemiology, and diagnostic evaluation. Prz Menopauzalny 2018;17(3):105–8.
[93] Nelson SM. Biomarkers of ovarian response: current and future applications. Fertil Steril 2013;99(4):963–9.
[94] Beck-Peccoz P, Persani L. Premature ovarian failure. Orphanet J Rare Dis 2006;1:9.
[95] Kokcu A. Premature ovarian failure from current perspective. Gynecol Endocrinol 2010;26(8):555–62.
[96] Tsigkou A, Marzotti S, Borges L, et al. High serum inhibin concentration discriminates autoimmune oophoritis from other forms of primary ovarian insufficiency. J Clin Endocrinol Metab 2008;93(4):1263–9.
[97] Corenblum B, Rowe T, Taylor PJ. High-dose, short-term glucocorticoids for the treatment of infertility resulting from premature ovarian failure. Fertil Steril 1993;59(5):988–91.
[98] Cowchock FS, McCabe JL, Montgomery BB. Pregnancy after corticosteroid administration in premature ovarian failure (polyglandular endocrinopathy syndrome). Am J Obstet Gynecol 1988;158(1):118–19.
[99] Simon A, Laufer N. Repeated implantation failure: clinical approach. Fertil Steril 2012;97(5):1039–43.
[100] Gleicher N, Kushnir VA, Darmon SK, et al. New PCOS-like phenotype in older infertile women of likely autoimmune adrenal etiology with high AMH but low androgens. J Steroid Biochem Mol Biol 2017;167:144–52.
[101] Bosdou JK, Venetis CA, Dafopoulos K, et al. Transdermal testosterone pretreatment in poor responders undergoing ICSI: a randomized clinical trial. Hum Reprod 2016;31(5):977–85.
[102] Noventa M, Vitagliano A, Andrisani A, et al. Testosterone therapy for women with poor ovarian response undergoing IVF: a meta-analysis of randomized controlled trials. J Assist Reprod Genet 2019;36(4):673–83.
[103] Montoya-Botero P, Rodriguez-Purata J, Polyzos NP. Androgen supplementation in assisted reproduction: where are we in 2019? Curr Opin Obstet Gynecol 2019;31(3):188–94.
[104] Liu TE, Wang S, Zhang L, et al. Growth hormone treatment of premature ovarian failure in a mouse model via stimulation of the Notch-1 signaling pathway. Exp Ther Med 2016;12(1):215–21.
[105] Adashi EY, Resnick CE, D'Ercole AJ, Svoboda ME, Van Wyk JJ. Insulin-like growth factors as intraovarian regulators of granulosa cell growth and function. Endocr Rev 1985;6(3):400–20.
[106] Blumenfeld Z, Dirnfeld M, Gonen Y, Abramovici H. Growth hormone co-treatment for ovulation induction may enhance conception in the co-treatment and succeeding cycles, in clonidine negative but not clonidine positive patients. Hum Reprod 1994;9(2):209–13.
[107] Albu D, Albu A. Is growth hormone administration essential for in vitro fertilization treatment of female patients with growth hormone deficiency? Syst Biol Reprod Med 2019;65(1):71–4.
[108] Suzuki N, Yoshioka N, Takae S, et al. Successful fertility preservation following ovarian tissue vitrification in patients with primary ovarian insufficiency. Hum Reprod 2015;30(3):608–15.
[109] Canonico M, Oger E, Conard J, et al. Obesity and risk of venous thromboembolism among postmenopausal women: differential impact of hormone therapy by route of estrogen administration. The ESTHER Study. J Thromb Haemost 2006;4(6):1259–65.
[110] Canonico M, Plu-Bureau G, Lowe GD, Scarabin PY. Hormone replacement therapy and risk of venous thromboembolism in postmenopausal women: systematic review and meta-analysis. BMJ. 2008;336(7655):1227–31.
[111] Langrish JP, Mills NL, Bath LE, et al. Cardiovascular effects of physiological and standard sex steroid replacement regimens in premature ovarian failure. Hypertension. 2009;53(5):805–11.
[112] White WB. Drospirenone with 17beta-estradiol in the postmenopausal woman with hypertension. Climacteric. 2007;10(Suppl 1):25–31.

CHAPTER

23

Polycystic ovarian syndrome and reproductive failure

Joseph Duero and Reshef Tal

Department of Obstetrics, Gynecology and Reproductive Sciences, Yale School of Medicine, New Haven, CT, United States

1 Introduction

Polycystic ovarian syndrome (PCOS) is a complex endocrine disorder affecting up to 10% of reproductive-age women [1]. First described by Stein and Leventhal in 1935 [2], PCOS is characterized by the presence of three primary features: clinical or biochemical hyperandrogenism, ovulatory dysfunction, and polycystic ovaries, of which two must be present for diagnosis according to the Rotterdam criteria [3,4]. Clinical presentations of PCOS vary by an individual due to the heterogeneity of the condition [5]. Hyperandrogenism often leads to acne, alopecia, and/or hirsutism [6]. Chronic oligo- or anovulation can cause amenorrhea or oligomenorrhea [7]. Women with PCOS also commonly have enlarged ovaries with greater stromal volume and increased follicular count due to arrest of follicular development [8]. Over half of PCOS women are obese [9], which can both be attributed to and also exacerbates the feature of insulin resistance that is found in 65%–70% of all PCOS patients, including 70%–80% of obese patients and 20%–25% of lean patients [10].

While the pathophysiology of PCOS is not well understood, androgen excess is a hallmark feature of the disorder, present in 60%–80% of PCOS women [11–15]. Hyperandrogenism is often diagnosed by clinical manifestations, most commonly hirsutism, and/or by abnormal circulating levels of androgens, including total testosterone, free testosterone, sex hormone-binding globulin (SHBG), androstenedione, and dehydroepiandrosterone (DHEA) [16]. In healthy women, androgens are produced by the adrenal glands and ovaries at similar levels [17]. Excess androgen production in the ovaries is considered a hallmark of PCOS and negatively impacts follicular development, leading to ovulatory dysfunction [18]. However, 20%–30% of PCOS patients also exhibit elevated adrenal androgen production [19]. Excess androgen production can occur via dysfunction of the hypothalamic–pituitary–gonadal/adrenal axes and the byproduct of insulin resistance and compensatory hyperinsulinemia [20].

Immunology of Recurrent Pregnancy Loss and Implantation Failure
DOI: https://doi.org/10.1016/B978-0-323-90805-4.00009-2

While the greatest barrier that PCOS patients face when attempting pregnancy is achieving ovulation, primarily due to ovarian dysfunction, hyperandrogenism also negatively impacts the endometrium and placenta. Through multiple mechanisms, excess androgens cause reduced endometrial receptivity [21] and placental dysfunction [22]. Even if conception is achieved, women with PCOS face a higher risk of early pregnancy loss or miscarriage [23–25].

In addition to hyperandrogenism, insulin resistance plays a major role in PCOS pathophysiology. Insulin resistance is associated with a higher risk of developing obesity, diabetes, and cardiovascular disorders [26,27]. Those risks are higher for patients already suffering from obesity. Over half of all PCOS patients are obese [9]. While insulin resistance in PCOS is independent of obesity, the added visceral adipose tissue in obese patients further desensitizes the body's response to insulin [28]. The primary mechanism by which insulin resistance and compensatory hyperinsulinemia are thought to lead to infertility is contributed by hyperandrogenism [29]. Insulin has been shown to have the ability to stimulate steroidogenesis in the ovaries through multiple mechanisms [30,31]. Insulin also inhibits SHBG production [32] and insulin growth factor binding protein-1 (IGFBP-1) production [33], which results in more free androgen and insulin, respectively. Women with PCOS display abnormalities of insulin action that are independent of obesity [34]. However, in obese PCOS patients, obesity and PCOS-related insulin resistance work in concert to negatively impact glucose metabolism [35], creating a vicious cycle. Specifically, regarding ovarian function, hyperinsulinemia causes arrest of folliculogenesis and prevents antral follicles from fully maturing [36]. Insulin resistance has also been shown to lower rates of implantation [37] by affecting endometrial receptivity [21] and causing placental abnormalities [22].

A growing body of evidence suggests that PCOS is a proinflammatory state and implicates chronic inflammation in reproductive failure is associated with this condition. Women with PCOS display elevated levels of proinflammatory cytokines that in turn contribute to insulin resistance and androgen production [38], and multiple inflammatory genes in adipose tissue, the ovaries, and the endometrium are altered in women with PCOS [39–42]. While gene expression abnormalities in adipose tissue can contribute to infertility through androgen-associated pathways, inflammation also directly affects ovarian, endometrial, and placental reproductive function. Altered inflammatory gene expression in the ovaries leads to the arrest of folliculogenesis [43]. PCOS is associated with increased oxidative stress and decreased antioxidant capacity [44]. Reactive oxygen species (ROS) are necessary for ovulation, but excessive ROS are cytotoxic, which are often present in PCOS and diminish oocyte quality [45,46]. Inflammatory genes are upregulated in multiple cell types, and inflammatory cytokines reduce endometrial receptivity through several mechanisms [21]. Similarly, chronic inflammation contributes to placental dysfunction [47]. While the chronic, low-grade inflammation associated with PCOS is independent of obesity [48], the increased adiposity of obese patients with PCOS exacerbates the inflammatory state through the NFκB pathway [49,50].

The effects of PCOS on the ovaries, uterus, and placenta culminate in reproductive dysfunction through reduced ability to ovulate, decreased endometrial receptivity, implantation failure, placental dysfunction, recurrent pregnancy loss, and various pregnancy complications. *Meta*-analyses of studies in which pregnancy outcomes were compared between women with PCOS and controls found that PCOS causes a significant increase in the risk of adverse pregnancy outcomes, including gestational diabetes, pregnancy-induced hypertension, preeclampsia, and preterm birth [51,52]. The presence of a link between PCOS and the rate of miscarriage has been inconsistently reported due to variation in diagnostic criteria and study design [53]. However, several *meta*-analyses have established a correlation between PCOS and elevated rates of miscarriage in patients undergoing IVF [53–55]. This chapter reviews the respective roles and interplay of hyperandrogenism, insulin resistance, and inflammation in PCOS pathophysiology related to reproductive failure.

2 Hyperandrogenism and reproductive failure in polycystic ovarian syndrome

Hyperandrogenism is diagnosed by clinical manifestations and/or biochemical evidence of hyperandrogenemia typically identified through elevated serum androgen levels. Testosterone is present in serum either as free testosterone or bound to proteins such as SHBG and albumin [56]. In normal conditions, 80% of testosterone is bound to SHBG, 19% to albumin, and 1% remain free in the circulation [57,58]. However, PCOS patients have lower concentrations of SHBG and elevated androgen production, resulting in greater concentrations of the bioactive free testosterone [59]. Therefore circulating free testosterone is one parameter assessed for a hyperandrogenism diagnosis. Other parameters include free androgen index, androstenedione, and DHEAS [3]. In addition to elevated androgen levels, PCOS patients also present a higher luteinizing hormone (LH) to follicle stimulating hormone (FSH) ratio (LH:FSH) [60]. Both excess androgen production and an elevated LH:FSH ratio are believed to contribute to infertility by interfering with ovarian and endometrial function [18,21,60]. Specific mechanisms by which hyperandrogenism affects reproductive functions are discussed in the following sections.

2.1 Causes of hyperandrogenism

The first of three proposed mechanisms leading to excess androgen production in PCOS is hypothalamic–pituitary–ovarian (HPO) axis dysfunction. In normal HPO axis function, the hypothalamus releases gonadotropin-releasing hormone (GnRH) that stimulates the pituitary gland to release LH and FSH [61]. The 2-cell, 2-gonadotropin concept, which was established to model ovarian estrogen production, consists of the following: LH stimulates the LH receptors of theca cells within the ovaries, which then activates a cAMP-mediated pathway to express cholesterol side-chain cleavage enzymes, including cytochrome P450 (CYP11A), 3β-hydroxysteroid dehydrogenase (3β-HSD), and 17α-hydroxylase/C17–20 lyase cytochrome P450 (CYP17). With those three enzymes, theca cells can convert cholesterol into androstenedione. Androstenedione then diffuses across the basal lamina to an adjacent granulosa cell. FSH stimulates granulosa cells, and through a cAMP-mediated mechanism, they express aromatase cytochrome P450 (CYP19) and 17β-hydroxysteroid dehydrogenase (17β-HSD). CYP19 converts androstenedione to estrone, followed by 17β-HSD-mediated conversion to estradiol [62]. Patients with PCOS commonly display an inappropriately high LH:FSH ratio due to both elevated LH and insufficient FSH [63,64]. One proposed mechanism by which this can occur is heightened GnRH pulsatility, which favors LH pulse frequency over that of FSH and leads to overstimulation of theca cells in relation to granulosa cells [65]. Theca cell overstimulation combined with an inability of granulosa cells to convert the appropriate amount of theca cell-derived androgen to estrogen due to insufficient FSH stimulation leads to hyperandrogenism [66].

The second mechanism of excess androgen production relates to the expression of key androgen-synthesizing enzymes. Overactive allele variants of the gene encoding CYP11a, which is responsible for side-chain cleavage in the synthesis pathway of androstenedione within theca cells, have been associated with elevated serum testosterone levels [67–72]. Similarly, dysregulated expression of CYP17, also involved in androstenedione production, is most likely a cause of ovarian hyperandrogenism in PCOS [73–81]. Downregulated CYP19, which is responsible for catalyzing the conversion of androgens to estrogens, also leads to excess androgen production due to insufficient conversion [82–86].

The third mechanism by which excess androgen is produced is through insulin-mediated biosynthesis of testosterone in theca cells [87–89]. In normal physiological function, the steroidogenic action of insulin on the ovaries is maintained at a healthy level. However, in PCOS patients who display insulin resistance, compensatory hyperinsulinemia results in high concentrations of insulin, which in turn stimulates excessive androgen production in theca cells [90,91]. To stimulate androgen production, insulin binds to its own receptor and activates a signaling cascade driven by inositol glycan messengers to transduce the stimulatory signal [92]. An important aspect of this pathway is that it remains independent of the insulin-mediated glucose uptake pathway. The implication is that in PCOS patients, who have an impaired glucose uptake pathway and consequently insulin resistance [93], the mechanism by which insulin stimulates steroidogenesis remains intact and becomes hyperactive due to compensatory hyperinsulinemia [94].

2.2 Hyperandrogenism and ovaries

Hyperandrogenism has detrimental effects on ovarian function by interfering with proper folliculogenesis. Folliculogenesis is regulated by granulosa cells, which are dysfunctional and lower in number in PCOS patients, causing abnormal follicular development [95]. PCOS patients have increased numbers of antral follicles; however, they are arrested at a premature developmental stage [96–98]. Androgens stimulate preantral follicles to grow [99]. Small follicles are further stimulated by androgens to convert theca cell-derived androgens to estrogens. However, in the antral follicle stage, androgens inhibit FSH-mediated androgen conversion to estrogen within granulosa cells, resulting in a hormonal imbalance and follicular arrest [100]. The presence of excess androgens in the ovaries has also been shown to cause follicular dysplasia [99,101,102].

Androgen levels are correlated with follicular development, where the lower level promotes folliculogenesis [103], but the high level causes the overrecruitment of follicles. Overrecruitment of follicles is associated with granulosa cell hypersecretion of anti-Müllerian hormone (AMH), an exclusive granulosa cell product implicated in PCOS ovarian dysfunction [104]. Several studies have shown that serum AMH is correlated with LH and androgen levels in women with PCOS [105–108]. LH increases AMH production fourfold in granulosa cells of PCOS ovaries but not of normal ovaries [109]. Moreover, androgens play a role in stimulating early (FSH independent) stages of follicular growth and may thus contribute to increased AMH production [110,111]. However, Carlsen et al. reported that 6-month androgen suppression with dexamethasone did not change AMH concentrations [112], suggesting other mechanisms likely responsible for the induction and maintenance of AMH production in PCOS women. Indeed, the follicular arrest in PCOS happens when granulosa cell AMH production is greatest (preantral and small antral stages). This explains why serum AMH concentration is elevated in PCOS women compared with women with normal ovaries [113–115]. AMH is likely related to the pathogenesis of anovulation in PCOS; [104] it inhibits FSH-induced follicular growth in addition to aromatase expression, leading to follicular arrest both directly by inhibiting growth and indirectly by suppressing androgen–estrogen conversion, causing excess androgen production and a deficiency of estrogen, which is required for follicle maturation [116,117]. Oligo-/anovulation develops due to chronic follicular arrest and prolonged impairment of the selection of a dominant follicle [118], impeding natural conception.

2.3 Hyperandrogenism and the endometrium

While excess androgens present major challenges to achieving pregnancy through anovulation and ovarian dysfunction, it also impedes conception by acting through the endometrium directly. In PCOS, hyperandrogenism reduces endometrial receptivity through multiple mechanisms. Excess androgen suppresses decidualization, which is required for implantation [119] and causes delayed decidualization, affecting the window of implantation [120]. During the window of implantation, endometrial stromal cells normally express the Wilms tumor suppressor gene (WT1) [121]. However, androgen receptors (ARs) are upregulated in response to hyperandrogenism, and AR, in turn, downregulates WT1 [122]. Bcl2, which is normally inhibited by WT1, becomes upregulated and activates p27, blocking the cell cycle and interfering with endometrial stromal cell function [122]. Additionally, epidermal growth factor (EGFR), which plays a role in regenerating the endometrium postmenses [123] and is directly regulated by WT1 [124], is downregulated as a consequence of AR-mediated WT1 suppression. Impairing the ability of the endometrium to regenerate due to EGFR downregulation causes reduced endometrial receptivity [21].

HOXA genes are also critical for endometrial receptivity and implantation [125–127]. Multiple studies have shown that PCOS is associated with endometrial HOXA-10 and HOXA-11 downregulation [128,129]. Whereas estrogen and progesterone stimulate upregulation of HOXA-10, androgens have the opposite effect [128,130]. Hyperandrogenism also directly influences the expression of various proteins within the endometrium. The cyclin-dependent inhibitor CDKN2A, which is important for blastocyst implantation, is downregulated in the presence of excess androgen [131]. Expression of the L-selectin ligand, which is expressed during the window of implantation, is also suppressed by testosterone [132]. Multiple types of cell-cell adhesion proteins, including $\alpha v\beta 3$-integrin, E-Cadherin, Mucin-1, claudin-4, and occludin, are downregulated by excess androgen, affecting endometrial permeability, selectivity of the endometrium, and directly affect receptivity [133–135]. Due to multiple mechanisms causing impaired endometrial receptivity, there is a higher risk of miscarriage for patients with hyperandrogenic PCOS [23–25,131,136].

2.4 Hyperandrogenism and the placenta

Women with PCOS who become pregnant display elevated circulating levels of androgens compared to pregnant women without PCOS regardless of fetal sex [137,138]. Since maternal androgens are not thought to cross the placenta [139], and the fetus is protected from excess androgen by placental aromatase-mediated conversion of androgens to estrogens, a potential mechanism through which maternal hyperandrogenism impacts fetal development is by altering placental and/or fetal steroidogenesis, resulting in excess feto-placental androgen [22,140–143]. This concept was supported by a study that showed upregulated 3β-hydroxysteroid dehydrogenase 1 activity and lower aromatase activity in placental tissue from PCOS women compared to non-PCOS women, suggesting upregulated androgen production and lower conversion of androgens to estrogen [142].

Multiple animal models have shown associations between hyperandrogenism and placental abnormalities. Maternal testosterone administration induces gestational hyperandrogenism [144–146]. Testosterone treatments in rats and sheep have also provided insight on the role of androgens in placental function, demonstrating a variety of placental effects, including decreased fetal and placental weight, decreased placental amino acid transport, increased expression of placental AR, decreased uterine artery blood flow, and advanced placental differentiation [139,141,147–153]. Animal models have also demonstrated a correlation between prenatal androgen levels and placental vasculopathy [154–156]. Excess androgens decrease the expression of placental proangiogenic factors [157,158] and increase the vascular tone of uterine arteries, in turn negatively affecting the primary uterine blood supply [147]. Preeclampsia, a leading cause of maternal and fetal morbidity thought to occur as a result of abnormal placental development, is also associated with excess androgen [159–161].

3 Insulin resistance and reproductive failure in polycystic ovarian syndrome

Insulin, synthesized in the β-cells of the pancreas, mediates glucose intake into cells and is critical for metabolic regulation [162]. In healthy individuals, insulin binds to its receptor on the cell surface and induces activation of a signaling cascade that stimulates GLUT4, an insulin-dependent transmembrane protein, to facilitate the entry of glucose into the cell [163]. In the face of insulin resistance, the compensatory response for the reduced cellular ability to uptake glucose produces more insulin, leading to hyperinsulinemia [164]. Approximately 65%–70% of all PCOS patients have insulin resistance and compensatory hyperinsulinemia, including 70%–80% of obese patients and 20%–25% of lean patients [10]. In addition, insulin resistance affects female fertility through both indirect and direct mechanisms that are discussed in the following sections.

3.1 Molecular basis for insulin resistance

Insulin resistance in PCOS is due to multiple defects in the downstream signaling pathway after insulin binds to its receptor [93]. The insulin receptor (IR) is a heterotetramer consisting of two pairs of identical dimers, IRα which contains the ligand-binding site, and IRβ, which contains a tyrosine phosphorylation site [165,166]. Under normal circumstances, once insulin binds to IRα, a conformational change is induced in the receptor that leads to tyrosine autophosphorylation of IRβ, which then interacts with a variety of intracellular proteins to transduce the message delivered by insulin [167]. However, in contrast to its tyrosine phosphorylation site, IRβ also contains a serine phosphorylation site that terminates insulin signaling upon phosphorylation [168–170]. In PCOS women, this serine site is excessively phosphorylated, blocking the transduction of insulin signaling without impacting the ability of insulin to bind to its receptor [93]. Over 50% of PCOS patients present this abnormality [171].

An additional postbinding defect proposed as a cause of insulin resistance in PCOS patients involves the signal-transducing molecules that interact with tyrosine-phosphorylated IRβ. Insulin receptor substrate (IRS) proteins are cytoplasmic proteins that transmit the insulin signal upon binding to tyrosine-phosphorylated IRβ [172]. Once IRβ is activated, it recruits IRS-1 and subsequently phosphorylates it at tyrosine residues, which is critical for signaling to continue. IRS-1 then activates phosphatidylinositol 3-kinase (PI3K), which is necessary for GLUT4-mediated glucose uptake [173]. Whereas tyrosine phosphorylation of IRS-1 begins signal transduction, serine phosphorylation of IRS-1 can either enhance or terminate transduction depending on the pattern of phosphorylation [172]. Women with PCOS have reduced PI3K activity, supporting the notion that IRS-1 serine phosphorylation is a potential cause of PCOS-related insulin resistance [93].

Multiple studies have also investigated the prevalence of polymorphisms in the IRS-1 gene and their potential role with regards to insulin resistance and PCOS. After an initial study brought to attention the possible link between a single nucleotide polymorphism in IRS-1, G972R, and PCOS [174], multiple subsequent studies have been reported, yielding conflicting results [173,175–181]. However, a *meta*-analysis reviewing the body of research into the potential link between polymorphisms in IRS-1 and PCOS found a significant correlation between the IRS-1 G972R variant and the risk of developing PCOS due to insulin resistance and compensatory hyperinsulinemia [182].

3.2 Insulin resistance and obesity

In PCOS, insulin resistance is independent of obesity [34], but increased adiposity exacerbates insulin resistance [35]. Obese patients have more adipose tissue that releases several metabolites that influence insulin secretion [183]. As a result, while lean PCOS patients can still experience insulin resistance, obese patients with PCOS present more severe insulin resistance and compensatory hyperinsulinemia [9,184]. The following three sections explain how hyperinsulinemia affects the ovaries, endometrium, and placenta, leading to reproductive failure.

3.3 Insulin resistance and the ovaries

Insulin affects ovarian function in multiple ways, the first of which is by contributing to hyperandrogenism and its effects on the ovaries. Insulin can directly stimulate steroidogenesis in the ovaries [30,31,185]. Insulin receptors are present on ovarian stromal cells, theca cells, and granulosa cells [185]. By binding to its receptor, insulin can stimulate P450c17 activity, resulting in the eventual production of excess testosterone [92]. Additionally, insulin has been shown to increase the amount of LH secreted by the pituitary gland, which in turn stimulates theca cell androgen production [31,186–190]. A third mechanism by which insulin exacerbates hyperandrogenism is by suppressing the synthesis of SHBG [32], allowing for more circulating testosterone to remain unbound and able to interfere with ovarian function.

In addition to exerting its effects via hyperandrogenism, insulin can directly affect the ovaries. In normal menstrual cycles, granulosa cells of the dominant follicle become sensitive to LH at a follicular diameter around 10 mm, whereas smaller and less mature follicles do not become sensitive [191,192]. Once stimulated by LH, LH-sensitive granulosa cells begin terminal differentiation and only undergo two more rounds of cell division [193]. In patients with PCOS, the excess insulin present in the ovarian environment causes premature sensitization of granulosa cells to LH. This results in premature terminal differentiation of granulosa cells, which in turn causes follicular arrest after the final two rounds of cell division and contributes to the larger number of small antral follicles present in patients with PCOS [36,194,195].

3.4 Insulin resistance and the endometrium

Results from in vitro fertilization studies have shown that oocytes obtained from patients with insulin-resistant PCOS are able to develop normally. However, the rate at which those patients achieved pregnancy was significantly lower than controls, suggesting that insulin resistance plays a role in endometrial receptivity and implantation [37,196].

The primary mechanism through which insulin resistance and compensatory hyperinsulinemia are thought to affect the endometrium is by interfering with the process of decidualization. Glucose metabolism is critical for proper endometrial decidualization [197–200]. Without proper decidualization, embryo implantation cannot occur [201]. PCOS patients with insulin resistance have altered glucose metabolism and therefore face a greater risk that their endometria will not properly undergo decidualization; Patients with PCOS have lower endometrial mRNA expression of GLUT4, the insulin-dependent transmembrane glucose transporter. This abnormality further contributes to the altered glucose metabolism of PCOS patients due to an impaired ability to sense glucose [202]. Lastly, insulin growth factor binding protein-1 (IGFBP-1), which is one of the most established biomarkers of decidualization and endometrial receptivity [203], has been shown to be downregulated in patients with PCOS, suggesting a defect in decidualization and consequently in endometrial receptivity [204].

3.5 Insulin resistance and the placenta

Insulin resistance has been implicated in multiple pregnancy complications, including miscarriage, pregnancy-induced hypertension, and fetal growth restriction [205]. The existence of those associations combined with the well-studied placental expression of the insulin receptor [206] suggests a placental role in insulin resistance-mediated pregnancy complications [207,208]. In the early stages of pregnancy, hyperglycemia potentially increases rates of miscarriage and preeclampsia, most likely due to impaired trophoblast invasion [22,209]. One study showed that insulin resistance leads to impaired insulin growth factor 1 (IGF-1) signaling in placental tissue. Impaired IGF-1 signaling decreases migration and trophoblast invasion into maternal spiral arteries [207], which is crucial for placental development and normal fetal growth [210]. Consistent with PCOS-associated insulin resistance, women with PCOS show lower levels of IGF-1 in early pregnancy compared to non-PCOS controls [211,212]. PCOS women are also at a higher risk of developing gestational diabetes due to insulin resistance [213–215]. Gestational diabetes may have multiple effects on placental function, including impaired vasculogenesis, reduced placental apoptosis, and increased ischemia [216–220].

4 Chronic inflammation and reproductive failures in polycystic ovarian syndrome

The inflammation process includes upregulation and release of chemokines, proinflammatory cytokines, and markers of inflammation. Studies have found significantly higher levels of the inflammatory markers, such as C-reactive protein (CRP), interleukin 18 (IL-18), tumor necrosis factor (TNF-α), and interleukin 6 (IL-6), in women with PCOS compared to BMI and age-matched controls [221–224]. CRP has been investigated thoroughly in studies that have evaluated chronic low-grade inflammation in PCOS. Numerous studies have shown an increased level of serum CRP in women with PCOS compared to matched controls [223,225–228]. Two *meta*-analyses concluded that CRP was elevated in women with PCOS by an average of 96% compared to controls [229,230]. Studies have also shown a correlation between IL-18 serum levels and PCOS [231–233]. IL-18 is a proinflammatory cytokine associated with insulin resistance and a greater risk of cardiovascular disease [221]. Both elevated CRP and IL-18 reported in PCOS women indicate a proinflammatory state, and inflammation is becoming recognized as a universal factor affecting the pathophysiology of PCOS [234–237].

Chronic inflammation in PCOS works in multiple ways to affect the ovaries, endometrium, and placenta. Inflammation is the last component of triads, including hyperandrogenism and insulin resistance, perpetuating abnormal hormonal and metabolic functions in PCOS patients and resulting in infertility [238] (Fig. 1). The specific mechanisms by which chronic inflammation contributes to reproductive failure are discussed in the following sections.

4.1 Mechanisms of inflammation in polycystic ovarian syndrome and interplay with insulin resistance and hyperandrogenism

There are several major causes of inflammation in women with PCOS. The first is altered gene expression of multiple proinflammatory cytokines in both ovaries and endometrium that promote inflammatory responses and contribute to the chronic, low-grade inflammatory state associated with PCOS [40–43]. The second cause is a result of the interplay between existing inflammatory markers and insulin resistance. Due to impaired glucose uptake, PCOS patients with insulin resistance induce an inflammatory response upon glucose ingestion [239]. Circulating mononuclear cells (MNCs) use glucose for glycolysis and synthesis of nicotinamide adenine dinucleotide phosphate (NADPH) [240]. Upon oxidation of NADPH by NADPH oxidase, a ROS is generated in the form of superoxide, which induces oxidative stress [241]. Oxidative stress activates the nuclear factor kappa B (NFκB), which in turn upregulates the expression and secretion of TNF-α and IL-6 by MNCs, which perpetuate the inflammatory response initiated by NFκB [242]. In addition to its role in inflammation, TNF-α contributes to insulin resistance. Upon binding to its receptor, TNF-α activates c-Jun amino-terminal kinase (JNK), which in turn causes phosphorylation of serine 307 of IRS-1, preventing tyrosine phosphorylation and subsequent activation of IRS-1 [243,244]. Through this mechanism insulin resistance and inflammation engage in cyclic reciprocal activation, creating a vicious cycle.

The third mechanism by which inflammation is caused in patients with PCOS is through obesity and abdominal adiposity. Over half of patients with PCOS are obese, and all patients with PCOS are at a greater risk of developing excess abdominal adiposity and/or obesity due to PCOS [9]. Hypoxia-related death of adipocytes due to adipose tissue expansion causes an influx of MNCs and MNC-derived macrophages, which are the primary sources of TNF-α and IL-6 [245]. In adipose tissue, TNF-α stimulates further proinflammatory cytokine release that contributes to the state of inflammation in obese patients with PCOS [246]. Although the chronic, low-grade inflammatory state observed in patients with PCOS is independent of obesity, excess abdominal adiposity exacerbates the hormonal and metabolic abnormalities associated with the condition [239].

Hyperandrogenism itself is another potential mechanism that promotes inflammation in PCOS. When hyperandrogenemia is induced acutely in normal-weight ovulatory women to the degree present in PCOS by treatment with an oral androgen, MNC-derived ROS generation, phosphorylated p47 protein content, activated NFκB, TNFα mRNA content, and TNFα secretion increase in the fasting state, and in response to glucose ingestion [247–249]. This suggests that an AR-dependent mechanism in MNC is responsible for this prooxidant, proinflammatory effect [250]. On the other hand, inflammation in PCOS may directly promote hyperandrogenism and ovarian dysfunction independent of insulin resistance. Macrophages derived from MNC are known to infiltrate the ovary [251]. The anti-inflammatory agents, statins and resveratrol, suppress proliferation of androgen-producing theca cells from rat and human polycystic ovaries, while TNFα stimulates [252,253]. Moreover, in vitro evidence shows that CYP17, the androgen-producing enzyme within ovarian theca cells, undergoes upregulation in response to proinflammatory stimuli and downregulation in response to resveratrol and statins [254]. Thus MNC recruited into the polycystic ovary likely promote a local inflammatory response that increases the number of theca cells and induces their steroidogenic capability to directly stimulate ovarian androgen production in women with PCOS [236].

4.2 Chronic inflammation and ovaries

As previously discussed, inflammation represents the third major player in a vicious cycle with hyperandrogenism and insulin resistance. The first way chronic inflammation affects the ovary is by promoting insulin resistance and hyperandrogenism. However, inflammation also directly influences ovarian function. Multiple inflammatory genes are upregulated in the PCOS ovaries, including TNF-α and IL-6 [39–42]. An increase in TNF-α in the ovaries has been linked to poor oocyte development and pregnancy outcomes in women with PCOS undergoing IVF treatment [255–257]. Another study with PCOS patients seeking IVF found that women with PCOS had significantly higher levels of both TNF-α and IL-6 in the serum and follicular fluid [258]. TNF-α has been shown to affect follicular maturation [259–261] and inhibit granulosa cell differentiation [262], suggesting that TNF-α negatively impacts folliculogenesis in excess. Specifically, in granulosa cells, other inflammatory genes that play roles in folliculogenesis are upregulated in patients with PCOS. IL-1β and IL-8, which normally recruit and activate leukocytes during ovulation [263], could cause premature invasion of leukocytes and disrupt the final maturation of follicles [43]. Moreover, the ovarian microenvironment in PCOS women has dysregulation of various angiogenic cytokines [264]. These include vascular endothelial growth factor [265–267], angiopoietins [268,269], transforming growth factor-β (TGF- β) [270,271], platelet-derived growth factor [269], and basic fibroblast growth factor [266], that have been shown to be dysregulated in serum and/or follicular fluid of PCOS patients. Such angiogenic dysregulation is thought to be contributory to the abnormal folliculogenesis as well as increased stromal vascularity and propensity to ovarian hyperstimulation syndrome [264]. Angiogenesis and inflammation have long been coupled together in many chronic inflammatory disorders, including psoriasis, rheumatoid arthritis, Crohn's disease, diabetes, obesity, and cancer [272]. It has been shown that clomiphene-resistance in PCOS is associated with decreased circulating levels of angiopoietin-2 and increased levels of the inflammatory markers CRP and CXCL16, a known chemoattractant of mononuclear cells, suggesting a potential interplay between angiogenesis and inflammation in altering the ovarian response to clomiphene [273]. Elucidating the underlying cellular and molecular mechanisms that tie these two processes together would be essential to understand their synergistic effect and develop new therapeutic approaches for the management of PCOS. In summary, the ovarian cytokine expression profile of PCOS patients is abnormal, and various proinflammatory cytokines have detrimental effects on folliculogenesis and oocyte quality.

4.3 Chronic inflammation and the endometrium

Similar to the ovaries, altered gene expression of proinflammatory cytokines interferes with endometrial function. IL-18 was found to be upregulated in the endometria of patients with PCOS compared to controls [274]. The levels of IL-6, IL-8, and other proinflammatory markers were also elevated in undecidualized stromal cells of PCOS patients, and their conditioned media had more chemotactic activity to immune cells (T cells, monocytes), suggesting interference with normal decidualization and subsequent endometrial receptivity [275]. IL-6 and TNF-α also interfere with endometrial receptivity by downregulating the expression of adiponectin, an insulin-sensitizing molecule, which causes impaired insulin signaling and glucose metabolism in the endometrium [239,276,277]. TNF-α also exacerbates inflammation in the endometrium by activating the NFκB pathway [278,279]. Other cytokines, including IL-1α were markedly downregulated with PCOS patients [280]. Similarly, a rat model of PCOS showed that decreased mRNA expression of IL-1 is associated with thinning of the endometrium and lower implantation success rates [280,281]. Women with PCOS also present decreased levels of stanniocalcin-1 (STC-1), a prosurvival factor protecting against inflammation, which could negatively impact the implantation microenvironment [21,282].

4.4 Chronic inflammation and the placenta

Should a woman with PCOS become pregnant despite impaired ovulatory function and endometrial receptivity, while inflammation can still negatively impact the pregnancy? The low-grade chronic inflammation persists during gestation and is exacerbated by pregnancy. Pregnant women with PCOS had higher white blood count, CRP, and ferritin levels at baseline and all gestational ages than controls, and these inflammatory markers were associated with adverse pregnancy and neonatal outcomes [283]. In addition, placentae of PCOS women have been demonstrated to have increased expression of phosphorylated STAT-3 [284]. The clinical consequences of increased placental STAT-3 remain to be established but may reflect a proinflammatory state similar to that observed in women with maternal obesity [284]. Placentae of obese women have increased infiltration of proinflammatory macrophages, which express high levels of IL-1, TNF-α, and IL-6, associated with increased expression of chemotactic cytokines and neutrophils in the maternal interstitial space and muscular layer of placental vessels [285]. These aberrant inflammatory pathways could be subsequently associated with immune and vascular dysfunction in placenta/decidua interactions inducing a significant impairment of decidual endovascular trophoblast invasion [286]. Indeed, multiple studies have demonstrated a link between upregulated inflammatory markers in PCOS and pregnancy complications, including gestational diabetes, preterm birth, preeclampsia, fetal growth restriction, and antepartum hemorrhage [283,287–289].

5 Conclusion

PCOS is a heterogeneous hormonal and metabolic condition characterized by hyperandrogenism, oligo-anovulation, and polycystic ovarian morphology. It is the most common endocrinopathy affecting reproductive-age women. Women with PCOS face higher risks of type II diabetes, cardiovascular disease, metabolic syndrome, obesity, reproductive dysfunction, and adverse pregnancy outcomes. Three main features of PCOS, namely hyperandrogenism, insulin resistance, and low-grade chronic inflammation, form a vicious cycle in which each feature feeds into the other two resulting in reproductive dysfunction. Women suffering from PCOS have abnormal ovarian and endometrial function due to that vicious cycle, causing infertility through multiple mechanisms. It is also evident that hyperandrogenism, insulin resistance, and low-grade chronic inflammation cause a series of endometrial and placental alterations that underlie the poor pregnancy and obstetric outcomes observed in these patients. While PCOS is independent of obesity, obesity aggravates PCOS by contributing to inflammation, PCOS-related insulin resistance, compensatory hyperinsulinemia, and consequently hyperandrogenism. Understanding the interplay between inflammatory, endocrine, and metabolic mechanisms responsible for the increased incidence of infertility, implantation failure, and obstetric complications associated with PCOS is essential for developing new therapies.

6 Summary and Recommendations

- PCOS affects up to 10% of reproductive-age women and is characterized by three hallmark features: hyperandrogenism, oligo-/anovulation, and polycystic ovaries.
- Hyperandrogenism, insulin resistance, and chronic inflammation form a vicious cycle that leads to reproductive dysfunction by negatively impacting the ovaries, uterus, and placenta.
- Ovarian aberrations in PCOS include abnormal follicular development, oligoovulation/anovulation, and reduced egg quality.
- Uterine abnormalities include altered endometrial gene expression, increased inflammation, and decreased endometrial receptivity to implantation.
- The effects of PCOS on the ovaries, uterus, and placenta culminate in reproductive dysfunction through reduced ability to ovulate, decreased endometrial receptivity, implantation failure, placental dysfunction, recurrent pregnancy loss, and various pregnancy complications, including preterm birth and preeclampsia.

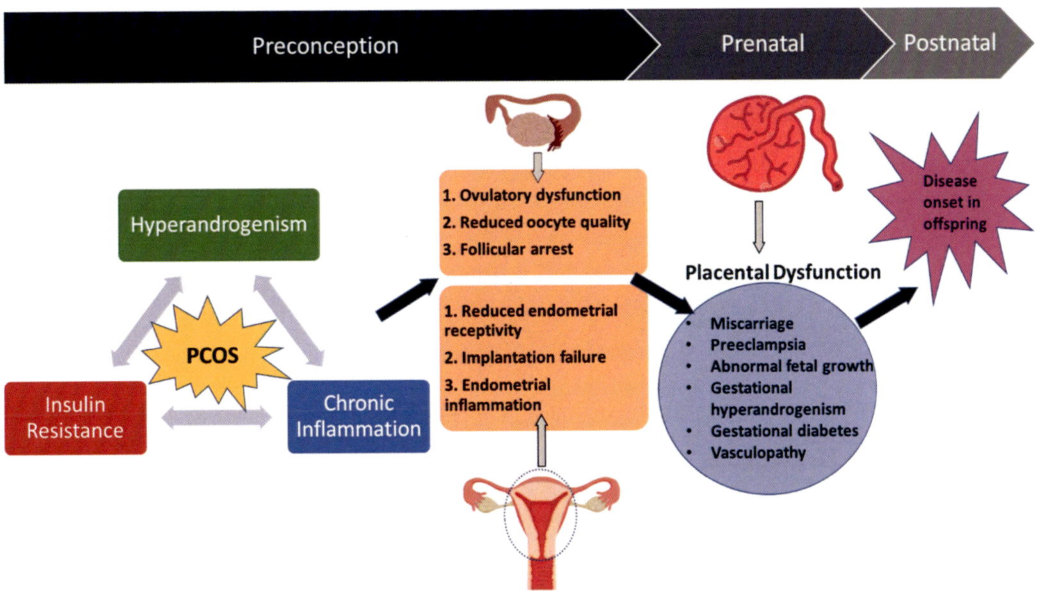

FIGURE 1 **Polycystic ovarian syndrome (PCOS) and reproductive dysfunction.** In PCOS, multiple proinflammatory cytokines are altered in both the ovary and endometrium, contributing to the chronic, low-grade inflammatory state. Three main features of PCOS, namely, hyperandrogenism, insulin resistance, and low-grade chronic inflammation, interplay to form a vicious cycle, resulting in reproductive dysfunction. Prior to conception, the effects on the ovary include ovulatory dysfunction, reduced oocyte quality, and follicular arrest; the effects on the endometrium include reduced endometrial receptivity, inflammation, and implantation failure. Prenatally, placental dysfunction further contributes to miscarriage, preeclampsia, abnormal fetal growth, gestational hyperandrogenism, gestational diabetes, and vasculopathy, which, in turn, lead to disease in the offspring.

- A growing body of evidence suggests that PCOS is a proinflammatory state and implicates chronic inflammation in reproductive failure associated with this condition.
- While PCOS is independent of obesity, obesity aggravates PCOS reproductive and endocrine dysfunction by contributing to inflammation, PCOS-related insulin resistance, compensatory hyperinsulinemia, and consequently hyperandrogenism.
- PCOS patients face a greater risk of developing cardiovascular and metabolic disorders in addition to reproductive dysfunction.

References

[1] Deswal R, Narwal V, Dang A, Pundir CS. The prevalence of polycystic ovary syndrome: a brief systematic review. J Hum Reprod Sci 2020;13(4):261–71. Available from: https://doi.org/10.4103/jhrs.JHRS_95_18.

[2] Stein I, Leventhal ML. Amenorrhea associated with bilateral polycystic ovaries. Am J Obstet Gynecol 1935;29:181–91.

[3] Rotterdam EA-SPcwg. Revised 2003 consensus on diagnostic criteria and long-term health risks related to polycystic ovary syndrome (PCOS). Hum Reprod 2004;19(1):41–7. Available from: https://doi.org/10.1093/humrep/deh098.

[4] Teede HJ, Misso ML, Costello MF, et al. Recommendations from the international evidence-based guideline for the assessment and management of polycystic ovary syndrome. Clin Endocrinol (Oxf) 2018;89(3):251–68. Available from: https://doi.org/10.1111/cen.13795.

[5] Acien P, Quereda F, Matallin P, et al. Insulin, androgens, and obesity in women with and without polycystic ovary syndrome: a heterogeneous group of disorders. Fertil Steril 1999;72(1):32–40. Available from: https://doi.org/10.1016/s0015-0282(99)00184-3.

[6] Witchel SF, Oberfield SE, Peña AS. Polycystic ovary syndrome: pathophysiology, presentation, and treatment with emphasis on adolescent girls. J Endocr Soc 2019;3(8):1545–73. Available from: https://doi.org/10.1210/js.2019-00078.

[7] van Hooff MH, Voorhorst FJ, Kaptein MB, Hirasing RA, Koppenaal C, Schoemaker J. Polycystic ovaries in adolescents and the relationship with menstrual cycle patterns, luteinizing hormone, androgens, and insulin. Fertil Steril 2000;74(1):49–58. Available from: https://doi.org/10.1016/s0015-0282(00)00584-7.

[8] Silfen ME, Denburg MR, Manibo AM, et al. Early endocrine, metabolic, and sonographic characteristics of polycystic ovary syndrome (PCOS): comparison between nonobese and obese adolescents. J Clin Endocrinol Metab 2003;88(10):4682–8. Available from: https://doi.org/10.1210/jc.2003-030617.

[9] Pasquali R. The impact of obesity on hyperandrogenism and polycystic ovary syndrome in premenopausal women. Clin Endocrinol (Oxf) 1993;39(1):1–16. Available from: https://doi.org/10.1111/j.1365-2265.1993.tb01744.x 1993.

[10] DeUgarte CM, Bartolucci AA, Azziz R. Prevalence of insulin resistance in the polycystic ovary syndrome using the homeostasis model assessment. Fertil Steril 2005;83(5):1454–60. Available from: https://doi.org/10.1016/j.fertnstert.2004.11.070.

[11] Balen AH, Conway GS, Kaltsas G, et al. Polycystic ovary syndrome: the spectrum of the disorder in 1741 patients. Hum Reprod 1995;10(8):2107–11. Available from: https://doi.org/10.1093/oxfordjournals.humrep.a136243.

[12] Chang WY, Knochenhauer ES, Bartolucci AA, Azziz R. Phenotypic spectrum of polycystic ovary syndrome: clinical and biochemical characterization of the three major clinical subgroups. Fertil Steril 2005;83(6):1717–23. Available from: https://doi.org/10.1016/j.fertnstert.2005.01.096.

[13] Conway GS, Honour JW, Jacobs HS. Heterogeneity of the polycystic ovary syndrome: clinical, endocrine and ultrasound features in 556 patients. Clin Endocrinol (Oxf) 1989;30(4):459–70. Available from: https://doi.org/10.1111/j.1365-2265.1989.tb00446.x.

[14] Hahn S, Tan S, Elsenbruch S, et al. Clinical and biochemical characterization of women with polycystic ovary syndrome in North Rhine-Westphalia. Horm Metab Res 2005;37(7):438–44. Available from: https://doi.org/10.1055/s-2005-870236.

[15] Orio Jr. F, Matarese G, Di Biase S, et al. Exon 6 and 2 peroxisome proliferator-activated receptor-gamma polymorphisms in polycystic ovary syndrome. J Clin Endocrinol Metab 2003;88(12):5887–92. Available from: https://doi.org/10.1210/jc.2002-021816.

[16] De Leo V, Musacchio MC, Cappelli V, Massaro MG, Morgante G, Petraglia F. Genetic, hormonal and metabolic aspects of PCOS: an update. Reprod Biol Endocrinol 2016;14(1):38. Available from: https://doi.org/10.1186/s12958-016-0173-x.

[17] Carmina E, Rosato F, Janni A, Rizzo M, Longo RA. Extensive clinical experience: relative prevalence of different androgen excess disorders in 950 women referred because of clinical hyperandrogenism. J Clin Endocrinol Metab 2006;91(1):2–6. Available from: https://doi.org/10.1210/jc.2005-1457.

[18] Abinaya S, Siva D, Sabitha R, Achiraman S. An overview of hyperandrogenism in PCOS and the prospective underlying factors. Res J Life Sci Bioinform Pharm Chem Sci 2019;1(5):179–86.

[19] Kumar A, Woods KS, Bartolucci AA, Azziz R. Prevalence of adrenal androgen excess in patients with the polycystic ovary syndrome (PCOS). Clin Endocrinol (Oxf) 2005;62(6):644–9. Available from: https://doi.org/10.1111/j.1365-2265.2005.02256.x.

[20] Diamanti-Kandarakis E, Dunaif A. Insulin resistance and the polycystic ovary syndrome revisited: an update on mechanisms and implications. Endocr Rev 2012;33(6):981–1030. Available from: https://doi.org/10.1210/er.2011-1034.

[21] Jiang N-X, Li X-L. The disorders of endometrial receptivity in PCOS and its mechanisms. Reprod Sci 2021;. Available from: https://doi.org/10.1007/s43032-021-00629-9 2021/05/27.

[22] Kelley AS, Smith YR, Padmanabhan V. A narrative review of placental contribution to adverse pregnancy outcomes in women with polycystic ovary syndrome. J Clin Endocrinol Metab 2019;104(11):5299–315. Available from: https://doi.org/10.1210/jc.2019-00383.

[23] Balen AH, Tan SL, MacDougall J, Jacobs HS. Miscarriage rates following in-vitro fertilization are increased in women with polycystic ovaries and reduced by pituitary desensitization with buserelin. Hum Reprod 1993;8 (6):959–64. Available from: https://doi.org/10.1093/oxfordjournals.humrep.a138174.

[24] Boomsma CM, Fauser BC, Macklon NS. Pregnancy complications in women with polycystic ovary syndrome. © Thieme Medical Publishers; 2008, p. 072–84.

[25] Sagle M, Bishop K, Ridley N, et al. Recurrent early miscarriage and polycystic ovaries. BMJ 1988;297 (6655):1027–8. Available from: https://doi.org/10.1136/bmj.297.6655.1027.

[26] Teede HJ, Hutchison S, Zoungas S, Meyer C. Insulin resistance, the metabolic syndrome, diabetes, and cardiovascular disease risk in women with PCOS. Endocrine. 2006;30(1):45–53.

[27] Zhao L, Zhu Z, Lou H, et al. Polycystic ovary syndrome (PCOS) and the risk of coronary heart disease (CHD): a meta-analysis. Oncotarget 2016;7(23):33715–21. Available from: https://doi.org/10.18632/oncotarget.9553.

[28] Gambineri A, Pelusi C, Vicennati V, Pagotto U, Pasquali R. Obesity and the polycystic ovary syndrome. Int J Obes 2002;26(7):883–96. Available from: https://doi.org/10.1038/sj.ijo.0801994 2002/07/01.

[29] Toprak S, Yonem A, Cakir B, et al. Insulin resistance in nonobese patients with polycystic ovary syndrome. Horm Res 2001;55(2):65–70. Available from: https://doi.org/10.1159/000049972.

[30] McGee E, Sawetawan C, Bird I, Rainey WE, Carr BR. The effects of insulin on 3β-hydroxysteroid dehydrogenase expression in human luteinized granulosa cells. J Soc Gynecol Investig: JSGI 1995;2(3):535–41.

[31] Nestler JE, Jakubowicz DJ. Decreases in ovarian cytochrome P450c17α activity and serum free testosterone after reduction of insulin secretion in polycystic ovary syndrome. N Engl J Med 1996;335(9):617–23.

[32] Plymate SR, Matej LA, Jones RE, Friedl KE. Inhibition of sex hormone-binding globulin production in the human hepatoma (Hep G2) cell line by insulin and prolactin. J Clin Endocrinol Metab 1988;67(3):460–4.

[33] Poretsky L, Chandrasekher Y, Bai C, Liu H, Rosenwaks Z, Giudice L. Insulin receptor mediates inhibitory effect of insulin, but not of insulin-like growth factor (IGF)-I, on IGF binding protein 1 (IGFBP-1) production in human granulosa cells. J Clin Endocrinol Metab 1996;81(2):493–6.

[34] Dunaif A, Segal KR, Futterweit W, Dobrjansky A. Profound peripheral insulin resistance, independent of obesity, in polycystic ovary syndrome. Diabetes 1989;38(9):1165–74. Available from: https://doi.org/10.2337/diab.38.9.1165.

[35] Dunaif A, Segal KR, Shelley DR, Green G, Dobrjansky A, Licholai T. Evidence for distinctive and intrinsic defects in insulin action in polycystic ovary syndrome. Diabetes 1992;41(10):1257–66. Available from: https://doi.org/10.2337/diab.41.10.1257.

[36] Jonard S, Robert Y, Cortet-Rudelli C, Pigny P, Decanter C, Dewailly D. Ultrasound examination of polycystic ovaries: is it worth counting the follicles? Hum Reprod 2003;18(3):598–603. Available from: https://doi.org/10.1093/humrep/deg115.
[37] Chang EM, Han JE, Seok HH, Lee DR, Yoon TK, Lee WS. Insulin resistance does not affect early embryo development but lowers implantation rate in in vitro maturation–in vitro fertilization–embryo transfer cycle. Clin Endocrinol 2013;79(1):93–9.
[38] Samy N, Hashim M, Sayed M, Said M. Clinical significance of inflammatory markers in polycystic ovary syndrome: their relationship to insulin resistance and body mass index. Dis Markers 2009;26(4):163–70. Available from: https://doi.org/10.3233/DMA-2009-0627.
[39] Escobar-Morreale HF, Calvo RM, Sancho J, San Millan JL. TNF-alpha and hyperandrogenism: a clinical, biochemical, and molecular genetic study. J Clin Endocrinol Metab 2001;86(8):3761–7. Available from: https://doi.org/10.1210/jcem.86.8.7770.
[40] Escobar-Morreale HF, Calvo RM, Villuendas G, Sancho J, San Millán JL. Association of polymorphisms in the interleukin 6 receptor complex with obesity and hyperandrogenism. Obes Res 2003;11(8):987–96.
[41] Peral B, San Millan JL, Castello R, Moghetti P, Escobar-Morreale HF. Comment: the methionine 196 arginine polymorphism in exon 6 of the TNF receptor 2 gene (TNFRSF1B) is associated with the polycystic ovary syndrome and hyperandrogenism. J Clin Endocrinol Metab 2002;87(8):3977–83. Available from: https://doi.org/10.1210/jcem.87.8.8715.
[42] Villuendas G, San Millan JL, Sancho J, Escobar-Morreale HF. The -597 G– > A and -174 G– > C polymorphisms in the promoter of the IL-6 gene are associated with hyperandrogenism. J Clin Endocrinol Metab 2002;87(3):1134–41. Available from: https://doi.org/10.1210/jcem.87.3.8309.
[43] Schmidt J, Weijdegård B, Mikkelsen AL, Lindenberg S, Nilsson L, Brännström M. Differential expression of inflammation-related genes in the ovarian stroma and granulosa cells of PCOS women. Mol Hum Reprod 2013;20(1):49–58. Available from: https://doi.org/10.1093/molehr/gat051.
[44] Fenkci V, Fenkci S, Yilmazer M, Serteser M. Decreased total antioxidant status and increased oxidative stress in women with polycystic ovary syndrome may contribute to the risk of cardiovascular disease. Fertil Steril 2003;80(1):123–7. Available from: https://doi.org/10.1016/s0015-0282(03)00571-5.
[45] Shkolnik K, Tadmor A, Ben-Dor S, Nevo N, Galiani D, Dekel N. Reactive oxygen species are indispensable in ovulation. Proc Natl Acad Sci USA 2011;108(4):1462–7. Available from: https://doi.org/10.1073/pnas.1017213108.
[46] Wu LL, Norman RJ, Robker RL. The impact of obesity on oocytes: evidence for lipotoxicity mechanisms. Reprod Fertil Dev 2011;24(1):29–34. Available from: https://doi.org/10.1071/RD11904.
[47] Palomba S, Russo T, Falbo A, et al. Macroscopic and microscopic findings of the placenta in women with polycystic ovary syndrome. Hum Reprod 2013;28(10):2838–47. Available from: https://doi.org/10.1093/humrep/det250.
[48] Çakıroğlu Y, Vural F, Vural B. The inflammatory markers in polycystic ovary syndrome: association with obesity and IVF outcomes. J Endocrinol Investigation 2016;39(8):899–907. Available from: https://doi.org/10.1007/s40618-016-0446-4 2016/08/01.
[49] Baldwin Jr. AS. Series introduction: the transcription factor NF-kappaB and human disease. J Clin Invest 2001;107(1):3–6. Available from: https://doi.org/10.1172/JCI11891.
[50] Wisse BE. The inflammatory syndrome: the role of adipose tissue cytokines in metabolic disorders linked to obesity. J Am Soc Nephrol 2004;15(11):2792–800. Available from: https://doi.org/10.1097/01.ASN.0000141966.69934.21.
[51] Boomsma CM, Eijkemans MJC, Hughes EG, Visser GHA, Fauser BCJM, Macklon NS. A meta-analysis of pregnancy outcomes in women with polycystic ovary syndrome. Hum Reprod Update 2006;12(6):673–83. Available from: https://doi.org/10.1093/humupd/dml036.
[52] Kjerulff LE, Sanchez-Ramos L, Duffy D. Pregnancy outcomes in women with polycystic ovary syndrome: a metaanalysis. Am J Obstet Gynecol 2011;204(6):558.e1–6. Available from: https://doi.org/10.1016/j.ajog.2011.03.021 2011/06/01/.
[53] Yu HF, Chen HS, Rao DP, Gong J. Association between polycystic ovary syndrome and the risk of pregnancy complications: a PRISMA-compliant systematic review and meta-analysis. Med (Baltim) 2016;95(51):e4863. Available from: https://doi.org/10.1097/MD.0000000000004863.
[54] Bahri Khomami M, Joham AE, Boyle JA, et al. Increased maternal pregnancy complications in polycystic ovary syndrome appear to be independent of obesity—a systematic review, meta-analysis, and meta-regression. Obes Rev 2019;20(5):659–74. Available from: https://doi.org/10.1111/obr.12829.

[55] Sha T, Wang X, Cheng W, Yan Y. A meta-analysis of pregnancy-related outcomes and complications in women with polycystic ovary syndrome undergoing IVF. Reprod Biomed Online 2019;39(2):281–93. Available from: https://doi.org/10.1016/j.rbmo.2019.03.203.

[56] Ashraf S, Nabi M, Rasool SUA, Rashid F, Amin S. Hyperandrogenism in polycystic ovarian syndrome and role of CYP gene variants: a review. Egypt J Med Hum Genet 2019;20(1):25. Available from: https://doi.org/10.1186/s43042-019-0031-4 2019/11/20.

[57] Chen MJ, Yang WS, Yang JH, Hsiao CK, Yang YS, Ho HN. Low sex hormone-binding globulin is associated with low high-density lipoprotein cholesterol and metabolic syndrome in women with PCOS. Hum Reprod 2006;21(9):2266–71. Available from: https://doi.org/10.1093/humrep/del175.

[58] Parker Jr. CR, Slayden SM, Azziz R, et al. Effects of aging on adrenal function in the human: responsiveness and sensitivity of adrenal androgens and cortisol to adrenocorticotropin in premenopausal and postmenopausal women. J Clin Endocrinol Metab 2000;85(1):48–54. Available from: https://doi.org/10.1210/jcem.85.1.6265.

[59] Deswal R, Yadav A, Dang AS. Sex hormone binding globulin - an important biomarker for predicting PCOS risk: a systematic review and meta-analysis. Syst Biol Reprod Med 2018;64(1):12–24. Available from: https://doi.org/10.1080/19396368.2017.1410591 2018/01/02.

[60] Malini NA, Roy George K. Evaluation of different ranges of LH:FSH ratios in polycystic ovarian syndrome (PCOS)—clinical based case control study. Gen Comp Endocrinol 2018;260:51–7. Available from: https://doi.org/10.1016/j.ygcen.2017.12.007 2018/05/01/.

[61] Conn PM, Huckle WR, Andrews WV, McArdle CA. The molecular mechanism of action of gonadotropin releasing hormone (GnRH) in the pituitary. In: Clark JH, editor. Proceedings of the 1986 Laurentian Hormone Conference. Academic Press; 1987, p. 29–68.

[62] Magoffin DA. Ovarian theca cell. Int J Biochem Cell Biol 2005;37(7):1344–9.

[63] Rebar R, Judd HL, Yen SS, Rakoff J, Vandenberg G, Naftolin F. Characterization of the inappropriate gonadotropin secretion in polycystic ovary syndrome. J Clin Invest 1976;57(5):1320–9. Available from: https://doi.org/10.1172/JCI108400.

[64] Yen SS, Vela P, Rankin J. Inappropriate secretion of follicle-stimulating hormone and luteinizing hormone in polycystic ovarian disease. J Clin Endocrinol Metab 1970;30(4):435–42. Available from: https://doi.org/10.1210/jcem-30-4-435.

[65] Waldstreicher J, Santoro NF, Hall JE, Filicori M, Crowley Jr. WF. Hyperfunction of the hypothalamic-pituitary axis in women with polycystic ovarian disease: indirect evidence for partial gonadotroph desensitization. J Clin Endocrinol Metab 1988;66(1):165–72. Available from: https://doi.org/10.1210/jcem-66-1-165.

[66] Wood JR, Nelson VL, Ho C, et al. The molecular phenotype of polycystic ovary syndrome (PCOS) theca cells and new candidate PCOS genes defined by microarray analysis. J Biol Chem 2003;278 (29):26380–90.

[67] Daneshmand S, Weitsman SR, Navab A, Jakimiuk AJ, Magoffin DA. Overexpression of theca-cell messenger RNA in polycystic ovary syndrome does not correlate with polymorphisms in the cholesterol side-chain cleavage and 17alpha-hydroxylase/C(17-20) lyase promoters. Fertil Steril 2002;77(2):274–80. Available from: https://doi.org/10.1016/s0015-0282(01)02999-5.

[68] Diamanti-Kandarakis E, Bartzis MI, Bergiele AT, Tsianateli TC, Kouli CR. Microsatellite polymorphism (tttta)(n) at -528 base pairs of gene CYP11alpha influences hyperandrogenemia in patients with polycystic ovary syndrome. Fertil Steril 2000;73(4):735–41. Available from: https://doi.org/10.1016/s0015-0282 (99)00628-7.

[69] Gharani N, Waterworth DM, Batty S, et al. Association of the steroid synthesis gene Cyp11a with polycystic ovary syndrome and hyperandrogenism. Hum Mol Genet 1997;6(3):397–402. Available from: https://doi.org/10.1093/hmg/6.3.397.

[70] Pusalkar M, Meherji P, Gokral J, Chinnaraj S, Maitra A. CYP11A1 and CYP17 promoter polymorphisms associate with hyperandrogenemia in polycystic ovary syndrome. Fertil Steril 2009;92(2):653–9. Available from: https://doi.org/10.1016/j.fertnstert.2008.07.016.

[71] Reddy KR, Deepika ML, Supriya K, et al. CYP11A1 microsatellite (tttta)n polymorphism in PCOS women from South India. J Assist Reprod Genet 2014;31(7):857–63. Available from: https://doi.org/10.1007/s10815-014-0236-x.

[72] Wang Y, Wu XK, Cao YX, et al. Microsatellite polymorphism of (tttta) n in the promoter of CYP11a gene in Chinese women with polycystic ovary syndrome. Zhonghua Yi Xue Za Zhi 2005;85(48):3396–400.

[73] Carey AH, Waterworth D, Patel K, et al. Polycystic ovaries and premature male pattern baldness are associated with one allele of the steroid metabolism gene CYP17. Hum Mol Genet 1994;3(10):1873–6. Available from: https://doi.org/10.1093/hmg/3.10.1873.
[74] Crocitto LE, Feigelson HS, Yu MC, Kolonel LN, Henderson BE, Coetzee GA. A polymorphism in intron 6 of the CYP17 gene. Clin Genet 1997;52(1):68–9. Available from: https://doi.org/10.1111/j.1399-0004.1997.tb02519.x.
[75] Diamanti-Kandarakis E, Bartzis MI, Zapanti ED, et al. Polymorphism T−>C (-34 bp) of gene CYP17 promoter in Greek patients with polycystic ovary syndrome. Fertil Steril 1999;71(3):431–5. Available from: https://doi.org/10.1016/s0015-0282(98)00512-3.
[76] Ehrmann DA, Rosenfield RL, Barnes RB, Brigell DF, Sheikh Z. Detection of functional ovarian hyperandrogenism in women with androgen excess. N Engl J Med 1992;327(3):157–62. Available from: https://doi.org/10.1056/NEJM199207163270304.
[77] Escobar-Morreale H, Serrano-Gotarredona J, Garcia-Robles R, Sancho J, Varela C. Mild adrenal and ovarian steroidogenic abnormalities in hirsute women without hyperandrogenemia: does idiopathic hirsutism exist? Metabolism. 1997;46(8):902–7.
[78] Gilling-Smith C, Willis DS, Beard RW, Franks S. Hypersecretion of androstenedione by isolated thecal cells from polycystic ovaries. J Clin Endocrinol Metab 1994;79(4):1158–65. Available from: https://doi.org/10.1210/jcem.79.4.7962289.
[79] Miyoshi Y, Iwao K, Ikeda N, Egawa C, Noguchi S. Genetic polymorphism in CYP17 and breast cancer risk in Japanese women. Eur J Cancer 2000;36(18):2375–9. Available from: https://doi.org/10.1016/s0959-8049(00)00334-8.
[80] Wickenheisser JK, Nelson-DeGrave VL, McAllister JM. Dysregulation of cytochrome P450 17α-hydroxylase messenger ribonucleic acid stability in theca cells isolated from women with polycystic ovary syndrome. J Clin Endocrinol Metab 2005;90(3):1720–7.
[81] Wickenheisser JK, Nelson-DeGrave VL, Quinn PG, McAllister JM. Increased cytochrome P450 17alpha-hydroxylase promoter function in theca cells isolated from patients with polycystic ovary syndrome involves nuclear factor-1. Mol Endocrinol 2004;18(3):588–605. Available from: https://doi.org/10.1210/me.2003-0090.
[82] Erickson GF, Hsueh AJ, Quigley ME, Rebar RW, Yen SS. Functional studies of aromatase activity in human granulosa cells from normal and polycystic ovaries. J Clin Endocrinol Metab 1979;49(4):514–19. Available from: https://doi.org/10.1210/jcem-49-4-514.
[83] Ito Y, Fisher CR, Conte FA, Grumbach MM, Simpson ER. Molecular basis of aromatase deficiency in an adult female with sexual infantilism and polycystic ovaries. Proc Natl Acad Sci USA 1993;90(24):11673–7. Available from: https://doi.org/10.1073/pnas.90.24.11673.
[84] Jakimiuk AJ, Weitsman SR, Brzechffa PR, Magoffin DA. Aromatase mRNA expression in individual follicles from polycystic ovaries. Mol Hum Reprod 1998;4(1):1–8. Available from: https://doi.org/10.1093/molehr/4.1.1.
[85] Mostafa RA, Al-Sherbeeny MM, Abdelazim IA, et al. Relation between aromatase gene CYP19 variation and hyperandrogenism in polycystic ovary syndrome Egyptian women. J Infertility Reprod Biol 2016;4(1):1–5.
[86] Zhang X-L, Zhang C-W, Xu P, et al. SNP rs2470152 in CYP19 is correlated to aromatase activity in Chinese polycystic ovary syndrome patients. Mol Med Rep 2012;5(1):245–9.
[87] Nahum R, Thong KJ, Hillier SG. Metabolic regulation of androgen production by human thecal cells in vitro. Hum Reprod 1995;10(1):75–81. Available from: https://doi.org/10.1093/humrep/10.1.75.
[88] Romero G, Gamez G, Huang LC, Lilley K, Luttrell L. Anti-inositolglycan antibodies selectively block some of the actions of insulin in intact BC3H1 cells. Proc Natl Acad Sci USA 1990;87(4):1476–80. Available from: https://doi.org/10.1073/pnas.87.4.1476.
[89] Romero G, Luttrell L, Rogol A, Zeller K, Hewlett E, Larner J. Phosphatidylinositol-glycan anchors of membrane proteins: potential precursors of insulin mediators. Science 1988;240(4851):509–11. Available from: https://doi.org/10.1126/science.3282305.
[90] Barbieri RL, Makris A, Randall RW, Daniels G, Kistner RW, Ryan KJ. Insulin stimulates androgen accumulation in incubations of ovarian stroma obtained from women with hyperandrogenism. J Clin Endocrinol Metab 1986;62(5):904–10. Available from: https://doi.org/10.1210/jcem-62-5-904.
[91] Jakubowicz DJ, Nestler JE. 17 Alpha-fydroxyprogesterone responses to leuprolide and serum androgens in obese women with and without polycystic ovary syndrome offer dietary weight loss. J Clin Endocrinol Metab 1997;82(2):556–60. Available from: https://doi.org/10.1210/jcem.82.2.3753.

[92] Nestler JE, Jakubowicz DJ, de Vargas AF, Brik C, Quintero N, Medina F. Insulin stimulates testosterone biosynthesis by human thecal cells from women with polycystic ovary syndrome by activating its own receptor and using inositolglycan mediators as the signal transduction system. J Clin Endocrinol Metab 1998;83 (6):2001–5. Available from: https://doi.org/10.1210/jcem.83.6.4886.
[93] Corbould A, Kim YB, Youngren JF, et al. Insulin resistance in the skeletal muscle of women with PCOS involves intrinsic and acquired defects in insulin signaling. Am J Physiol Endocrinol Metab 2005;288(5): E1047–54. Available from: https://doi.org/10.1152/ajpendo.00361.2004.
[94] Nestler JE. Insulin regulation of human ovarian androgens. Hum Reprod 1997;12(Suppl 1):53–62. Available from: https://doi.org/10.1093/humrep/12.suppl_1.53.
[95] Wang M, Liu M, Sun J, et al. MicroRNA-27a-3p affects estradiol and androgen imbalance by targeting Creb1 in the granulosa cells in mouse polycytic ovary syndrome model. Reprod Biol 2017;17(4):295–304.
[96] Erickson GF, Magoffin DA, Garzo VG, Cheung AP, Chang RJ. Granulosa cells of polycystic ovaries: are they normal or abnormal? Hum Reprod 1992;7(3):293–9.
[97] Hsueh AJ. 6 Paracrine mechanisms involved in granulosa cell differentiation. Clin Endocrinol Metab 1986;15(1):117–34.
[98] Mason HD, Willis DS, Beard RW, Winston R, Margara R, Franks S. Estradiol production by granulosa cells of normal and polycystic ovaries: relationship to menstrual cycle history and concentrations of gonadotropins and sex steroids in follicular fluid. J Clin Endocrinol Metab 1994;79(5):1355–60.
[99] Rodrigues J, Navarro P, Zelinski M, Stouffer R, Xu J. Direct actions of androgens on the survival, growth and secretion of steroids and anti-Müllerian hormone by individual macaque follicles during three-dimensional culture. Hum Reprod 2015;30(3):664–74.
[100] Harlow C, Shaw H, Hillier S, Hodges J. Factors influencing follicle-stimulating hormone-responsive steroidogenesis in marmoset granulosa cells: effects of androgens and the stage of follicular maturity. Endocrinology. 1988;122(6):2780–7.
[101] Pierre A, Taieb J, Giton F, et al. Dysregulation of the anti-Müllerian hormone system by steroids in women with polycystic ovary syndrome. J Clin Endocrinol Metab 2017;102(11):3970–8.
[102] Xu F, Liu R, Cao X. Hyperandrogenism stimulates inflammation and promote apoptosis of cumulus cells. Cell Mol Biol 2017;63(10):64–8.
[103] Ware VC. The role of androgens in follicular development in the ovary. I. A quantitative analysis of oocyte ovulation. J Exp Zool 1982;222(2):155–67.
[104] Garg D, Tal R. The role of AMH in the pathophysiology of polycystic ovarian syndrome. Reprod BioMed Online 2016;33(1):15–28. Available from: https://doi.org/10.1016/j.rbmo.2016.04.007.
[105] Eldar-Geva T, Margalioth EJ, Gal M, et al. Serum anti-Mullerian hormone levels during controlled ovarian hyperstimulation in women with polycystic ovaries with and without hyperandrogenism. Hum Reprod 2005;20(7):1814–19. Available from: https://doi.org/10.1093/humrep/deh873.
[106] Homburg R, Ray A, Bhide P, et al. The relationship of serum anti-Mullerian hormone with polycystic ovarian morphology and polycystic ovary syndrome: a prospective cohort study. Hum Reprod 2013;28 (4):1077–83. Available from: https://doi.org/10.1093/humrep/det015.
[107] Pigny P, Merlen E, Robert Y, et al. Elevated serum level of anti-mullerian hormone in patients with polycystic ovary syndrome: relationship to the ovarian follicle excess and to the follicular arrest. J Clin Endocrinol Metab 2003;88(12):5957–62. Available from: https://doi.org/10.1210/jc.2003-030727.
[108] Tal R, Seifer DB, Khanimov M, Malter HE, Grazi RV, Leader B. Characterization of women with elevated antimüllerian hormone levels (AMH): correlation of AMH with polycystic ovarian syndrome phenotypes and assisted reproductive technology outcomes. Am J Obstet Gynecol 2014;211(1):59.e1–8. Available from: https://doi.org/10.1016/j.ajog.2014.02.026 2014/07/01/.
[109] Pellatt L, Hanna L, Brincat M, et al. Granulosa cell production of anti-Mullerian hormone is increased in polycystic ovaries. J Clin Endocrinol Metab 2007;92(1):240–5. Available from: https://doi.org/10.1210/jc.2006-1582.
[110] Vendola KA, Zhou J, Adesanya OO, Weil SJ, Bondy CA. Androgens stimulate early stages of follicular growth in the primate ovary. J Clin Invest 1998;101(12):2622–9. Available from: https://doi.org/10.1172/JCI2081.
[111] Weil S, Vendola K, Zhou J, Bondy CA. Androgen and follicle-stimulating hormone interactions in primate ovarian follicle development. J Clin Endocrinol Metab 1999;84(8):2951–6. Available from: https://doi.org/10.1210/jcem.84.8.5929.

[112] Carlsen SM, Vanky E, Fleming R. Anti-Mullerian hormone concentrations in androgen-suppressed women with polycystic ovary syndrome. Hum Reprod 2009;24(7):1732–8. Available from: https://doi.org/10.1093/humrep/dep074.
[113] Fallat ME, Siow Y, Marra M, Cook C, Carrillo A. Mullerian-inhibiting substance in follicular fluid and serum: a comparison of patients with tubal factor infertility, polycystic ovary syndrome, and endometriosis. Fertil Steril 1997;67(5):962–5. Available from: https://doi.org/10.1016/s0015-0282(97)81417-3.
[114] Laven JS, Mulders AG, Visser JA, Themmen AP, De Jong FH, Fauser BC. Anti-Mullerian hormone serum concentrations in normoovulatory and anovulatory women of reproductive age. J Clin Endocrinol Metab 2004;89(1):318–23. Available from: https://doi.org/10.1210/jc.2003-030932.
[115] Mulders AG, Laven JS, Eijkemans MJ, de Jong FH, Themmen AP, Fauser BC. Changes in anti-Mullerian hormone serum concentrations over time suggest delayed ovarian ageing in normogonadotrophic anovulatory infertility. Hum Reprod 2004;19(9):2036–42. Available from: https://doi.org/10.1093/humrep/deh373.
[116] Hayes E, Kushnir V, Ma X, et al. Intra-cellular mechanism of anti-Müllerian hormone (AMH) in regulation of follicular development. Mol Cell Endocrinol 2016;433:56–65. Available from: https://doi.org/10.1016/j.mce.2016.05.019 2016/09/15/.
[117] Visser JA, Durlinger AL, Peters IJ, et al. Increased oocyte degeneration and follicular atresia during the estrous cycle in anti-Mullerian hormone null mice. Endocrinology 2007;148(5):2301–8.
[118] Franks S, Stark J, Hardy K. Follicle dynamics and anovulation in polycystic ovary syndrome. Hum Reprod Update 2008;14(4):367–78. Available from: https://doi.org/10.1093/humupd/dmn015.
[119] Lee MH, Yoon JA, Kim HR, et al. Hyperandrogenic milieu dysregulates the expression of insulin signaling factors and glucose transporters in the endometrium of patients with polycystic ovary syndrome. Reprod Sci 2020;27(8):1637–47. Available from: https://doi.org/10.1007/s43032-020-00194-7.
[120] Younas K, Quintela M, Thomas S, et al. Delayed endometrial decidualisation in polycystic ovary syndrome; the role of AR-MAGEA11. J Mol Med 2019;97(9):1315–27.
[121] Barcena C, Oliva E. WT1 expression in the female genital tract. Adv Anat Pathol 2011;18(6):454–65. Available from: https://doi.org/10.1097/PAP.0b013e318234aaed.
[122] Gonzalez D, Thackeray H, Lewis PD, et al. Loss of WT1 Expression in the endometrium of infertile PCOS patients: a hyperandrogenic effect? J Clin Endocrinol Metab 2012;97(3):957–66. Available from: https://doi.org/10.1210/jc.2011-2366.
[123] Ejskjaer K, Sorensen BS, Poulsen SS, Mogensen O, Forman A, Nexo E. Expression of the epidermal growth factor system in human endometrium during the menstrual cycle. Mol Hum Reprod 2005;11(8):543–51. Available from: https://doi.org/10.1093/molehr/gah207.
[124] Jafferji I, Bain M, King C, Sinclair JH. Inhibition of epidermal growth factor receptor (EGFR) expression by human cytomegalovirus correlates with an increase in the expression and binding of Wilms' tumour 1 protein to the EGFR promoter. J Gen Virol 2009;90(Pt 7):1569–74. Available from: https://doi.org/10.1099/vir.0.009670-0.
[125] Akbas GE, Taylor HS. HOXC and HOXD gene expression in human endometrium: lack of redundancy with HOXA paralogs. Biol Reprod 2004;70(1):39–45.
[126] McGinnis W, Krumlauf R. Homeobox genes and axial patterning. Cell 1992;68(2):283–302. Available from: https://doi.org/10.1016/0092-8674(92)90471-n.
[127] Taylor HS. Transcriptional regulation of implantation by HOX genes. Rev Endocr Metab Disord 2002;3(2):127–32. Available from: https://doi.org/10.1023/a:1015454828489.
[128] Cermik D, Selam B, Taylor HS. Regulation of HOXA-10 expression by testosterone in vitro and in the endometrium of patients with polycystic ovary syndrome. J Clin Endocrinol Metab 2003;88(1):238–43.
[129] Kara M, Ozcan SS, Aran T, Kara O, Yilmaz N. Evaluation of endometrial receptivity by measuring HOXA-10, HOXA-11, and leukemia inhibitory factor expression in patients with polycystic ovary syndrome. Gynecol Minim Invasive Ther 2019;8(3):118.
[130] He H, Li T, Yin D, et al. HOXA10 expression is decreased by testosterone in luteinized granulosa cells in vitro. Mol Med Rep 2012;6(1):51–6.
[131] Rahman TU, Ullah K, Guo M-X, et al. Androgen-induced alterations in endometrial proteins crucial in recurrent miscarriages. Oncotarget 2018;9(37):24627.

[132] Mokhtar MH, Giribabu N, Salleh N. Testosterone decreases the number of implanting embryos, expression of pinopode and L-selectin ligand (MECA-79) in the endometrium of early pregnant rats. Int J Env Res Public Health 2020;17(7). Available from: https://doi.org/10.3390/ijerph17072293.
[133] Mokhtar M, Giribabu N, Salleh N. Testosterone down-regulates expression of αVβ3-integrin, E-cadherin and mucin-1 during uterine receptivity period in rats. Sains Malaysiana 2018;47(10):2509–17.
[134] Mokhtar MH, Giribabu N, Salleh N. Testosterone reduces tight junction complexity and down-regulates expression of claudin-4 and occludin in the endometrium in ovariectomized, sex-steroid replacement rats. vivo 2020;34(1):225–31.
[135] Quezada S, Avellaira C, Johnson MC, Gabler F, Fuentes A, Vega M. Evaluation of steroid receptors, coregulators, and molecules associated with uterine receptivity in secretory endometria from untreated women with polycystic ovary syndrome. Fertil Steril 2006;85(4):1017–26.
[136] Su Y, Shi H. High androgen level causes recurrent miscarriage and impairs endometrial receptivity. Tropical J Pharm Res 2019;18(7):1547–52.
[137] Glintborg D, Jensen RC, Bentsen K, et al. Testosterone levels in third trimester in polycystic ovary syndrome: odense child cohort. J Clin Endocrinol Metab 2018;103(10):3819–27. Available from: https://doi.org/10.1210/jc.2018-00889.
[138] Sir-Petermann T, Maliqueo M, Angel B, Lara HE, Perez-Bravo F, Recabarren SE. Maternal serum androgens in pregnant women with polycystic ovarian syndrome: possible implications in prenatal androgenization. Hum Reprod 2002;17(10):2573–9. Available from: https://doi.org/10.1093/humrep/17.10.2573.
[139] Sathishkumar K, Elkins R, Chinnathambi V, Gao H, Hankins GD, Yallampalli C. Prenatal testosterone-induced fetal growth restriction is associated with down-regulation of rat placental amino acid transport. Reprod Biol Endocrinol 2011;9:110. Available from: https://doi.org/10.1186/1477-7827-9-110.
[140] Barry JA, Hardiman PJ, Siddiqui MR, Thomas M. Meta-analysis of sex difference in testosterone levels in umbilical cord blood. J Obstet Gynaecol 2011;31(8):697–702. Available from: https://doi.org/10.3109/01443615.2011.614971.
[141] Beckett EM, Astapova O, Steckler TL, Veiga-Lopez A, Padmanabhan V. Developmental programing: impact of testosterone on placental differentiation. Reproduction 2014;148(2):199–209. Available from: https://doi.org/10.1530/REP-14-0055.
[142] Maliqueo M, Lara HE, Sanchez F, Echiburu B, Crisosto N, Sir-Petermann T. Placental steroidogenesis in pregnant women with polycystic ovary syndrome. Eur J Obstet Gynecol Reprod Biol 2013;166(2):151–5. Available from: https://doi.org/10.1016/j.ejogrb.2012.10.015.
[143] Palomba S, Falbo A, Russo T, Tolino A, Orio F, Zullo F. Pregnancy in women with polycystic ovary syndrome: the effect of different phenotypes and features on obstetric and neonatal outcomes. Fertil Steril 2010;94(5):1805–11. Available from: https://doi.org/10.1016/j.fertnstert.2009.10.043.
[144] Domonkos E, Borbelyova V, Kolatorova L, et al. Sex differences in the effect of prenatal testosterone exposure on steroid hormone production in adult rats. Physiol Res 2017;66(Suppl 3):S367–74. Available from: https://doi.org/10.33549/physiolres.933722.
[145] Padmanabhan V, Veiga-Lopez A. Sheep models of polycystic ovary syndrome phenotype. Mol Cell Endocrinol 2013;373(1-2):8–20. Available from: https://doi.org/10.1016/j.mce.2012.10.005.
[146] Shah AB, Nivar I, Speelman DL. Elevated androstenedione in young adult but not early adolescent prenatally androgenized female rats. PLoS One 2018;13(5):e0196862. Available from: https://doi.org/10.1371/journal.pone.0196862.
[147] Chinnathambi V, Blesson CS, Vincent KL, et al. Elevated testosterone levels during rat pregnancy cause hypersensitivity to angiotensin II and attenuation of endothelium-dependent vasodilation in uterine arteries. Hypertension 2014;64(2):405–14. Available from: https://doi.org/10.1161/HYPERTENSIONAHA.114.03283.
[148] Cleys ER, Halleran JL, Enriquez VA, et al. Androgen receptor and histone lysine demethylases in ovine placenta. PLoS One 2015;10(2):e0117472. Available from: https://doi.org/10.1371/journal.pone.0117472.
[149] Fornes R, Maliqueo M, Hu M, et al. The effect of androgen excess on maternal metabolism, placental function and fetal growth in obese dams. Sci Rep 2017;7(1):8066. Available from: https://doi.org/10.1038/s41598-017-08559-w.
[150] Fornes R, Manti M, Qi X, et al. Mice exposed to maternal androgen excess and diet-induced obesity have altered phosphorylation of catechol-O-methyltransferase in the placenta and fetal liver. Int J Obes (Lond) 2019;43(11):2176–88. Available from: https://doi.org/10.1038/s41366-018-0314-8.

[151] Gopalakrishnan K, Mishra JS, Chinnathambi V, et al. Elevated testosterone reduces uterine blood flow, spiral artery elongation, and placental oxygenation in pregnant rats. Hypertension 2016;67(3):630–9. Available from: https://doi.org/10.1161/HYPERTENSIONAHA.115.06946.
[152] Hu M, Richard JE, Maliqueo M, et al. Maternal testosterone exposure increases anxiety-like behavior and impacts the limbic system in the offspring. Proc Natl Acad Sci USA 2015;112(46):14348–53. Available from: https://doi.org/10.1073/pnas.1507514112.
[153] Sun M, Maliqueo M, Benrick A, et al. Maternal androgen excess reduces placental and fetal weights, increases placental steroidogenesis, and leads to long-term health effects in their female offspring. Am J Physiol Endocrinol Metab 2012;303(11):E1373–85. Available from: https://doi.org/10.1152/ajpendo.00421.2012.
[154] Kumar S, Gordon GH, Abbott DH, Mishra JS. Androgens in maternal vascular and placental function: implications for preeclampsia pathogenesis. Reproduction 2018;156(5):R155–67. Available from: https://doi.org/10.1530/REP-18-0278.
[155] Metzler VM, de Brot S, Robinson RS, et al. Androgen dependent mechanisms of pro-angiogenic networks in placental and tumor development. Placenta 2017;56:79–85. Available from: https://doi.org/10.1016/j.placenta.2017.02.018.
[156] Williams CJ, Chu A, Jefferson WN, et al. Epithelial membrane protein 2 (EMP2) deficiency alters placental angiogenesis, mimicking features of human placental insufficiency. J Pathol 2017;242(2):246–59. Available from: https://doi.org/10.1002/path.4893.
[157] Fornes R, Hu M, Maliqueo M, et al. Maternal testosterone and placental function: effect of electroacupuncture on placental expression of angiogenic markers and fetal growth. Mol Cell Endocrinol 2016;433:1–11. Available from: https://doi.org/10.1016/j.mce.2016.05.014.
[158] Maliqueo M, Echiburu B, Crisosto N. Sex steroids modulate uterine-placental vasculature: implications for obstetrics and neonatal outcomes. Front Physiol 2016;7:152. Available from: https://doi.org/10.3389/fphys.2016.00152.
[159] Brown SH, Eather SR, Freeman DJ, Meyer BJ, Mitchell TW. A lipidomic analysis of placenta in preeclampsia: evidence for lipid storage. PLoS One 2016;11(9):e0163972. Available from: https://doi.org/10.1371/journal.pone.0163972.
[160] Mastrogiannis DS, Spiliopoulos M, Mulla W, Homko CJ. Insulin resistance: the possible link between gestational diabetes mellitus and hypertensive disorders of pregnancy. Curr Diab Rep 2009;9(4):296–302. Available from: https://doi.org/10.1007/s11892-009-0046-1.
[161] Scioscia M, Gumaa K, Rademacher TW. The link between insulin resistance and preeclampsia: new perspectives. J Reprod Immunol 2009;82(2):100–5. Available from: https://doi.org/10.1016/j.jri.2009.04.009.
[162] Fu Z, Gilbert ER, Liu D. Regulation of insulin synthesis and secretion and pancreatic Beta-cell dysfunction in diabetes. Curr Diab Rev 2013;9(1):25–53.
[163] Ducluzeau PH, Fletcher LM, Vidal H, Laville M, Tavare JM. Molecular mechanisms of insulin-stimulated glucose uptake in adipocytes. Diab Metab 2002;28(2):85–92.
[164] Crofts CAP, Zinn C, Wheldon M, Schofield G. Hyperinsulinemia: a unifying theory of chronic disease. Diabesity 2015;1(4):34–43.
[165] Goldfine ID. The insulin receptor: molecular biology and transmembrane signaling. Endocr Rev 1987;8(3):235–55.
[166] Ullrich A, Bell J, Chen EY, et al. Human insulin receptor and its relationship to the tyrosine kinase family of oncogenes. Nature. 1985;313(6005):756–61.
[167] Bremer AA, Miller WL. The serine phosphorylation hypothesis of polycystic ovary syndrome: a unifying mechanism for hyperandrogenemia and insulin resistance. Fertil Steril 2008;89(5):1039–48. Available from: https://doi.org/10.1016/j.fertnstert.2008.02.091 2008/05/01/.
[168] Bollage G, Roth R, Beaudoin J, Mochly-Rosen D, Doshland Jr. DE. Protein kinase C directly phosphorylates the insulin receptor in vitro and reduces its protein-tyrosine kinase activity. Proc Natl Acad Sci USA 1986;83:5822–4.
[169] Stadtmauer L, Rosen O. Increasing the cAMP content of IM-9 cells alters the phosphorylation state and protein kinase activity of the insulin receptor. J Biol Chem 1986;261(7):3402–7.
[170] Takayama S, White M, Kahn C. Phorbol ester-induced serine phosphorylation of the insulin receptor decreases its tyrosine kinase activity. J Biol Chem 1988;263(7):3440–7.

[171] Dunaif A, Xia J, Book C-B, Schenker E, Tang Z. Excessive insulin receptor serine phosphorylation in cultured fibroblasts and in skeletal muscle. A potential mechanism for insulin resistance in the polycystic ovary syndrome. J Clin Investig 1995;96(2):801–10.

[172] Gual P, Le Marchand-Brustel Y, Tanti J-F. Positive and negative regulation of insulin signaling through IRS-1 phosphorylation. Biochimie 2005;87(1):99–109. Available from: https://doi.org/10.1016/j.biochi.2004.10.019 2005/01/01/.

[173] Ertunc D, Tok EC, Aktas A, Erdal EM, Dilek S. The importance of IRS-1 Gly972Arg polymorphism in evaluating the response to metformin treatment in polycystic ovary syndrome. Hum Reprod 2005;20(5):1207–12. Available from: https://doi.org/10.1093/humrep/deh747.

[174] Sir-Petermann T, Perez-Bravo F, Angel B, Maliqueo M, Calvillan M, Palomino A. G972R polymorphism of IRS-1 in women with polycystic ovary syndrome. Diabetologia. 2001;44(9):1200–1.

[175] Dilek S, Ertunc D, Tok EC, Erdal EM, Aktas A. Association of Gly972Arg variant of insulin receptor substrate-1 with metabolic features in women with polycystic ovary syndrome. Fertil Steril 2005;84 (2):407–12.

[176] Dravecká I, Lazúrová I, Habalová V. The prevalence of Gly972Arg and C825T polymorphisms in Slovak women with polycystic ovary syndrome and their relation to the metabolic syndrome. Gynecol Endocrinol 2010;26(5):356–60.

[177] El Mkadem SA, Lautier C, Macari F, et al. Role of allelic variants Gly972Arg of IRS-1 and Gly1057Asp of IRS-2 in moderate-to-severe insulin resistance of women with polycystic ovary syndrome. Diabetes. 2001;50(9):2164–8.

[178] Lin T-C, Yen J-M, Gong K-B, et al. Abnormal glucose tolerance and insulin resistance in polycystic ovary syndrome amongst the Taiwanese population-not correlated with insulin receptor substrate-1 Gly972Arg/Ala513Pro polymorphism. BMC Med Genet 2006;7(1):1–8.

[179] Pappalardo M, Vita R, Di Bari F, Le Donne M, Trimarchi F, Benvenga S. Gly972Arg of IRS-1 and Lys121Gln of PC-1 polymorphisms act in opposite way in polycystic ovary syndrome. J Endocrinol Investig 2017;40 (4):367–76.

[180] Sir-Petermann T, Angel B, Maliqueo M, et al. Insulin secretion in women who have polycystic ovary syndrome and carry the Gly972Arg variant of insulin receptor substrate-1 in response to a high-glycemic or low-glycemic carbohydrate load. Nutrition. 2004;20(10):905–10.

[181] Valdés P, Cerda A, Barrenechea C, Kehr M, Soto C, Salazar LA. No association between common Gly972Arg variant of the insulin receptor substrate-1 and polycystic ovary syndrome in Southern Chilean women. Clinica Chim Acta 2008;390(1-2):63–6.

[182] Ioannidis A, Ikonomi E, Dimou NL, Douma L, Bagos PG. Polymorphisms of the insulin receptor and the insulin receptor substrates genes in polycystic ovary syndrome: a Mendelian randomization meta-analysis. Mol Genet Metab 2010;99(2):174–83. Available from: https://doi.org/10.1016/j.ymgme.2009.10.013 2010/02/01/.

[183] Vettor R, Lombardi A, Fabris R, et al. Substrate competition and insulin action in animal models. Int J Obes 2000;24(2):S22–4.

[184] Ciaraldi TP, Morales AJ, Hickman MG, Odom-Ford R, Olefsky JM, Yen SS. Cellular insulin resistance in adipocytes from obese polycystic ovary syndrome subjects involves adenosine modulation of insulin sensitivity. J Clin Endocrinol Metab 1997;82(5):1421–5.

[185] Poretsky L, Cataldo NA, Rosenwaks Z, Giudice LC. The insulin-related ovarian regulatory system in health and disease. Endocr Rev 1999;20(4):535–82.

[186] Adashi EY, Hsueh AJ, Yen SS. Insulin enhancement of luteinizing hormone and follicle-stimulating hormone release by cultured pituitary cells. Endocrinology 1981;108(4):1441–9. Available from: https://doi.org/10.1210/endo-108-4-1441.

[187] Dunaif A, Scott D, Finegood D, Quintana B, Whitcomb R. The insulin-sensitizing agent troglitazone improves metabolic and reproductive abnormalities in the polycystic ovary syndrome. J Clin Endocrinol Metab 1996;81(9):3299–306. Available from: https://doi.org/10.1210/jcem.81.9.8784087.

[188] Nestler JE, Jakubowicz DJ. Lean women with polycystic ovary syndrome respond to insulin reduction with decreases in ovarian P450c17 alpha activity and serum androgens. J Clin Endocrinol Metab 1997;82 (12):4075–9. Available from: https://doi.org/10.1210/jcem.82.12.4431.

[189] Soldani R, Cagnacci A, Yen SS. Insulin, insulin-like growth factor I (IGF-I) and IGF-II enhance basal and gonadotrophin-releasing hormone-stimulated luteinizing hormone release from rat anterior pituitary cells in vitro. Eur J Endocrinol 1994;131(6):641–5. Available from: https://doi.org/10.1530/eje.0.1310641.

[190] Velazquez EM, Mendoza S, Hamer T, Sosa F, Glueck CJ. Metformin therapy in polycystic ovary syndrome reduces hyperinsulinemia, insulin resistance, hyperandrogenemia, and systolic blood pressure, while facilitating normal menses and pregnancy. Metabolism 1994;43(5):647–54. Available from: https://doi.org/10.1016/0026-0495(94)90209-7.

[191] Hillier S. Current concepts of the roles of follicle stimulating hormone and luteinizing hormone in folliculogenesis. Hum Reprod 1994;9(2):188–91.

[192] Willis DS, Watson H, Mason HD, Galea R, Brincat M, Franks S. Premature response to luteinizing hormone of granulosa cells from anovulatory women with polycystic ovary syndrome: relevance to mechanism of anovulation. J Clin Endocrinol Metab 1998;83(11):3984–91.

[193] McNatty K, Smith DM, Osathanondh R, Ryan K. The human antral follicle: functional correlates of growth and atresia. EDP Sciences; 1979, p. 1547–58.

[194] Franks S, Gilling-Smith C, Watson H, Willis D. Insulin action in the normal and polycystic ovary. Endocrinol Metab Clin North Am 1999;28(2):361–78. Available from: https://doi.org/10.1016/S0889-8529(05)70074-8 1999/06/01/.

[195] Jonard S, Dewailly D. The follicular excess in polycystic ovaries, due to intra-ovarian hyperandrogenism, may be the main culprit for the follicular arrest. Hum Reprod Update 2004;10(2):107–17. Available from: https://doi.org/10.1093/humupd/dmh010.

[196] Qi J, Wang W, Zhu Q, et al. Local cortisol elevation contributes to endometrial insulin resistance in polycystic ovary syndrome. J Clin Endocrinol Metab 2018;103(7):2457–67. Available from: https://doi.org/10.1210/jc.2017-02459.

[197] Frolova AI, Moley KH. Glucose transporters in the uterus: an analysis of tissue distribution and proposed physiological roles. Reproduction 2011;142(2):211–20. Available from: https://doi.org/10.1530/REP-11-0114.

[198] Frolova AI, Moley KH. Quantitative analysis of glucose transporter mRNAs in endometrial stromal cells reveals critical role of GLUT1 in uterine receptivity. Endocrinology 2011;152(5):2123–8. Available from: https://doi.org/10.1210/en.2010-1266.

[199] Kim ST, Moley KH. Regulation of facilitative glucose transporters and AKT/MAPK/PRKAA signaling via estradiol and progesterone in the mouse uterine epithelium. Biol Reprod 2009;81(1):188–98. Available from: https://doi.org/10.1095/biolreprod.108.072629.

[200] von Wolff M, Ursel S, Hahn U, Steldinger R, Strowitzki T. Glucose transporter proteins (GLUT) in human endometrium: expression, regulation, and function throughout the menstrual cycle and in early pregnancy. J Clin Endocrinol Metab 2003;88(8):3885–92. Available from: https://doi.org/10.1210/jc.2002-021890.

[201] Okada H, Tsuzuki T, Murata H. Decidualization of the human endometrium. Reprod Med Biol 2018;17(3):220–7.

[202] Zhai J, Liu C-X, Tian Z-R, Jiang Q-H, Sun Y-P. Effects of metformin on the expression of GLUT4 in endometrium of obese women with polycystic ovary syndrome. Biol Reprod 2012;87(2):29 1-5.

[203] Kim JJ, Fazleabas AT. Uterine receptivity and implantation: the regulation and action of insulin-like growth factor binding protein-1 (IGFBP-1), HOXA10 and forkhead transcription factor-1 (FOXO-1) in the baboon endometrium. Reprod Biol Endocrinol 2004;2(1):34. Available from: https://doi.org/10.1186/1477-7827-2-34 2004/06/16.

[204] Giudice LC, Dsupin BA, Irwin JC. Steroid and peptide regulation of insulin-like growth factor-binding proteins secreted by human endometrial stromal cells is dependent on stromal differentiation. J Clin Endocrinol Metab 1992;75(5):1235–41. Available from: https://doi.org/10.1210/jcem.75.5.1385468.

[205] Frasca F, Pandini G, Sciacca L, et al. The role of insulin receptors and IGF-I receptors in cancer and other diseases. Arch Physiol Biochem 2008;114(1):23–37. Available from: https://doi.org/10.1080/13813450801969715.

[206] Westermeier F, Saez T, Arroyo P, et al. Insulin receptor isoforms: an integrated view focused on gestational diabetes mellitus. Diabetes Metab Res Rev 2016;32(4):350–65. Available from: https://doi.org/10.1002/dmrr.2729.

[207] Mayama R, Izawa T, Sakai K, Suciu N, Iwashita M. Improvement of insulin sensitivity promotes extravillous trophoblast cell migration stimulated by insulin-like growth factor-I. Endocr J 2013;60(3):359–68. Available from: https://doi.org/10.1507/endocrj.ej12-0241.

[208] Solomon CG, Seely EW. Brief review: hypertension in pregnancy: a manifestation of the insulin resistance syndrome? Hypertension 2001;37(2):232–9. Available from: https://doi.org/10.1161/01.hyp.37.2.232.

[209] Gabbay-Benziv R, Baschat AA. Gestational diabetes as one of the "great obstetrical syndromes"–the maternal, placental, and fetal dialog. Best Pract Res Clin Obstet Gynaecol 2015;29(2):150–5. Available from: https://doi.org/10.1016/j.bpobgyn.2014.04.025.

[210] Huppertz B. Traditional and new routes of trophoblast invasion and their implications for pregnancy diseases. Int J Mol Sci 2019;21(1). Available from: https://doi.org/10.3390/ijms21010289.

[211] Jakubowicz DJ, Essah PA, Seppala M, et al. Reduced serum glycodelin and insulin-like growth factor-binding protein-1 in women with polycystic ovary syndrome during first trimester of pregnancy. J Clin Endocrinol Metab 2004;89(2):833–9. Available from: https://doi.org/10.1210/jc.2003-030975.

[212] Morris DV, Falcone T. The relationship between insulin sensitivity and insulin-like growth factor-binding protein-1. Gynecol Endocrinol 1996;10(6):407–12. Available from: https://doi.org/10.3109/09513599609023605.

[213] de Wilde MA, Veltman-Verhulst SM, Goverde AJ, et al. Preconception predictors of gestational diabetes: a multicentre prospective cohort study on the predominant complication of pregnancy in polycystic ovary syndrome. Hum Reprod 2014;29(6):1327–36. Available from: https://doi.org/10.1093/humrep/deu077.

[214] Gozukara YM, Aytan H, Ertunc D, et al. Role of first trimester total testosterone in prediction of subsequent gestational diabetes mellitus. J Obstet Gynaecol Res 2015;41(2):193–8. Available from: https://doi.org/10.1111/jog.12525.

[215] Powe CE. Early pregnancy biochemical predictors of gestational diabetes mellitus. Curr Diab Rep 2017;17(2):12. Available from: https://doi.org/10.1007/s11892-017-0834-y.

[216] Belkacemi L, Kjos S, Nelson DM, Desai M, Ross MG. Reduced apoptosis in term placentas from gestational diabetic pregnancies. J Dev Orig Health Dis 2013;4(3):256–65. Available from: https://doi.org/10.1017/S2040174413000068.

[217] Madazli R, Tuten A, Calay Z, Uzun H, Uludag S, Ocak V. The incidence of placental abnormalities, maternal and cord plasma malondialdehyde and vascular endothelial growth factor levels in women with gestational diabetes mellitus and nondiabetic controls. Gynecol Obstet Invest 2008;65(4):227–32. Available from: https://doi.org/10.1159/000113045.

[218] Meng Q, Shao L, Luo X, et al. Expressions of VEGF-A and VEGFR-2 in placentae from GDM pregnancies. Reprod Biol Endocrinol 2016;14(1):61. Available from: https://doi.org/10.1186/s12958-016-0191-8.

[219] Wang K, Zheng J. Signaling regulation of fetoplacental angiogenesis. J Endocrinol 2012;212(3):243–55. Available from: https://doi.org/10.1530/JOE-11-0296.

[220] Zhou J, Ni X, Huang X, et al. Potential role of hyperglycemia in fetoplacental endothelial dysfunction in gestational diabetes mellitus. Cell Physiol Biochem 2016;39(4):1317–28. Available from: https://doi.org/10.1159/000447836.

[221] Rudnicka E, Suchta K, Grymowicz M, et al. Chronic low grade inflammation in pathogenesis of PCOS. Int J Mol Sci 2021;22(7):3789.

[222] Boulman N, Levy Y, Leiba R, et al. Increased C-reactive protein levels in the polycystic ovary syndrome: a marker of cardiovascular disease. J Clin Endocrinol Metab 2004;89(5):2160–5. Available from: https://doi.org/10.1210/jc.2003-031096.

[223] Rudnicka E, Kunicki M, Suchta K, Machura P, Grymowicz M, Smolarczyk R. Inflammatory markers in women with polycystic ovary syndrome. Biomed Res Int 2020;2020:4092470. Available from: https://doi.org/10.1155/2020/4092470.

[224] Xiong YL, Liang XY, Yang X, Li Y, Wei LN. Low-grade chronic inflammation in the peripheral blood and ovaries of women with polycystic ovarian syndrome. Eur J Obstet Gynecol Reprod Biol 2011;159(1):148–50. Available from: https://doi.org/10.1016/j.ejogrb.2011.07.012.

[225] Kelly CC, Lyall H, Petrie JR, Gould GW, Connell JM, Sattar N. Low grade chronic inflammation in women with polycystic ovarian syndrome. J Clin Endocrinol Metab 2001;86(6):2453–5. Available from: https://doi.org/10.1210/jcem.86.6.7580.

[226] Orio Jr F, Palomba S, Cascella T, et al. The increase of leukocytes as a new putative marker of low-grade chronic inflammation and early cardiovascular risk in polycystic ovary syndrome. J Clin Endocrinol Metab 2005;90(1):2–5.

[227] Souza dos Santos AC, Soares NP, Costa EC, de Sá JCF, Azevedo GD, Lemos TMAM. The impact of body mass on inflammatory markers and insulin resistance in polycystic ovary syndrome. Gynecol Endocrinol 2015;31(3):225–8.

[228] Tola EN, Yalcin SE, Dugan N. The predictive effect of inflammatory markers and lipid accumulation product index on clinical symptoms associated with polycystic ovary syndrome in nonobese adolescents and younger aged women. Eur J Obstet Gynecol Reprod Biol 2017;214:168–72.

[229] Escobar-Morreale HF, Luque-Ramírez M, González F. Circulating inflammatory markers in polycystic ovary syndrome: a systematic review and metaanalysis. Fertil Steril 2011;95(3):1048–58 e2.
[230] Toulis KA, Goulis DG, Mintziori G, et al. Meta-analysis of cardiovascular disease risk markers in women with polycystic ovary syndrome. Hum Reprod update 2011;17(6):741–60.
[231] Deligeoroglou E, Vrachnis N, Athanasopoulos N, et al. Mediators of chronic inflammation in polycystic ovarian syndrome. Gynecol Endocrinol 2012;28(12):974–8.
[232] Escobar-Morreale HCF, Botella-Carretero JI, Villuendas G, Sancho J, San Millán JL. Serum interleukin-18 concentrations are increased in the polycystic ovary syndrome: relationship to insulin resistance and to obesity. J Clin Endocrinol Metab 2004;89(2):806–11.
[233] Kaya C, Pabuccu R, Berker B, Satıroglu H. Plasma interleukin-18 levels are increased in the polycystic ovary syndrome: relationship of carotid intima-media wall thickness and cardiovascular risk factors. Fertil Steril 2010;93(4):1200–7.
[234] Boots CE, Jungheim ES. Inflammation and human ovarian follicular dynamics. Thieme Medical Publishers; 2015, p. 270–5.
[235] Duleba AJ, Dokras A. Is PCOS an inflammatory process? Fertil Steril 2012;97(1):7–12.
[236] González F. Nutrient-induced inflammation in polycystic ovary syndrome: role in the development of metabolic aberration and ovarian dysfunction. Thieme Medical Publishers; 2015, p. 276–86.
[237] Spritzer PM, Lecke SB, Satler F, Morsch DM. Adipose tissue dysfunction, adipokines, and low-grade chronic inflammation in polycystic ovary syndrome. Reproduction. 2015;149(5):R219–27.
[238] Shorakae S, Ranasinha S, Abell S, et al. Inter-related effects of insulin resistance, hyperandrogenism, sympathetic dysfunction and chronic inflammation in PCOS. Clin Endocrinol (Oxf) 2018;89(5):628–33. Available from: https://doi.org/10.1111/cen.13808.
[239] González F. Inflammation in polycystic ovary syndrome: underpinning of insulin resistance and ovarian dysfunction. Steroids 2012;77(4):300–5. Available from: https://doi.org/10.1016/j.steroids.2011.12.003 2012/03/10/.
[240] Tan AS, Ahmed N, Berridge MV. Acute regulation of glucose transport after activation of human peripheral blood neutrophils by phorbol myristate acetate, fMLP, and granulocyte-macrophage colony-stimulating factor. Blood, J Am Soc Hematol 1998;91(2):649–55.
[241] Chanock SJ, El Benna J, Smith RM, Babior BM. The respiratory burst oxidase. J Biol Chem 1994;269 (40):24519–22.
[242] Löms Ziegler-Heitbrock H-W. The biology of the monocyte system. Eur J Cell Biol 1989;49(1):1–12.
[243] Aguirre V, Uchida T, Yenush L, Davis R, White MF. The c-Jun NH(2)-terminal kinase promotes insulin resistance during association with insulin receptor substrate-1 and phosphorylation of Ser(307). J Biol Chem 2000;275(12):9047–54. Available from: https://doi.org/10.1074/jbc.275.12.9047.
[244] Chen G, Goeddel DV. TNF-R1 signaling: a beautiful pathway. Science 2002;296(5573):1634–5. Available from: https://doi.org/10.1126/science.1071924.
[245] Weisberg SP, McCann D, Desai M, Rosenbaum M, Leibel RL, Ferrante AW. Obesity is associated with macrophage accumulation in adipose tissue. J Clin Investig 2003;112(12):1796–808.
[246] Fain J, Bahouth S, Madan A. TNF α release by the nonfat cells of human adipose tissue. Int J Obes 2004;28 (4):616–22.
[247] Gonzalez F, Nair KS, Daniels JK, Basal E, Schimke JM. Hyperandrogenism sensitizes mononuclear cells to promote glucose-induced inflammation in lean reproductive-age women. Am J Physiol Endocrinol Metab 2012;302(3):E297–306. Available from: https://doi.org/10.1152/ajpendo.00416.2011.
[248] Gonzalez F, Nair KS, Daniels JK, Basal E, Schimke JM, Blair HE. Hyperandrogenism sensitizes leukocytes to hyperglycemia to promote oxidative stress in lean reproductive-age women. J Clin Endocrinol Metab 2012;97(8):2836–43. Available from: https://doi.org/10.1210/jc.2012-1259.
[249] Gonzalez F, Sia CL, Bearson DM, Blair HE. Hyperandrogenism induces a proinflammatory TNFalpha response to glucose ingestion in a receptor-dependent fashion. J Clin Endocrinol Metab 2014;99(5):E848–54. Available from: https://doi.org/10.1210/jc.2013-4109.
[250] Gonzalez F. Nutrient-induced inflammation in polycystic ovary syndrome: role in the development of metabolic aberration and ovarian dysfunction. Semin Reprod Med 2015;33(4):276–86. Available from: https://doi.org/10.1055/s-0035-1554918.

[251] Best CL, Pudney J, Welch WR, Burger N, Hill JA. Localization and characterization of white blood cell populations within the human ovary throughout the menstrual cycle and menopause. Hum Reprod 1996;11 (4):790–7. Available from: https://doi.org/10.1093/oxfordjournals.humrep.a019256.
[252] Sancho-Tello M, Perez-Roger I, Imakawa K, Tilzer L, Terranova PF. Expression of tumor necrosis factor-alpha in the rat ovary. Endocrinology 1992;130(3):1359–64. Available from: https://doi.org/10.1210/endo.130.3.1537297.
[253] Spaczynski RZ, Arici A, Duleba AJ. Tumor necrosis factor-alpha stimulates proliferation of rat ovarian theca-interstitial cells. Biol Reprod 1999;61(4):993–8. Available from: https://doi.org/10.1095/biolreprod61.4.993.
[254] Ortega I, Villanueva JA, Wong DH, et al. Resveratrol potentiates effects of simvastatin on inhibition of rat ovarian theca-interstitial cells steroidogenesis. J Ovarian Res 2014;7:21. Available from: https://doi.org/10.1186/1757-2215-7-21.
[255] Carlberg M, Nejaty J, Fröysa B, Guan Y, Söder O, Bergqvist A. Elevated expression of tumour necrosis factor α in cultured granulosa cells from women with endometriosis. Hum Reprod 2000;15(6):1250–5. Available from: https://doi.org/10.1093/humrep/15.6.1250.
[256] Cianci A, Calogero AE, Palumbo MA, et al. Relationship between tumour necrosis factor α and sex steroid concentrations in the follicular fluid of women with immunological infertility. Hum Reprod 1996;11(2):265–8.
[257] Lee KS, Joo BS, Na YJ, Yoon MS, Choi OH, Kim WW. Relationships between concentrations of tumor necrosis factor-α and nitric oxide in follicular fluid and oocyte quality. J Assist Reprod Genet 2000;17(4):222–8.
[258] Amato G, Conte M, Mazziotti G, et al. Serum and follicular fluid cytokines in polycystic ovary syndrome during stimulated cycles. Obstet Gynecol 2003;101(6):1177–82. Available from: https://doi.org/10.1016/S0029-7844(03)00233-3 2003/06/01/.
[259] Adashi EY, Resnick CE, Packman JN, Hurwitz A, Payne DW. Cytokine-mediated regulation of ovarian function: tumor necrosis factor a inhibits gonadotropin-supported progesterone accumulation by differentiating and luteinized murine granulosa cells. Am J Obstet Gynecol 1990;162(4):889–99.
[260] Roby K, Terranova P. Tumor necrosis factor alpha alters follicolar steroidogenesis in vitro. Endocrinology 1988;123(6):2952–4.
[261] Zolti M, Meirom R, Shemesh M, et al. Granulosa cells as a source and target organ for tumor necrosis factor-α. FEBS Lett 1990;261(2):253–5.
[262] Darbon J, Oury F, Laredo J, Bayard F. Tumor necrosis factor-α inhibits follicle-stimulating hormone-induced differentiation in cultured rat granulosa cells. Biochem Biophys Res Commun 1989;163(2):1038–46.
[263] Wang LJ, Brannstrom M, Pascoe V, Norman RJ. Cellular composition of primary cultures of human granulosa-lutein cells and the effect of cytokines on cell proliferation. Reprod Fertil Dev 1995;7(1):21–6. Available from: https://doi.org/10.1071/rd9950021.
[264] Tal R, Seifer DB, Arici A. The emerging role of angiogenic factor dysregulation in the pathogenesis of polycystic ovarian syndrome. Semin Reprod Med 2015;33(3):195–207. Available from: https://doi.org/10.1055/s-0035-1552582.
[265] Agrawal R, Sladkevicius P, Engmann L, et al. Serum vascular endothelial growth factor concentrations and ovarian stromal blood flow are increased in women with polycystic ovaries. Hum Reprod 1998;13(3):651–5. Available from: https://doi.org/10.1093/humrep/13.3.651.
[266] Artini PG, Monti M, Matteucci C, Valentino V, Cristello F, Genazzani AR. Vascular endothelial growth factor and basic fibroblast growth factor in polycystic ovary syndrome during controlled ovarian hyperstimulation. Gynecol Endocrinol 2006;22(8):465–70. Available from: https://doi.org/10.1080/09513590600906607.
[267] Tal R, Seifer DB, Grazi RV, Malter HE. Follicular fluid placental growth factor is increased in polycystic ovarian syndrome: correlation with ovarian stimulation. Reprod Biol Endocrinol 2014;12(1):1–7. Available from: https://doi.org/10.1186/1477-7827-12-82 82.
[268] Tal R, Seifer DB, Grazi RV, Malter HE. Angiopoietin-1 and angiopoietin-2 are altered in polycystic ovarian syndrome (PCOS) during controlled ovarian stimulation. Vasc Cell 2013;5(1):18. Available from: https://doi.org/10.1186/2045-824X-5-18.
[269] Scotti L, Parborell F, Irusta G, et al. Platelet-derived growth factor BB and DD and angiopoietin1 are altered in follicular fluid from polycystic ovary syndrome patients. Mol Reprod Dev 2014;81(8):748–56. Available from: https://doi.org/10.1002/mrd.22343.

[270] Raja-Khan N, Kunselman AR, Demers LM, Ewens KG, Spielman RS, Legro RS. A variant in the fibrillin-3 gene is associated with TGF-β and inhibin B levels in women with polycystic ovary syndrome. Fertil Steril 2010;94(7):2916–19. Available from: https://doi.org/10.1016/j.fertnstert.2010.05.047.

[271] Tal R, Seifer DB, Shohat-Tal A, Grazi RV, Malter HE. Transforming growth factor-β1 and its receptor soluble endoglin are altered in polycystic ovary syndrome during controlled ovarian stimulation. Fertil Steril 2013;100(2):538–43. Available from: https://doi.org/10.1016/j.fertnstert.2013.04.022.

[272] Costa C, Incio J, Soares R. Angiogenesis and chronic inflammation: cause or consequence? Angiogenesis 2007;10(3):149–66. Available from: https://doi.org/10.1007/s10456-007-9074-0.

[273] Wang L, Qi H, Baker PN, et al. Altered circulating inflammatory cytokines are associated with anovulatory polycystic ovary syndrome (PCOS) women resistant to clomiphene citrate treatment. Med Sci Monit 2017;23:1083–9. Available from: https://doi.org/10.12659/msm.901194.

[274] Long X, Li R, Yang Y, Qiao J. Overexpression of IL-18 in the proliferative phase endometrium of patients with polycystic ovary syndrome. Reprod Sci 2017;24(2):252–7. Available from: https://doi.org/10.1177/1933719116653681.

[275] Piltonen TT, Chen JC, Khatun M, et al. Endometrial stromal fibroblasts from women with polycystic ovary syndrome have impaired progesterone-mediated decidualization, aberrant cytokine profiles and promote enhanced immune cell migration in vitro. Hum Reprod 2015;30(5):1203–15. Available from: https://doi.org/10.1093/humrep/dev055.

[276] Fasshauer M, Kralisch S, Klier M, et al. Adiponectin gene expression and secretion is inhibited by interleukin-6 in 3T3-L1 adipocytes. Biochem Biophys Res Commun 2003;301(4):1045–50.

[277] García V, Oróstica L, Poblete C, et al. Endometria from obese PCOS women with hyperinsulinemia exhibit altered adiponectin signaling. Hormone Metab Res 2015;47(12):901–9.

[278] Orostica L, Astorga I, Plaza-Parrochia F, et al. Proinflammatory environment and role of TNF-α in endometrial function of obese women having polycystic ovarian syndrome. Int J Obes 2016;40(11):1715–22.

[279] Oróstica L, García P, Vera C, García V, Romero C, Vega M. Effect of TNF-α on molecules related to the insulin action in endometrial cells exposed to hyperandrogenic and hyperinsulinic conditions characteristics of polycystic ovary syndrome. Reprod Sci 2018;25(7):1000–9.

[280] Jasper MJ, Tremellen KP, Robertson SA. Reduced expression of IL-6 and IL-1α mRNAs in secretory phase endometrium of women with recurrent miscarriage. J Reprod Immunol 2007;73(1):74–84.

[281] Zhao DM, Shan YH, Li FH, Jiang L, Qu QL. Correlation between endometrial receptivity with expressions of IL-1 and VEGF in rats with polycystic ovary syndrome. Eur Rev Med Pharmacol Sci 2019;23(13):5575–80. Available from: https://doi.org/10.26355/eurrev_201907_18291.

[282] Khatun M, Arffman RK, Lavogina D, et al. Women with polycystic ovary syndrome present with altered endometrial expression of stanniocalcin-1†. Biol Reprod 2019;102(2):306–15. Available from: https://doi.org/10.1093/biolre/ioz180.

[283] Palomba S, Falbo A, Chiossi G, et al. Low-grade chronic inflammation in pregnant women with polycystic ovary syndrome: a prospective controlled clinical study. J Clin Endocrinol Metab 2014;99(8):2942–51. Available from: https://doi.org/10.1210/jc.2014-1214.

[284] Maliqueo M, Sundstrom Poromaa I, Vanky E, et al. Placental STAT3 signaling is activated in women with polycystic ovary syndrome. Hum Reprod 2015;30(3):692–700. Available from: https://doi.org/10.1093/humrep/deu351.

[285] Aye IL, Lager S, Ramirez VI, et al. Increasing maternal body mass index is associated with systemic inflammation in the mother and the activation of distinct placental inflammatory pathways. Biol Reprod 2014;90(6):129. Available from: https://doi.org/10.1095/biolreprod.113.116186.

[286] Roberts KA, Riley SC, Reynolds RM, et al. Placental structure and inflammation in pregnancies associated with obesity. Placenta 2011;32(3):247–54. Available from: https://doi.org/10.1016/j.placenta.2010.12.023.

[287] Du M, Basu A, Fu D, et al. Serum inflammatory markers and preeclampsia in type 1 diabetes: a prospective study. Diabetes Care 2013;36(7):2054–61. Available from: https://doi.org/10.2337/dc12-1934.

[288] Sacks GP, Seyani L, Lavery S, Trew G. Maternal C-reactive protein levels are raised at 4 weeks gestation. Hum Reprod 2004;19(4):1025–30. Available from: https://doi.org/10.1093/humrep/deh179.

[289] Wolf M, Sandler L, Hsu K, Vossen-Smirnakis K, Ecker JL, Thadhani R. First-trimester C-reactive protein and subsequent gestational diabetes. Diabetes Care 2003;26(3):819–24. Available from: https://doi.org/10.2337/diacare.26.3.819.

CHAPTER

24

Metabolic control of immune responses in women with recurrent pregnancy loss and recurrent implantation failure

Yiqiu Wei[1,*], *Songchen Cai*[2,*], *Jinli Ding*[1], *Yong Zeng*[2], *Tailang Yin*[1] *and Lianghui Diao*[2]

[1]Reproductive Medicine Center, Renmin Hospital of Wuhan University, Wuhan, Hubei, P.R. China [2]Shenzhen Key Laboratory for Reproductive Immunology of Peri-implantation, Shenzhen Zhongshan Institute for Reproduction and Genetics, Shenzhen Zhongshan Urology Hospital, Shenzhen, P.R. China

1 Introduction

Metabolism and immunity are central mechanisms for the body in maintaining tissue homeostasis. The role played by these two pathways during pregnancy has been documented in extensive studies. At the beginning of pregnancy, immune cells confer "immune privilege" to the embryo through adaptive metabolic reprogramming to support normal fetal development until delivery. Given that the metabolism shapes the cell function, inadequate metabolic response of immune cells to the allograft might be the pathological condition detrimental to placentation and pregnancy outcomes [1]. Recurrent pregnancy loss (RPL) and recurrent implantation failures (RIFs) are the two most frustrating conditions for patients and physicians. Although RIF and RPL occur at different reproductive stages, the laboratory results often exhibit comparable features, so these two adverse reproductive conditions are often evaluated and treated by the same group of physicians or at the same department.

* Yiqiu Wei and Songchen Cai contributed equally to this work and shared the first authorship.

Immunology of Recurrent Pregnancy Loss and Implantation Failure
DOI: https://doi.org/10.1016/B978-0-323-90805-4.00021-3

The most prominent cause of RPL is chromosomal abnormalities of the embryo, followed by neuroendocrine, immune, and metabolic factors [2]. However, the definition, incidence, and intervention of RIF remain under debate [3,4]. Among all etiologies, the endocrine, metabolic, and immune factors are classified as maternal factors. Previous studies usually interpreted immunological pathogenesis from two perspectives: alloimmunity and autoimmunity. The pathogenesis of alloimmunity is mainly based on the concept "paradox of the fetal allograft." Once the pregnancy is established, the maternal immune system actively adapts to the exposure of fetal antigens to achieve the state of immune homeostasis [5]. Dysregulation of immune homeostasis associated with pregnancy contains Th1/Th2 cell imbalance [6], Treg/Th17 cell imbalance [7], increased natural killer (NK) cell activity [8], M1/M2 cell imbalance [9], abnormal expressions of cytokines and their receptors [10]. The autoimmune disorders often manifest as an overactive immune response to autoantigens, producing autoantibodies or attacking intrauterine fetus and placental vessels, leading to an embryo/fetal loss. The autoantibodies include antiphospholipid antibodies, antiinsulin receptor antibodies, antithyroid antibodies, etc. With the latest advances in reproductive immunology, the researchers understand that pregnancy is more like cancer but not transplantation [11]. Breakthroughs in cancer immunometabolism have also led to revisiting the immunometabolic landscape during pregnancy and the impact of nutrition and metabolic disorders on adverse pregnancy outcomes [12].

As the embryonic factors are addressed in more detail elsewhere, this chapter focuses on the etiology, mechanisms, evaluation, and treatment of RPL and RIF from an immunological and metabolic perspective.

2 Definition and prevalence

The prevalence of immunometabolic diseases in RPL and RIF have been summarized in Table 1. Evidence suggests that many immunometabolic diseases are associated with the occurrence of RPL and RIF, such as antiphospholipid syndrome (APS), insulin resistance (IR), and polycystic ovary syndrome (PCOS), autoimmune thyroid disease (AITD), hyperprolactinemia, and other autoimmune-related diseases.

TABLE 1 Prevalence of immunological and metabolic diseases in recurrent pregnancy loss and recurrent implantation failure.

Prevalence in women of	Childbearing age	Infertile women	RPL	RIF
Antiphospholipid syndrome (APS)	2.7%–7%	20%–25%	5%–15%	9%–20%
Insulin resistance (IR)	10%	20%	20%–30%	20%–30%
Polycystic ovary syndrome (PCOS)	8%–13%	30%–40%	27%–40%	27%–40%
Autoimmune thyroid disease (AITD)	6%–20%	15%–30%	17%–33%	15%–30%
Hyperprolactinemia	0.4%	5%	36%	-

Note: The references are cited in the following context.

2.1 Antiphospholipid syndrome in recurrent pregnancy loss and recurrent implantation failure

Antiphospholipid syndrome (APS), also known as Hughes syndrome, is an immune disorder characterized by positive antiphospholipid antibodies (aPLs) in the laboratory evaluation, with clinical indications of increased thrombotic risk and pregnancy loss [13]. Among RPL patients, the prevalence of APS following Sapporo criteria was 2.5%, while the single positive rate for aPLs was 10.7%, and the recurrent positive rate was 4.5% [2]. However, it is rare to find RIF patients with persistent aPL. The value of aPL testing in RIF patients is still under debate, which need additional data to make a firm recommendation [14].

2.2 Insulin resistance and polycystic ovary syndrome in recurrent pregnancy loss and recurrent implantation failure

IR is a pathological condition when the body fails to respond well to insulin and cannot efficiently uptake glucose for metabolism. Hence, the hypoglycemic effect of insulin cannot be exerted. Hyperglycemia stimulates the secretion of excess insulin in the body, which is called hyperinsulinemia. Although PCOS diagnosis does not require IR, it seems to cause or exacerbate hyperandrogenemia in many patients, and treatment with insulin sensitizers can reduce androgen levels. A recent study revealed that IR adversely affected in vitro fertilization (IVF) outcomes in lean women without PCOS [15]. Although the underlying mechanism of correlation between IR and RPL remains uncertain in patients without PCOS. However, current studies have found that about 70% of women with PCOS have IR [16]. Furthermore, 27%–40% of women with RPL are reported to have PCOS [16,17].

2.3 Autoimmune thyroid disease in recurrent pregnancy loss and recurrent implantation failure

AITD is one of the most common organ-specific autoimmune diseases, including autoimmune thyroiditis and Graves' disease. Thyroid dysfunction and thyroid autoimmunity (TAI) are often present together in women of childbearing age and may be related to ovulation dysfunction and poor pregnancy outcomes [18]. Among unselected women, the prevalence varies from 6% to 20% [19,20], and even higher among women with RPL, ranging from 17% to 33% [21,22]. However, the effect of Graves' disease, which leads to hyperthyroidism, on RPL or RIF has not been reported.

2.4 Hyperprolactinemia in recurrent pregnancy loss and recurrent implantation failure

Hyperprolactinemia is defined as abnormally high serum prolactin (PRL) level, usually defined as higher than 25 μg/L (530 mIU/L) in women [23,24]. Several physiological states, including pregnancy, breastfeeding, stress, exercise, and sleep, can cause prolactin elevation [24]. Certain prescription medicines such as calcium-channel blockers and methyldopa, tricyclic, and selective serotonin reuptake inhibitor antidepressants, opiates, and estrogen, can also increase prolactin levels [25]. Other causes include germline loss-of-function mutation, idiopathic hyperprolactinemia, hypothyroidism, chest wall injury, estrogen administration, chronic renal failure, PCOS, and macroprolactinoma [26]. Excessive PRL acts on the corresponding receptors in the hypothalamus through feedback, inhibits pituitary gonadotropin secretion, and causes anovulation and amenorrhea. Clinical observation found that the concentration of prolactin in women with RPL was significantly higher, and the rate of hyperprolactinemia in patients with RPL was about 36% [27], which indicated that there was a causal relationship between hyperprolactinemia and the occurrence of miscarriage, especially in the women with RPL who have no other apparent reasons [28,29].

2.5 Other immunological and metabolic factors in recurrent pregnancy loss and recurrent implantation failure

Systemic lupus erythematosus (SLE) is an autoimmune disease that is more prevalent in women and often associated with APS, leading to an increased risk of miscarriage [30]. Sjogren's syndrome [31], rheumatoid arthritis [32], and systemic sclerosis [33] also have a certain degree of associations with the development of pregnancy complications, but it is less evident for RPL or RIF. Vitamin D (VitD) deficiency is thought to be one of the underlying causes of autoimmune diseases, and the decreased VitD level was demonstrated in patients with TAI [34]. In turn, VitD deficiency is also associated with infertility and pregnancy losses, suggesting a possible interaction among TAI, VitD, and adverse pregnancy outcomes.

3 Underlying immunometabolic pathologies related to recurrent pregnancy loss and recurrent implantation failure

It has been well documented that a successful pregnancy depends on adequate education of the maternal immune system by trophoblast cells and a proper reciprocal dialog at the maternal–fetal interface [35]. Immune cells that resident the uterus exhibit tissue-specific phenotypes and functions involved in immune regulation at different stages of human reproduction [5]. How immune cells adapt themselves to support pregnancy through immunometabolism has been progressively emphasized in recent years. Cell metabolism is a housekeeping process to maintain cell survival and a highly adaptive regulatory force for cell destiny [36]. The metabolic pathways that mediate different immune cell functions can be reprogrammed, and this plasticity can then alter the immune cell function [37]. In this part, we elaborate on the changes of immunometabolism and its mechanism in women with

3.1 Antiphospholipid syndrome

Antiphospholipid antibody-mediated pregnancy loss is associated with placental vascular thrombosis, decidual vascular lesions, intervillous fibrin deposition, and placental bed infarction [38–40]. In a study of women with RPL, 56% of obstetric APS patients had higher peripheral NK cell (pNK) levels [41], suggesting that NK cells may be involved in APS-induced RPL. The cytotoxicity and percentage of pNK cells change dynamically during pregnancy [42]. There is a weak association between the RPL and the number of pNK according to the European Society of Human Reproduction and Embryology (ESHRE) justification [43]. Studies have shown the complexity and plasticity of NK cells, making the phenotypic definition of NK cell subsets incongruous. The functional subsets are classically dichotomized into $CD56^{dim}CD16^{+}$ NK and $CD56^{bright}CD16^{-}$ NK. The majority of pNK is $CD56^{dim}CD16^{+}$ NK with greater cytotoxic functions, while many uterine NK is $CD56^{bright}CD16^{-}$ NK with greater cytokine-producing capabilities. Distinct metabolic profiles might serve as reliable fingerprints to identify functional fates of NK cells. For example, glucose-driven glycolysis and oxidative metabolism are required for cytotoxic capacity, and hypoxic and glycolysis-limiting conditions are favored for the regulatory function [44], suggesting higher glycolysis and oxidative metabolism of pNK in obstetric APS patients.

The immune metabolism that immune cells undergo during pregnancy includes increased glycolysis, decreased oxidative phosphorylation, increased fatty acid (FA) synthesis, and other switches in energy use patterns [45]. This process usually enhances glycolysis and mitochondrial function by activating the mammalian target of rapamycin (mTOR) [46,47]. APS-associated vascular lesions have been demonstrated to be associated with abnormal mTOR signaling [48]. mTOR is also involved in regulating aPL-induced tissue factor (TF) and IL-8 expression in monocytes as well as platelet activation/aggregation [49,50]. In addition, TF/FVIIa/PAR2 signal transduction mediates neutrophil activation and fetal death in APS. Hence, as rate-limiting enzyme inhibitors of cholesterol synthesis in the liver, statins may be an excellent candidate to treat women with aPLs-induced pregnancy complications [51], since the immunopathology of this process was suggested to be regulated by inhibiting cholesterol synthesis.

3.2 Insulin resistance

The occurrence of IR may be related to the production of autoantibodies against the autoantigen of the insulin receptor, resulting in the inability of insulin to bind to its receptor and play a role in regulating blood glucose. NK cells are critical macrophage polarization and IR regulators on adipocytes with obesity-induced stress [52]. Metabolic activation is essential for NK cells to produce effector proteins (including IFN-γ and granzyme B) and to form correct immune synapses with target cells [53]. During the activation of NK cells, glycolysis and mitochondrial function are enhanced by activating mTOR [46]. FA administration and peroxisome proliferation activated receptor agonists inhibited mTOR-mediated glycolysis, effector function, and metabolism of NK cells [53].

Adipose tissue inflammation is implicated in associations between obesity and its multiple complications [54]. The decreased number of Treg cells observed in obese people contributes to the systemic IR because Treg cells usually improve glucose tolerance and insulin sensitivity [55]. Short-chain FA can affect the steady-state of Th17 and Treg that may lead to unexpected pregnancy loss [56]. In addition, the increase in the $CD4^{+}/CD8^{+}$ ratio is positively correlated with tumor necrosis factor-α (TNF-α) levels [57]. It is worth noting that triglyceride (TG) levels induce high levels of plasma TNF-α by increasing the production of very-low-density lipoprotein particles [58]. The increased expression of TNF-α contributes to the systemic IR because they usually improve glucose tolerance and insulin sensitivity [55]. Moreover, increased levels of CC chemokine ligand 5 (CCL5) and CC chemokine receptor 5 (CCR5) in peripheral blood mononuclear cell and adipose tissue are associated with hyperandrogenemia and IR in women with PCOS [59]. In contrast, B cells in adipose tissue may promote IR by activating pro-inflammatory macrophages and T cells and releasing resistin [60,61].

In addition to lipid metabolism, glucose metabolism was disordered in women affected by PCOS [62]. Six proteomic biomarkers were reported to predict the risk of preterm delivery in women with PCOS, of which increased pyruvate kinase isozymes M1/M2 (PKM1/M2) were closely associated with defected glucose metabolism [63]. Activation and terminal complement pathway components are increased in patients with PCOS, especially in the presence of IR and obesity [64]. A study has demonstrated the disturbance of the metabolism-immune axis in $CD8^{+}PD\text{-}1^{+}$ T cells in type 2 diabetes, while metformin treatment rescued the antigen-specific signaling and cytokine production by increasing the aerobic glycolysis activity [65]. In addition, insulin receptor-deficient T cells exhibit reduced inflammatory potential and poor protective immunity, suggesting that insulin receptors directly regulate T cell function via controlling cellular metabolism [66]. The metabolic disorder caused by IR disrupts the balance of various immune cells closely involved in RPL and RIF and are worthy of careful consideration.

3.3 Autoimmune thyroid disease

TAI is the leading cause of hypothyroidism. Even in women with normal thyroid function, a link between thyroid autoantibodies, especially thyroid peroxidase antibodies (TPOAb), and adverse pregnancy outcomes have been reported [67]. In many women with unexplained RPL, positive TPOAb predicts a decrease in the live birth rate [68]. It was reported that thyroid autoantibodies were detected in 22.5% of women with a history of RPL and 14.5% of healthy controls with a significant difference [69]. At present, the mechanism of RPL caused by thyroid autoantibodies remains to be elucidated. Currently, three hypotheses have been proposed [70,71]. The first hypothesis is that autoantibodies potentially impair thyroid function, resulting in metabolic abnormalities during pregnancy. The second hypothesis is that thyroid antibody disrupts the placental barrier, leading to abnormal rejection of the placenta and fetus. The third hypothesis is that thyroid autoantibodies cause abnormalities in the maternal autoimmune system, producing a similar "rejection" of the fetus by cellular and humoral immune activation. Indeed, Th1-directed changes in peripheral blood innate immunity occur in women with impaired TAI, suggesting that abnormal T cells may underlie TAI and infertility [71]. Women with AITD have up to 40% increased hyperactivity and migration of cytotoxic NK cells that alter uterine immune and hormonal responses [71]. Hyperactivity and increased NK cells have also been associated with infertility and are more common in TAI. In women with TAI, increased percentages of NK cells and NKT-like cells, as well as increased NK cell cytotoxicity, indicate an altered immune status and may impair pregnancy outcomes [72].

3.4 Other immune factors

The studies on how other immunometabolic factors affect RPL and RIF are rare. The causes of miscarriage due to autoimmune diseases represented by SLE are similar to those of APS. In addition to aCLs and anti-β2 GP1, anti-PT antibodies and anti-Anx V antibodies in patients with SLE may play an important role in RPL [73]. The immunomodulatory properties of vitamin D are attributed to its effects on cells of the innate and adaptive systems, including macrophages, dendritic cells [74], and T and B lymphocytes cells [75], all of which contain vitamin D3 receptors. In vitro studies have shown that CD4 activation promotes the Th2 phenotype (production of IL-4 and IL-5) and suppresses Th1 activity (production of interferon-gamma and IL-2) in vitamin D-expressing T cells [75].

4 Evaluation

For the diagnosis of RPL and RIF regarding immune and metabolic disorders, the presence of the above clinical conditions should be considered first, followed by further analysis of laboratory tests based on clinical symptoms. APS should be considered when the patient has unexplained recurrent thrombosis, recurrent fetal loss, eclampsia, preeclampsia, thrombocytopenia, lower extremity ulcers, or reticulocytosis [76]. Laboratory tests for antiphospholipid antibodies, such as anticardiolipin antibodies (aCL), anti-β2 GP1, or LA, are recommended. Medium to high titers of antiphospholipid antibodies may indicate the presence of APS [76].

Obesity, dyslipidemia, atherosclerosis, coagulation disorders, hypertension, diabetes, hyperuricemia, premeal hypoglycemic symptoms, and acanthosis nigricans indicate the need for the IR evaluation. Hyperinsulinemic-euglycemic clamp is widely considered the "gold standard" for measuring peripheral insulin sensitivity. However, this method is costly, time-consuming, and invasive, making it clinically impractical. Oral glucose tolerance test and fasting-based surrogates, as well as the Quantitative Insulin Sensitivity Check Index, can be used as an alternative for large-scale clinical studies [77]. Additionally, hypertriglyceridemia and high free FAs can further exacerbate IR, which requires more attention. Women with scanty menstruation, amenorrhea, and hirsutism are indicated to evaluate PCOS. Hyperandrogenism is a key indicator, while other conditions such as congenital adrenal hyperplasia, Cushing's syndrome, and androgen-secreting tumors need to be ruled out in the diagnosis of PCOS [78].

Thyroid functional examinations are required when symptoms associated with hypothyroidism such as goiter, cold intolerance, constipation, dry skin, and myxedema are present. The results of thyroid function tests for TAI may vary at different stages of the disease. In the early stages, triiodothyronine (T3) and thyroxine (T4) increase while TSH decreases. In the later stage, the opposite may happen. TPOAb or thyroglobulin antibodies (TgAb) may be present in the serum of patients with autoimmune thyroiditis. In addition, it is important to note that hypothyroidism can generally coexist with hypercholesterolemia and hypertriglyceridemia [79].

5 Treatment and reproductive outcomes

Interventions targeting key enzymes or key molecules in metabolic pathways can effectively manage human autoimmune and metabolic diseases [36]. Many drugs targeting metabolism and immunity have been approved by the US Food and Drug Administration (Table 2) (Fig. 1).

To date, several immunometabolic modulating drugs have been utilized or are expected to be released for women with metabolic disorders associated with adverse pregnancy outcomes. However, there are many enigmas regarding the safety and efficacy of immunometabolism-targeted drugs (Fig. 2). More evidence is needed to apply immunosuppressive and metabolic agents to aid in pregnancy success in clinical practice.

5.1 Treatment of antiphospholipid syndrome

The combination of aspirin and low molecular weight heparin (LMWH) is the standard therapy for APS. Studies have also shown that intravenous immunoglobulin (IVIG)

TABLE 2 Current Food and Drug Administration-approved drugs targeting metabolic and immune responses.

Drugs	Targets	Mechanisms	FDApregnancycategories
Metformin	AMPK	Increase FAO; Inhibit Complex I; Decrease mitochondrial ROS	B
Orlistat	Gastrointestinal lipase	Inhibition of fat hydrolysis	B
Sirolimus	mTOR	Inhibit lymphokine-driven T cell proliferation	C
Tacrolimus	mTOR (FKBP12)	Inhibit the activation of T cells	C
Hydroxychloroquine	Unclear	Inhibit inflammation and anticoagulation; Decrease cholesterol and LDL	C
Immunoglobulin	Immunoglobulin receptor	Induce antiinflammatory cytokines Inhibit monocyte and macrophage activation Suppress the proliferation of B cells and modulate antibody production Suppress autoreactive B cells and neutralize cytokines Expand T regulatory cells Induce apoptosis in DCs, Th1 and Th17 cells Suppress NK cell levels and cytotoxicity Inhibit DC maturation and differentiation	C
Pravastatin Simvastatin Atorvastatin	HMG-CoA reductase	Inhibit cholesterol synthesis	X

Abbreviations: *AMPK*: AMP-activated protein kinase; *FAO*: fatty acid oxidation; *ROS*: reactive oxygen species; *mTOR*: mammalian target of rapamycin; *FKBP12*: FK506-binding protein 12; *LDL*: low-density lipoprotein; *HMG-CoA*: 3-hydroxy-3-methylglutaryl-coenzymeA.

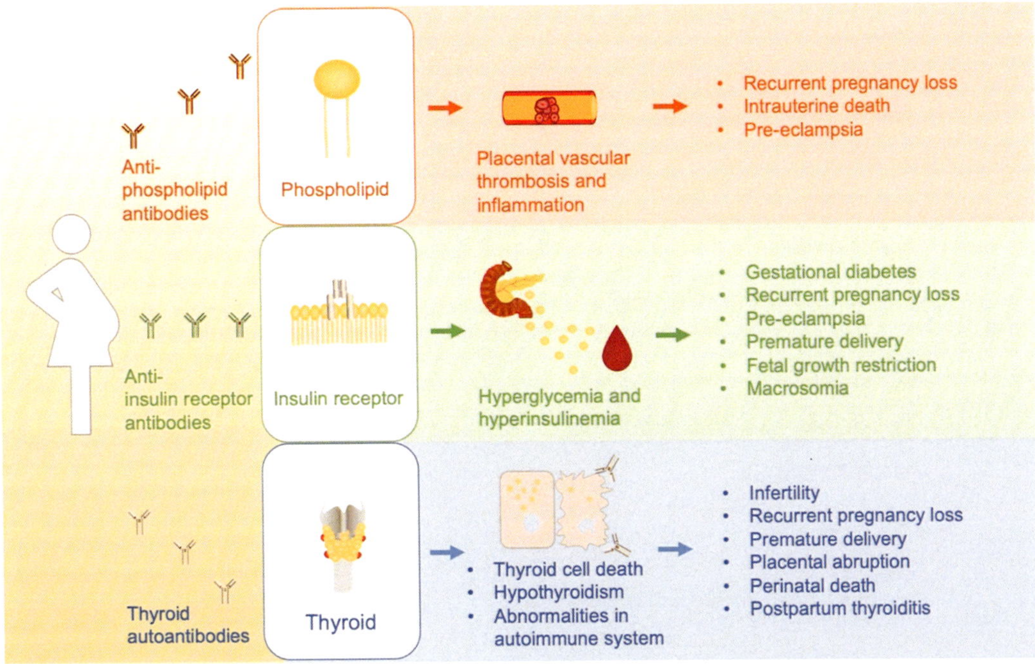

FIGURE 1 **Schematic of immunometabolism pathologies related to reproduction.** The figure illustrates the potential pathological mechanisms of three typical immunometabolic disorders leading to adverse pregnancy outcomes. Upper part: Antiphospholipid antibodies bind indirectly to phospholipids via phospholipid-binding proteins, damaging the vascular epithelium and inducing thrombosis. When thrombosis occurs in the placenta, it decreases placental vascular circulation, resulting in fetal loss or death. Middle part: Antiinsulin receptor antibodies bind to insulin receptors on cell membranes and reduce their insulin sensitivity, thereby preventing the appropriate functioning of insulin secreted by the pancreas, resulting in hyperglycemia and hyperinsulinemia, which can lead to a series of metabolic disorders and adverse pregnancy outcomes. The lower part: autoantibodies bind to thyroid cells and disrupt the normal functioning of the thyroid gland. Consequently, hypothyroidism and abnormalities in the maternal autoimmune system lead to adverse pregnancy outcomes.

reduces fetal loss rate in patients with APS [80,81]. Aspirin and LMWH modulate the immune response in patients with RPL and RIF by inhibiting inflammatory factors, binding integrins, and participating in the complement response. Unfortunately, approximately 20%–30% of women with APS still fail to deliver liveborn infants with standard treatment. These cases are defined as "refractory obstetric APS," and other emerging approaches such as IVIG or hydroxychloroquine need to be considered. Hydroxychloroquine, which has immune regulatory and anticoagulant properties, promotes trophoblast migration and reduces the binding of aPLs to trophoblasts [82].

Statins have antiinflammatory and antithrombotic effects in aPL-positive patients. Although clinical data remain insufficient, statins may play an adjunctive role in patients with APS in whom anticoagulant therapy is not practical [82]. There are some adverse effects associated with statins in pregnant women. However, the potential benefit of statins in pregnant women has prompted a new evaluation of the risk-benefit ratio of statins usage during pregnancy [83].

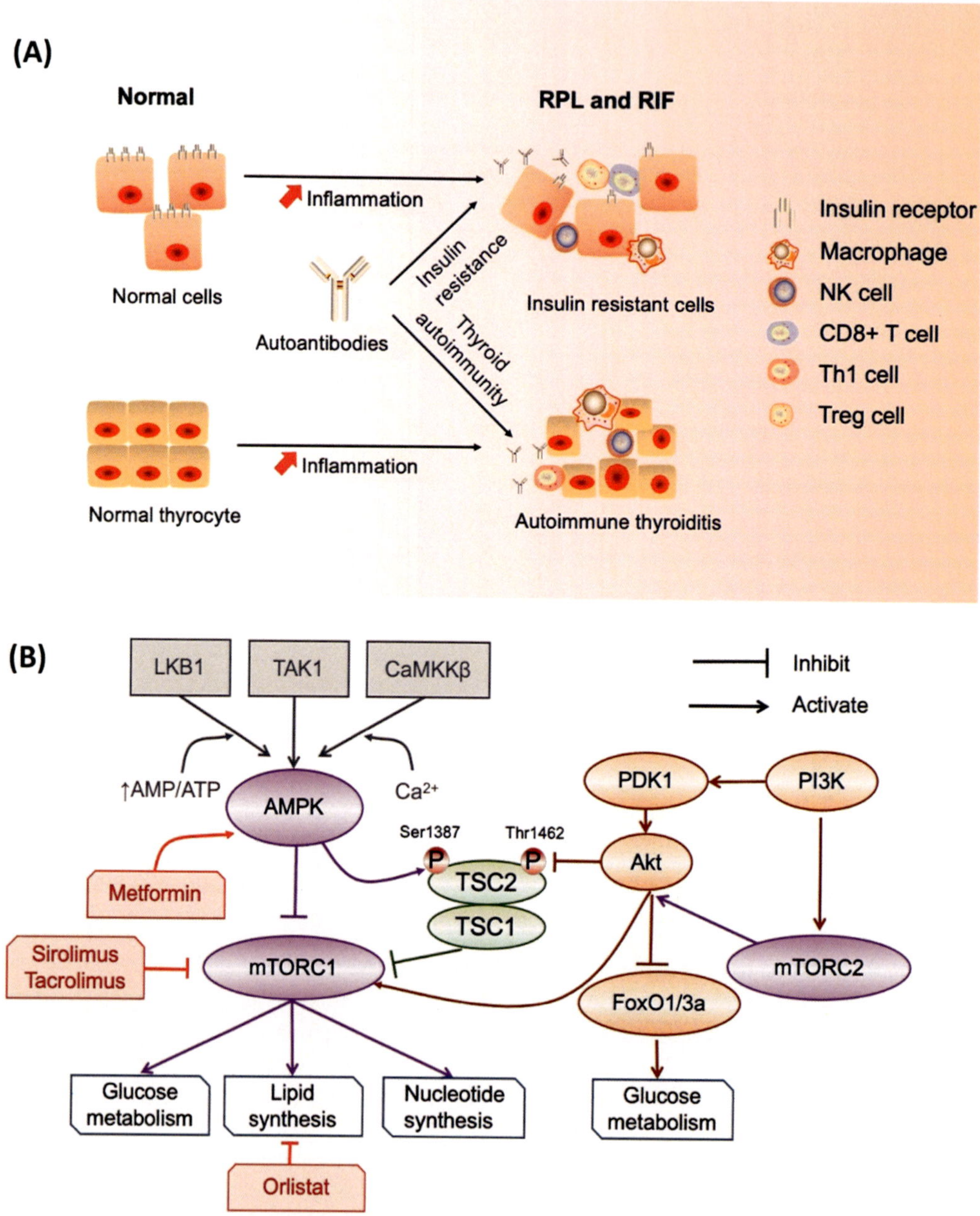

FIGURE 2 **Evaluation and the treatment target of the immune and metabolic causes for recurrent pregnancy loss (RPL) and recurrent implantation failure (RIF).** (A) The cellular level immune and metabolic changes in normal pregnant women and RPL/RIF women; (B) Related metabolic pathways and possible treatment targets. Abbreviation: *mTOR*, mammalian target of rapamycin; *AMPK*, adenosine 5′monophosphate-activated protein kinase; *LKB1*, liver kinase B1; *TAK1*, transforming growth factor beta-activated kinase 1; *CaMKKb*, Ca^{2+}/calmodulin-dependent protein kinase kinase beta; *TSC*, tuberous sclerosis complex; *PI3K*, phosphoinositide 3-kinase; *Akt*, serine-threonine kinase; *PDK1*, phosphoinositide-dependent kinase-1; *FoxO1/3a*, forkhead box protein O1/3a.

Sirolimus, also known as rapamycin, inhibits the mTOR complex (mTORC), which may be an option for patients with aPL-positive microvascular disease who are resistant to the standard treatment protocol [50]. During pregnancy, decidual $CD8^+$ T cells exhibit transient dysfunction with high IFN- γ expression and low expression of perforin and granzyme B [84]. It seems somewhat paradoxical that rapamycin can inhibit T cell proliferation inhibiting mTORC1 and mTORC2. However, it promotes immunity in the form of T cell memory. It can promote the differentiation of memory $CD8^+$ T cells and memory Treg cells, which may contribute to the maintenance of pregnancy [85,86]. However, its therapeutic efficacy and safety during pregnancy have not been established.

IVIG may improve pregnancy outcomes by modulating Th1/Th2 cell balance and Treg/Th17 cell balance while reducing NK cell cytotoxicity. Side effects of IVIG are rare and mostly mild and transient. IVIG infusion should be given preconceptionally in women with positive cellular immune markers, such as increased NK cell cytotoxicity, NK cell counts, or increased Th1/Th2 ratios [87].

In summary, early identification of risk factors associated with poor pregnancy outcomes may be the best strategy for developing additional treatment criteria, such as immunosuppressants, IVIG, and plasma exchange [88].

5.2 Treatment of insulin resistance and polycystic ovary syndrome

Metabolic abnormalities play a crucial role in the development and progression of IR in patients with PCOS. Metformin combined with clomiphene induces ovulation and improves the clinical pregnancy rate in PCOS patients [89]. In addition, metformin improves endovascular permeability and alleviates spontaneous abortion in patients with PCOS [90,91]. A randomized controlled trial (RCT) of metformin in PCOS treatment found that the metformin group significantly reduced late abortion and preterm delivery [92]. Orlistat is a pancreatic lipase inhibitor, which can negatively regulate the FA synthesis and effectively interfere with metabolism in patients with PCOS. Both metformin and orlistat improved body weight, ovulation rate, and androgen levels in obese anovulatory PCOS patients [93].

In immunometabolism, metformin participates in T cell differentiation and induces T cell differentiation into Treg cells and memory T cells [94]. In addition, metformin can reduce estradiol and progesterone-induced decidualization of endometrial stromal cells by regulating the secretion of various cytokines, inhibiting the expression of MMP-2 and MMP-9, activating p38-MAPK signal transduction, and reducing progesterone receptor (PGR) expression [95]. As a metabolic stress factor, metformin can also activate the AMPK energy sensor and enhance the production of $CD8^+$ T cells [96]. Orlistat administration can attenuate FA metabolic reprogramming and inhibit the FABP5 induced lymph node metastasis in cervical cancer [97]. It may have protective and therapeutic effects on the metabolic dysfunction-associated fatty liver disease by activating the nuclear factor erythroid-2-related factor 2 (Nrf2) signal pathway in the rodent model [98]. Nevertheless, it is unclear whether metformin and orlistat can affect pregnancy by regulating immunometabolism.

5.3 Treatment of autoimmune thyroid disease

A *meta*-analysis study suggested that women with RPL should screen/treat major thyroid diseases rather than TAI [99]. Levothyroxine supplementation and IVIG treatment have been reported to prevent miscarriages in women with AITD. Exogenous supplementation with synthetic thyroid hormone analogs to reduce the risk of miscarriage in women with AITD is based on consideration of the association between AITD and hypothyroidism in pregnancy. However, a *meta*-analysis of RCT reported that levothyroxine was not associated with improved clinical pregnancy outcomes in women with positive TPOAb [100], suggesting the presence of other mechanisms rather than hormonal deficiency for miscarriages. Based on current evidence, it is recommended that serum TSH should be carefully monitored in AITD women during pregnancy. Moreover, in women with euthyroid hypothyroidism (TSH $\geq$2.5 and $<$4.3), ATA should be evaluated.

6 Summary and recommendations

- The etiology of RIF and RPL is multifactorial, with the contribution of immune and metabolic disorders accounting for a substantial proportion of their pathogenesis.
- Lifestyle changes, social support, and medication controls are available to improve the subsequent pregnancy outcome in these women.
- Notably, metabolic modulation targeting immune cells is one of the proven tools to improve pregnancy outcomes in RPL and RIF, but it needs to be administered after a precise diagnosis.
- For novel immunomodulatory and metabolic treatments that have proven effective in other areas, special attention should be paid to the safety of the treatment for the mother and infants, including long-term effects.

Fundings

The writing of this chapter was made possible through funding by the Shenzhen Natural Science Foundation (JCYJ20190813161801676).

References

[1] Cha J, Sun X, Dey SK. Mechanisms of implantation: strategies for successful pregnancy. Nat Med 2012;18 (12):1754–67.

[2] Sugiura-Ogasawara M, Ozaki Y, Katano K, Suzumori N, Kitaori T, Mizutani E. Abnormal embryonic karyotype is the most frequent cause of recurrent miscarriage. Hum Reprod 2012;27(8):2297–303.

[3] Shaulov T, Sierra S, Sylvestre C. Recurrent implantation failure in IVF: a canadian fertility and andrology society clinical practice guideline. Reprod Biomed Online 2020;41(5):819–33.

[4] Garneau AS, Young SL. Defining recurrent implantation failure: a profusion of confusion or simply an illusion? Fertil Steril 2021;116(6):1432–5.

[5] Schumacher A, Sharkey DJ, Robertson SA, Zenclussen AC. Immune cells at the fetomaternal interface: how the microenvironment modulates immune cells to foster fetal development. JI. 2018;201(2):325–34.

[6] Raghupathy R, Makhseed M, Azizieh F, Hassan N, Al-Azemi M, Al-Shamali E. Maternal Th1- and Th2-type reactivity to placental antigens in normal human pregnancy and unexplained recurrent spontaneous abortions. Cell Immunol 1999;196(2):122–30.
[7] Qian J, Zhang N, Lin J, et al. Distinct pattern of Th17/Treg cells in pregnant women with a history of unexplained recurrent spontaneous abortion. Biosci Trends 2018;12(2):157–67.
[8] King K, Smith S, Chapman M, Sacks G. Detailed analysis of peripheral blood natural killer (NK) cells in women with recurrent miscarriage. Hum Reprod 2010;25(1):52–8.
[9] Tsao FY, Wu MY, Chang YL, Wu CT, Ho HN. M1 macrophages decrease in the deciduae from normal pregnancies but not from spontaneous abortions or unexplained recurrent spontaneous abortions. J Formos Med Assoc 2018;117(3):204–11.
[10] Jasper MJ, Tremellen KP, Robertson SA. Reduced expression of IL-6 and IL-1alpha mRNAs in secretory phase endometrium of women with recurrent miscarriage. J Reprod Immunol 2007;73(1):74–84.
[11] Patil R, Patil SA, Beaman KD, Patil SA. Indole molecules as inhibitors of tubulin polymerization: potential new anticancer agents, an update (2013–2015). Future Med Chem 2016;8(11):1291–316.
[12] Thiele K, Diao L, Arck PC. Immunometabolism, pregnancy, and nutrition. Semin Immunopathol 2018;40(2):157–74.
[13] Garcia D, Erkan D. Diagnosis and management of the antiphospholipid syndrome Longo DL, (ed.) N Engl J Med 2018;378(21):2010–21.
[14] Mascarenhas M, Jeve Y, Polanski L, et al. Management of recurrent implantation failure: British Fertility Society policy and practice guideline. Hum Fertil 2021;1–25.
[15] Wang H, Zhang Y, Fang X, Kwak-Kim J, Wu L. Insulin resistance adversely affect IVF outcomes in lean women without PCOS. Front Endocrinol (Lausanne) 2021;12:734638.
[16] Goodarzi MO, Korenman SG. The importance of insulin resistance in polycystic ovary syndrome. Fertil Steril 2003;80(2):255–8.
[17] Rai R, Backos M, Rushworth F, Regan L. Polycystic ovaries and recurrent miscarriage—a reappraisal. Hum Reprod 2000;15(3):612–15.
[18] Thangaratinam S, Tan A, Knox E, Kilby MD, Franklyn J, Coomarasamy A. Association between thyroid autoantibodies and miscarriage and preterm birth: meta-analysis of evidence. BMJ. 2011;342(may09 1):d2616.
[19] Stagnaro-Green A, Glinoer D. Thyroid autoimmunity and the risk of miscarriage. Best Pract Res Clin Endocrinol Metab 2004;18(2):167–81.
[20] Poppe K, Velkeniers B, Glinoer D. The role of thyroid autoimmunity in fertility and pregnancy. Nat Clin Pract Endocrinol Metab 2008;4(7):394–405.
[21] Ticconi C, Giuliani E, Veglia M, Pietropolli A, Piccione E, Di Simone N. Thyroid autoimmunity and recurrent miscarriage. Am J Reprod Immunol 2011;66(6):452–9.
[22] Sarkar D. Recurrent pregnancy loss in patients with thyroid dysfunction. Indian J Endocrinol Metab 2012;16 (Suppl 2):S350–1.
[23] Chahal J, Schlechte J. Hyperprolactinemia. Pituitary. 2008;11(2):141–6.
[24] Melmed S, Casanueva FF, Hoffman AR, et al. Diagnosis and treatment of hyperprolactinemia: an endocrine society clinical practice guideline. J Clin Endocrinol & Metab 2011;96(2):273–88.
[25] Molitch ME. Drugs and prolactin. Pituitary. 2008;11(2):209–18.
[26] Prabhakar VKB, Davis JRE. Hyperprolactinaemia. Best Pract & Res Clin Obstet & Gynaecol 2008;22(2):341–53.
[27] Bussen S, Sütterlin M, Steck T. Endocrine abnormalities during the follicular phase in women with recurrent spontaneous abortion. Hum Reprod 1999;14(1):18–20.
[28] Ando N, Gorai I, Hirabuki T, Onose R, Hirahara F, Minaguchi H. [Prolactin disorders in patients with habitual abortion]. Nihon Sanka Fujinka Gakkai Zasshi 1992;44(6):650–6.
[29] Hirahara F, Andoh N, Sawai K, Hirabuki T, Uemura T, Minaguchi H. Hyperprolactinemic recurrent miscarriage and results of randomized bromocriptine treatment trials. Fertil Steril 1998;70(2):246–52.
[30] Chen D, Lao M, Zhang J, et al. Fetal and maternal outcomes of planned pregnancy in patients with systemic lupus erythematosus: a retrospective multicenter study. J Immunol Res 2018;2018:2413637.
[31] De Carolis S, Salvi S, Botta A, et al. The impact of primary Sjogren's syndrome on pregnancy outcome: our series and review of the literature. Autoimmun Rev 2014;13(2):103–7.
[32] Ota K, Dambaeva S, Lee J, Gilman-Sachs A, Beaman K, Kwak-Kim J. Persistent high levels of IgM antiphospholipid antibodies in a patient with recurrent pregnancy losses and rheumatoid arthritis. Am J Reprod Immunol 2014;71(3):286–92.

[33] Blagojevic J, AlOdhaibi KA, Aly AM, et al. Pregnancy in systemic sclerosis: results of a systematic review and metaanalysis. J Rheumatol 2020;47(6):881–7.

[34] Twig G, Shina A, Amital H, Shoenfeld Y. Pathogenesis of infertility and recurrent pregnancy loss in thyroid autoimmunity. J Autoimmun 2012;38(2–3):J275–81.

[35] Gellersen B, Brosens JJ. Cyclic decidualization of the human endometrium in reproductive health and failure. Endocr Rev 2014;35(6):851–905.

[36] Jung J, Zeng H, Horng T. Metabolism as a guiding force for immunity. Nat Cell Biol 2019;21(1):85–93.

[37] Pearce EL, Pearce EJ. Metabolic pathways in immune cell activation and quiescence. Immunity. 2013;38 (4):633–43.

[38] Rand JH, Wu XX, Andree HA, et al. Pregnancy loss in the antiphospholipid-antibody syndrome–a possible thrombogenic mechanism. N Engl J Med 1997;337(3):154–60.

[39] Kwak-Kim J, Agcaoili MSL, Aleta L, et al. Management of women with recurrent pregnancy losses and antiphospholipid antibody syndrome. Am J Reprod Immunol 2013; Published online March.

[40] Pantham P, Abrahams VM, Chamley LW. The role of anti-phospholipid antibodies in autoimmune reproductive failure. Reproduction. 2016;151(5):R79–90.

[41] Perricone C, De Carolis C, Giacomelli R, et al. High levels of NK cells in the peripheral blood of patients affected with anti-phospholipid syndrome and recurrent spontaneous abortion: a potential new hypothesis. Rheumatol (Oxf) 2007;46(10):1574–8.

[42] Gabrilovac J, Zadjelović J, Osmak M, Suchanek E, Zupanović Z, Boranić M. NK cell activity and estrogen hormone levels during normal human pregnancy. Gynecol Obstet Invest 1988;25(3):165–72.

[43] The ESHRE Guideline Group on RPL, Bender Atik R, Christiansen OB, et al. ESHRE guideline: recurrent pregnancy loss. Hum Reprod Open 2018;2018(2):hoy004.

[44] Poznanski SM, Ashkar AA. What defines NK cell functional fate: phenotype or metabolism? Front Immunol 2019;10:1414.

[45] Loftus RM, Finlay DK. Immunometabolism: cellular metabolism turns immune regulator. J Biol Chem 2016;291(1):1–10.

[46] Donnelly RP, Loftus RM, Keating SE, et al. mTORC1-dependent metabolic reprogramming is a prerequisite for NK cell effector function. J Immunol 2014;193(9):4477–84.

[47] Chi H. Regulation and function of mTOR signalling in T cell fate decisions. Nat Rev Immunol 2012;12 (5):325–38.

[48] Canaud G, Bienaimé F, Tabarin F, et al. Inhibition of the mtorc pathway in the antiphospholipid syndrome. N Engl J Med 2014;371(4):303–12.

[49] Xia L, Zhou H, Wang T, et al. activation of mTOR is involved in anti-β2GPI/β2GPI-induced expression of tissue factor and IL-8 in monocytes. Thromb Res 2017;157:103–10.

[50] Hollerbach A, Müller-Calleja N, Ritter S, et al. Platelet activation by antiphospholipid antibodies depends on epitope specificity and is prevented by mTOR inhibitors. Thromb Haemost 2019;119(7):1147–53.

[51] Redecha P, Franzke CW, Ruf W, Mackman N, Girardi G. Neutrophil activation by the tissue factor/Factor VIIa/PAR2 axis mediates fetal death in a mouse model of anti-phospholipid syndrome. J Clin Invest 2008;118(10):3453–61.

[52] Wensveen FM, Jelenčić V, Valentić S, et al. NK cells link obesity-induced adipose stress to inflammation and insulin resistance. Nat Immunol 2015;16(4):376–85.

[53] Michelet X, Dyck L, Hogan A, et al. Metabolic reprogramming of natural killer cells in obesity limits antitumor responses. Nat Immunol 2018;19(12):1330–40.

[54] Zeng Q, Sun X, Xiao L, Xie Z, Bettini M, Deng T. A unique population: adipose-resident regulatory T cells. Front Immunol 2018;9:2075.

[55] Bapat SP, Myoung Suh J, Fang S, et al. Depletion of fat-resident Treg cells prevents age-associated insulin resistance. Nature. 2015;528(7580):137–41.

[56] Smith PM, Howitt MR, Panikov N, et al. The microbial metabolites, short-chain fatty acids, regulate colonic treg cell homeostasis. Science. 2013;341(6145):569–73.

[57] Wang K, Wang DC, Feng YQ, Xiang-Feng L. Changes in cytokine levels and CD4 + /CD8 + T cells ratio in draining lymph node of burn wound. J Burn Care Res 2007;28(5):747–53.

[58] Deng Y, Scherer PE. Adipokines as novel biomarkers and regulators of the metabolic syndrome. Ann N Y Acad Sci 2010;1212:E1–19.

[59] Juan CC, Chen KH, Chen CW, et al. Increased regulated on activation, normal T-cell expressed and secreted levels and cysteine-cysteine chemokine receptor 5 upregulation in omental adipose tissue and peripheral blood mononuclear cells are associated with testosterone level and insulin resistance in polycystic ovary syndrome. Fertil Steril 2021;116(4):1139–46.

[60] Nishimura S, Manabe I, Nagasaki M, et al. CD8 + effector T cells contribute to macrophage recruitment and adipose tissue inflammation in obesity. Nat Med 2009;15(8):914–20.

[61] Lauterbach MAR, Wunderlich FT. Macrophage function in obesity-induced inflammation and insulin resistance. Pflug Arch - Eur J Physiol 2017;469(3–4):385–96.

[62] Kim JY, Song H, Kim H, et al. Transcriptional profiling with a pathway-oriented analysis identifies dysregulated molecular phenotypes in the endometrium of patients with polycystic ovary syndrome. J Clin Endocrinol & Metab 2009;94(4):1416–26.

[63] Galazis N, Docheva N, Nicolaides KH, Atiomo W. Proteomic biomarkers of preterm birth risk in women with polycystic ovary syndrome (PCOS): a systematic review and biomarker database integration. PLoS One 2013;8(1):e53801.

[64] Lewis RD, Narayanaswamy AK, Farewell D, Rees DA. Complement activation in polycystic ovary syndrome occurs in the postprandial and fasted state and is influenced by obesity and insulin sensitivity. Clin Endocrinol (Oxf) 2021;94(1):74–84.

[65] Nojima I, Eikawa S, Tomonobu N, et al. Dysfunction of CD8 + PD-1 + T cells in type 2 diabetes caused by the impairment of metabolism-immune axis. Sci Rep 2020;10(1):14928.

[66] Tsai S, Clemente-Casares X, Zhou AC, et al. Insulin receptor-mediated stimulation boosts T cell immunity during inflammation and infection. Cell Metab 2018;28(6):922–34 e4.

[67] Bussen S, Steck T. Thyroid autoantibodies in euthyroid non-pregnant women with recurrent spontaneous abortions. Hum Reprod 1995;10(11):2938–40.

[68] Bliddal S, Feldt-Rasmussen U, Rasmussen ÅK, et al. Thyroid peroxidase antibodies and prospective live birth rate: a cohort study of women with recurrent pregnancy loss. Thyroid. 2019;29(10):1465–74.

[69] Muller AF, Verhoeff A, Mantel MJ, De Jong FH, Berghout A. Decrease of free thyroxine levels after controlled ovarian hyperstimulation. J Clin Endocrinol Metab 2000;85(2):545–8.

[70] Cavagna M, Mantese JC. Biomarkers of endometrial receptivity—a review. placenta. 2003;24(suppl b):s39–47.

[71] kim ny, cho hj, kim hy, et al. thyroid autoimmunity and its association with cellular and humoral immunity in women with reproductive failures: thyroid autoimmunity and reproductive failures. Am J Reprod Immunol 2011;65(1):78–87.

[72] Miko E, Meggyes M, Doba K, et al. Characteristics of peripheral blood NK and NKT-like cells in euthyroid and subclinical hypothyroid women with thyroid autoimmunity experiencing reproductive failure. J Reprod Immunol 2017;124:62–70.

[73] Bizzaro N, Tonutti E, Villalta D, Tampoia M, Tozzoli R. Prevalence and clinical correlation of anti-phospholipid-binding protein antibodies in anticardiolipin-negative patients with systemic lupus erythematosus and women with unexplained recurrent miscarriages. Arch Pathol Lab Med 2005;129(1):61–8.

[74] Penna G, Adorini L. 1α,25-dihydroxyvitamin D $_3$ inhibits differentiation, maturation, activation, and survival of dendritic cells leading to impaired alloreactive T cell activation. J Immunol 2000;164(5):2405–11.

[75] Boonstra A, Barrat FJ, Crain C, Heath VL, Savelkoul HFJ, O'Garra A. 1α,25-dihydroxyvitamin D3 has a direct effect on naive CD4 $^+$ T cells to enhance the development of Th2 cells. J Immunol 2001;167(9):4974–80.

[76] Sammaritano LR. Anti-phospholipid syndrome. Best Pract & Res Clin Rheumatol 2020;34(1):101463.

[77] Otten J, Ahrén B, Olsson T. Surrogate measures of insulin sensitivity vs the hyperinsulinaemic-euglycaemic clamp: a meta-analysis. Diabetologia. 2014;57(9):1781–8.

[78] Fauser BCJM, Tarlatzis BC, Rebar RW, et al. Consensus on women's health aspects of polycystic ovary syndrome (PCOS): the Amsterdam ESHRE/ASRM-sponsored 3rd PCOS consensus workshop group. Fertil Steril 2012;97(1):28–38 e25.

[79] Alexander EK, Pearce EN, Brent GA, et al. Guidelines of the American thyroid association for the diagnosis and management of thyroid disease during pregnancy and the postpartum. Thyroid 2017;27(3):315–89 2017.

[80] Heilmann L, von Tempelhoff GF, Pollow K. Antiphospholipid syndrome in obstetrics. Clin Appl Thromb Hemost 2003;9(2):143–50.

[81] Ziakas PD, Pavlou M, Voulgarelis M. Heparin treatment in antiphospholipid syndrome with recurrent pregnancy loss: a systematic review and meta-analysis. Obstet & Gynecol 2010;115(6):1256–62.

[82] Sevim E, Willis R, Erkan D. Is there a role for immunosuppression in anti-phospholipid syndrome? Hematol Am Soc Hematol Educ Program 2019;2019(1):426–32.
[83] Vahedian-Azimi A, Makvandi S, Banach M, Reiner Ž, Sahebkar A. Fetal toxicity associated with statins: a systematic review and meta-analysis. Atherosclerosis. 2021;327:59–67.
[84] Liu L, Huang X, Xu C, et al. Decidual CD8 + T cells exhibit both residency and tolerance signatures modulated by decidual stromal cells. J Transl Med 2020;18(1):221.
[85] Powell JD, Delgoffe GM. The mammalian target of rapamycin: linking T cell differentiation, function, and metabolism. Immunity. 2010;33(3):301–11.
[86] Araki K, Turner AP, Shaffer VO, et al. mTOR regulates memory CD8 T-cell differentiation. Nature. 2009;460 (7251):108–12.
[87] Yang X, Meng T. Is there a role of intravenous immunoglobulin in immunologic recurrent pregnancy loss? Bayry J, (ed.) J Immunol Res 2020;2020:1–14.
[88] De Carolis S, Tabacco S, Rizzo F, et al. Anti-phospholipid syndrome: an update on risk factors for pregnancy outcome. Autoimmun Rev 2018;17(10):956–66.
[89] Practice Committee of the American Society for Reproductive Medicine. Electronic address: ASRM@asrm.org, practice committee of the American society for reproductive medicine. Role of metformin for ovulation induction in infertile patients with polycystic ovary syndrome (PCOS): a guideline. Fertil Steril 2017;108 (3):426–41.
[90] Attia GR, Rainey WE, Carr BR. Metformin directly inhibits androgen production in human thecal cells. Fertil Steril 2001;76(3):517–24.
[91] Jakubowicz DJ, Iuorno MJ, Jakubowicz S, Roberts KA, Nestler JE. Effects of metformin on early pregnancy loss in the polycystic ovary syndrome. J Clin Endocrinol Metab 2002;87(2):524–9.
[92] Løvvik TS, Carlsen SM, Salvesen Ø, et al. Use of metformin to treat pregnant women with polycystic ovary syndrome (PregMet2): a randomised, double-blind, placebo-controlled trial. Lancet Diabetes Endocrinol 2019;7(4):256–66.
[93] Metwally M, Amer S, Li TC, Ledger WL. An RCT of metformin vs orlistat for the management of obese anovulatory women. Hum Reprod 2009;24(4):966–75.
[94] Pereira FV, Melo ACL, Low JS, et al. metformin exerts antitumor activity via induction of multiple death pathways in tumor cells and activation of a protective immune response. Oncotarget. 2018;9(40):25808–25.
[95] Xiong F, Xiao J, Bai Y, Zhang Y, Li Q, Lishuang X. Metformin inhibits estradiol and progesterone-induced decidualization of endometrial stromal cells by regulating expression of progesterone receptor, cytokines and matrix metalloproteinases. Biomed Pharmacother 2019;109:1578–85.
[96] Pearce EL, Walsh MC, Cejas PJ, et al. Enhancing CD8 T-cell memory by modulating fatty acid metabolism. Nature. 2009;460(7251):103–7.
[97] Zhang C, Liao Y, Liu P, et al. FABP5 promotes lymph node metastasis in cervical cancer by reprogramming fatty acid metabolism. Theranostics. 2020;10(15):6561–80.
[98] Zakaria Z, Othman ZA, Bagi Suleiman J, Jalil NAC, Ghazali WSW, Mohamed M. Protective and therapeutic effects of orlistat on metabolic syndrome and oxidative stress in high-fat diet-induced metabolic dysfunction-associated fatty liver disease (MAFLD) in rats: role on Nrf2 activation. Vet Sci 2021;8(11):274.
[99] Dong AC, Morgan J, Kane M, Stagnaro-Green A, Stephenson MD. Subclinical hypothyroidism and thyroid autoimmunity in recurrent pregnancy loss: a systematic review and meta-analysis. Fertil Steril 2020;113 (3):587–600 e1.
[100] Wang X, Zhang Y, Tan H, et al. Effect of levothyroxine on pregnancy outcomes in women with thyroid autoimmunity: a systematic review with meta-analysis of randomized controlled trials. Fertil Steril 2020;114 (6):1306–14.

CHAPTER

25

Endometriosis and reproductive failures

Gentaro Izumi, Kaori Koga and Yutaka Osuga

Department of Obstetrics and Gynecology, Graduate School of Medicine, The University of Tokyo, Tokyo, Japan

1 Introduction

Endometriosis is characterized by the implantation and growth of endometrial tissues outside the uterus, such as ovaries, tubes, and other pelvic peritoneal organs, including the peritoneal wall. It is a common gynecological condition in women of reproductive age and causes dysmenorrhea and chronic pelvic pain. Endometriosis is known to cause infertility, and large-scale studies have suggested the prevalence rate of the condition to be 25%–40% in patients with infertility [1].

The correlation between endometriosis and reproductive failure is still under discussion. Several studies have suggested that endometriosis can damage the endometrial function, leading to recurrent pregnancy loss and implantation failure. A *meta*-analysis of in vitro fertilization (IVF) outcomes showed lower implantation rates in patients with endometriosis than in control patients [2]. Endometriosis does not affect embryo aneuploidy rates but implantation rates, patients with endometriosis are thought to have implantation failure. However, a recent study showed that live birth rates after frozen embryo transfers of euploid blastocysts of endometriosis patients were similar to those of controls [3]. Using the Society for Assisted Reproductive Technologies Database, Senapati et al. analyzed the IVF outcome data and concluded that isolated endometriosis was not a risk factor for early pregnancy losses, but endometriosis with tubal factor was associated with increased risk of early pregnancy loss [4]. Patients with endometriosis and concomitant infertility diagnoses such as tubal, unexplained infertility, and all other diagnoses, had significantly decreased live birth rates than infertile women without endometriosis [4]. These recent studies suggest that endometriosis, when associated with other reproductive tract alterations, has the lowest chance of live birth. Therefore, treatment for endometriosis-related infertility may enhance the chance of pregnancy and live birth in

Immunology of Recurrent Pregnancy Loss and Implantation Failure
DOI: https://doi.org/10.1016/B978-0-323-90805-4.00004-3

patients with implantation failure or recurrent pregnancy loss. Thus, this chapter discusses the treatment options for endometriosis-related infertility that are considered effective.

2 Mechanism of endometriosis-related reproductive failure

Numerous studies have reported mechanisms by which endometriosis causes infertility. One major mechanism is tubal dysfunction of oocyte pick-up and embryo transportation induced by fibrosis, adhesion, and chronic inflammation of the tube. Endometriosis is also thought to alter endometrial receptivity, which can cause reproductive failure.

The mechanism of endometriosis-related endometrial dysfunction remains unclear. However, several studies have proposed hypotheses (Fig. 1). Aberrant gene expression in the endometrium of patients with endometriosis has been reported for genes that play an important role in implantation in several studies [5]. In addition, progesterone resistance in the endometrium of endometriosis patients causes an unopposed estrogen state that is not suitable for implantation [6]. These abnormal molecular and hormonal dysfunctions are thought to impair endometrial receptivity in patients with endometriosis.

3 Immune aspect of endometriosis-related pregnancy failure

3.1 Chronic endometritis

Recently, a strong association between chronic endometritis and endometriosis has been reported [7]. Chronic endometritis is a persistent inflammation of the uterine endometrium, which is often induced by altered endometrial microbiota. As a higher incidence rate of chronic endometritis has been reported among patients with implantation failures

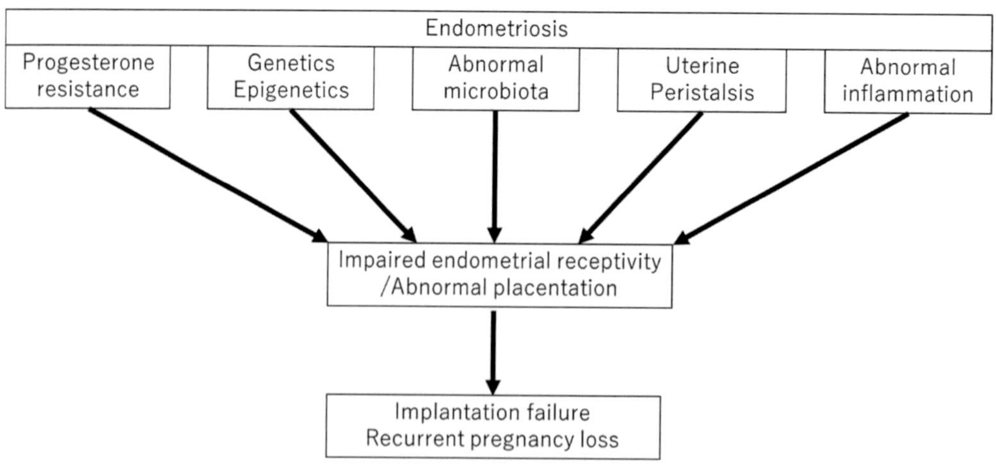

FIGURE 1 Mechanism of endometriosis-related implantation failure and recurrent pregnancy loss.

and recurrent pregnancy loss [8,9], some endometriosis-related reproductive failures reported in the past might be associated with the coexistent chronic endometritis.

3.2 Cytokines

Abnormal inflammatory status or immune cell dysfunctions have been reported in endometriosis patients. In the peritoneal fluid of endometriosis patients, various cytokine levels were higher than in control patients, including interleukin (IL)-1, IL-6, IL-8, IL-10, and tumor necrosis factor (TNF)-α [10]. Immune cell dysfunction may be a possible cause of these cytokine dysregulations. Macrophages from endometriosis patients are characterized by higher activation status [11] and producing increased proinflammatory cytokines [12]. Mast cells, neutrophils, and dendritic cells are also reported to secrete various proinflammatory cytokines, participating in endometriosis pathologies [13,14]. These proinflammatory cytokines are necessary for fertility; however, aberrant levels of these cytokines are thought to be harmful for reproduction. For instance, IL-6 plays an essential role in trophoblast attachment and implantation; however, its activity must be well regulated to avoid implantation failure or recurrent miscarriages [15].

3.3 Immune cell and surveillance against retrograde menstruation

Inappropriate immune response against retrograde menstruation is an important mechanism of endometriosis development. Sampson proposed that endometriosis is caused by the implantation of endometrial tissue from retrograde menstruation [16], and this theory has been widely accepted. However, retrograde menstruation is observed in most women of reproductive age. For endometriosis lesion formation, immune cells play an important role in enhancing endometrial cell survival, promoting their implantation, and inducing angiogenesis in endometriosis lesions [17]. Natural killer (NK) cell cytotoxicity [18,19] and phagocytic activity of macrophages [20,21] are reduced in endometriosis patients, helping endometrial cell survival. Macrophages are a key source of vascular endothelial growth factor [22], which promotes the growth of blood vessels in the endometriosis lesion. However, the pathogenetic step of endometriotic lesions is similar to that of trophoblast implantation. Thus, the treatment for immune cell dysfunctions in women with endometriosis may inhibit embryo implantation, making immune therapy for endometriosis-related infertility difficult.

4 Treatment options

4.1 Hormonal treatment

As hormonal treatment is commonly used to relieve the pain symptoms of endometriosis, a Cochrane review reported that hormonal treatments for ovarian suppression, such as gonadotropin-releasing hormone (GnRH) agonists, oral contraception, and danazol, are not effective in improving fertility in infertile women with endometriosis [23]. Thus, hormonal treatment should not be offered to women with endometriosis who are trying to conceive, as these treatments postpone pregnancy and cause side effects.

4.2 Hysterosalpingography

Hysterosalpingography is a diagnostic test used to assess tubal patency and is a part of infertility evaluation in several cases. In addition, it has been suggested that tubal flushing with an oil-soluble contrast medium directly increases pregnancy rates in the months after hysterosalpingography [24]. Johnson et al. reported that the fertility-enhancing effect of oil-soluble contrast medium in endometriosis patients was higher than that in patients without confirmed endometriosis [25]. The mechanism of the fertility-enhancing effect of oil-based contrast media is still unclear. The proposed mechanism of the enhancing fertility effect includes (1) flushing debris in fallopian tubes, (2) modulation of peritoneal immune profiles, such as macrophages, dendritic cells, and regulatory T cells [26–28], and (3) manipulating endometrial receptivity [29]. In fact, a fertility-enhancing effect was reported in patients without oil-based contrast medium spilling from the fallopian tubes, which supports the therapeutic effect of this medium for implantation failure [30]. In contrast, one RCT reported that hysterosalpingography before IVF cycles did not improve live birth rates in patients with endometriosis or recurrent implantation failures [31].

4.3 Immune therapy

Several immune therapies have been tried; however, there is still not enough evidence to recommend immune therapy as a standard treatment. Two studies reported that corticosteroid use in the IVF cycle might be effective for endometriosis patients with positive autoantibodies [32,33]. Onalan et al. reported that the pregnancy rate and live birth rate were higher in the TNF-α antagonist treatment group than controls in a retrospective analysis of the TNF-α antagonist treatment during the IVF cycle [34]. However, no randomized controlled studies are available to confirm the benefit of these drugs. Further studies are needed.

4.4 Laparoscopic surgery

Laparoscopic surgery is both diagnostic and therapeutic for endometriosis. However, surgical treatment of infertility associated with endometriosis has a limited effect on endometriosis-related pregnancy failure. A *meta*-analysis concluded that laparoscopic excision and ablation of endometriosis reduced overall pain and increased pregnancy rates [35]. However, there is limited evidence to support the improvement of pregnancy loss. A randomized controlled study (RCT) showed that laparoscopic surgery for minimal or mild endometriosis did not affect fetal loss [36]. Further, in patients undergoing IVF treatment, no differences were found in pregnancy rate and occurrence of miscarriage between the surgery and expectant management groups [37]. There is much evidence that surgery is usually the treatment of choice. However, little evidence supports the superiority of surgical treatment over assisted reproductive technology for reproductive outcomes.

4.5 In vitro fertilization protocol options

IVF is an important and the most effective treatment for endometriosis-related infertility. However, there is currently no established IVF protocol for patients with endometriosis. Non-randomized studies have reported that the protocols using GnRH agonist and antagonist for controlled ovarian stimulation had similar effects on IVF outcomes in patients with endometriosis [38,39].

Several studies have evaluated the effects of medical treatment before ovarian stimulation to improve the live birth rate. In 2002, Surrey et al. reported that a 3-month course of GnRH agonist (leuprolide acetate 3.75 mg i.m. every 28 days, three times) increased ongoing pregnancy rates [40]. Noteworthy, they also reported an improvement in the implantation rate after GnRH agonist administration. Several similar studies have been published, and a *meta*-analysis published in 2006 concluded that pituitary downregulation for three or more months before the IVF cycle for endometriosis patients increased the live birth rate [41]. However, recent studies showed no difference in pregnancy rate between the GnRH agonist pretreatment group and the control group [42]. A *meta*-analysis published in 2019 concluded that compared to no pretreatment, long-term GnRH agonist therapy before IVF/intracytoplasmic sperm injection in women with endometriosis affects the live birth rate and overall complication rate [43]. As the criteria (such as diagnosis of endometriosis, stages of endometriosis, and coexistence of endometrioma) of these RCTs varied, these results are considered very low-quality evidence. Several studies have challenged the pretreatment with progestins, and as in the case of GnRH agonists, these results are controversial as well. Tamura et al. reported that pretreatment with dienogest, a fourth-generation progestin, did not affect the live-birth rate in patients with endometriosis undergoing IVF cycles [44]. In contrast, Barra et al. reported that pretreatment with dienogest improved IVF outcomes [45]. Interestingly, all embryos used in previous studies were frozen; [44] however, in the latter, fresh embryos were transferred [45]. From these data, it is speculated that pretreatment with GnRH agonists or progestins immediately before embryo transfer may contribute to enhancing fertility in patients with endometriosis. Well designed clinical trials are needed in the future.

4.6 Fresh versus frozen embryo transfer

The IVF outcomes between fresh embryo transfer and frozen embryo transfer in patients with endometriosis have also been compared in several studies. Wu et al. retrospectively compared matched cohorts in terms of the live birth and implantation rates and found both to be higher in the frozen embryo transfer group than in the fresh embryo transfer group [46]. Bourdon et al. also reported that the frozen embryo transfer group showed higher live birth and implantation rates than the fresh embryo transfer group in patients with endometriosis [47]. In contrast, Tan et al. reported that there was no benefit of the frozen embryo transfer IVF strategy in patients with endometriosis. Both studies were retrospectively matched cohort studies. Further studies, especially well-designed RCTs, are needed on this matter. Frozen embryo transfer after pretreatment with a GnRH agonist or progestins might be a treatment option for patients with endometriosis.

5 Summary and recommendations

- It is still controversial whether endometriosis is a cause of implantation failure and recurrent pregnancy loss. However, several studies have suggested that endometriosis decreases the receptivity of the endometrium.
- There is no specific treatment for implantation failure or recurrent pregnancy loss in patients with endometriosis. However, treatments for endometriosis-related infertility are also effective for these patients.
- IVF is the most effective fertility treatment for endometriosis. Frozen embryo transfer and pretreatment with GnRH agonists or progestins may enhance fertility.

References

[1] Ozkan S, Murk W, Arici A. Endometriosis and infertility: epidemiology and evidence-based treatments. Ann N Y Acad Sci 2008;1127:92–100.
[2] Barnhart K, Dunsmoor-Su R, Coutifaris C. Effect of endometriosis on in vitro fertilization. Fertil Steril 2002;77 (6):1148–55.
[3] Bishop LA, et al. Endometriosis does not impact live-birth rates in frozen embryo transfers of euploid blastocysts. Fertil Steril 2021;115(2):416–22.
[4] Senapati S, et al. Impact of endometriosis on in vitro fertilization outcomes: an evaluation of the Society for assisted reproductive technologies database. Fertil Steril 2016;106(1):164–71 e1.
[5] Macer ML, Taylor HS. Endometriosis and infertility: a review of the pathogenesis and treatment of endometriosis-associated infertility. Obstet Gynecol Clin North Am 2012;39(4):535–49.
[6] Lessey BA, Kim JJ. Endometrial receptivity in the eutopic endometrium of women with endometriosis: it is affected, and let me show you why. Fertil Steril 2017;108(1):19–27.
[7] Takebayashi A, et al. The association between endometriosis and chronic endometritis. PLoS One 2014;9(2): e88354.
[8] Kimura F, et al. Review: chronic endometritis and its effect on reproduction. J Obstet Gynaecol Res 2019;45 (5):951–60.
[9] McQueen DB, Bernardi LA, Stephenson MD. Chronic endometritis in women with recurrent early pregnancy loss and/or fetal demise. Fertil Steril 2014;101(4):1026–30.
[10] Wu MY, Ho HN. The role of cytokines in endometriosis. Am J Reprod Immunol 2003;49(5):285–96.
[11] Lousse JC, et al. Increased activation of nuclear factor-kappa B (NF-kappaB) in isolated peritoneal macrophages of patients with endometriosis. Fertil Steril 2008;90(1):217–20.
[12] Montagna P, et al. Peritoneal fluid macrophages in endometriosis: correlation between the expression of estrogen receptors and inflammation. Fertil Steril 2008;90(1):156–64.
[13] Binda MM, Donnez J, Dolmans MM. Targeting mast cells: a new way to treat endometriosis. Expert Opin Ther Targets 2017;21(1):67–75.
[14] Izumi G, et al. Involvement of immune cells in the pathogenesis of endometriosis. J Obstet Gynaecol Res 2018;44(2):191–8.
[15] Prins JR, Gomez-Lopez N, Robertson SA. Interleukin-6 in pregnancy and gestational disorders. J Reprod Immunol 2012;95(1–2):1–14.
[16] Sampson JA. Metastatic or embolic endometriosis, due to the menstrual dissemination of endometrial tissue into the venous circulation. Am J Pathol 1927;3(2):93–110 43.
[17] Giudice LC, Kao LC. Endometriosis. Lancet 2004;364(9447):1789–99.
[18] Thiruchelvam U, Wingfield M, O'Farrelly C. Natural killer cells: key players in endometriosis. Am J Reprod Immunol 2015;74(4):291–301.
[19] Kang YJ, et al. An increased level of IL-6 suppresses NK cell activity in peritoneal fluid of patients with endometriosis via regulation of SHP-2 expression. Hum Reprod 2014;29(10):2176–89.
[20] Chuang PC, et al. Inhibition of CD36-dependent phagocytosis by prostaglandin E2 contributes to the development of endometriosis. Am J Pathol 2010;176(2):850–60.

[21] Wu MH, et al. Suppression of annexin A2 by prostaglandin E(2) impairs phagocytic ability of peritoneal macrophages in women with endometriosis. Hum Reprod 2013;28(4):1045–53.

[22] McLaren J, et al. Vascular endothelial growth factor is produced by peritoneal fluid macrophages in endometriosis and is regulated by ovarian steroids. J Clin Invest 1996;98(2):482–9.

[23] Hughes E, et al. Ovulation suppression for endometriosis. Cochrane Database Syst Rev 2007;(3):CD000155.

[24] Wang R, et al. Tubal flushing for subfertility. Cochrane Database Syst Rev 2020;10:CD003718.

[25] Johnson NP, et al. The FLUSH trial–flushing with lipiodol for unexplained (and endometriosis-related) subfertility by hysterosalpingography: a randomized trial. Hum Reprod 2004;19(9):2043–51.

[26] Johnson JV, Montoya IA, Olive DL. Ethiodol oil contrast medium inhibits macrophage phagocytosis and adherence by altering membrane electronegativity and microviscosity. Fertil Steril 1992;58(3):511–17.

[27] Sawatari Y, Horii T, Hoshiai H. Oily contrast medium as a therapeutic agent for infertility because of mild endometriosis. Fertil Steril 1993;59(4):907–11.

[28] Izumi G, et al. Oil-soluble contrast medium (OSCM) for hysterosalpingography modulates dendritic cell and regulatory T cell profiles in the peritoneal cavity: a possible mechanism by which OSCM enhances fertility. J Immunol 2017;198(11):4277–84.

[29] Johnson NP, et al. Lipiodol alters murine uterine dendritic cell populations: a potential mechanism for the fertility-enhancing effect of lipiodol. Fertil Steril 2005;83(6):1814–21.

[30] Court KA, et al. Establishment of lipiodol as a fertility treatment—prospective study of the complete innovative treatment data set. Aust N Z J Obstet Gynaecol 2014;54(1):13–19.

[31] Reilly SJ, et al. The IVF-LUBE trial—a randomized trial to assess lipiodol((R)) uterine bathing effect in women with endometriosis or repeat implantation failure undergoing IVF. Reprod Biomed Online 2019;38 (3):380–6.

[32] Kim CH, et al. The immunotherapy during in vitro fertilization and embryo transfer cycles in infertile patients with endometriosis. J Obstet Gynaecol Res 1997;23(5):463–70.

[33] Dmowski WP, et al. The effect of endometriosis, its stage and activity, and of autoantibodies on in vitro fertilization and embryo transfer success rates. Fertil Steril 1995;63(3):555–62.

[34] Onalan G, Tohma YA, Zeyneloglu HB. Effect of etanercept on the success of assisted reproductive technology in patients with endometrioma. Gynecol Obstet Invest 2018;83(4):358–64.

[35] Bafort C, et al. Laparoscopic surgery for endometriosis. Cochrane Database Syst Rev 2020;10:CD011031.

[36] Marcoux S, Maheux R, Berube S. Laparoscopic surgery in infertile women with minimal or mild endometriosis. Canadian collaborative group on endometriosis. N Engl J Med 1997;337(4):217–22.

[37] Benschop L, et al. Interventions for women with endometrioma prior to assisted reproductive technology. Cochrane Database Syst Rev 2010;(11):CD008571.

[38] Drakopoulos P, et al. Does the type of GnRH analogue used, affect live birth rates in women with endometriosis undergoing IVF/ICSI treatment, according to the rAFS stage? Gynecol Endocrinol 2018;34(10):884–9.

[39] Rodriguez-Purata J, et al. Endometriosis and IVF: are agonists really better? Analysis of 1180 cycles with the propensity score matching. Gynecol Endocrinol 2013;29(9):859–62.

[40] Surrey ES, et al. Effect of prolonged gonadotropin-releasing hormone agonist therapy on the outcome of in vitro fertilization-embryo transfer in patients with endometriosis. Fertil Steril 2002;78(4):699–704.

[41] Sallam HN, et al. Long-term pituitary down-regulation before in vitro fertilization (IVF) for women with endometriosis. Cochrane Database Syst Rev 2006;(1):CD004635.

[42] Decleer W, et al. RCT to evaluate the influence of adjuvant medical treatment of peritoneal endometriosis on the outcome of IVF. Hum Reprod 2016;31(9):2017–23.

[43] Georgiou EX, et al. Long-term GnRH agonist therapy before in vitro fertilisation (IVF) for improving fertility outcomes in women with endometriosis. Cochrane Database Syst Rev 2019;2019(11).

[44] Tamura H, et al. The clinical outcome of dienogest treatment followed by in vitro fertilization and embryo transfer in infertile women with endometriosis. J Ovarian Res 2019;12(1):123.

[45] Barra F, et al. Pretreatment with dienogest in women with endometriosis undergoing IVF after a previous failed cycle. Reprod Biomed Online 2020;41(5):859–68.

[46] Wu J, et al. Fertility and neonatal outcomes of freeze-all vs. fresh embryo transfer in women with advanced endometriosis. Front Endocrinol (Lausanne) 2019;10:770.

[47] Bourdon M, et al. The deferred embryo transfer strategy improves cumulative pregnancy rates in endometriosis-related infertility: a retrospective matched cohort study. PLoS One 2018;13(4):e0194800.

CHAPTER

26

Laboratory approaches for reproductive failure: immunological biomarkers for reproductive failures

Katharine Wolf, Kenneth Beaman, Svetlana Dambaeva and Alice Gilman-Sachs

Clinical Immunology Laboratory, Faculty of Microbiology and Immunology, Center for Cancer Biology, Infection and Immunology, The Chicago Medical School, Rosalind Franklin University of Medicine and Science, Chicago, IL, United States

1 Genomic evaluation

1.1 Human leukocyte antigen-C and natural killer cells

The key question of pregnancy asks why the fetus is not rejected by the mother's immune system. The expression of human leukocyte antigen (HLA) genes within the major histocompatibility complex class I (MHC-I) proteins that regulate the acceptance or rejection of transplanted organs were believed to also play a role in pregnancy. However, early data suggested that the key transplantation antigens, HLA-A and -B, were not involved in the maternal immune response to the fetus during normal pregnancy [1–3]. There is, nevertheless, a relationship between MHC class I loci and reproduction. In pregnancy, HLA-C, -G, and -E molecules regulate maternal–fetal immune responses [4,5].

While traditional transplantation responses do not apply to the fetal placental units as anticipated, natural killer (NK) cells appear to be a different matter. Variation in uterine NK activity has come to the forefront of our interest in the fetal placental unit, and the uterine cyclic variation is observed during the normal menstrual cycle [6,7]. Over 6800 HLA-C alleles have been recorded in the HLA sequence database [8]. These alleles can be divided into two groups, C1 and C2, based on the amino acid sequence at residue number 80. The C1 group encodes for arginine, and the C2 group encodes lysine [4]. Since HLA-C is polymorphic, genotyping of both maternal and paternal HLA-C alleles would provide

Immunology of Recurrent Pregnancy Loss and Implantation Failure
DOI: https://doi.org/10.1016/B978-0-323-90805-4.00027-4

information on possible HLA-C group combination/combinations of the fetus and thus could be useful for the evaluation of recurrent pregnancy loss (RPL).

Fetal HLA-C binds a group of ligands found on NK cells known as killer immunoglobulin-like receptors (KIRs) (Fig. 1). Like other ligands of HLA, KIR genes are genetically linked to each other and can be activating or inhibitory of NK cell activities. Activation or inhibition of NK cell activities depends on the combination of fetal HLA-C allele and maternal KIR allele. For the sake of simplicity, KIR genotypes can be divided into two groups: AA and Bx. AA genotypes are composed of inhibitory KIR alleles, while Bx genotypes include one or more activating KIR alleles. The KIR and HLA-C genotype regulate NK activation during pregnancy. When the HLA-C2 interacts with the AA genotype, insufficient activation and subsequent lack of support for placentation occur. Other HLA-C and AA combinations have a lesser effect on pregnancy outcomes. Hence, the negative predictive value of the AA/HLA-C C2 genotype is an important clinical test in the evaluation of RPL [9].

The variation of NK cell activity observed between KIR/HLA-C combination cohorts is, therefore, due to differences in types of activation rather than activation or inhibition. Despite their name, uterine NK cells are not universally destructive, but in the case of pregnancy, they can promote immunologic tolerance of trophoblasts, induce angiogenesis, and secrete growth factors that support the growing embryo [10–12]. Two important findings come from investigations into NK/HLA interaction: (1) uterine NK cell activation is good, not bad, for supporting pregnancy and (2) therapeutic intervention is possible based on these mechanistic findings for RPL [13,14]. Genomic evaluation should therefore focus on the maternal and paternal HLA-C alleles and maternal KIR alleles. The genetic typing of HLA ligands and receptors appears to be a valuable tool in the diagnosis of RPL. While the standard HLA typing needed for transplantations may give only limited information for the evaluation of RPL, the analysis of the HLA-C along with KIR genotypes reveals useful diagnostic information.

1.2 Prevalence of abnormal genomic test

The inhibitory or activating alleles for KIRs, as well as representative diagrams for the AA and 3 Bx genotypes, are shown in Fig. 2. Dambaeva et al. conducted a study of North American women with RPL compared to a corresponding North American population. Among these women, the AA genotype was the single most common genotype with a frequency of 32.7%. The remaining women were distributed among 27 Bx genotypes, with no single genotype being more common in frequency than 13%. There was no significant difference between women with RPL and the reference population for any KIR genotype. Similar

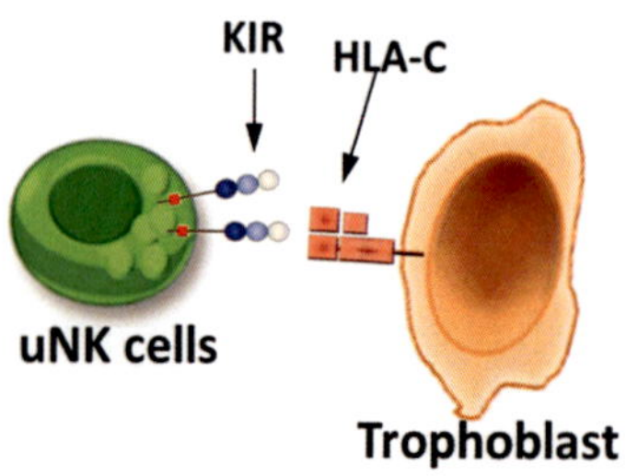

FIGURE 1 The recognition of the placental trophoblast cells is done through human leukocyte antigen (HLA) C and killer immunoglobulin-like receptor. It may also be associated with HLA E. This recognition and activation of natural killer cells are necessary for vascularization and growth of the human placenta.

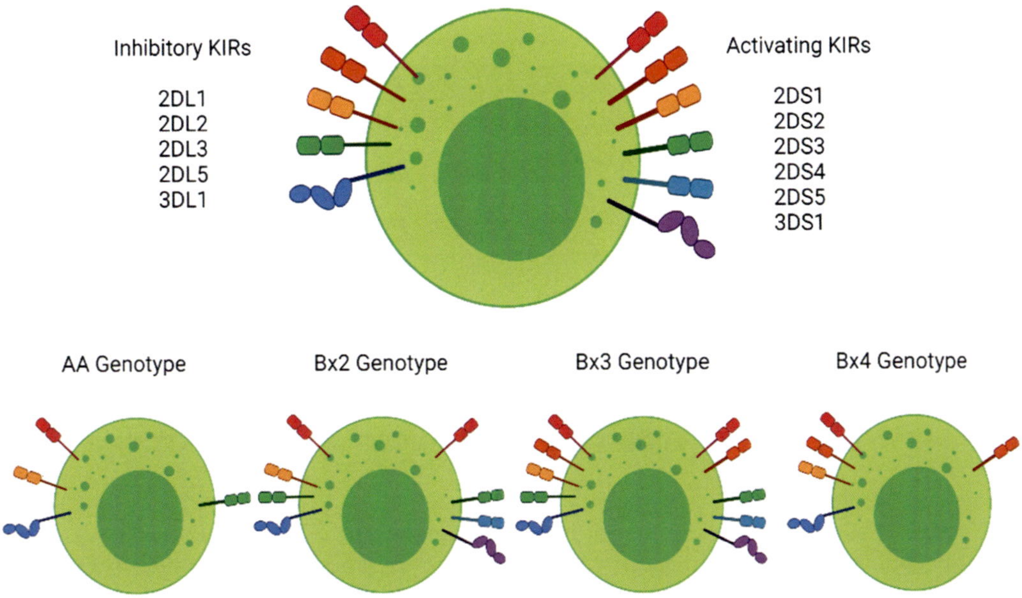

FIGURE 2 The AA or any of the 27 Bx genotypes for killer immunoglobulin-like receptors depend on the number of inhibitory or activating alleles. The figure shows some, but not all, inhibitory or activating KIRS. Representative images of the AA genotype and three Bx genotypes are shown.

to the frequency of KIRs, there was no significant difference in the allelic frequencies of HLA-C C1 or C2 between women with RPL, their partners, or the reference population [15]. Similarly, there is no significant difference between women with RPL and the reference population in frequencies HLA-C allelic combinations: C1 homozygous, C2 homozygous, or C1/C2 heterozygous frequencies are similar in both populations.

Significant differences are observed when genetic combinations of particular KIR to HLA-C alleles are tested for [9,15,16]. Women with RPL who express the activating Bx receptor, KIR2DS1, are significantly more likely to carry the HLA-C2 allele (Fig. 3). This combination (KIR2DS1 and cognate HLA-C2) is involved in NK cell education/licensing and result in the prevalence of hypofunctional NK cells in individuals carrying both genes [17]. An increased frequency of HLA-C2 is also observed among the partners of women with RPL. Finally, as stated above, a higher frequency of inhibitory KIR (AA genotype) with HLA-C2 combinations has been reported in women with RPL [18–21].

2 Local decidual measurements

2.1 Endometrial immune profile testing

2.1.1 Procedure

The EIP can be performed with an endometrial biopsy sample obtained in either a hormone replacement therapy cycle in preparation of a frozen embryo transfer or a natural

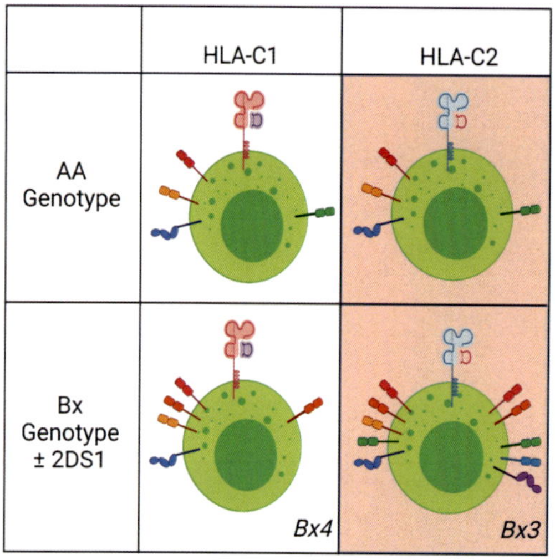

FIGURE 3 Some genotypes (highlighted in red) are significantly enriched in women with recurrent pregnancy loss (RPL) compared to the referenced population. Women with RPL who have the AA genotype are significantly more likely to also have human leukocyte antigen (HLA)-C2. Women with RPL who have the activating receptor 2DS1 (as an example women with Bx3 genotype) are significantly more likely to have HLA-C2.

cycle. Endometrial biopsy is performed according to a standard procedure. About 30 mg of endometrial tissue is enough for analysis.

1. **Hormone replacement therapy cycle:** Upon initiation of an HRT cycle, the biopsy is taken after five full days of progesterone treatment or the expected day of frozen embryo transfer [3,22,23]. The day for the first intake of progesterone is considered progesterone at day zero; the day of the biopsy is progesterone plus day 5. Therefore, if you refer to the first day as progesterone plus day one, then the day of the biopsy is progesterone plus day 6.
2. **Natural cycle:** The biopsy is taken 7–9 days after the luteinizing hormone surge. The day of the luteinizing hormone surge is considered as luteinizing hormone plus day 0, and the biopsy is taken at LH + 7–9. Therefore, the best way to identify the luteinizing hormone surge is with the urinary luteinizing hormone tests.

2.1.2 *Assay profile*

The histological measurements are limited by the number of sections taken and the quantitation of findings. While it can give spatial consideration between cells, its quantitation depends on the individual and varies between the investigators. Hence, a better quantitation technique is a molecular measurement [24–27]. Both cDNA and mRNA sequencing eliminates the human variation and the limit on the number of factors measured. The total sequence approach was the first to be promulgated. Although it is expensive and time-consuming, it has been reported to have good predictive outcomes.

The EIP is a test to evaluate the molecular immune expression of the endometrium to see if it is appropriately prepared to promote successful pregnancy. It is a simple, less expensive approach. In order for pregnancy to be successful, the endometrium must undergo two processes: decidualization and placentation. Decidualization is accompanied by substantial recruitment of maternal immune cells into the decidua, with approximately 70% of decidual lymphocytes being NK cells in the early pregnancy. In this assay, CD56 mRNA expression is measured, which correlates with the number of CD56 NK cells in a sample [6,28,29]. During placentation, embryonic extravillous trophoblastic cells invade the decidua and actively participate in spiral arteries remodeling, resulting in increased blood supply to the developing embryo/fetus.

Women experiencing repeated implantation failure (RIF) have been shown to have an excessive or decreased number of uterine NK cells. An abnormal change of endometrial levels of interleukin (IL)-15 and IL-18 as well as TNF weak inducer of apoptosis (TWEAK) and its receptor, Fn-14 (fibroblast growth factor-inducible molecule), often occurs when compared with fertile women. IL-15 and IL-18 are cytokines involved in the activation of uterine NK cells. TWEAK is a pro-angiogenic and antiinflammatory cytokine acting via Fn14 [13,20,21]. The balance between activating vs controlling signals can be measured as IL-18/TWEAK and IL-15/Fn14 ratios. An abundance of uterine NK cells is assessed via CD56 mRNA transcript quantification. The ratios, along with uterine NK cell evaluation, have been shown to provide the molecular tools that help to determine whether the endometrium is receptive to the embryo for implantation and placentation.

2.1.3 *Evaluation*

Women who have a history of RPL or RIF as well as women with a history of a thin endometrial thickness (≤ 6 mm) are candidates for this test. Although not as yet tested, there are theoretical reasons to believe that women tested before starting an assisted reproduction cycle with EIP (endometrial immune profile) and treated appropriately would experience a higher live birth rate than women not screened and treated [14,20,21,28].

2.2 Endometrial decidualization score

In studies by Dambaeva and colleagues, multiple growth factors and cytokines were studied [27] and reported that 6 of the observed factors were often abnormal in women with reproductive failures when compared with normal fertile controls. They identified predictive factors that could be modified by medical intervention to enhance the possibility of a successful pregnancy outcome. When all the factors were analyzed together by ROC analysis, the predictive probabilities of endometrial decidualization score (EDS) were enhanced. Three factors, including the transcription factor FOX01, the Serum- and Glucocorticoid-inducible Kinase 1 (SGK1), the sodium channel SCNN1A, could be enhanced by increased progesterone levels. Likewise, the glucose transporter, Solute Carrier Family 2 Member 1 (SLC2A1), is known to be upregulated in decidualized endometrium [30]. Like the work of Ledee, endometrial mRNA expression of the NK cell marker granzyme B (GZMB) and cytokine IL-15 were altered in women with IVF failures and early pregnancy losses. In EDS, these markers were combined, providing an excellent predictive value of pregnancy outcomes. Like the HLA-C and KIR typings, these findings could be incorporated into a treatment that can intervene and prevent pregnancy losses.

3 Peripheral blood cell analysis

While measuring immune parameters in the blood to detect changes in the pregnant uterus is not ideal, it is often the best we can do. Someone once said: it is like seeing fire engines going down the road with red lights and sirens. It might be an emergency, or it could be a parade. Nevertheless, it can provide important information in ascertaining the success or failure of the pregnancy. For over 20 years, reproductive immunologists have assessed pregnancy by examining T and B cells and nearly that long evaluating the NK cell populations [24,25,31].

3.1 T helper (Th) 1/Th2 cell ratio

Original evaluations for RPL were testing of the immune response to women who were treated with paternal lymphocyte immunizations. This treatment was aimed at improving pregnancy outcomes by desensitizing a woman to paternal antigens; however, it is no longer used in the United States due to a concern of stem cell regulation. Nevertheless, it spawned a review of the findings, which suggested that an immunoglobulin or noncytotoxic immune response was perhaps important for a successful pregnancy. Subsequent studies began to reveal that an immune response dominated by cytotoxic or Th1 types cytokines in the peripheral blood was not conducive to successful pregnancy outcomes. Several studies have shown that an optimal balance between Th1 and Th2 production is key for a proper immune response during pregnancy, and an increased Th1/Th2 ratio is linked to reproductive failures [32]. Hence, the Th1/Th2 test for peripheral blood is a useful diagnostic indicator in patients determined to have RPL [28].

While an increase in Th1 cytokines (TNF-α and IFN-γ) leads to pregnancy losses, these cytokines appear important in the local uteroplacental immune response at the appropriate time and concentrations [32]. However, these cytokines being expressed in numerous peripheral blood monocytes during early pregnancy are not indicative of a successful pregnancy outcome [28]. In studies examining the expression of PD1 in pregnancy and comparing RPL with normal pregnancy, the Th1/Th2 paradigm was shown to correspond with other markers such as Th17 and T regulatory cells. While these markers lack the statistically relevant attributes to be useful prognostic markers in a clinical setting, they do strongly support the Th1/Th2 paradigm [33].

3.2 Natural killer cell levels and cytotoxicities

Another important marker in both the peripheral blood and at the placental interface is the NK cell. The measurement of NK cells has evolved from early cell counting in peripheral blood to the analysis of their functional activity, namely to measurements of cytotoxicity. All in all, the evaluation can be most informative in detecting abnormalities in placental growth. This is true for early pregnancy, before and during implantation, and in the first trimester of fetal growth.

Originally NK cells were thought to be mononuclear cells that mediate cytolysis against cells that lack MHC Class I or II molecules. Now it is known that they also regulate the immune system, mediate natural resistance against tumor cells and produce cytokines. In the cytotoxic cell assay, mononuclear cells (containing natural kill cells) are incubated for 2 hours with target K562 cells that have been prestained with PKH-2, a lipophilic dye that binds to the cell membrane and fluoresces green [34]. If the green-K562 cells are killed by NK cells, they become permeabilized, and propidium iodide dye can enter the cell through the pores that are created by cytolysis, and the K562 target cells fluoresce bright red/orange and green. The percentage of killed target cells that contain propidium iodide can then be measured by flow cytometry. When lymphocytes are incubated with target cells as well as IgG, Intralipid, or prednisolone, inhibition of cytotoxicity may be seen.

4 Autoantibodies in recurrent pregnancy loss

4.1 Antiphospholipid antibodies

Autoantibodies such as antiphospholipid antibodies, antithyroid antibodies, or antinuclear antibodies are associated with RPL and can be easily detected and measured [18,35,36]. The antiphospholipid syndrome (APS) is a thromboembolic disease that presents with antibodies to phospholipids, particularly cardiolipin, which is the most relevant of these various autoantibodies. With the development of sensitive methods for the detection of various antiphospholipid antibodies (among which anticardiolipin antibodies are the most frequently assayed), the clinical associations of these antibodies with thromboembolic disease, thrombocytopenia, SLE, RPL, and obstetrical complications have become apparent. The association of a biological false-positive test for syphilis, lupus anticoagulant, and anticardiolipin antibodies were demonstrated in women with SLE and then RPL. A solid-phase ELISA assay for anticardiolipin antibodies and other antiphospholipid antibodies was developed, which is both sensitive and semi-quantitative.

Antiphospholipid antibodies have been characterized as heterogeneous subpopulations of cross-reacting autoantibodies that react with at least six different phospholipids, that is, cardiolipin, phosphatidylserine, phosphatidylethanolamine, phosphatidic acid, phosphatidylglycerol, phosphatidylinositol, etc. In 2006, these antibodies were shown to bind to a protein, beta-2-glycoprotein I (GPI), a carrier protein for phospholipids. They also bind to a complex of phospholipids and this protein. Assays have been devised to detect both these individual autoantibodies and autoantibodies specific for this complex. These assays have been standardized and are tested according to the CLIA 1988 regulations. Antibodies of the IgM, IgG, and IgA classes have been found for all of these antibodies in pregnant women.

The assay for antiphospholipid antibodies is a well-characterized ELISA assay. Purified β_2GPI antigen is bound to the wells of a polystyrene microwell plate under conditions that will preserve the antigen in its native state. Pre-diluted controls and diluted patient sera are added to separate wells, allowing any β_2GPI IgG antibodies present to bind to the immobilized antigen. Unbound sample is washed away, and an enzyme-labeled antihuman IgG conjugate is added to each well. A second incubation allows the enzyme-labeled antihuman IgG to bind to any patient antibodies, which have become attached to the microwells. After washing away any unbound enzyme-labeled antihuman IgG, the remaining enzyme activity is measured by adding a chromogenic substrate and measuring the intensity of the color that develops. The assay can be evaluated spectrophotometrically by measuring and comparing the color intensity that develops in the patient wells with that of a five-point calibration curve.

Patients with RPL present with laboratory results and pathology different than found in patients with classical APS syndrome, that is, levels of autoantibodies are lower and overt thrombosis is rarely seen. The relationship between these autoantibodies is thought to be through the anticoagulation/coagulation complex. Antiphospholipid antibodies can bind to GPI found on surfaces of trophoblast cells and endothelial cells and induce proinflammatory and antimigration profiles, which prevent implantation or placental growth [36]. Also, antiphospholipid antibodies can bind to complement and result in activation, subsequently interfering with the growth of the placenta by binding to the decidua and inhibiting the invasion of the trophoblast cells. Complement activation may also recruit neutrophils, which adds another layer to placental insufficiency, restricted intrauterine growth, and fetal loss. Thus, in addition to thrombosis first postulated as the role of these antibodies in pregnancy loss, these other factors have been postulated and shown to occur. Administration of anticoagulants such as heparin with or without aspirin has been used to prevent pregnancy loss when administered early in the pregnancy but not necessarily late in the pregnancy.

4.2 Anti nuclear antibodies (ANA)

Patients with RPL should also be tested for ANA, and if elevated titers are found, antibodies should be further differentiated to rule out a Sjögren's syndrome (SS-A/RO and SS-B/La) or lupus erythematosus. A high titer of a positive ANA result usually occurs in many autoimmune disorders such as systemic lupus erythematosus (SLE), mixed connective tissue disease (MCTD), rheumatoid arthritis (RA), Sjogren's syndrome (SS), and progressive systemic sclerosis (PSS). A high titer of a positive ANA result usually occurs in several autoimmune disorders such as SLE, mixed connective tissue disease (MCTD), RA, SS, and progressive systemic sclerosis (PSS).

Each of the various rheumatic diseases generally exhibits a particular ANA profile. Hence, identification of the specificity of a patient's ANAs can be extremely useful in the differential diagnosis of these diseases. Anti-Sm is a highly specific serological marker found in 20%–30% of patients with SLE. Antibodies to RNP are frequently detected in several rheumatic diseases; however, the presence of a high titer response to this antigen is usually considered diagnostic of mixed connective tissue disease. Antibodies to SS-A/Ro and SS-B/La are observed with the highest frequency in SS, although these antigen specificities are also found in a significant percentage of patients with SLE. Furthermore, there is an association between SS-A/Ro antibodies and congenital fetal heart block. All these autoantibodies (RNP, Sm, SSA/Ro, and SS-B/La) are detected by ELISA.

The mechanisms used by anti-DNA antibodies to negatively affect pregnancy outcome and result in RPL probably occur by processes identified for other pathogenic antibodies. These antibodies may bind to placental cells or other tissues and activate complement, behave as immune complexes, attract neutrophils or monocytes to a region of the maternal–fetal interface, or attract lymphoid or plasmacytoid dendritic cells, which can activate other cells to secrete proinflammatory cytokines [23]. These processes may affect and inhibit the growth of the placenta or the embryo.

4.3 Other autoantibodies

Other autoantibodies have been associated with RPL. Testing for IgA antibodies against tissue Transglutaminase (tTG-IgA test) can be considered for women with a history of food sensitivity followed by a biopsy if positive to rule out celiac disease.

5 Endocrine workup

An endocrine workup determining TSH levels is recommended in women with RPL [23,35]. If TSH levels are found to be abnormal, T3, T4, and thyroid autoantibody concentrations must be determined. Circulating thyroid autoantibodies have been widely implicated in the etiology of autoimmune thyroid disease, and both thyroglobulin and thyroid peroxidase autoantibodies are measured routinely in clinical practice. Serum autoantibodies to thyroid microsomal antigen(s) are commonly found in patients with thyroid autoimmune diseases. In addition, thyroid disorders have been associated with preterm delivery and pregnancy losses. A recent review indicated that anti-TPO antibodies could be correlated with a decreased live birth rate. Therefore, a workup for thyroid levels and autoantibodies is recommended for women with autoimmune thyroid disorders [23].

6 Other assays

Many other assays to manage RPL have been used, such as genotyping the various hereditary thrombophilia genes (FV Leiden, MTHFR, or PAI-1), quantitating other TH subsets such as T regulatory cells or TH17 cells, identifying TNF-alpha levels following administration of TNF inhibitors, and measuring extracellular vesicles to monitor RPL [37].

7 Summary and recommendations

Pregnancy loss is a complex immunological, genetic, physiological process, and the outcomes of the various tests for women with this sad outcome should be better understood to know how to treat RPL:

- Over 50% of women may remain unexplained after currently recommended testing for RPL and RIF.
- A significant proportion of women with RPL and RIF have immune abnormalities.
- KIR and HLA-C genotyping provide information on NK cells contribution to pregnancy support.
- Peripheral blood NK cell levels, cytotoxicities, and Th1/Th2 ratio testings aid to select women with immune etiology RPL and RIF.
- Autoantibody screening for APA, ANA, ATA, and antiovarian antibodies are often helpful to investigate underlying autoimmune etiologies for RPL and RIF.

References

[1] Beaman KD, Dambaeva S, Katara GK, Kulshrestha A, Gilman-Sachs A. The immune response in pregnancy and in cancer is active and supportive of placental and tumor cell growth not their destruction. Gynecol Oncol 2017;145(3):476–80. Available from: https://doi.org/10.1016/j.ygyno.2017.04.019.

[2] Beaman KD, Jaiswal MK, Katara GK, et al. Pregnancy is a model for tumors, not transplantation. Am J Reprod Immunol N Y N 1989 2016;76(1):3–7. Available from: https://doi.org/10.1111/aji.12524.

[3] Moffett A, Chazara O, Colucci F. Maternal allo-recognition of the fetus. Fertil Steril 2017;107(6):1269–72. Available from: https://doi.org/10.1016/j.fertnstert.2017.05.001.

[4] Felker AM, Croy BA. Uterine natural killer cell partnerships in early mouse decidua basalis. J Leukoc Biol 2016;100(4):645–55. Available from: https://doi.org/10.1189/jlb.1HI0515-226R.

[5] Croy BA, Zhang J, Tayade C, Colucci F, Yadi H, Yamada AT. Analysis of uterine natural killer cells in mice. Meth Mol Biol Clifton NJ 2010;612:465–503. Available from: https://doi.org/10.1007/978-1-60761-362-6_31.

[6] Trundley A, Moffett A. Human uterine leukocytes and pregnancy. Tissue Antigens 2004;63(1):1–12. Available from: https://doi.org/10.1111/j.1399-0039.2004.00170.x.

[7] Tukwasibwe S, Nakimuli A, Traherne J, et al. Variations in killer-cell immunoglobulin-like receptor and human leukocyte antigen genes and immunity to malaria. Cell Mol Immunol 2020;17(8):799–806. Available from: https://doi.org/10.1038/s41423-020-0482-z.

[8] Robinson J, Barker DJ, Georgiou X, Cooper MA, Flicek P, Marsh SGE. IPD-IMGT/HLA database. Nucleic Acids Res 2020;48(D1):D948–55. Available from: https://doi.org/10.1093/nar/gkz950.

[9] Hiby SE, Walker JJ, O'Shaughnessy KM, et al. Combinations of maternal KIR and fetal HLA-C genes influence the risk of preeclampsia and reproductive success. J Exp Med 2004;200(8):957–65. Available from: https://doi.org/10.1084/jem.20041214.

[10] Fu B, Zhou Y, Ni X, et al. Natural killer cells promote fetal development through the secretion of growth-promoting factors. Immunity 2017;47(6):1100–13. Available from: https://doi.org/10.1016/j.immuni.2017.11.018 e6.

[11] Lima PDA, Tu MM, Rahim MMA, Peng AR, Croy BA, Makrigiannis AP. Ly49 receptors activate angiogenic mouse DBA$^+$ uterine natural killer cells. Cell Mol Immunol 2014;11(5):467–76. Available from: https://doi.org/10.1038/cmi.2014.44.

[12] Sojka DK, Yang L, Yokoyama WM. Uterine natural killer cells. Front Immunol 2019;10:960. Available from: https://doi.org/10.3389/fimmu.2019.00960.

[13] Petitbarat M, Rahmati M, Sérazin V, et al. TWEAK appears as a modulator of endometrial IL-18 related cytotoxic activity of uterine natural killers. PLoS ONE 2011;6(1):e14497. Available from: https://doi.org/10.1371/journal.pone.0014497.

[14] Lédée N, Gridelet V, Ravet S, et al. Impact of follicular G-CSF quantification on subsequent embryo transfer decisions: a proof of concept study. Hum Reprod Oxf Engl 2013;28(2):406–13. Available from: https://doi.org/10.1093/humrep/des354.

[15] Dambaeva SV, Lee DH, Sung N, et al. Recurrent pregnancy loss in women with killer cell immunoglobulin-like receptor KIR2DS1 is associated with an increased HLA-C2 allelic frequency. Am J Reprod Immunol 2016;75(2):94–103. Available from: https://doi.org/10.1111/aji.12453.

[16] Yang X, Yang E, Wang WJ, et al. Decreased HLA-C1 alleles in couples of KIR2DL2 positive women with recurrent pregnancy loss. J Reprod Immunol 2020;142:103186. Available from: https://doi.org/10.1016/j.jri.2020.103186.

[17] Pittari G, Liu XR, Selvakumar A, et al. NK cell tolerance of self-specific activating receptor KIR2DS1 in individuals with cognate HLA-C2 ligand. J Immunol Baltim Md 1950 2013;190(9):4650–60. Available from: https://doi.org/10.4049/jimmunol.1202120.

[18] Kwak-Kim J, Agcaoili MSL, Aleta L, et al. Management of women with recurrent pregnancy losses and antiphospholipid antibody syndrome. Am J Reprod Immunol N Y N 1989 2013;69(6):596–607. Available from: https://doi.org/10.1111/aji.12114.

[19] Huhn O, Ivarsson MA, Gardner L, et al. Distinctive phenotypes and functions of innate lymphoid cells in human decidua during early pregnancy. Nat Commun 2020;11(1):381. Available from: https://doi.org/10.1038/s41467-019-14123-z.

[20] Lédée N, Petitbarat M, Chevrier L, et al. The uterine immune profile may help women with repeated unexplained embryo implantation failure after in vitro fertilization. Am J Reprod Immunol N Y N 1989 2016;75(3):388–401. Available from: https://doi.org/10.1111/aji.12483.

[21] Lédée N, Prat-Ellenberg L, Chevrier L, et al. Uterine immune profiling for increasing live birth rate: a one-to-one matched cohort study. J Reprod Immunol 2017;119:23–30. Available from: https://doi.org/10.1016/j.jri.2016.11.007.
[22] El-Azzamy H, Dambaeva SV, Katukurundage D, et al. Dysregulated uterine natural killer cells and vascular remodeling in women with recurrent pregnancy losses. Am J Reprod Immunol N Y N 1989 2018;80(4): e13024. Available from: https://doi.org/10.1111/aji.13024.
[23] Vomstein K, Feil K, Strobel L, et al. Immunological risk factors in recurrent pregnancy loss: guidelines vs current state of the art. J Clin Med 2021;10(4):869. Available from: https://doi.org/10.3390/jcm10040869.
[24] Beaman KD, Ntrivalas E, Mallers TM, Jaiswal MK, Kwak-Kim J, Gilman-Sachs A. Immune etiology of recurrent pregnancy loss and its diagnosis. Am J Reprod Immunol N Y N 1989 2012;67(4):319–25. Available from: https://doi.org/10.1111/j.1600-0897.2012.01118.x.
[25] Davies ML, Dambaeva SV, Katukurundage D, et al. Predicting NK cell subsets using gene expression levels in peripheral blood and endometrial biopsy specimens. Am J Reprod Immunol N Y N 1989 2017;78(3). Available from: https://doi.org/10.1111/aji.12730.
[26] Shreeve N, Depierreux D, Hawkes D, et al. The CD94/NKG2A inhibitory receptor educates uterine NK cells to optimize pregnancy outcomes in humans and mice. Immunity 2021;54(6):1231–44. Available from: https://doi.org/10.1016/j.immuni.2021.03.021 e4.
[27] Dambaeva S, Bilal M, Schneiderman S, et al. Decidualization score identifies an endometrial dysregulation in samples from women with recurrent pregnancy losses and unexplained infertility. FS Rep 2021;2(1):95–103. Available from: https://doi.org/10.1016/j.xfre.2020.12.004.
[28] Sojka DK, Yang L, Plougastel-Douglas B, Higuchi DA, Croy BA, Yokoyama WM. Cutting edge: local proliferation of uterine tissue-resident NK cells during decidualization in mice. J Immunol Baltim Md 1950 2018;201 (9):2551–6. Available from: https://doi.org/10.4049/jimmunol.1800651.
[29] Turco MY, Moffett A. Development of the human placenta. Dev Camb Engl 2019;146(22):dev163428. Available from: https://doi.org/10.1242/dev.163428.
[30] von Wolff M, Ursel S, Hahn U, Steldinger R, Strowitzki T. Glucose transporter proteins (GLUT) in human endometrium: expression, regulation, and function throughout the menstrual cycle and in early pregnancy. J Clin Endocrinol Metab 2003;88(8):3885–92. Available from: https://doi.org/10.1210/jc.2002-021890.
[31] Kwak-Kim JYH, Chung-Bang HS, Ng SC, et al. Increased T helper 1 cytokine responses by circulating T cells are present in women with recurrent pregnancy losses and in infertile women with multiple implantation failures after IVF. Hum Reprod Oxf Engl 2003;18(4):767–73. Available from: https://doi.org/10.1093/humrep/deg156.
[32] Wang W, Sung N, Gilman-Sachs A, Kwak-Kim JT. Helper (Th) cell profiles in pregnancy and recurrent pregnancy losses: Th1/Th2/Th9/Th17/Th22/Tfh cells. Front Immunol 2020;11:2025. Available from: https://doi.org/10.3389/fimmu.2020.02025.
[33] Wang WJ, Salazar Garcia MD, Deutsch G, et al. PD-1 and PD-L1 expression on T-cell subsets in women with unexplained recurrent pregnancy losses. Am J Reprod Immunol N Y N 1989 2020;83(5):e13230. Available from: https://doi.org/10.1111/aji.13230.
[34] Salazar MD, Wang WJ, Skariah A, et al. Post-hoc evaluation of peripheral blood natural killer cell cytotoxicity in predicting the risk of recurrent pregnancy losses and repeated implantation failures. J Reprod Immunol 2022;150:103487. Available from: https://doi.org/10.1016/j.jri.2022.103487.
[35] He Q, Zhang Y, Qiu W, Fan J, Zhang C, Kwak-Kim J. Does thyroid autoimmunity affect the reproductive outcome in women with thyroid autoimmunity undergoing assisted reproductive technology. Am J Reprod Immunol N Y N 1989 2020;84(6):e13321. Available from: https://doi.org/10.1111/aji.13321.
[36] Pantham P, Abrahams VM, Chamley LW. The role of anti-phospholipid antibodies in autoimmune reproductive failure. Reprod Camb Engl 2016;151(5):R79–90. Available from: https://doi.org/10.1530/REP-15-0545.
[37] Rajaratnam N, Ditlevsen NE, Sloth JK, Bæk R, Jørgensen MM, Christiansen OB. Extracellular vesicles: an important biomarker in recurrent pregnancy loss? J Clin Med 2021;10(12):2549. Available from: https://doi.org/10.3390/jcm10122549.

Further reading

Bilal MY, Katara G, Dambaeva S, Kwak-Kim J, Gilman-Sachs A, Beaman KD. Clinical molecular genetics evaluation in women with reproductive failures. Am J Reprod Immunol N Y N 1989 2021;85(4):e13313. Available from: https://doi.org/10.1111/aji.13313.

Index

Note: Page numbers followed by "*f*" and "*t*" refer to figures and tables, respectively.

B

C

D

E

G

H

J

K

L

M

N

O

S

T

Zeitfracht Medien GmbH
Ferdinand-Jühlke-Straße 7
99095 Erfurt, Deutschland
produktsicherheit@kolibri360.de